Sources and Studies in the History of Mathematics and Physical Sciences

For other titles published in this series, go to
http://www.springer.com/series/4142

Hans Fischer

A History of the Central Limit Theorem

From Classical to Modern Probability Theory

Springer

Hans Fischer
Katholische Universität Eichstätt-Ingolstadt
Mathematisch-Geographische Fakultät
D-85072 Eichstätt
Germany
hans.fischer@ku-eichstaett.de

ISBN 978-0-387-87856-0 e-ISBN 978-0-387-87857-7
DOI 10.1007/978-0-387-87857-7
Springer New York Dordrecht Heidelberg London

Library of Congress Control Number: 2010935663

Mathematics Subject Classification (2010): Primary: 60-03; Secondary: 01A55, 01A60, 26-03, 42-03, 60F05, 62-03

Printed on acid-free paper

Springer is part of Springer Science+Business Media (www.springer.com)

Preface

This book is an entirely revised, in some parts considerably shortened, in other parts enlarged, version of a study which was originally published in German with the title "Die verschiedenen Formen und Funktionen des zentralen Grenzwertsatzes in der Entwicklung von der klassischen zur modernen Wahrscheinlichkeitsrechnung" at Shaker Verlag (Aachen, Germany) in 2000. I thank Shaker for giving me the permission to republish large parts of that work in an English translation. This mainly concerns chapters 1, 3, 4, and 5 of the present book. Most parts of this book have been translated from German or directly written in English by the author himself, except for chapters 1 and 8, as well as the first part of chapter 5 up to section 5.1.2 inclusively, section 6.1, and section 7.2, which were translated by Gavin Bruce from the author's German drafts. Brent Runyan and Andreas Ellwanger corrected those portions of text which are less mathematical.

During the long time of preparing the original and thereafter the present version of the book, many people gave me assistance and advice. As representative of all the persons who considerably helped me with the 2000 book, I thank Ivo Schneider in the first place for his comprehensive and continuous support. Ulrich Oppel kindly gave many valuable hints about mathematical details. The idea to publish an English version was primarily inspired by Reinhard Siegmund-Schultze, and I thank him very much for his encouragement. Bo Isenberg drew my attention to the characterization of modernity through the notion of contingency and introduced me to this subject. I am very grateful to Walter Purkert for discussing questions concerning Hausdorff's reception of Lyapunov's work with me. Günther Wirsching helped me with his friendly and continued interest in the progress of this book, and with numerous discussions on stochastic and historical issues. René Grothmann carefully reworked the pictures and graphics enclosed in this book. Vladimir Andrievskii provided me with copies of sources that were very difficult to access. Without Fritz Heberlein's and Peter Zimmermann's professional aid I would not have overcome the frequent LaTeX problems which occurred during the making of this book. Finally, I thank Jesper Lützen for accepting my book in the Springer series *Sources and Studies*.

I am very grateful to "Maximilian-Bickhoff-Universitätsstiftung" (Eichstätt), which granted financial support for the editing and correction process of a considerable part of the text.

To the following persons and institutions I am indebted for providing me with details on documents, copies of unpublished material, or allowing me to publish texts and pictures: Professor Menso Folkerts (Universität München), Archive of Eidgenössische Technische Hochschule Zürich, Archive of Berlin-Brandenburgische Akademie der Wissenschaften, Picture Library of The Royal Society (London), General Collections of MIT Museum, Archive of University of Minnesota, Archive of New York University, Archive of Mathematisches Forschungsinstitut Oberwolfach, Library of University of Bonn, Library of University of Greifswald, and Oldenbourg-Verlag München.

I consider it a favorable coincidence that this book appears 200 years after the first publication of Laplace's approximation of the distributions of sums of large numbers of independent random variables by normal distributions, which can be interpreted as a first fairly general central limit theorem. This achievement decisively influenced the course of history of probability theory up to modern times. Therefore, this book is also intended to serve as a modest contribution to the 200th anniversary of one of Laplace's most outstanding findings.

Eichstätt, April 2010 *Hans Fischer*

Contents

List of Figures

Abbreviations and Denotations

Abbreviations

CLT	Central limit theorem
TAP	*Théorie analytique des probabilités*

Mathematical Denotations

$\mathbb{N}$	Set of natural numbers $\{1, 2, 3, \ldots\}$
$\mathbb{N}_0$	$\mathbb{N} \cup \{0\}$
$\mathbb{R}$	Set of real numbers
$\mathbb{R}^+$	Set of positive real numbers
$\mathbb{Z}$	Set of integers
$\mathbb{Q}$	Set of rational numbers
$\mathbb{P}_r$	Set of polynomials with real coefficients and maximum degree r
$[a; b]$	Closed interval
$]a; b[$	Open interval
$C^m(I)$	Set of functions $I \to \mathbb{R}$ (I an interval) for which the mth derivative in I exists; $m = 0$ corresponds to continuity
$L^p(I)$	Set of function classes (each class consisting of all functions $f : I \to \mathbb{R}$, I an interval, which only differ from each other in a Lebesgue null set of arguments) such that $\int_I \lvert f(x)\rvert^p dx$ $(0 < p < \infty)$ exists for each representative of each class
$L^\infty(I)$	Set of function classes (each class consisting of all functions $f : I \to \mathbb{R}$, I an interval, which only differ from each other in a Lebesgue null set of arguments) such that for each class a representative exists that is uniformly bounded in I
$\boldsymbol{x}$	Column vector with coordinates $x_1, \ldots, x_n$
$\lvert\boldsymbol{x}\rvert$	$\sqrt{x_1^2 + \cdots + x_n^2}$
$\log x$	Natural logarithm of $x \in \mathbb{R}^+$
$\operatorname{sgn} x$	Sign-function of $x \in \mathbb{R}$ with values $-1, 0, 1$
$\phi_{\mu,\sigma^2}(t)$	Normal density $\frac{1}{\sigma\sqrt{2\pi}} e^{\frac{(t-\mu)^2}{2\sigma^2}}$ $(\mu \in \mathbb{R}, \sigma \in \mathbb{R}^+)$

$\Phi_{\mu,\sigma^2}(x)$	Normal distribution function $\int_{-\infty}^{x} \phi_{\mu,\sigma^2}(t)dt$
Φ	Standard normal distribution function ($= \Phi_{0,1}(x)$)
$L_{\alpha,\beta}$	Class of stable laws with parameters α, β
$F \star G$	Convolution of the distribution functions F and G, $F \star G(x) := \int_{-\infty}^{x} F(x-t)dG(t)$
$F^{n\star}$	n-fold convolution $F \star F \star \cdots \star F$
$\varphi_X(\gamma)$	Lévy-dispersion of the random variable X with respect to the probability level $\gamma \in]0;1[$
$f_X(l)$	Lévy-concentration of the random variable X with respect to the interval length $l > 0$.

Let u and $v > 0$ be functions of x, and let $x \to a$. Then:

$$\left.\begin{array}{l} u = \mathrm{O}(v) \\ u = \mathrm{o}(v) \\ u \sim v \end{array}\right\} \text{ if } \frac{u}{v} \left\{\begin{array}{l} \text{remains bounded} \\ \to 0 \\ \to 1. \end{array}\right.$$

Chapter 1
Introduction

The term "central limit theorem" most likely traces back to Georg Pólya. As he recapitulated at the beginning of a paper published in 1920, it was "generally known that the appearance of the Gaussian probability density[1] e^{-x^2}" in a great many situations "can be explained by one and the same limit theorem," which plays "a central role in probability theory" [Pólya 1920, 171]. Laplace had discovered the essentials of this fundamental theorem in 1810, and with the designation "central limit theorem of probability theory," which was even emphasized in the paper's title, Pólya gave it the name that has been in general use ever since.

These days the term "central limit theorem" is associated with a multitude of statements having to do with the convergence of probability distributions of functions of an increasing number of one- or multi-dimensional random variables[2] or even more general random elements (with values in Banach spaces or more general spaces) to a normal distribution[3] (or related distributions). In an effort to reduce ambiguity—and in view of historic developments—the denotation "central limit theorem" in the present examination will usually refer only to the "classical" case, which deals with the asymptotic equality of distributions of sums of independent or weakly dependent random variables and of a normal distribution. Yet even by this definition, which accords with Pólya's view, "central limit theorem" actually amounts to a collective term for the entire group of theorems about the convergence of distribution functions, densities or discrete probabilities of sums of random

---

[1] Pólya was a bit careless with his phraseology here. Naturally he knew that e^{-x^2} is not a probability density (the norming factor $\frac{1}{\sqrt{\pi}}$ is missing).

[2] Whenever the term "random variable" is used in this book with no additional information, it always refers to one-dimensional real-valued random variables.

[3] A one-dimensional normal distribution with expected value μ and variance σ^2 is characterized by the distribution function $\Phi_{\mu,\sigma^2}(x) := \frac{1}{\sigma\sqrt{2\pi}} \int_{-\infty}^{x} e^{-\frac{(t-\mu)^2}{2\sigma^2}}\, dt$. In the case where $\mu=0$ and $\sigma=1$, one speaks of the "standard normal distribution." By definition, a random vector $(X_1, \ldots, X_n)$ is normally distributed if and only if, for all vectors $(t_1, \ldots, t_n) \in \mathbb{R}^n$, the random variable $\sum_{k=1}^{n} t_k X_k$ follows a one-dimensional normal distribution. The term *the* normal distribution serves to designate the class of all normal distributions.

H. Fischer, *A History of the Central Limit Theorem*, Sources and Studies in the History of Mathematics and Physical Sciences, DOI 10.1007/978-0-387-87857-7_1,

variables. One theorem from this group—which was admittedly only a very specific case—had already been derived by de Moivre in 1733.

Strictly speaking, therefore, one should not really refer to *the* central limit theorem in connection with sums of independent random variables, but rather to *a* central limit theorem on a case-by-case basis. Nevertheless, after ca. 1810 when Laplace utilized normal distributions to present approximations that were valid in decidedly general situations, the statement regarding the universal existence of a "law of frequency" similar to e^{-x^2} for sums of large numbers of independent random variables took on the status of a natural law in the eyes of almost all 19th-century probabilists. This law would serve as a leitmotif for the theory of errors and the field of distribution statistics, which began with Quetelet. Thus it is no wonder that Pólya, who in a sense was rooted in this 19th-century tradition, would talk about *the* central limit theorem. The development of this theorem from a basic idea in the natural and social sciences (see Fig. 1.1 for an example) into an autonomous mathematical theorem—or more correctly into an entire group of such theorems—is the subject matter of this book.

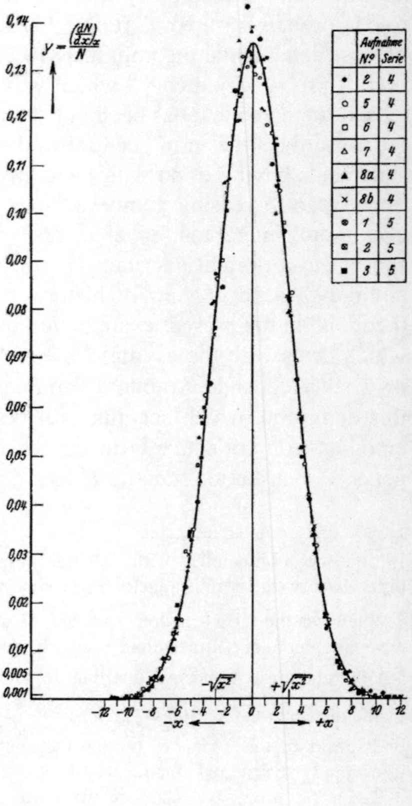

Fig. 1.1 Empirical frequency curve of the deflections of a torsion balance according to its Brownian movement [Kappler 1931]; the deflection at a certain time can be considered as caused by a large number of molecular impacts during a certain time interval just before; the normal distribution may therefore be interpreted as a consequence of the central limit theorem; for an exact stochastic discussion see [von Mises 1931, 512–518]

1.1 Different Versions of Central Limit Theorems

From approximately 1810 to 1935, the period primarily examined in this work, limit theorems for sums of (finite-dimensional) random vectors were also discussed sporadically, but the focus of interest lay on central limit theorems for sums of independent and—in the 20th century—weakly dependent one-dimensional random variables. A differentiation is made today between central limit theorems for normed sums and for triangular arrays on the one hand, and between integral and local limit theorems on the other.

The most historically important version of the central limit theorem (hereafter abbreviated "CLT") is the one pertaining to an integral limit theorem for normed sums: Let (X_k) be a sequence of independent (or weakly dependent) random variables on a common probability space. Under particular conditions on X_k, there exist sequences $(a_n > 0)$ and (b_k) such that

$$\forall r \in \mathbb{R} : P\left(\frac{\sum_{k=1}^{n}(X_k - b_k)}{a_n} \leq r\right) \rightarrow \Phi(r) \quad (n \rightarrow \infty). \tag{1.1}$$

Φ stands for the distribution function of the standard normal distribution. In the case of $a_n^2 = \operatorname{Var} \sum_{k=1}^{n} X_k$ and $b_k = \mathrm{E}X_k$, one speaks of "classical norming."[4]

Corresponding local limit theorems emerge from a representation

$$P\left(\frac{\sum_{k=1}^{n}(X_k - b_k)}{a_n} \leq x\right) = \int_{t \leq x} f_n(t) d\mu_n(t)$$

with a suitable sequence of σ-finite measures (μ_n), which are defined on the system of real Borel sets, and with a sequence of nonnegative functions (f_n). This is to determine the conditions under which pointwise or uniform convergence of (f_n) to the density function[5] of the standard normal distribution occurs. Important special cases include:

— The X_k are continuous random variables, μ_n is the Lebesgue measure for every n, and f_n is the density function associated with $\frac{\sum_{k=1}^{n}(X_k - b_k)}{a_n}$;

— the X_k take on only the values $a + ih$ $(i = 0; \pm 1, \pm 2, \ldots)$,[6]

$$f_n(x) := P\left(\sum_{k=1}^{n} X_k = na + zh\right)\frac{a_n}{h}$$

$$\text{for } x \in \left]\frac{na + (z - 0,5)h - \sum_{k=1}^{n} b_k}{a_n}; \frac{na + (z + 0,5)h - \sum_{k=1}^{n} b_k}{a_n}\right] \quad (z \in \mathbb{Z}),$$

---

[4] The terms $\frac{\sum_{k=1}^{n}(X_k - b_k)}{a_n}$ are referred to as "normed sums." Often both the a_n and the b_k are called "norming constants" or something similar (see, e.g., [Feller 1945, 818; Le Cam 1986, 79]), although, technically, only the constants a_n merit this name.

[5] The term "density function" is used in the present book only for densities with respect to the Lebesgue measure. A random variable with a density function is designated "continuous."

[6] Random variables of this kind are designated "lattice distributed."

and μ_n is a discrete measure that respectively assigns the weight $\frac{h}{a_n}$ to the point $\frac{na+zh-\sum_{k=1}^{n} b_k}{a_n}$.

In the case of the CLT for a triangular array, a double sequence of random variables

$$\begin{array}{l} Y_{11} \\ Y_{21}, Y_{22} \\ Y_{31}, Y_{32}, Y_{33} \\ Y_{41}, Y_{42}, Y_{43}, Y_{44} \\ \text{etc.} \end{array}$$

is considered, or more generally

$$\begin{array}{l} Y_{11}, \ldots, Y_{1m_1} \\ Y_{21}, \ldots\ldots, Y_{2m_2} \\ Y_{31}, \ldots\ldots\ldots\ldots, Y_{3m_3} \\ Y_{41}, \ldots\ldots\ldots\ldots\ldots\ldots, Y_{4m_4} \\ \text{etc.} \end{array}$$

with $m_n \to \infty$. Within each row, the random variables are assumed to be independent or weakly dependent. The integral form of the CLT examines the convergence of distributions of the row sums to a normal distribution, i.e., it determines under what conditions the following is true:

$$\forall r \in \mathbb{R} : P\left(\sum_{k=1}^{m_n} Y_{nk} \le r\right) \to \Phi(r) \quad (n \to \infty). \tag{1.2}$$

As with the procedure for normed sums, local versions of the CLT can also be considered for triangular arrays.[7] Apparently, by the definition $Y_{nk} = \frac{X_k - b_k}{a_n}$ the CLT for normed sums (1.1) becomes a special case of the CLT for triangular arrays (1.2) if the norming constants a_n and b_k are considered given. However, the problem of normed sums also brings the question of how to find suitable a_n and b_k.

With certain modifications, the problems in question can also be transferred to CLTs for random vectors.

Chebyshev [1887/90] was the first to formulate a statement involving the CLT (1.1) for a sequence of independent random variables using classical norming, and he attempted to prove his assertion under certain conditions. In the late 19th and early 20th centuries, mathematicians were mostly considering relations that were equivalent to (1.1):

---

[7] The phrase "triangular array" apparently did not gain prevalence in literature before the Second World War. Feller [1971] mentions "triangular arrays" at several points in his popular textbook (1st edition, 1966). Beginning with papers by Bernshtein and Lindeberg in the year 1922, row sums related to double sequences of random variables $(Y_{nk})_{n \in \mathbb{N}, 1 \le k \le m_n}$ were studied well before the war in exactly the way described, and so it seems appropriate to generally employ the "modern" term "triangular array" in this book in order to simplify the discussion.

$$\forall a \le b \in \mathbb{R} : P\left(a \le \frac{\sum_{k=1}^{n}(X_k - b_k)}{a_n} \le b\right) \to \Phi(b) - \Phi(a) \quad (n \to \infty)$$

or

$$\forall a < b \in \mathbb{R} : P\left(a < \frac{\sum_{k=1}^{n}(X_k - b_k)}{a_n} < b\right) \to \Phi(b) - \Phi(a) \quad (n \to \infty).^8$$

The formulation of integral limit theorems in exactly the form (1.1) is likely attributable to von Mises [1919a]. Local limit theorems explicitly appear in Nekrasov's work [1898]. The treatment of nonclassical normings began in earnest in the 1920s with [Bernshtein 1922; 1926] and with [Lévy 1925b]. The idea of considering triangular arrays is found in [Bernshtein 1922] and [Lindeberg 1922b;c], but was really intensively pursued in the 1930s only. The contemplation of CLTs for sums of dependent random variables started with Markov [1907/10]. From the publication of Chebyshev's paper until the mid-1930s, the CLT (1.1) for independent random variables was constantly in the foreground of interest.

Before Chebyshev, people were not actually studying limit theorems but approximations of probability densities, individual probabilities or probabilities that a sum of random variables lay "between" predetermined limits, in an absolute or relative sense. The corresponding statements can be interpreted from the standpoint of both normed sums in classical norming and triangular arrays. One always presupposed—often tacitly—the independence of the summands in question. Approximations for densities (in the case of continuous random variables) or discrete probabilities (in the case of lattice distributed random variables), and approximations for integral probabilities were considered equivalent. For this reason, it is fairly pointless to differentiate between integral forms and local forms of the CLT in the era before Chebyshev.

1.2 Objectives and Focus of the Present Examination

The history of the CLT as a universal law begins with Laplace; all relevant studies in the 18th century, starting with [de Moivre 1733], essentially contained only approximations of the binominal distribution and their scope of application remained narrow. Laplace's finding of 1810, according to which the additive coaction of a large number of independent random variables generally leads to probabilities that can, at least approximately, be calculated according to the normal distribution, substantially expanded the numerical possibilities of probability theory, especially in the discussion of mass phenomena. Laplace's CLT was trend-setting for the development of stochastics in the 19th century as a discipline that was

[8] It is not clear at all from the wording of almost all of the 19th-century authors, including Chebyshev, whether the version with "$<$" or with "$\le$" was meant (or perhaps both of them together). Where there is doubt, statements such as "the sum lies between the limits ..." are interpreted to mean "$\le$."

defined foremost by its applications. A number of the mathematical methods used in connection with the problem of approximating probabilities for sums of independent random variables were so important to analysis that, starting in the mid-19th century, the central "limit" theorem occasionally also served to illustrate these methods from a primarily mathematical standpoint. The CLT became a mathematically discrete object which was examined for its own sake around the turn of the century as a result of the papers by Lyapunov. During the period between the world wars, "the" CLT fulfilled an important integrative role in the process of developing the discipline of modern probability theory. Around 1935 this process reached a first conclusion, which also corresponded with the "definitive" solution of limit theorems (1.1) and (1.2), at least for independent summands, in the sense of establishing necessary and sufficient conditions. By 1940, the scope of analysis of the CLT for independent summands was expanded to nonnormal—stable, and more general infinitely divisible—limit laws. The monograph *Limit Distributions for Sums of Independent Random Variables* by Gnedenko and Kolmogorov, originally published in Russian in 1949, was dedicated to this complex of problems and represented one of the most important works of probability theory in the 20th century. In order to contrast this development, which in a sense corresponds to a "direct" line of evolution of the CLT since Laplace, the present book also addresses other CLT-related topics which occurred only shortly before and after World War II: generalizations in the direction of martingales, and in the direction of stochastic processes and random elements in metric spaces.

In accordance with the largely auxiliary role the CLT played in the 19th century, space is also given in this book to the applications of approximate normal distributions, e.g., in the theory of errors and in distribution statistics. Reducing the presentation to the analytical content of the individual papers would not have done justice to the significance of the CLT in classical probability theory and the ensuing era. In this respect, the present disquisition also encompasses various aspects of the history of the theory of errors and of statistics in general.

Particular attention is paid to analytical methods. Above all, this pertains to the method of characteristic functions established by Laplace and the method of moments which Chebyshev and Markov preferred. If possible, the analytical procedures are presented and discussed in accordance with the respective contemporary standards. In particular, the treatment of infinitely large and small numbers in the 19th century, which arises in conjunction with sums of "infinitely many" random variables, shall be depicted as authentically as possible.

Extensive reconstruction is also necessary in the case of some modern articles, especially those by Lévy, in order to be able to elaborate upon the original ideas. Commencing with [Gnedenko & Kolmogorov 1949], the textbook literature that was established after the Second World War provided proofs for many of the theorems proposed in the 1920s and 1930s, which did not fully align with the original accounts.

When we look for coherent concepts in the history of the CLT, we first encounter the basic idea that would determine thinking from Laplace until modern times, namely, that the accumulation of many (small) random variables results in a normal distribution. This, though, seems to be the only main thread running through the entire history of CLTs. All the other ways people viewed CLTs at various points in time were just too different. Yet a number of "local" leitmotifs can be detected in certain periods. For instance, questions surrounding the theory of errors dominated dealings with CLTs from Laplace until the 1920s. In the early 1920s, the view of probability theory as a subfield of analysis, which began with Laplace, and which was thereafter promoted by Dirichlet and particularly stressed by Chebyshev and Markov, had a bearing on young analysts like von Mises, Lévy, and Lindeberg and their activities involving limit theorems of probability theory.

"Local" continuity and "global" fissures are likewise evident in the analytical methods associated with the CLT. Even the method of characteristic functions, which seems to have persisted unchanged at its core since Laplace, and which remains one of the most vital tools associated with sums of independent random variables, enjoys a consistent track record only when examined superficially. If it was Laplace's method of approximation to "functions of large numbers" that ultimately allowed Lyapunov to devise proofs that were increasingly sophisticated and adapted to the analytical standards of the day, then it was Lévy's theorem of the "continuous correspondence" between distributions and associated characteristic functions that determined the use of characteristic functions since the early 1920s. In addition, several other at least intermittent threads can be traced: Chebyshev's method of moments, the method first employed by Crofton to add a further "auxiliary random variable" to the sum of random variables considered, or the method of truncated random variables that is attributed to Markov.

The transition undergone by the CLT between 1810 and the outbreak of the Second World War, from a "natural law," whose universality was scarcely challenged, to an entity whose scope of application could be precisely explored using purely mathematical methods, corresponds to the development of probability theory from its "classical" genesis to its "modern" shape.

I would like to employ Lorraine Daston's interpretation of the term "classical probability theory" as it is explained in detail in her *Classical Probability in the Enlightenment* [1988]. According to Daston [1988, xi–xii], "classical probability theory" was just a discipline of mathematics in a wider sense, that built upon a consensus of what was "reasonable," and that should assist "common sense" by providing the resource of calculation during decision-making. Daston places the phase of classical probability theory in the period stretching from about 1650 (although there was as yet no discussion of "probabilities" at that time) up until the time of Laplace and his successors, such as Poisson. However, in specifically stressing the analytical relevance of those problems that were linked to his approximation through normal distribution, Laplace had already begun to abandon the standpoint of classical probability theory characterized by Daston, that of viewing stochastic problems almost exclusively in terms of practical applicability.

When I use the adjective "modern," it is oriented primarily to Herbert Mehrtens's description as it emerges from his book *Moderne—Sprache—Mathematik* [1990]. According to Mehrtens, modern mathematics equates to working on a language that establishes relationships between abstract concepts without making reference to physical or even ideally existing objects. It is not concerns about "imaginativeness," "usefulness," or even "beauty" that determine the value of results within modern mathematics, but rather their consistency within freely and autonomously selected rules. Mathematical truth is determined entirely by the rules of this language. In contrast to the natural sciences, mathematics does not point to anything; it always invariably leads back to itself. Yet, in his examination Mehrtens concentrates on the positions taken by various mathematicians toward fundamental questions and speaks less about their attitudes with respect to problem areas outside of this field. For instance, he mentions at various points (e.g., [1990, 187 f; 270 f.]) that articles by avowed "counter-modernists" such as Henri Poincaré or Luitzen Brouwer could be "extremely modern," but he scarcely analyzes this contradiction between basic attitude and pragmatic work. Practically all proponents of "modern" probability theory between the world wars also exhibit a discrepancy between their "counter-modern" attitudes regarding fundamentals, as evidenced in their speeches about mathematics, and their "modern" work on the further development of stochastic problems, where a successive departure from external performance criteria was occurring particularly with regard to outer-mathematical applicability.

Mehrtens's approach can be supplemented by attempts from the field of sociology to characterize "modernity" as an epoch with an "unusual degree of contingency" [Luhmann 1992, 93].[9] If we follow Luhmann's definition [1992, 96], then:

> Anything is contingent that is neither necessary nor impossible.

According to Luhmann [1992, 47], however, contingency is not a reservoir of arbitrary possibilities, but rather possesses an "order with bound alternatives." Luhmann (e.g., [1992, 100–103]) describes mechanisms which, in social systems (including scientific systems, art systems, state systems, legal systems), lead to contingency as well as to a self-reference that permits the "self-limitation" of the contingency of each system. When applied to Mehrtens's "language of mathematics," this would mean that rather than speaking about things lying outside the realm of mathematics, an internally mathematical consideration of language forms appears and continuously seeks new limits for mathematics in the area of its contingency, limits which are determined only by the inner consistency of the language rules.

With his theory of the contingency of (social) systems, Luhmann is attempting to nullify the difference between the modern and the postmodern. Obviously drawing directly upon the "founder" of the postmodern perspective, Jean-François Lyotard [1979/84], and his observation of the "grand narratives"—the ideologies that shaped the modern period—Luhmann [1992, 42] writes:

> If we understand "postmodern" to mean the lack of a unified cosmography, a universally applicable rationality, or even just a collective attitude toward the world and society, then this results from the structural conditions to which contemporary society delivers itself.

By contrast, though with a certain amount of caution, Mehrtens [1990, 318–326] recognizes a delineation in mathematics between the modern—he sees

[9] Can also be found in [Blumenberg 1987, 57] or [Makropoulos 1997, 34] in a form similar to Luhmann. My thanks to Bo Isenberg (Malmö) for drawing my attention to this aspect of modernity.

Bourbakism and its ideology of "mathematical structures" as both the apex and end of this period—and the postmodern, which is concerned with "heterogeneous individual problems."

In the 20th-century history of the CLT, patterns actually can be discerned which conform to both Mehrtens's and Luhmann's approaches. The principle of autonomy and of self-limitation that is independent of external criteria is voiced in connection with the CLT beginning with the papers by Lyapunov (1900/1901). The contingency of the CLT manifests itself in the establishment of ever more general conditions and in generalizations to nonnormal limit distributions and to the weakening of independence. The results achieved before World War II involving limit distributions of independent random variables were condensed by Gnedenko and Kolmogorov [1949] using consistent analytical methods in such a way that, in a certain sense, one may speak of a "grand narrative" in the history of the CLT. The consistency shown here was superseded after the war by a variety of other problems, and so one may actually speak of a "postmodern" development. On the other hand, though, this development can also be regarded as an extrapolation of approaches introduced before the Second World War and conforming to the contingency of the CLT.

1.3 The Development of Analysis in the 19th Century

One main focus of this study consists in illuminating the history of the CLT before the backdrop of changes in analysis. In fact, during the final third of the 19th century the disparate analytical concepts and procedures that might possibly exist side by side in works by the very same author earlier in the century were gradually replaced entirely by "Weierstrassian rigor," and that implies "epsilonic" analysis in the modern sense. Significant to the history of the CLT during the transition from classical to modern probability theory are essentially three ways of analytical procedure in the 19th century: algebraic analysis, calculation with infinitely small and large "quantities," and finally the aforementioned "epsilonic" concept, which can already be found in connection with individual problems before Weierstrass, e.g., in the work of Dirichlet or Cauchy.[10]

Algebraic analysis was based on algebraic manipulations of series expansions and sought to completely exclude examinations of limits or considerations of infinite quantities. Faith in the analytical sense of each series expansion and in the unlimited scope of formulae thus formed the basis of this concept, which Euler and Lagrange in particular promoted. The algorithmic-formal approaches in algebraic analysis were not without a certain elegance, the attractiveness of which inspired Laplace in particular to undertake similar research, especially in light of his theory of generating functions.[11] In many cases, however, it remained at least open to interpretation what value—in the truest sense of the word—the results achieved by this calculus actually had.

[10] An excellent overview of the various concepts of analysis in the 19th century is provided by [Laugwitz 1999, 46–63] in which a total of five approaches are presented.

[11] See the general explanations in [Laplace 1814/20/86, XXXVII–XL].

Arguing on the basis of infinitesimal considerations was quite popular in the first half of the 19th century, for example with Fourier, Poisson, and Cauchy.[12] Of course, these authors also pragmatically applied other analytical ways of thinking from case to case—for Cauchy, this included a nascent epsilonics. Even if some passages initially suggest an epsilontic interpretation, the discussion of those sources from the standpoint of infinitesimal analysis may be insightful and inspiring.[13] One has to take care, however, not to confuse 19th-century reasoning with concepts and proofs of modern nonstandard analysis. Therefore, there is still a quite controversial discussion on the interpretation of sources in analysis, especially those by Cauchy.[14] The infinitesimal approach in connection with the CLT can be found in the work of Poisson (Sect. 2.2.4), Hagen (Sect. 3.2.1), or Bessel (Sect. 3.2.2). What was especially attractive about infinitesimal considerations was likely the possibility of intuitively grasping problems relating to "limit formulae." However, this intuitiveness also left something of a gap in proof, which did not go unnoticed in the course of rigorous evaluation at the time. Aside from this, it was also probably a lack of clarity in the basic principles[15] that ultimately led to infinitesimal considerations being displaced by epsilonics.

In his Berlin lectures in the 1860s, Weierstrass advocated the complete reduction of analytical considerations to those involving the properties of real numbers without including infinitely small or large quantities. A construct such as this required a clear concept of real numbers (Weierstrass also attempted one, see [Dugac 1978, 364–366]) and pushed work with inequalities to the fore. A particularly important aid for advanced investigations was the mean value theorem of differential calculus, which in many cases was able to replace the series expansions of algebraic analysis. Weierstrassian analysis did retain some vocabulary of infinitesimal analysis at first, but it was founded—at least ostensibly—upon purely finitistic views.[16]

Most authors in the 19th century dealt with the concepts presented in a fairly pragmatic way and differently from case to case. This also applied to the CLT. In Chebyshev's work, for example, one recognizes aspects of algebraic analysis (in his dealings with continued fractions), the use of infinitely small numbers, and the "modern" perception of limits. Overall, though, the transition to modern analysis exerted a substantial influence on the history of the CLT through the first decades of the 20th century. We have this analytical reorientation to thank not only for the formulation as a limit theorem that more precisely defined the statement about an approximation that becomes "exact" when there are "infinitely many" summands.

[12] See, e.g., [Laugwitz 1990].

[13] This applies, for example, to Cauchy's well-known "error"—the assertion that a convergent series of functions in a neighborhood of x results in a function that is continuous in x if only each summand is continuous in x. Actually, Cauchy's concept of convergence "everywhere" (in other words, also for numbers that are "infinitely close" to x) includes what today is called uniform convergence in a neighborhood of x; see [Laugwitz 1986, 78 f.].

[14] See [Schubring 2005, 431–433] for a survey.

[15] There is also a case example of this in connection with the CLT; see Sect. 3.2.1.

[16] Of course, it is open to discussion whether statements such as "for all $n > n_0$" reintroduce the (potentially) infinite through the back door.

Rather, the working methods of modern analysis were gradually carried over to work on the CLT. This process was deeply involved in the transition from classical to modern probability theory.

1.4 Literature on the History of the Central Limit Theorem

In accordance with its paramount importance in the evolution of probability theory, the CLT has already garnered some historical attention, as is shown by the following overview of works devoted to this subject.

Adams's book *The Life and Times of the Central Limit Theorem* (2nd edn.) [2009] gives in its first part—which is intended for more of a general audience and which is essentially identical to the first edition from 1974—insight into developments from de Moivre to Lyapunov. The second part contains additional material by other authors related to the period from 1900 to ca. 1935 (English translations of Lyapunov's 1900 and 1901 papers as well as reprints of [Feller 1945] and [Le Cam 1986]). A thorough account of analytical-technical aspects, especially for the period from Laplace to Cauchy, is provided by [Hald 1998, 303–350] in the chapter "Early History of the Central Limit Theorem" of his monumental *History of Mathematical Statistics from 1750 to 1930.* Hald [2002] likewise dealt exhaustively with the history of series expansions as they can be associated with the CLT. Additional information on this topic is found in his presentation of the *History of Parametric Statistical Inference* [2007]. Schneider [1987b] primarily illuminated the "intellectual background" of stochastic limit theorems in the 18th and 19th centuries. The Russian contributions to the CLT in the 19th and early 20th centuries are highlighted, for example, by Maistrov [1974, 188–224] and Gnedenko & Sheynin [1992, 247–268]. A multitude of details surrounding the history of the CLT up to the time of Markov can be found in many of the articles Sheynin has written—mainly in the *Archive for History of Exact Sciences*—on the achievements of important probability theorists and statisticians of the 18th, 19th, and 20th centuries. The two monographs by Sheynin [1996a; 2005b] about the history of stochastics have a summarizing character. Of particular value for the present study were the volumes of Russian sources [Sheynin 2004a;b; 2005a; Nekrasov 2004] that Sheynin translated into English and published.[17]

The years following the turn of the century have heretofore been accorded with distinctly less regard than the preceding period as far as the CLT was concerned. At least, two major phases have already been examined at some detail: Firstly, and just recently [Siegmund-Schultze 2006], the time around 1920 with the eminent and groundbreaking papers by Pólya and von Mises[18] and, secondly, the development in the 1930s by Le Cam's article "The Central Limit Theorem around 1935" [1986] (reprinted in [Adams 2009], see above), which gives an excellent survey, in

[17] All these sources are also available at `http://www.sheynin.de`.

[18] Siegmund-Schulze at some places refers to the German edition [Fischer 2000] of the present book.

particular of the works by Feller and Lévy. Volume V of the *Collected Works* of Hausdorff [2006] also contains plenty of material relating to the history of the CLT, above all in the 1920s.

Despite the wealth of historical information about the CLT, a coherent overview, in particular one covering the years after 1900, still seems desirable. Moreover, the extant literature on the history of probability theory and statistics concentrates mainly on purely stochastic and—in the 20th century—measure-theoretical aspects. By contrast, the CLT was closely linked to specific methods of classical analysis until well into the first decades of the 20th century and was therefore part of a probability theory that could be perceived as a subfield of analysis.

1.5 Terminology and Notation

One particular difficulty in completing studies on the history of probability theory and statistics consists in the fact that, for the sake of succinctness and clarity, some 18th and 19th century contributions must be presented in the modern terminology to which the reader is accustomed. In particular, this relates to the use of stochastic terms like "random variable," "variance," or "estimated values," and to the shortened notation of linear equation systems in matrix form.

The notion of "random variable" as it is employed in modern probability theory was introduced by Kolmogorov in the 1930s, but this term can still be used largely intuitively. Laplace [1781] himself devised a formula for those probabilities that a sum of "quantités variables" can assume. In 1829, Poisson developed approximations to probabilities that the sum of the "values" ("valeurs") that a (!) "thing" ("chose") receives in various independent experiments remains between certain limits. Hauber [1830], likely motivated by Poisson, emphasized the difference between "undetermined quantities" ("unbestimmte Größen") themselves and the "values" that they each can receive with a particular probability. Chebyshev [1867; 1887/90] clearly differentiated between "quantités" and the different "values" they can take, but in his notation he usually made no distinction between these "random variables" themselves and their concrete values. Nekrasov [1898] examined limit theorems for probabilities that a sum of "random magnitudes"[19] ("sluchainye velichiny") will take a given value. In his papers on the CLT around the turn of the century, Lyapunov proceeded in a way very similar to Chebyshev. In error theory, however, the prevalent practice in the 19th century was not to make a difference between errors in the sense of random variables and their concrete values, neither in notation nor in the way of speaking. Before the backdrop of a parlance with origins in the 18th century, it is therefore appropriate to speak of "random variables" when discussing contributions of classical probability theory already, and not always to explicitly distinguish between random variables and their values. This is likewise

[19] Sheynin's translation, see [Nekrasov 2004, 12].

true for the use of terms like "estimated value" or "compensation value" in the historical discussion of the method of least squares.

The term "variance," which was presumably coined by Ronald Alymer Fisher (see [Hald 1998, 461]), did not become widespread until after the Second World War. The associated concept essentially traces back to the approximation of distributions of sums of independent random variables in the tradition of Laplace, where the variance appears as a coefficient of the first nontrivial term in series expansions, and to the discussion by Laplace and Gauss regarding possible measures for quantifying the mean variation of errors. If I were to use only the various names that cropped up in the 19th century to account for variance or for particular fractions of it, such as simply "factor" (Laplace), square of the "mean error" (Gauss) or "mean of the squares of the differences of the errors from their mean" ("moyenne des carrés des différences des erreurs à leur moyenne," Bienaymé), it would lead to an inconsistent and complicated text. When it comes to establishing the mathematical essence, the denotation "variance" will always be used. The same holds for the use of "expectation."

The modern short form $A\boldsymbol{x} = \boldsymbol{d}$ with $A \in \mathbb{R}^{n,m}$, $\boldsymbol{x} \in \mathbb{R}^m$, $\boldsymbol{d} \in \mathbb{R}^n$ is the simplified notation for systems of equations that are written as follows:

$$\begin{aligned} &ax + by + cz + \text{etc.} = d \\ &a'x + b'y + c'z + \text{etc.} = d' \\ &a''x + b''y + c''z + \text{etc.} = d'' \\ &\text{etc.} \end{aligned}$$

This does not refer to the treatment of linear systems of equations in the sense of a matrix calculus as became common in the 20th century only.[20]

Having said that, my writing does very often include historical denotations which are no longer conventional, mainly in instances when they represent consistent usage. This applies, for example, to the terms "error law" or "frequency law" for the densities of observation errors or other consistent random variables. In accordance with Gauss [1809, 241], a function φ was called an error or frequency law in the 19th century when $\varphi(x)dx$ signified the probability of an error lying between x and $x + dx$.

The phrase "distribution function" is—even today—managed differently by different authors. In the years before World War II, we see the distribution function V of a random variable in the sense of $V(x) = P(X \leq x)$ (von Mises, Pólya, Lindeberg, Feller), $V(x) = P(X < x) + \frac{1}{2}P(X = x)$ (Lévy) or $V(x) = P(X < x)$ (Kolmogorov). However, in practically all cases it is irrelevant which definition was intended.[21] Unless otherwise stipulated, I will be using von Mises's interpretation of "distribution function."

---

[20] Farebrother [1999] provides a history of the theory of errors that gives special consideration to the problems of linear algebra in the original notation.

[21] One exception is the inversion formula for characteristic functions; see Sect. 5.2.3.4. Furthermore, in some estimates it is necessary to consider whether a "$<$" or a "$\leq$" is correct.

A chart of frequently used mathematical notations is located in the index of acronyms at the beginning of this book. Many terms and abbreviations which are not assumed to be generally known or whose usage is inconsistent are explained when they make their first substantial appearance in the main text. References to these explanations are contained in the "Subject Index" (in boldface).

The transliteration of Russian words is according to the Library of Congress system (except that hard and soft signs are omitted).

1.6 The Prehistory: De Moivre's Theorem

As has already been mentioned, de Moivre's approximations to binomial distributions did not do justice to the universality that characterizes *the* CLT. For the sake of completeness, but also to demonstrate what tremendous progress Laplace's approximations of 1810 represented, de Moivre's 1733 paper should be recognized as a sort of "0th chapter," as it were, in the history of the CLT.

Abraham de Moivre (1667–1754) set himself the task of refining the main theorem of *ars conjectandi* [1713] by Jakob Bernoulli, known today as "Bernoulli's Law of Large Numbers." Specifically, Bernoulli had shown—in a derivation that is still rigorous by today's standards—that

$$\lim_{n\to\infty} P(|h_n - p| \le \varepsilon) = 1 \quad \forall \varepsilon > 0,$$

if h_n represents the relative frequency of a particular event occurring with the probability $p = \frac{a}{b}$ $(a, b \in \mathbb{N})$ in a series of n identical and independent trials (see, e.g., [Stigler 1986, 67–69]). Bernoulli gave an estimate of n such that, for any assigned $0 < \eta < 1$ and $\varepsilon = \frac{1}{b}$—where b can be chosen to be arbitrarily large and thus ε to be arbitrarily small—the inequality $P(|h_n - p| \le \varepsilon) \ge \eta$ holds. Although this estimate was better than those one infers today within the framework of a popular elementary exercise involving the Bienaymé–Chebyshev inequality,[22] it resulted in such considerable values for n, even if ε was relatively large, that an improvement had to seem worth pursuing. Conversely, to answer the question of the value of ε for a predetermined n such that $P(|h_n - p| \le \varepsilon)$ is still sufficiently close to 1, Bernoulli's estimate could only deliver unrealistically large values for ε. De Moivre, who was interested in precisely this question, thus had to develop a far more precise approximation to the binomial distribution than Bernoulli had achieved.

---

[22] See the very illustrative examples in [Barth & Haller 1994, 71/70; 273/82] in which the minimum values of n according to Bernoulli and Bienaymé–Chebyshev for $p = \frac{3}{5}$ and various η and ε are compared to each other.

In a seven-page offprint[23] entitled *Approximatio ad summam terminorum binomii* $(a+b)^n$ *in seriem expansi*, which was circulated among close friends and students, and of which just three copies survive today (see [Schneider 1968, 295]), de Moivre concisely described his method for the special case of $p = \frac{1}{2}$: To approximate the probability

$$P(Z = \left[\frac{n}{2}\right] + i) = 2^{-n}\binom{n}{\left[\frac{n}{2}\right]+i}$$

that exactly $\left[\frac{n}{2}\right] + i$ "successes" Z will be achieved for a large number n of trials, de Moivre first provided the approximations

$$\frac{\binom{n}{[\frac{n}{2}]}}{2^n} \approx \frac{2}{\sqrt{2\pi n}} \quad \text{and} \log \frac{\binom{n}{[\frac{n}{2}]+i}}{\binom{n}{[\frac{n}{2}]}} \approx -2\frac{i^2}{n}. \tag{1.3}$$

Tools to assist in making the approximations included the power series for $\log(1+x)$ as well as the approximation to $n!$ that today is named after James Stirling but was actually developed jointly by him and de Moivre in friendly competition around 1730. From (1.3) follows

$$P(Z = \left[\frac{n}{2}\right] + i) \approx \frac{2}{\sqrt{2\pi n}} e^{-2\frac{i^2}{n}}.$$

If you like, this statement can be "translated" into the modern "local" limit theorem

$$\frac{B(n; \frac{1}{2}; k)\sqrt{2n\pi}}{2e^{-\frac{2(k-\frac{n}{2})^2}{n}}} \to 1,$$ [24]

but we must not forget that de Moivre's objective was not a limit theorem, and certainly not a local one, but rather an approximation to $P(|h_n - \frac{1}{2}| \leq \varepsilon)$ and $P(|Z - \left[\frac{n}{2}\right]| \leq t)$. Furthermore, he did not have a concept at his disposal to adequately match the idea of the exponential function. De Moivre approximated the probability $P(|Z - \left[\frac{n}{2}\right]| \leq t) = \sum_{i=-t}^{t} P(Z = \left[\frac{n}{2}\right] + i)$ according to

$$P(|Z - \left[\frac{n}{2}\right]| \leq t) \approx 2\frac{2}{\sqrt{2\pi n}} \sum_{i=0}^{t} e^{-2\frac{i^2}{n}}$$

$$\approx \frac{4}{\sqrt{2\pi}} \int_0^{t/\sqrt{n}} e^{-2y^2} dy = \frac{4}{\sqrt{2\pi n}} \int_0^t e^{-2\frac{x^2}{n}} dx. \tag{1.4}$$

---

[23] An English translation of the offprint was later included, with minor amendments, in the second edition of de Moivre's textbook *The Doctrine of Chances* (1738). A reprint of the relevant passage from the third edition (1756) of the *Doctrine* can be found in [Schneider 1988, 125–134]. For in-depth discussions of de Moivre's method, see [Schneider 1968, 292–300; 1995], [Stigler 1986, 70–77], and [Hald 1998, 17–21].

[24] In general, $B(n; p; k) := \binom{n}{k} p^k (1-p)^{n-k}$.

In contrast to this modern representation and fully in the tradition of the Newtonian form of the infinitesimal calculus, de Moivre indicated the exponential function and the associated integrals only by series expansions. Partly by summation of the first members of his series and partly with the help of approximative integration, de Moivre on the basis of the integral "limit theorem" (1.4) arrived at values for $P(|h_n - \frac{1}{2}| \leq \varepsilon)$ when $\varepsilon = \frac{1}{2\sqrt{n}}, \frac{1}{\sqrt{n}}$, and $\frac{3}{2\sqrt{n}}$. In this last case, a value of approximately 0.99874 resulted for the probability that was sought. Therefore, the question of the required order of magnitude for ε with a predetermined probability of ≈ 1 was answered by explicitly indicating numerical values at least for $p = \frac{1}{2}$.

The fact that de Moivre's approach was very cumbersome compared to the use of Leibniz's integral notation does not mean that his solution to the derivation which Laplace [1812/20/86, 280–284] later provided using basically the same analytical methods and which comes very close to the modern version, was inferior. Moreover, de Moivre made observations about the procedure for a general probability of success $p \neq \frac{1}{2}$ that, though incomplete, was basically viable.[25]

Probabilities for sums of independent random variables played a not insubstantial role in the probability theory of the 18th century, as in problems involving games of chance (e.g., with regard to sums of dice rolls) and in the field of the theory of errors, which began to emerge around 1750. Using the method of generating functions, a rudimentary form of which can already be found in the *ars conjectandi* and which de Moivre substantially expanded around 1730 [Seal 1949, 209–211], it was possible to establish formulae for probabilities and density functions of sums of independent identically distributed random variables if the distribution of the individual summands could be expressed by simple algebraic terms.[26] However, even with a number of random variables that was still fairly small it became impossible to numerically and analytically evaluate the results obtained in this way. Although Daniel Bernoulli succeeded in 1780 in introducing an approximation method that was completely different from the de Moivrian approach (see Sect. 3.2.1), its scope of application remained limited to binomial distributions and thus to distributions of sums of two-valued random variables. In the 18th century, it was impossible to get significantly beyond de Moivre's "limit" theorems.[27]

---

[25] Schneider [1995] was able to reconstruct how de Moivre must have recognized that the approximation for $P(|h_n - \frac{1}{2}| \leq \varepsilon)$ (h_n is the relative frequency of success in a Bernoulli process with a success probability of $\frac{1}{2}$) was equal to the approximation for $P(|h'_n - p| \leq 2\varepsilon\sqrt{p(1-p)})$ (h'_n is the relative frequency of success in a Bernoulli process with a success probability p). However, likely due to contemporary publication practices, de Moivre did not explicitly publish this finding.

[26] A detailed survey of the respective works from the 18th century can be found in [Sheynin 1973]; see also [Hald 1998, Chapt. 2].

[27] This observation also applies to Lagrange's approximation of the multinomial distribution (ca. 1775), which was derived analogously to de Moivre's approximation of the binomial, see Sect. 3.3.2.4.

Chapter 2
The Central Limit Theorem from Laplace to Cauchy: Changes in Stochastic Objectives and in Analytical Methods

In 1812, Pierre-Simon de Laplace (1749–1827) published the first edition of his *Théorie analytique des probabilités* (henceforth simply abbreviated by *TAP*).[1] With its typical problems, stochastic models, and analytic methods this book would considerably influence probability theory and mathematical statistics right until the beginning of the 20th century.

Until Laplace and his successors, classical probability consisted mainly in the sum of its applications to physical, social, and moral problems. However, as Laplace already pointed out in the concise preface to the first edition of his *TAP*, probability was also important for mathematics in a narrower sense. In many problems referring to stochastic models depending on a large number of trials, probabilities could only be expressed by formulae too complicated for direct numerical evaluation. Thus, for a reasonable application of many of the results of probability calculus, particular considerations were needed to obtain useful approximations of the "formulae of large numbers." In the aforementioned preface, Laplace called these problems "the most delicate, the most difficult, and the most useful" of the entire theory. He expressed his hope that discussion of these problems would catch the attention of other "geometers." Therefore, in addition to the qualitative feature of applicability, which was characteristic for classical probability theory, a new, purely mathematical aspect emerged: the relevance of specific analytical methods of probability theory.

Laplace had been intensely dealing with the "delicate problems" of probability just described from the very beginning of his scientific career. In his 1781 "Mémoire sur les probabilités," one can already find "in nuce" almost all of the problems of *TAP*, which can be roughly divided into two categories: "sums of random variables"

[1] For a description of the origin and the major contents of this book, see [Stigler 2005; Sheynin 2005b, 99–110]. An English translation by Richard Pulskamp of the second, probabilistic, part of the *TAP* is available at
`http://www.cs.xu.edu/math/Sources/Laplace/index.html`.

H. Fischer, *A History of the Central Limit Theorem*, Sources and Studies in the History of Mathematics and Physical Sciences, DOI 10.1007/978-0-387-87857-7_2,

and "inverse probabilities."[2] The first category includes, for example, the a priori probabilities of profit and loss in certain games of chance, or of the arithmetic mean of observations being subject to random errors; the latter for instance deals with the a posteriori probabilities that the ratio of the chances of a boy's and a girl's birth is within a given interval centered around the ratio of the corresponding observed values. By 1774, Laplace had already developed useful approximation methods for those a posteriori probabilities depending on a large number of observations. He did not succeed in adapting this method to a priori probabilities until 1810, however. Only then, with a "tricky" modification of the method of generating functions, did he achieve any usable results on approximations of probabilities of sums of independent random variables, which, from the modern point of view, are subsumed under the rubric of the "central limit theorem." It was the CLT which considerably shaped the contents and methods of the *TAP* and significantly influenced the development of probability and error theory during the 19th century.

As we have already seen (Sect. 1.4), the history of the CLT, as far as the contributions of Laplace and his successors are concerned, has already been studied in fair detail. Therefore, a main focus in the present section will be on those questions which still seem to be open: Which changes in the probabilistic and analytical context of the CLT occurred between ca. 1810 and 1850; how did these changes come about, and how have these changes influenced analytical style and methods in the treatment of this theorem?

2.1 Laplace's Central "Limit" Theorem

As already noticed, Laplace's CLT was the result of an almost forty years' effort. In the following, we will describe the historical development of Laplace's treatment of sums of independent random variables, his methods for finding appropriate approximation formulae, and the major applications of his finally achieved CLT.

---

[2] Inverse probabilities are conditional probabilities $P(H|B)$ for certain "hypothetic" causes H which may have entailed the observed results B. ($P(H|B)$ is considered as "inverse" to $P(B|H)$.) The probabilities $P(H|B)$ can be interpreted as if they quantify conclusions from an observation B to its causation H "a posteriori." If there are n possible causes H_j ($j = 1, \ldots, n$), and if the $P(H_j)$ are known, then, by virtue of Bayes's formula:

$$P(H_k|B) = \frac{P(B|H_k)P(H_k)}{\sum_{j=1}^{n} P(B|H_j)P(H_j)}, \quad k = 1, \ldots, n.$$

Since the probabilities $P(H_j)$ are unknown in most cases, one is often forced to the "subjective" assumption of the H_j being equiprobable. If, conversely, a certain probability distribution is—more or less arbitrarily—presupposed, then any probabilities derived therefrom can be interpreted as "a priori probabilities."

2.1.1 Sums of Independent Random Variables

Sums of independent random variables had played an important role in Laplace's probabilistic work from the very beginning.[3] In this context, the problem of calculating the probability distribution of the sum of angles of inclination, which were assumed to be determined randomly, as well as the related problem of calculating the probabilities of the deviations between the arithmetic mean of data which were afflicted by observational errors and the underlying "true value," became especially important. In one of his first published papers, Laplace [1776] had already set out to determine the probability that the sum of the angles of inclination of comet orbits (or the arithmetic mean of these angles respectively) is within given limits. He assumed that all angles, which had to be measured against the ecliptic, were distributed randomly according to a uniform distribution between 0° and 90° (and also tacitly presupposed that all angles were stochastically independent). Laplace succeeded in calculating these probabilities for an arbitrary number of comets via induction (with a minor mistake which was subsequently corrected in [Laplace 1781]). In this 1781 paper, Laplace even introduced a general—however very intricate—method, based on convolutions of density functions, in order to exactly determine the probability that a sum of independent random variables ("quantités variables," as Laplace put it) was within given limits.[4] In the most simple case, each of the n variables had the same rectangular distribution between 0 and h. For the probability P that the sum of those variables was between a and b with $0 \leq a < b \leq nh$, Laplace obtained (in modern notation)

$$P = \frac{1}{h^n n!}\left(\sum_{i=0}^{N}\binom{n}{i}(-1)^i(b-ih)^n - \sum_{i=0}^{M}\binom{n}{i}(-1)^i(a-ih)^n\right), \tag{2.1}$$

where $N = \min(n, [\frac{b}{h}])$ and $M = \min(n, [\frac{a}{h}])$. Formulae of this kind were too complicated for a direct numerical evaluation if the number of random variables exceeded a relatively small value. The arithmetic mean of the actual observed angles of inclination of the then known 63 comets was $46°16'$. Through the use of (2.1) alone, Laplace was unable to address the hypothesis that the comets' planes of motion resulted at "random." At this stage of his mathematical work, however, Laplace could not develop usable approximations.

[3] For a comprehensive biography also dealing with Laplace's probabilistic work, see [Gillispie 1997]. Detailed discussions of Laplace's contributions to probability and statistics can be found in [Sheynin 1976; 1977; 2005b; Stigler 1986; Hald 1998]. The web site already referred to in footnote 1 contains English translations of most works in probability theory by Laplace.

[4] See [Sheynin 1973, 219 f.] and [Hald 1998, 56–60] for descriptions of this method.

2.1.2 Laplace's Method of Approximating Integrals, and "Algebraic Analysis"

Beginning with his "Mémoire sur la probabilité des causes" [1774], Laplace developed techniques for approximating integrals depending on a "great number," such as, for example, the Gamma function $\Gamma(s+1) = \int_0^\infty e^{-x} x^s dx$ with the "great number" s. The basic idea of this "Laplacian method of approximation" is as follows: Let the integrand $f(x)$ depend on a very large parameter such that the function f has a single, very sharp peak, with the consequence that appreciable contributions to the entire integral result only from a small interval around this maximum. Then it can be expected that the function f is asymptotically equal to a function of the form $f(a)e^{-\alpha(x-a)^{2k}\pm\cdots}$ $(\alpha > 0)$ if f attains its maximum at $x = a$. Based on this idea, the Laplacian method consists of appropriate series expansions around the abscissa of the maximum. In the case of the Gamma function, Laplace started with

$$\Gamma(s+1) = \int_0^\infty e^{-x} x^s dx = \int_{-s}^\infty e^{(-z+s)}(z+s)^s dz.$$

The maximum $M = e^{-s}s^s$ of the integrand is attained at $x = s$, or equivalently $z = 0$. Laplace [1785, 258 f.; 1812/20/86, 128–131] set

$$e^{-s}e^{-z}(z+s)^s = Me^{-t^2(z)}$$

and expanded $t^2 = -\log(e^{-z}(1+z/s)^s)$ into a series of powers of z. Conversely, he also expanded z into a series of powers of t, and obtained the following expansion after transforming the variable of integration from z to t:

$$\begin{aligned}\Gamma(s+1) &= M\int_{-\infty}^\infty e^{-t^2}\sqrt{2s}\left(1+\frac{4t}{3\sqrt{2s}}+\frac{t^2}{6s}+\cdots\right)dt\\ &= s^{s+1/2}e^{-s}\sqrt{2\pi}\left(1+\frac{1}{12s}+\frac{1}{288s^2}+\cdots\right). \qquad (2.2)\end{aligned}$$

For many probabilistic formulae, Laplace's method of approximation worked extremely well. For the problem of sums of (independent) random variables, however, it was only at a rather late stage of his mathematical work that Laplace developed techniques based on which suitable approximations could be deduced.

In the above-mentioned article of 1774, Laplace treated approximation problems in an analytical style closely related to that of Euler. Laplace discussed the behavior of the peak with an "infinitely large" parameter, carefully considering "infinitely" large or small quantities. In his later work, however, he abandoned the "Eulerian" style of calculating with infinite quantities of different gradations and, influenced by Lagrange's algebraic analysis, developed a special algebraic-algorithmic style dealing primarily with formal series expansions, as we have just seen in connection with the Gamma function. Laplace's deduction of the CLT was likewise written in this style.

2.1.3 The Emergence of Characteristic Functions and the Deduction of Approximating Normal Distributions

Laplace for the first time exemplified his approach to the CLT in the "Mémoire sur les approximations des formules qui sont fonctions des très grands nombres et sur leur application aux probabilités" [1810a]. Crucial for this success in approximating distributions of sums of independent random variables by normal distributions was his modification of generating functions. Let me demonstrate the essentials of his approach to the CLT[5] in the special case of identically distributed random variables $X_1, \ldots, X_n$, which have zero means and which take the values $\frac{k}{m}$ (m a given natural number, $k = -m, -m+1, \ldots, m-1, m$) with the respective probabilities p_k.[6] For the calculation of the probability P_j that $\sum_{l=1}^{n} X_l$ has the value $\frac{j}{m}$ $(-nm \le j \le nm)$, Laplace made use of the generating function $T(t) = \sum_{k=-m}^{m} p_k t^k$. Due to the mutual independence of the X_l's—which was usually only tacitly presupposed by Laplace—P_j is equal to the coefficient of t^j in $[T(t)]^n$ after carrying out the multiplication. The direct execution of this method—its general principle going back to de Moivre, see [Seal 1949]—leads at best to very complicated algebraic terms for P_j. Laplace, however, introduced the trick of substituting the variable t by $\mathrm{e}^{\mathrm{i}x}$ ($\mathrm{i} = \sqrt{-1}$). Thus, he introduced the (now so-called) characteristic functions in a special case.

From

$$\frac{1}{2\pi}\int_{-\pi}^{\pi} \mathrm{e}^{-\mathrm{i}tx}\mathrm{e}^{\mathrm{i}sx}\,dx = \delta_{ts} \quad (t, s \in \mathbb{Z}) \tag{2.3}$$

it follows that

$$P(j) = \frac{1}{2\pi}\int_{-\pi}^{\pi} \mathrm{e}^{-\mathrm{i}jx}\left[\sum_{k=-m}^{m} p_k \mathrm{e}^{\mathrm{i}kx}\right]^n dx.$$

The last integral above was at least formally accessible to Laplace's method of approximation. There was, however, a certain modification necessary, as Laplace did not consider an expansion of the whole integrand around its maximum at $x = 0$, but only of the power with exponent n (equal to the characteristic function). By expanding $\mathrm{e}^{\mathrm{i}kx}$ into powers of x one gets

[5] The most important sources for Laplace's treatment of the CLT are [Laplace 1810a; 1811], and the fourth chapter of the *TAP*.

[6] The following explanation differs, as far as terminology and further details are concerned, from Laplace's exposition. Unlike Laplace, we only consider, for the sake of simplicity, random variables with values within the interval [−1; 1]. For paraphrases in Laplace's original style see [Sheynin 1977, 10–16] and [Fischer 2000, 29–33]. Hald [1998, 303–317] gives a thorough account on Laplace's analytical approach to the CLT.

$$P(j) = \frac{1}{2\pi}\int_{-\pi}^{\pi} \mathrm{e}^{-\mathrm{i}jx}\left[\sum_{k=-m}^{m} p_k \mathrm{e}^{\mathrm{i}kx}\right]^n dx$$
$$= \frac{1}{2\pi}\int_{-\pi}^{\pi} \mathrm{e}^{-\mathrm{i}jx}\left[\sum_{k=-m}^{m} p_k\left(1 + \mathrm{i}kx - \frac{k^2x^2}{2} - \frac{\mathrm{i}k^3x^3}{6} + \cdots\right)\right]^n dx.$$

Taking into consideration that $\sum_{k=-m}^{m} p_k k = 0$, and with the substitution $m^2\sigma^2 = \sum_{k=-m}^{m} p_k k^2$, we obtain

$$P(j) = \frac{1}{2\pi}\int_{-\pi}^{\pi} \mathrm{e}^{-\mathrm{i}jx}\left[1 - \frac{m^2\sigma^2x^2}{2} - \mathrm{i}Ax^3 + \cdots\right]^n dx,$$

where A is a constant depending on $\sum_{k=-m}^{m} p_k k^3$. The formal expansion of

$$\log\left[1 - \frac{m^2\sigma^2x^2}{2} - \mathrm{i}Ax^3 + \cdots\right]^n =: \log z$$

into a series of powers of x leads to

$$\log z = -\frac{m^2\sigma^2nx^2}{2} - \mathrm{i}Anx^3 + \cdots,$$

and therefrom to

$$z = \mathrm{e}^{-\frac{m^2\sigma^2nx^2}{2} - \mathrm{i}Anx^3 + \cdots} = \mathrm{e}^{-\frac{m^2\sigma^2nx^2}{2}}\left(1 - \mathrm{i}Anx^3 + \cdots\right).$$

After transforming the variable of integration according to $x = \frac{y}{\sqrt{n}}$, the result is

$$P(j) = \frac{1}{2\pi\sqrt{n}}\int_{-\pi\sqrt{n}}^{\pi\sqrt{n}} \mathrm{e}^{-\mathrm{i}j\frac{y}{\sqrt{n}}}\mathrm{e}^{-\frac{m^2\sigma^2y^2}{2}}\left(1 - \frac{\mathrm{i}Ay^3}{\sqrt{n}} + \cdots\right) dy.$$

For an approximation with a "very large" n we ignore, like Laplace, all series terms with a power of $\sqrt{n}$ in the denominator, and at the same time, set the limits of integration equal to $\pm\infty$. In this way we get

$$P(j) \approx \frac{1}{2\pi\sqrt{n}}\int_{-\infty}^{\infty} \mathrm{e}^{-\mathrm{i}j\frac{y}{\sqrt{n}}}\mathrm{e}^{-\frac{m^2\sigma^2y^2}{2}} dy,$$

where the last integral is, as Laplace showed in different ways, equal to

$$\frac{1}{m\sigma\sqrt{2\pi n}}\mathrm{e}^{-\frac{j^2}{2m^2\sigma^2n}}. \tag{2.4}$$

Summing up (2.4) for $\frac{j}{m} \in [r_1\sqrt{n}; r_2\sqrt{n}]$, which can be approximated by integration ($dx \approx \frac{1}{\sqrt{n}}$), leads to the result

$$P(r_1\sqrt{n} \le \sum X_l \le r_2\sqrt{n}) \approx \sum_{j\in[mr_1\sqrt{n};mr_2\sqrt{n}]} \frac{1}{m\sigma\sqrt{2\pi n}} \mathrm{e}^{-\frac{j^2}{2m^2\sigma^2 n}}$$
$$\approx \int_{mr_1}^{mr_2} \frac{1}{m\sigma\sqrt{2\pi}} \mathrm{e}^{-\frac{x^2}{2m^2\sigma^2}}\,dx = \int_{r_1}^{r_2} \frac{1}{\sigma\sqrt{2\pi}} \mathrm{e}^{-\frac{x^2}{2\sigma^2}}\,dx,$$

which corresponds to the integral form of the CLT. With only one exception (see Sect. 2.1.5.3) Laplace dealt with independent identically distributed and bounded random variables with densities.[7] To this aim he at first considered the range of values of those random variables discrete (as described above), and then he assumed m "infinitely large."

Nowhere in his work did Laplace state a general theorem which would have corresponded to the CLT in today's sense. He only treated particular problems concerning the approximation of probabilities of sums or linear combinations of a great number of random variables (in many cases errors of observation, see Sect. 2.1.5.2) by methods which in principle corresponded to the procedure described above. In modern notation, Laplace's most general version of the CLT [Laplace 1812/20/86, 335–338] was as follows: Let $\epsilon_1, \dots, \epsilon_n$ be a large number of independent errors of observation, each having the same density with mean μ and variance σ^2. If $\lambda_1, \dots, \lambda_n$ are constant multipliers and $a > 0$, then

$$P\left(\left|\sum_{j=1}^{n} \lambda_j(\epsilon_j - \mu)\right| \le a\sqrt{\sum_{j=1}^{n} \lambda_j^2}\right) \approx \frac{2}{\sigma\sqrt{2\pi}} \int_0^a \mathrm{e}^{-\frac{x^2}{2\sigma^2}}\,dx. \qquad (2.5)$$

The special case of a CLT for the binomial distribution Laplace [1812/20/86, 280–284] on the basis of Stirling's formula treated in a particular section of his *TAP* by methods which are in principle due to de Moivre and still employed in modern textbooks.

2.1.4 The "Rigor" of Laplace's Analysis

From Laplace's point of view, approximating an analytical expression depending on a great number n meant transforming it into a series expansion with terms whose order of magnitude decreased sufficiently fast with increasing n. The greater the number of calculated terms and the faster these terms decrease, the better the approximation. Laplace did not determine absolute or relative errors of approximations, but instead put his trust, according to the leitmotif of algebraic analysis, in the power of series expansions.

In the case of Laplace's CLT, the series terms seem to decrease with ascending powers of $\frac{1}{\sqrt{n}}$ (or even of $\frac{1}{n}$ if the individual random variables have a symmetric distribution). Apparently, it was Laplace's point of view to trust in the quality

[7] Laplace [1810a, 326 f.; 1812/20/86, 313 f.] hinted, though in a quite vague manner only, also at the possibility of analogous considerations concerning unbounded random variables.

of his approximations already because of those decreasing series terms. In the *Essai philosophique sur les probabilités*, whose first edition appeared in 1814 and served as a "popular" introduction to the *Théorie analytique*, Laplace [1814/20/86, XXXIX] wrote of his approximations:

> (...) the series converges the faster the more complicated the formula is, such that the procedure is more precise the more it becomes necessary.

However, some authors did, if rather rarely, object to Laplace's specific approach to approximations. A first hint came from Adrien Marie Legendre as early as 1811. In his *Exercises du calcul intégral* [1811, 290 f.] he discussed the approximation formula

$$n! \approx \frac{\sqrt{2\pi}n^{n+1/2}}{\exp(n)} \exp\left(E(n)\right), \quad E(n) = \sum_{k=1}^{m} \frac{B_{2k}}{2k(2k-1)n^{2k-1}}, \tag{2.6}$$

which can (with slight modifications) be traced back to de Moivre and Stirling around 1730 (see [Schneider 1968, 266–276]). The B_{2k} are the (Jakob) Bernoullian numbers; Leonhard Euler had already shown in 1739 that, from a certain index, these numbers grow faster than any geometric sequence [Schneider 1968, 276]. But only Legendre clearly addressed the divergence of the series $E(s)$ and the resulting difficulties for its analytical treatment. Laplace's series (2.2) was, as apparent from its first terms, equivalent to (2.6). (An exact proof for the equality of both series expansions, however, was not given during the 19th century.) From Legendre's description [1811, 343–348] of Laplace's account it became therefore plausible that the Laplacian method of approximation could lead in the general case to (in Legendre's own words) "semi-convergent expansions" only. Thus, for critical mathematicians, Laplace's treatment of the CLT became suspicious as well. How could it be justified neglecting series terms of "higher order," if the series was possibly divergent?

In 1844, Robert Leslie Ellis tried to discuss Laplace's reasoning regarding the CLT in a modified form (see [Hald 1998, 333–335]). He also explicitly analyzed the example of mutually independent random variables with the common density function $f(x) = \frac{1}{2}e^{-|x|}$. Referring to his—only quite formal—manipulations with series expansions in treating this particular case, he wrote at the end of his explanations [1844, 215]:

> But some doubt may perhaps remain, whether such an approximation to the *form* of the function P [the probability to be approximated], if such an expression may be used, is also an approximation to its numerical value (...)

A similar assessment of Laplace's series expansions was given by Cauchy in [1853g'] (see Sect. 2.5.6).

In 1856 Anton Meyer[8] submitted a proof of the CLT in the special case of two-valued random variables to the Academy in Brussels. Meyer's proof was not based

[8] Meyer was the author of a rather influential treatise of probability and error theory [Meyer 1874], which was also translated into German [Meyer 1874/79] and constitutes an important source for the state of the art at the beginning of the last quarter of the 19th century.

on the usual procedure which can be traced back to de Moivre, and which had also been elaborated in Laplace's *Théorie analytique*. He instead used Laplace's modification of generating functions. There exists a brief report by Jean Baptiste Brasseur on Meyer's article (which itself seems to have been lost). Brasseur [1856] hoped that Meyer's method would lead to a more exact discussion of the neglect of the "terms of higher order of smallness." Meyer's paper was accepted for publication, however on condition that a better examination of the "convergence of the series" be made. The publication failed, Meyer died the following year.

2.1.5 The Central Limit Theorem as a Tool of Good Sense

The examples of Ellis, Cauchy, and Meyer show that, in the middle of the 19th century, Laplace's methods of deducing approximative normal distributions for sums of random variables were considered to be unrigorous by some authors. Such criticism was quite rare, but this was in part due to the status of probability theory within mathematics during the 19th century. As Lorraine Daston [1988] explained, probability theory, at least until the middle of the 19th century, was not a discipline of mathematics in a narrower sense, but rather part of a "mathesis mixta." The value of probabilistic research was determined less by internal mathematical criteria, but rather by the quality of its application to "real" situations. Laplace's CLT met the latter point in an excellent manner. The results of all applications of this theorem matched with "good sense" and thus confirmed Laplace's well-known saying [1814/20/86, CLIII] that

> Basically, probability is only good sense reduced to a calculus.

We shall test this claim with three prominent applications of CLT: the comet problem (already mentioned above), the problem of foundation of the method of least squares, and the problem of risk in games of chance.

2.1.5.1 The Comet Problem

In 1810, Laplace could base his examinations of the "randomness" of the orbits of comets on the observation of 97 comets. Under the hypothesis of a uniform distribution for the angles of inclination between 0 G and 100 G (centesimal degrees, corresponding to $0°$ and $90°$) and with aid of the CLT, he calculated the probability that the arithmetic mean of all angles falls within a certain interval around "50 G." The mean of the observed values was 51.87663 G, and thus Laplace considered the interval [50 G − 1.87663 G; 50 G + 1.87663 G]. The probability of this interval was only around 0.5. Therefore, there was a considerable probability that, presupposing a uniform distribution, the mean of all angles deviated from 50 G even more than the observed mean. Laplace [1810a, 316] followed that there did not exist any "primitive cause" which affected the specific positions of comet orbits. Thus, Laplace, by using probabilistic methods, succeeded in confirming the prior assertion of Achille Pierre de Séjour (stated in *Essai sur les comètes* 1775) which he had already referred to in his first pertinent contribution [Laplace 1776, 280].

In contrast, an analogous calculation regarding the 10 planets (and planetoids) known at that time, which could be carried out with the "exact" formula (2.1) of 1776/1781, showed that the position of their orbits depended on a common "cause" [Laplace 1810a, 307 f.]. Such considerations were important regarding the currently so-called Kant–Laplace nebular-hypothesis. Stigler [1986, 137 f.] and Hald [1998, 303–306], both referring to the first, although very specific and purely algebraic, applications of the tricky substitution $t^x = e^{x\varpi\sqrt{-1}}$ in generating functions discussed by Laplace in [1785, 267–270], maintain that Laplace had already discovered "his" CLT by the 1780s. However, the relevance of this theorem for astronomical issues, intensively studied by Laplace between 1785 and 1810, was likely to have led to the publication of pertinent results as soon as possible. Thus, Laplace presumably did not develop his method for deriving approximate normal distributions for sums of independent random variables much earlier than around 1810.

The problem whether orbits of comets and planets depended on "primitive causes" was only one of several opportunities when Laplace searched for "regular causes" in nature. Other examples, treated similarly as the comets and planets issue, such as the daily changes of air pressure between mornings and evenings, or the slight deviations to the east during the free fall of bodies, can be found in the fifth chapter of Laplace's *TAP*.[9]

2.1.5.2 The Foundation of the Method of Least Squares

The most prominent application of the method of least squares[10] during the 19th century was as follows:
Let d_i $(i = 1, \ldots, s)$ be observed values, a_{ij} $(j = 1, \ldots, t, t < s)$ given coefficients, and ξ_j "elements" to be determined such that

$$d_i + \epsilon_i = \sum_{j=1}^{t} a_{ij}\xi_j \quad (i = 1, \ldots, s), \tag{2.7}$$

where the ϵ_i are unknown, mutually independent errors of observation. Laplace named the equations (2.7) "equations of condition" ("equations de condition"). The problem was to estimate the ξ_j as precisely as possible after observing the d_i. According to the method of least squares, first published by Legendre in 1805, estimators x_j for the ξ_j can be obtained by virtue of the principle

[9] For a survey of the pertinent work of Laplace see [Hald 1998, 431–443].

[10] There exists a good deal of historical literature on the method of least squares. For detailed discussions of the error theoretic development during the 18th and 19th centuries see [Stigler 1986; Hald 1998; Farebrother 1999]. The most important original sources can be found (mainly in German translation) in [Schneider 1988].

$$\sum_{i=1}^{s}\left(d_i - \sum_{j=1}^{t} a_{ij}x_j\right)^2 = \min, \tag{2.8}$$

from which the t equations

$$\sum_{i=1}^{s}\sum_{j=1}^{t} a_{ik}a_{ij}x_j = \sum_{i=1}^{s} a_{ik}d_i \quad (k = 1, \ldots, t)$$

follow. Thus, the method of least squares belongs to those methods which combine the equations of condition after setting $\epsilon_i = 0$ linearly to a new system of t equations in t unknowns. In modern matrix-notation, this means: Given the system of equations of condition

$$\boldsymbol{d} + \boldsymbol{\epsilon} = A\boldsymbol{\xi}$$

for the vector of unknown elements $\boldsymbol{\xi} = (\xi_1, \ldots, \xi_t)^T$ with

$$A = (a_{ij}) \in \mathbb{R}^{s,t}\ (s > t),\ \boldsymbol{d} = (d_1, \ldots, d_s)^T,\ \boldsymbol{\epsilon} = (\epsilon_1, \ldots, \epsilon_s)^T,$$

the goal is to find a system of "multipliers" $B \in \mathbb{R}^{t,s}$ such that the vector of solutions $\boldsymbol{x}$ of the equation system

$$B\boldsymbol{d} = BA\boldsymbol{x}$$

is in a certain sense "optimal" with regard to the "true" $\boldsymbol{\xi}$. Choosing $B = A^T$, one gets exactly the same values for the coordinates of $\boldsymbol{x}$ which result from the condition (2.8), that is, from the method of least squares.

In the special case of "direct observations" of one single element ξ, that means, in the case where the equations of condition have the particular form

$$d_i + \epsilon_i = \xi \quad (i = 1, \ldots, s),$$

the method of least squares yields the arithmetic mean $x = \sum_{i=1}^{s} d_i/s$ as an estimator for ξ. This property rather frequently played an important role in foundational discussions on least squares during the 19th century.

Legendre [1805] had only given an intuitive justification of least squares, which did not use any probabilistic arguments. In 1809 Carl Friedrich Gauss succeeded in showing that the least squares estimators x_j according to (2.8) are equal to the estimators meeting the condition of being "most probable," a condition which is now called the "maximum-likelihood-principle" (see Sect. 3.1). For this justification of giving preference to the method of least squares, Gauss presupposed that the errors of observation were identically normally distributed (with expectation 0).

The joint occurrence of normal distributions in Gauss's argument and in Laplace's CLT possibly motivated the latter to give a new foundation of least squares in the case of a large number of equations of condition (see [Stigler 1986, 143] for a discussion of this "Gauss–Laplace Synthesis"). Laplace [1811] showed that the method of least squares was "optimal" according to certain criteria, which suggested to him calling this method later, in the *TAP*, the "most advantageous."

If one takes the recapitulating description in the *Essai philosophique* (the introduction to the *TAP*) as a standard, the "most advantageous" method was, according to Laplace [1814/20/86, LXII], the method in which "one and the same error of the results is less probable than with any other procedure." A sensible translation of this sentence into modern mathematical language is: the estimator x' for a true value ξ according to the "most advantageous" method has, in comparison with all estimators x'' obtained by competing methods, the following property:

$$P(|\xi - x'| \geq a) < P(|\xi - x''| \geq a) \text{ for all } a > 0. \tag{2.9}$$

Laplace (e.g., [1812/20/86, 348]) claimed to have proven that the method of least squares would be, in this sense, the "most advantageous," at least among those methods which combine a large number of observational equations linearly into a set of equations with (if possible) a uniquely determined system of solutions.

In his foundation of the method of least squares, Laplace [1811, 387–398; 1812/20/86, 318–327] treated first the simplest case of equations of condition with a single element ξ:

$$a_1\xi = d_1 + \epsilon_1, \ldots, a_s\xi = d_s + \epsilon_s$$

(a_i given coefficients, d_i observations, ϵ_i mutually independent errors with zero means). Laplace estimated ξ in the form

$$x = \frac{\sum_{i=1}^{s} b_i d_i}{\sum_{i=1}^{s} b_i a_i},$$

$b_1, \ldots, b_s$ being indeterminate constants at first. The difference between the true value ξ and the estimator x became therefore

$$\xi - x = \frac{\sum_{i=1}^{s} b_i \epsilon_i}{\sum_{i=1}^{s} b_i a_i}. \tag{2.10}$$

In order to determine the "most advantageous" multipliers b_i, Laplace tried to calculate the probability law for linear forms $\sum_{i=1}^{s} b_i \epsilon_i$, s being a great number. For each error he assumed the same symmetric density function which vanished beyond a bounded interval. In his work of 1810 Laplace had already deduced an approximating normal distribution for the sum of a large number of identically distributed errors, a result which at first served only for a rather theoretical discussion of arithmetic means. Now, Laplace used an analogous analytical approach to the linear combination, with the following result (represented in modern notation):

$$P(-r\sqrt{s} \leq \sum b_i \epsilon_i \leq r\sqrt{s}) \approx \frac{\sqrt{2s}}{\sigma\sqrt{\pi \sum b_i^2}} \int_0^r \mathrm{e}^{-\frac{su^2}{2\sigma^2 \sum b_i^2}} du,$$

σ^2 being the variance common to all errors. Setting $r\sqrt{s} = c\sigma\sqrt{2\sum b_i^2}$, Laplace for $\xi - x$ according to (2.10) deduced:[11]

$$P\left(\left|\frac{\sum b_i\epsilon_i}{\sum a_i b_i}\right| \le \frac{c\sigma\sqrt{2\sum b_i^2}}{|\sum a_i b_i|}\right) = P\left(|\xi - x| \le \frac{c\sigma\sqrt{2\sum b_i^2}}{|\sum a_i b_i|}\right)$$
$$\approx \frac{2}{\sqrt{\pi}}\int_0^c \mathrm{e}^{-t^2}dt. \quad (2.11)$$

Laplace now proceeded, without giving any explanations, as if the approximation (2.11) was, presupposing a large number s, even exact. This was one of the crucial points of his foundation of least squares. As we will see below, Cauchy's criticism of exactly this point would later become a major motivation for his own "rigorous proof" of the CLT. Also Gauss, at several places of his work, critically pointed out that, strictly speaking, Laplace's argumentation was only valid for the unrealistic situation of an "infinitely large" number of observations.[12]

On the basis of the assumption of an exact normal distribution, Laplace required that one choose the multipliers b_i according to the condition that for any probability level (depending only on c) the "limits of error" $\pm\frac{c\sigma\sqrt{2\sum b_i^2}}{|\sum a_i b_i|}$ should be minimal. Because the modulus of these limits is minimal if and only if $b_i = ka_i$, with constant $k \neq 0$, this condition in fact leads to the least squares estimator $x = \frac{\sum a_i d_i}{\sum a_i^2}$. The criterion of "minimal limits" is equivalent to condition (2.9), which was discussed only in Laplace's *Essai philosophique*.

Laplace [1811, 401–409; 1812/20/86, 327–332] also tried to apply his reasoning to the simultaneous treatment of more than one element. To achieve this, he developed a rudimentary form of the multidimensional CLT, from which he, however, passed on to a one-dimensional consideration. A truly complete multidimensional solution of this problem, by an explicit consideration of confidence ellipsoids, was only reached by Bienaymé [1852]. Presupposing mutually independent errors of observation $\epsilon_1, \ldots, \epsilon_n$, each having the same density f with mean 0, Bienaymé by further developing Laplace's techniques derived a series expansion for the density $p(\boldsymbol{\tau})$ of the multi-dimensional linear combination $\boldsymbol{\Delta} := \sum_{i=1}^{s}\boldsymbol{\alpha}_i\epsilon_i$ with fixed $\boldsymbol{\alpha}_i \in \mathbb{R}^t$ $(t \le n)$. His result was equivalent to

$$p(\boldsymbol{\tau}) = \frac{1}{(2\pi)^{\frac{t}{2}}\sigma^t N}\exp\left(-\frac{1}{2\sigma^2}\sum_{j,k=1}^{t} a_{jk}\tau_j\tau_k\right)(1 - R(\boldsymbol{\tau})),$$

[11] In order to deduce the following approximation, Laplace in his *TAP* would have been able to apply equation (2.5), which was even derived for errors with an asymmetrical density, if he had set $\mu = 0$, $a = c\sigma\sqrt{2}$, and $\lambda_i = b_i/\sum a_i b_i$ there. As the *TAP* was largely a compilation of earlier work, he simply copied the argumentation from his 1811 paper, which was based on symmetric errors. And only in the subsequent section of the *TAP* did he establish the relation (2.5), however without any comment on its possible use for discussing least squares.

[12] See, for example, [Gauss 1821, 99; 1823, 18].

where σ^2 denotes the variance common to all errors, (a_{jk}) the inverse matrix of $(A_{i\ell}) \in \mathbb{R}^{t,t}$ with $A_{i\ell} = \sum_{r=1}^{s} \alpha_{r,i}\alpha_{r,\ell}$,[13] N^2 the determinant of the matrix $(A_{i\ell})$, and $R(\boldsymbol{\tau})$ an infinite series of terms, each depending on moments of f[14] and tending to 0 as $n \to \infty$ (see [Heyde & Seneta 1977, 66–71; Hald 1998, 501–504]).

By around 1810, several methods of dealing with observational data were available, but the method of least squares was apparently the most useful in the general case. Thus, it was reasonable to champion least squares even without a probabilistic discussion. Yet the CLT "proved" that, at least under "natural" assumptions, this method was superior to other procedures. From Laplace's point of view, his asymptotic discussion of least squares completely confirmed the established opinion of astronomers and geodesists. Thus, on the one hand, his CLT was a tool of good sense, and its rigor was not to be scrutinized. On the other hand, it became plausible that, in the time after Laplace, critical discussions of the superiority of least squares also questioned the validity of the applied normal approximations, and thus of the CLT itself.

2.1.5.3 Benefits from Games of Chance

As a general rule, Laplace considered independent identically distributed random variables with densities. A rare exception from this rule can be found in his discussion of the "benefits depending on the probability of future events" (chapter IX of *TAP*). Laplace [1812/20/86, 428–432] dealt with a particular sequence of games with only two outcomes for each single game: "gain" and "loss." He assumed that the respective probabilities of gain and loss were possibly different from game to game. According to these assumptions, Laplace based his analysis on a large number s of single games (tacitly considered as being independent) with results $X_1, \ldots, X_s$, where each X_i could take the values ν_i (gain) and $-\mu_i$ (loss) with probabilities q_i and $1 - q_i$, respectively. Proceeding in a way analogous to his treatment of sums of observational errors, he achieved the result that

$$P\left(\left|\sum X_i - \sum(q_i\nu_i - (1-q_i)\mu_i)\right| \le r\sqrt{2\sum q_i(1-q_i)(\nu_i+\mu_i)^2}\right)$$
$$\approx \frac{2}{\sqrt{\pi}}\int_0^r e^{-t^2}\,dt.$$

Laplace argued that $\sum(q_i\nu_i - (1-q_i)\mu_i)$ was of the order of magnitude s if each summand was "a little" greater than 0, whereas $r\sqrt{2\sum q_i(1-q_i)(\nu_i+\mu_i)^2}$ was of the order $\sqrt{s}$ only. Therefore, for arbitrarily large $r > 0$ and sufficiently large s, even

$$\sum(q_i\nu_i - (1-q_i)\mu_i) - r\sqrt{2\sum q_i(1-q_i)(\nu_i+\mu_i)^2}$$

[13] $\alpha_{r,i}$ designating the i-th coordinate of the vector $\boldsymbol{\alpha}_r$.

[14] Bienaymé explicitly calculated those terms which depend on moments up to the 4th order.

became greater than 0.[15] Laplace followed that an "infinitely large and certain" total gain would be accumulated if only $q_i \nu_i - (1 - q_i)\mu_i > 0$ for all $1 \leq i \leq s$. By this application of the CLT, Laplace provided the basis for a theory of risk, which in turn would even play an important role in the history of the CLT during the 1920s (see Sect. 5.2.8.1).

2.2 Poisson's Modifications

Among all contributions of the 19th century in connection with Laplace's CLT aiming at a more comprehensible presentation or at modifications of the Laplacian methods according to contemporary analytical standards, the two approaches [1824; 1829] by Siméon Denis Poisson (1781–1840) had a special influence on the contributions of later authors. Poisson shared Laplace's view on the status of probability theory in the classical sense.[16] Concerning moral problems, however, Poisson generalized Laplace's stochastic models to a considerable extent, and he did not share Laplace's cautious attitude toward these issues. Poisson's idea of all processes in the physical and moral world being governed by distinct mathematical laws is in line with his attempts toward a more exact mathematical analysis. Accordingly, the consequences for CLT were twofold: Firstly, Poisson formulated and proved this theorem generally for "choses," thus creating an early concept of random variables, and secondly, he tried to discuss the validity of this theorem, mainly through counterexamples.

2.2.1 Poisson's Concept of Random Variable

In the first [1824] of the above-mentioned articles, Poisson treated sums and linear combinations of observational errors with different (not necessarily symmetrical) distributions, followed by a discussion of the Laplacian foundation of least squares. In the second article of 1829, he took up the issue from a far more general point of view. There, Poisson investigated asymptotic behavior of the distribution of a sum of functions (!) of the values of a "thing" ("chose"), where in several independent experiments these values were obtained with possibly different probabilities. The additional complication of considering a "function" essentially served to cover both sums of random values and of powers of these values within the same theorem. From today's point of view, all these quantities would plainly be described as random variables. Thus, Poisson's concept of the values of a "thing" was directed primarily

---

[15] Apparently, Laplace tacitly assumed the existence of positive constants a, b such that $q_i \nu_i - (1 - q_i)\mu_i > a$ and $(\nu_i + \mu_i)^2 < b$ for all i.

[16] Poisson's work in probability is well described in [Sheynin 1978; Bru 1981; Hald 1998; Sheynin 2005b].

toward the most important applications, and was still far away from the modern conception of abstract "random variable," as explained by Kolmogorov [1933a].[17]

2.2.2 Poisson's Representation of the Probabilities of Sums

In his discussion of sums of independent random variables, Poisson normally assumed that each variable X_n took values within the interval $[a;b]$ ($-a$ and b could be even infinitely large) and had a density function f_n, which was introduced by $f_n(x) = F_n'(x)$, where $F_n(x) = P(X_n \le x)$. In a manner similar to Laplace's approach, Poisson started his analysis with discrete random variables. Unlike Laplace, however, he did not consider probabilities of single discrete values but immediately calculated, partly through combinatorial considerations, the probability that the sum $S_s = X_1 + \cdots + X_s$ would be within certain limits. Through the strict use of infinitesimal quantities in the transition from discrete to continuous random variables, he [1829, 5; 1824, 275; 286] established the formula

$$P(c-\varepsilon \le S_s \le c+\varepsilon) = \frac{1}{\pi}\int_{-\infty}^{\infty}\left(\prod_{n=1}^{s}\int_a^b f_n(x)\mathrm{e}^{\alpha x\sqrt{-1}}dx\right)\mathrm{e}^{-\alpha c\sqrt{-1}}\sin(\varepsilon\alpha)\frac{d\alpha}{\alpha}. \quad (2.12)$$

The justification of this formula was incomplete, even from a contemporary point of view. But Poisson [1824, 276] examined the validity of (2.12) in the special case $s = 1$. By interchanging the order of integration he concluded from (2.12) that

$$P(c-\varepsilon \le X_1 \le c+\varepsilon) = \frac{1}{\pi}\int_a^b\int_{-\infty}^{\infty}\left(\mathrm{e}^{(x-c)\alpha\sqrt{-1}}\sin(\varepsilon\alpha)\frac{d\alpha}{\alpha}\right)f_1(x)dx. \quad (2.13)$$

By virtue of the addition theorems for sine and cosine, and the well-known formula[18]

$$\int_0^{\infty}\frac{\sin(kx)}{x}dx = \frac{\pi}{2} \quad (k>0),$$

he showed that

$$\int_{-\infty}^{\infty}\mathrm{e}^{(x-c)\alpha\sqrt{-1}}\sin(\varepsilon\alpha)\frac{d\alpha}{\alpha} = \begin{cases}\pi \text{ for } x\in\,]c-\varepsilon;c+\varepsilon[\\ 0 \text{ for } x\notin[c-\varepsilon;c+\varepsilon].\end{cases} \quad (2.14)$$

[17] Poisson's approach to random variables was taken up and further developed soon afterwards by Carl Friedrich Hauber [1830], in his "Theorie der mittleren Werthe" ("Theory of Mean Values"), in an interesting attempt to develop a concept of far-reaching generality for random variables, which were named "unbestimmte Größen" ("indetermined quantities"). Many properties of expectations and variances of sums or products of independent random variables which today belong to the standards of each elementary theory of random variables, were explicitly stated and proven for the first time by Hauber.

[18] For a history of this formula, which can be essentially traced back to Euler and still plays an important role in several branches of analysis, see [Fischer 2007].

The required result

$$P(c-\varepsilon \le X_1 \le c+\varepsilon) = \int_{c-\varepsilon}^{c+\varepsilon} f_1(x)dx$$

followed immediately from (2.13) and (2.14). In turn, it must have been within Poisson's scope to establish (2.12) by means of (2.14), even in the general case of arbitrary s. But only Dirichlet and Cauchy, as we will see below, directly used the jump function in (2.14) for elegant derivations of formulae equivalent to (2.12) for the probabilities of sums. Dirichlet at least was most probably motivated by Poisson's discussion of (2.13) and (2.14).

Dealing with the general case, Poisson set

$$\int_a^b f_n(x)\cos(\alpha x)dx =: \rho_n \cos\varphi_n, \qquad \int_a^b f_n(x)\sin(\alpha x)dx =: \rho_n \sin\varphi_n, \quad (2.15)$$

and

$$R := \rho_1 \cdots \rho_s, \quad \psi := \varphi_1 + \cdots + \varphi_s. \quad (2.16)$$

Using $R(-\alpha) = R(\alpha)$ and $\psi(-\alpha) = -\psi(\alpha)$, he concluded from (2.12):

$$P(c-\varepsilon \le S_s \le c+\varepsilon) = \frac{2}{\pi}\int_0^\infty R\cos(\psi - c\alpha)\sin(\varepsilon\alpha)\frac{d\alpha}{\alpha}. \quad (2.17)$$

In his article of 1824, Poisson dealt with the case of an "infinitely large" s by calculating with infinitely small and infinitely large quantities. In his article of 1829, however, series expansions constituted the analytical background for an approximation with "large" (but not infinite) s. Afterwards, Poisson apparently preferred the second version (described in detail by Hald [1998, 317–327]), which was also adopted into his major probabilistic work, the *Recherches sur la probabilité des jugements en matière criminelle et en matière civile* [1837].

2.2.3 The Role of the Central Limit Theorem in Poisson's Work

As we will see in the following, Poisson's work on the CLT was based on Laplace's ideas on the one hand; on the other hand, however, Poisson's discussion of new analytical aspects paved the way toward a more rigorous treatment of the CLT.

2.2.3.1 Poisson's Version of the Central Limit Theorem

Poisson's results concerning the CLT can be summarized in modern terminology essentially as follows:
Let $X_1, \ldots, X_s$ be a great number of independent random variables with density functions which decrease sufficiently fast (Poisson did not specify exactly how fast)

as their arguments tend to $\pm\infty$. It is supposed that for the absolute values $\rho_n(\alpha)$ of the characteristic functions of X_n (see (2.15)) there exists a function $r(\alpha)$ independent of n with $0 \le r(\alpha) < 1$ for all $\alpha \neq 0$ such that

$$\rho_n(\alpha) \le r(\alpha). \tag{2.18}$$

Then, for arbitrary γ, γ',

$$P\left(\gamma \le \frac{\sum_{n=1}^{s}(X_n - \mathrm{E}X_n)}{\sqrt{2\sum_{n=1}^{s} \mathrm{Var}X_n}} \le \gamma'\right) \approx \frac{1}{\sqrt{\pi}}\int_{\gamma}^{\gamma'} \mathrm{e}^{-u^2}du, \tag{2.19}$$

where the approximation becomes all the better the larger s is, and the difference between the left and the right side becomes "infinitely small" with "infinite" s. Strictly speaking, Poisson's analysis could be used for arbitrary γ, γ', though he explicitly expressed end results in the sense of (2.19) only for the special case $\gamma = -\gamma' < 0$.

Poisson was convinced that this CLT was also valid for discrete random variables. In this case one could, according to Poisson [1837, 274 f.], assume that the values $c_1, \ldots, c_\nu$ of a random variable of this kind were subject to the respective probabilities $\gamma_1, \ldots, \gamma_\nu$ which were represented by $\gamma_i = \int_{c_i-\delta}^{c_i+\delta} f(z)dz$ with an "infinitely small" quantity δ and a "discontinuous" density function f.[19]

As with Laplace, the CLT for Poisson was an important tool of classical probability, but not an autonomous mathematical theorem. Unlike Laplace, however, Poisson pointed out essential presuppositions "en passant," such as the above-mentioned condition (2.18) for characteristic functions, and he discussed counterexamples to an overall validity of asymptotic normal distributions for sums. The most prominent of these counterexamples [Poisson 1824, 278] concerns the sum of identically distributed random variables with the probability density

$$f(x) = \frac{1}{\pi(1+x^2)},$$

for which the direct evaluation of (2.12) shows that

$$P\left(c - \varepsilon \le \sum X_n \le c + \varepsilon\right) = \frac{1}{\pi}\arctan\left(\frac{2\varepsilon s}{s^2 + c^2 - \varepsilon^2}\right).$$

Therefore in this case, even for large s, an approximate normal distribution can not be reached. Poisson [1824, 280] pointed out, however, that such cases of very slowly decreasing densities would not occur in practice, because all errors of observation were uniformly bounded in reality. Random variables with the density function f would later play an important role in Cauchy's critical discussion of least squares (see Sect. 2.5.2). In fact, such random variables are now called "Cauchy-distributed."

[19] Poisson at this place used the adjective "discontinuous" in the traditional sense, as being inaccessible to a representation through a uniform algebraic term.

The significance of his condition for characteristic functions (2.18) Poisson [1824, 289–291] illustrated by two similar examples, where neither the assertion of the CLT was true nor this condition was met: He considered linear combinations $\sum_{n=1}^{s} \gamma_n \epsilon_n$ of identically distributed errors obeying the law

$$f(x) = \mathrm{e}^{-2|x|}.$$

Using the formula (2.12) he showed that, for an "infinitely large" s,

$$P(-c \leq \sum \gamma_n \epsilon_n \leq c) = \frac{1 - \mathrm{e}^{-2c}}{1 + \mathrm{e}^{2c}} \quad \text{if} \quad \gamma_n = \frac{1}{n},$$

and

$$P(-c \leq \sum \gamma_n \epsilon_n \leq c) = 1 - \frac{4}{\pi} \arctan(\mathrm{e}^{-2c}) \quad \text{if} \quad \gamma_n = \frac{1}{2n-1}.$$

According to Poisson, in the first example we have

$$\rho_1(\alpha) \cdots \rho_s(\alpha) = \frac{1}{(1 + \frac{\alpha^2}{4})(1 + \frac{\alpha^2}{4 \cdot 4}) \cdots (1 + \frac{\alpha^2}{4s^2})} \to \frac{\pi\alpha}{\mathrm{e}^{\frac{1}{2}\pi\alpha} - \mathrm{e}^{-\frac{1}{2}\pi\alpha}},$$

whereas in the second

$$\rho_1(\alpha) \cdots \rho_s(\alpha) = \frac{1}{(1 + \frac{\alpha^2}{4})(1 + \frac{\alpha^2}{4 \cdot 9}) \cdots (1 + \frac{\alpha^2}{4(2s-1)^2})} \to \frac{2}{\mathrm{e}^{\frac{\pi\alpha}{4}} + \mathrm{e}^{-\frac{\pi\alpha}{4}}}.$$

2.2.3.2 Poisson's Law of Large Numbers

Regarding error theory, Poisson hardly made any modifications to the Laplacian discussion of least squares based on the CLT. Yet the discussion of (in modern terms) stochastic convergence of mean values and relative frequencies, respectively, which did not play a too dominant role in Laplace's work, became vital for Poisson and his major probabilistic work, the *Recherches*. Like Laplace, Poisson based such considerations on the CLT.

The approximate stability of arithmetic means or relative frequencies, quite often observed within different sequences of random experiments of the same kind, was so important for Poisson's probabilistic approach that he coined the term "law of large numbers" for this fact. In the introduction of his *Recherches*, he characterized this law as follows:

> The phenomena of any kind are subject to a general law, which one can call the *Law of Large Numbers*. It consists in the fact, that, if one observes very large numbers of phenomena of the same kind depending on constant or irregularly changeable causes, however not progressively changeable, but one moment in the one sense, the other moment in the other sense; one finds ratios of these numbers which are almost constant [Poisson 1837, 7].

It must be emphasized that Poisson's interpretation of "law of large numbers" is different from the modern definition of this term.

For a “proof” of his law of large numbers, Poisson [1837, 139–143, 277 f.] introduced a special two-stage model of causation for the occurrence of an event (or, more generally, for the occurrence of a special value of a “chose”), and he established two auxiliary theorems on stochastic convergence: the first concerning the arithmetic means of non-identically distributed random variables, the second concerning the relative frequencies of an event which generally does not occur with constant probability. He based these theorems, which are equivalent to the *now* so-called “laws of large numbers,” on his general CLT (for comprehensive historical accounts see [Bru 1981, 69–75] and [Hald 1998, 577–580]). A distinct deviation of the relative frequencies with which a certain event had occurred in different sequences of observations respectively, possibly gave rise to the assumption that these sequences originated from different systems of causation. In the third part of his *Recherches*, Poisson gave a probabilistic discussion of the significance of such hypotheses in the context of conviction rates, and he essentially used the CLT for calculating the respective probabilities (see [Stigler 1986, 186–194] for a detailed discussion).

Poisson’s law of large numbers (in its original form) was heavily criticized during the 19th century. Among these discussions, two crucial points became subject of debates: the practical meaning of Poisson’s causation system was scrutinized (mainly by Bienaymé, see [Stigler 1986, 185; Heyde & Seneta 1977, 46–49]), and the analytical rigor of the deduction of the “auxiliary” CLT was questioned. Chebyshev [1846, 17] criticized that Poisson’s analysis was only “approximative,” and did not provide exact “error limits.” In this way he showed a—still rather vague—unease with Poisson’s analytical approach. One can, however, interpret Chebyshev’s criticism as an indication of the shift from “classical” probability, chiefly determined by its applications, toward a “new mathematical” probability. Perhaps, Chebyshev’s objections resulted from Poisson’s (as well as Laplace’s) procedure of neglecting “higher” series terms without giving any justification for that. Yet, if this was the case, Chebyshev did possibly not realize that Poisson had given an—at least indirect—justification of this procedure with his first, infinitistic approach.

2.2.4 Poisson’s Infinitistic Approach

Poisson’s discussions of 1824 and 1829 on the CLT were essentially equivalent. The first account, however, clarified the fundamentals of Laplace’s method of approximations as applied to the CLT much more directly, and, as we will see below, paved the way for a more “rigorous” treatment of asymptotic normal distributions for sums of independent random variables. For a discussion of the essentials of Poisson’s “first” approach it is sufficient to confine the description to the special case of identically distributed random variables with a density f_1 vanishing beyond the finite interval $[a;b]$.

From (2.15), (2.16), (2.17) one gets with

$$\rho := \rho_1 = \sqrt{\left(\int_a^b f_1(x)\cos(\alpha x)dx\right)^2 + \left(\int_a^b f_1(x)\sin(\alpha x)dx\right)^2},$$

and $\varphi := \varphi_1$:

$$P(c - \varepsilon \leq S_s \leq c + \varepsilon) = \frac{2}{\pi}\int_0^\infty \rho^s \cos(s\varphi - c\alpha)\sin(\varepsilon\alpha)\frac{d\alpha}{\alpha}. \tag{2.20}$$

Poisson [1824, 279–281] carefully justified that $0 < \rho < 1$ if $\alpha \neq 0$. The following excerpt illustrates his notion of "infinite" quantities and his handling of these quantities in connection with an asymptotic representation of $P(c - \varepsilon \leq S_s \leq c + \varepsilon)$:

> We want to consider the number s infinitely large, such that the following formulae are rigorously true at this limit, and the more approximated, the larger s is. Now, from the quantity ρ being less than 1 if the variable α is not $= 0$ it follows that at the limit $s = \infty$ the power ρ^s attains finite values only for infinitely small values of this variable, and becomes infinitely small if α has a finite value [Poisson 1824, 280].

Poisson expressed in this text the contemporary view of the meaning of "approximation": Approximation formulae had to be "rigorously true" at the "limit." Moreover, he considered, as can be inferred from his phrasing "limit," an "infinite" quantity not as actually infinite. On the other hand, he treated infinitely small quantities as belonging to the common system of numbers.[20] This ambivalence in the attitude toward the infinite is typical for the "infinitesimal" period in the first half of the 19th century, which led away from the priority of algebraic analysis.

On the basis of the above-cited comment and on account of $\cos(\alpha x) \approx 1 - \frac{\alpha^2 x^2}{2}$ and $\sin(\alpha x) \approx \alpha x$ for "infinitely small" α, Poisson could deduce—at least for a finite interval $[a; b]$:

$$\rho^s \approx \begin{cases} (1 - h^2\alpha^2)^s & \text{for an "infinitely small" } \alpha \\ 0 & \text{otherwise,} \end{cases}$$

$$\text{where } h^2 := \frac{1}{2}\left(\int_a^b x^2 f_1(x)dx - \left(\int_a^b x f_1(x)dx\right)^2\right).$$

The sign $\approx$ (not explicitly used by Poisson) is used here to indicate an "infinitely close" position of one value to another. Poisson [1824, 281] set $\alpha =: y/\sqrt{s}$, "where the new variable y can attain finite values." Taking into account that $\rho \approx 1$ and $\int_a^b f_1(x)\sin(\alpha x)dx \approx \int_a^b f_1(x)\alpha x dx$ for $\alpha \approx 0$, he concluded that $\sin\varphi \approx k\alpha$ for $\alpha \approx 0$, where k is the expectation of the random variables. As a result of $\sin\varphi \approx \varphi$ for $\sin\varphi \approx 0$ it followed that $\varphi \approx k\alpha$ for $\alpha \approx 0$.

In this way Poisson obtained for "infinitely large" s on account of $(1 - \frac{h^2y^2}{s})^s \approx \mathrm{e}^{-h^2y^2}$:

$$P(c - \varepsilon \leq S_s \leq c + \varepsilon) \approx \frac{2}{\pi}\int_0^\infty \mathrm{e}^{-h^2y^2}\cos[(ks - c)\frac{y}{\sqrt{s}}]\sin\frac{\varepsilon y}{\sqrt{s}}\frac{dy}{y}. \tag{2.21}$$

---

[20] For Poisson's general preference to infinitesimals see [Schubring 2005, 455 f.].

For Poisson's inference from (2.20) to (2.21) further explanations would have been necessary. The only comment which Poisson gave in this context was in relation to (2.21):

> Strictly speaking, one is allowed to attribute to the variable y only finite values; because of the exponential factor $e^{-h^2y^2}$, however, one can expand the respective integral into the infinite, without a considerable error [Poisson 1824, 282].

From a rigorous point of view, one can deduce from (2.20) only that, for an arbitrarily large but finite Y,

$$P(c-\varepsilon \leq S_s \leq c+\varepsilon) \approx \frac{2}{\pi}\int_0^Y e^{-h^2y^2}\cos[(ks-c)\frac{y}{\sqrt{s}}]\sin\frac{\varepsilon y}{\sqrt{s}}\frac{dy}{y}+ \\ +\frac{2}{\pi}\int_{\frac{Y}{\sqrt{s}}}^{\infty}\rho^s\cos(s\varphi-c\alpha)\sin(\varepsilon\alpha)\frac{d\alpha}{\alpha}.$$

Apparently, for Poisson it was a matter of course, which did not need any special justification, that

$$\int_{\frac{Y}{\sqrt{s}}}^{\infty}\rho^s\cos(s\varphi-c\alpha)\sin(\varepsilon\alpha)\frac{d\alpha}{\alpha}\approx 0$$

for an "infinitely large" s.

From (2.21) one could infer, with the aid of the relation

$$\frac{1}{y}\cos[(ks-c)\frac{y}{\sqrt{s}}]\sin\frac{\varepsilon y}{\sqrt{s}}=\frac{1}{\pi\sqrt{s}}\int_{-\varepsilon}^{\varepsilon}\cos[(ks-c+z)\frac{y}{\sqrt{s}}]dz,$$

and consequently

$$\frac{2}{\pi}\int_0^{\infty}e^{-h^2y^2}\cos[(ks-c)\frac{y}{\sqrt{s}}]\sin\frac{\varepsilon y}{\sqrt{s}}\frac{dy}{y} \\ =\frac{1}{\pi\sqrt{s}}\int_{-\varepsilon}^{\varepsilon}\left(\int_0^{\infty}e^{-h^2y^2}\cos[(ks-c+z)\frac{y}{\sqrt{s}}]dy\right)dz,$$

that

$$P(c-\varepsilon\leq S_s\leq c+\varepsilon)\approx\frac{1}{2h\sqrt{\pi s}}\int_{-\varepsilon}^{\varepsilon}e^{-\frac{(ks-c+z)^2}{4h^2s}}dz. \tag{2.22}$$

Setting $c=ks$ and $\varepsilon=2hr\sqrt{s}$ in (2.22), Poisson finally obtained the result:

$$P(ks-2hr\sqrt{s}\leq S_s\leq ks+2hr\sqrt{s})\approx\frac{2}{\sqrt{\pi}}\int_0^r e^{-t^2}dt.$$

Poisson's discussion of sums of non-identically distributed random variables followed the model just described for identically distributed random variables. For the validity of his deductions in the general case, Poisson made a further condition explicit which was equivalent to (2.18).

2.2.5 Approximation by Series Expansions

The essential goal of Poisson's paper from 1829 on the CLT was to approximate the probabilities of a sum of a large number of random variables X_n, whose densities f_n vanish beyond a finite interval $[a;b]$, by a series expansion rather than to derive a "limiting formula." By the argument that the product $\rho_1(\alpha)\cdots\rho_s(\alpha)$, where

$$\rho_n = \sqrt{\left(\int_a^b f_n(x)\cos(\alpha x)dx\right)^2 + \left(\int_a^b f_n(x)\sin(\alpha x)dx\right)^2},$$

attained values significantly different from zero for very small α only, Poisson justified that it was possible to cut off that series expansion after its first terms.

With the abbreviations $\int_a^b xf_n(x)dx =: k_n$, $\int_a^b x^2 f_n(x)dx =: k_n', \ldots$, and the designations (2.15), Poisson [1829, 8 f.] derived the series expansions

$$\rho_n \cos\varphi_n = 1 - \frac{\alpha^2}{2}k_n' + \frac{\alpha^4}{4!}k_n''' - \cdots,$$

$$\rho_n \sin\varphi_n = \alpha k_n - \frac{\alpha^3}{3}k_n'' + \cdots.$$

Because of $|k_n| < |b| + |a|, |k_n'| < (|b| + |a|)^2, \ldots$ these series are convergent. Poisson [1829, 8] expressed the opinion that this convergence guaranteed the respective left sides being actually represented by the series expansions on the right. In this way, Laplace's formal calculations according to his method of approximation, which in many cases led to divergent series, were substituted by an explicit discussion of convergence.

By use of the series expansions for $\rho_n \cos\varphi_n$ and $\rho_n \sin\varphi_n$, series expansions for R, ψ (see (2.16)), and for $\cos(\psi - c\alpha)$ in powers of α were accomplished such that, because of (2.17),

$$P(c - \varepsilon \leq S_s \leq c + \varepsilon) = \frac{2}{\pi}\int_0^\infty e^{-\alpha^2 hs}\left(1 + \alpha^4 ls + \cdots\right) \times$$
$$\times \left(\cos[(ks - c)\alpha] + \alpha^3 gs\sin[(ks - c)\alpha] + \cdots\right)\sin(\varepsilon\alpha)\frac{d\alpha}{\alpha}.$$

In this formula k, h, g, l denote quantities depending on the moments of the single random variables; the absolute values of these quantities have upper bounds independent of s, as Poisson proved. In particular, $k = \frac{\Sigma \mathrm{E}X_n}{s}$ and $h = \frac{\Sigma \mathrm{Var}X_n}{2s}$ ensued. On the basis of these considerations Poisson apparently believed to have given an additional justification for the neglect of those terms which are, after having carried out the substitution $\alpha =: \beta/\sqrt{s}$, divided by a power of s larger than $s^{1/2}$. In this way, the approximation

$$P(c-\varepsilon \le S_s \le c+\varepsilon) \approx \frac{2}{\pi}\int_0^\infty e^{-\beta^2 h}\cos\frac{(ks-c)\beta}{\sqrt{s}}\sin\frac{\varepsilon\beta}{\sqrt{s}}\frac{d\beta}{\beta}+ \\ +\frac{2g}{\pi\sqrt{s}}\int_0^\infty e^{-\beta^2 h}\sin\frac{(ks-c)\beta}{\sqrt{s}}\sin\frac{\varepsilon\beta}{\sqrt{s}}\beta^2 d\beta \quad (2.23)$$

was reached [Poisson 1829, 9].

In his 1929 paper, Poisson's further proceeding was rather complicated. A considerably simplified approach was given in his book [1837, 270 f.]: Poisson in (2.23) set $c = ks$ and $\varepsilon = 2\gamma\sqrt{hs}$, with the result

$$P(ks-2\gamma\sqrt{hs} \le S_s \le ks+2\gamma\sqrt{hs}) \approx \frac{2}{\pi}\int_0^\infty e^{-\beta^2 h}\sin(2\beta\gamma\sqrt{h})\frac{d\beta}{\beta}. \quad (2.24)$$

Essentially making use of

$$\int_{-\infty}^{\infty} e^{-x^2}\cos(\alpha x)dx = \sqrt{\pi}e^{-\frac{\alpha^2}{4}},$$

Poisson showed that the integral in (2.24) was equal to

$$\frac{2}{\sqrt{\pi}}\int_0^\gamma e^{-u^2}du.$$

Poisson's preference for the just-described approach to the CLT by means of explicit series expansions might have been mainly caused by the fact that this method gave additional correction terms of the order $s^{-1/2}$ and less for "large" (but not infinite) s, and therefore was considered to be more general than the "simple" approximation by the normal distribution only. For the subsequent development of the CLT, Poisson's "infinitistic" approach seems to have been more influential, however.

2.3 The Central Limit Theorem After Poisson

During the time after Poisson, two crucial changes occurred in the development of probability theory. Firstly, probability eventually lost one of its major branches, the application to moral sciences. Secondly, the movement toward a purely mathematical view of stochastics, which in a certain sense had already begun with Laplace, gained momentum. The development of the CLT was connected with both fields, as we will see in the cases of Cauchy's and Dirichlet's contributions.

2.3.1 Toward a New Conception of Mathematics

Both Cauchy and Dirichlet are seen as representatives of a new mathematical conception emerging after 1800 which was generally accepted during the last third of the 19th century. The essentials of this new point of view can be summarized as follows: A separation of mathematics from its ontological relation to the physical and moral world was beginning to form, as stated by Kline [1972, 619 f.]. In [Laugwitz 1999, 187–191] this development is described as a transition from the consideration of the "contents" to the discussion of the "scope" of "concepts." The role of counterexamples in this context changed from irrelevant "curiosities" toward boundary posts indicating the limits of the specific concepts. Poisson, for example, still understood his examples of nonconvergence to the normal distribution in the sense of singular exceptions, which do not occur "in practice." Without external criteria, such as applicability, however, mathematics experienced an increased need to reflect on its internal logical consistency, as pointed out by Mehrtens [1990]. In this sense, Poisson's main counterexample would become especially important for Cauchy's critique of the method of least squares.

The framework of the growing abstraction of mathematics during the 19th century can only be roughly described in this exposition. An excellent survey is given by Schneider [1981a]. There were changes in the employment of mathematicians (from 18th-century academies to universities), which helped to promote pure mathematics.[21] The computational potentialities of analysis seemed to become gradually exhausted, so a turn to the discussion of analytical fundamentals or even to other, temporarily neglected disciplines, such as synthetic geometry, became plausible. The intellectual background was perhaps even more decisive. After the political upheavals due to the French Revolution, the confidence of the Enlightenment in a common standard of rationality began to vanish. The commonly accepted unity of mathematics and good sense began to drift apart (this process is exactly described by Daston [1988, 370–386], for the field of probability theory). The growing re-examination of basic definitions after 1800 can be considered as a reaction to the decline of the idea of self-evident "natural" standards.

The resulting changes toward "mathematical rigor" are not to be confused with changes in analytical style and methods. As several authors have pointed out since Lakatos [1966] and Spalt [1981], analytic reasoning during the first half of the 19th century using the language of infinitesimals was not fundamentally less rigorous than the application of epsilontic methods.[22] The decline of algebraic analysis, however, was closely connected to the new standards of mathematical rigor. This was also an essential point in the history of the CLT.

Certainly, the changes described above did not happen overnight. Cauchy and Dirichlet still worked a good deal in the tradition of problem solving of the 18th century. In the case of the CLT, however, the "new mathematics" can clearly be seen in the contributions of both authors.

[21] For more material on this topic see [Mehrtens, Bos, & Schneider 1981; Schubring 2005, Chapt. VII].

[22] Especially regarding Cauchy's work, the discussion is still quite controversial, see Sect. 1.3.

2.3.2 Changes in the Status of Probability Theory

Several subjects of classical probability were heavily attacked after Laplace's death. His personal authority, however, remained unharmed. This criticism was mainly directed toward applications of probability to human decisions, for example at court trials. Especially Poisson's work in this field caused a broad disapproval of the claim of classical probability for universal applicability, at least in France.[23] Daston [1988, 384] has pointed out that, as a consequence, a shift from the focus on the individual man toward the probability of mass phenomena occurred. Naturally, the CLT was also an excellent tool for the latter field. A further consequence was that a more critical awareness replaced the "natural" and often only tacit presuppositions of classical probability also in "unsuspicious" applications, such as error theory. In this way, error theory became the discipline of probability being subject to the most far-reaching mathematization. Some sources showed a rather abstract view of error theory and gave rise to demanding analytical discussions. This development was responsible for Cauchy's "rigorous" proof of the CLT during his dispute with Bienaymé over the priority of the method of least squares, as we will see below.

At several occasions during his work, Laplace had already pointed out the extreme relevance of his analytic methods of probability theory, especially his methods for approximating integrals depending on large numbers. Thus, from the analytical point of view, statements now interpreted as probabilistic limit theorems became appendages of the theory of definite integrals. Based on this idea, Dirichlet rather frequently gave courses on probability theory during the 1830s and 1840s, in which he directly referred to Laplacian methods, however with considerable modifications toward a "new" analytical rigor, from which his "rigorous" proof of the CLT (discussed in detail below) resulted. In this context, the CLT reached a quality different from the framework of classical probability theory. It was no longer only a tool for applications beyond mathematics, but also became a subject within (pure) mathematics, albeit with a mainly auxiliary character (serving as an illustration of the theory of definite integrals).

2.3.3 The Rigorization of Laplace's Idea of Approximation

As we have seen in the discussion of Poisson's deduction of an approximate normal distribution for sums of independent random variables, the following basic idea (for the sake of simplicity described only for identically distributed random variables with symmetric density function f on $[-a; a]$) was pursued: The probability P that a sum of s random variables of this kind has values within $[b-c; b+c]$ is (cf. formula (2.12)):

---

[23] There is also a German example: Jakob Fries's *Versuch einer Kritik der Prinzipien der Wahrscheinlichkeitsrechnung* [1842], which was based on Kant's philosophy, and met with Gauss's approval; see [Fischer 2004].

$$P = \frac{2}{\pi}\int_0^{\infty}\left(\int_{-a}^{a} f(x)\cos(\alpha x)dx\right)^s \cos(b\alpha)\sin(c\alpha)\frac{d\alpha}{\alpha}.$$

As expressed in the infinitesimal style of the first half of the 19th century, the power

$$\left(\int_{-a}^{a} f(x)\cos(\alpha x)dx\right)^s$$

with the "infinitely large" exponent s attains values which differ essentially from 0 only for "infinitely small" α. The whole integrand is, as a function of α, similar to a bell-shaped function, whose maximum peak becomes sharper and sharper as s increases. This circumstance gives rise to the conjecture that for "infinitely large" s the whole range $]-\infty;\infty[$ of the integral with respect to α can, with only an "infinitely small" error, be reduced to an "infinitely small" neighborhood of $\alpha = 0$. It was exactly the latter point which was used by Poisson (and many of his imitators) without any detailed justification. But, why should it be impossible for the value of the integral of an "infinitely" small function to be considerably large if the domain of integration itself is unbounded? This unsolved problem corresponded, in the end, to the unjustified neglect of higher terms in the approach via series expansions, and was most probably responsible for the already described unease (see Sect. 2.1.4) associated with Laplace's deduction of the CLT.

A more exact analysis of the CLT, which explicitly referred to the basic idea of the Laplacian method of approximation, had to show that for r in a specified range of "infinite smallness" the integral

$$\int_r^{\infty}\left(\int_{-a}^{a} f(x)\cos(\alpha x)dx\right)^s \cos(b\alpha)\sin(c\alpha)\frac{d\alpha}{\alpha}$$

would in fact become "infinitely small" for "infinitely large" values of s. As we have seen, Poisson's analysis had already shown that r had to be of an order around $1/\sqrt{s}$. Corresponding considerations were to be applied in the general case of non-identically and non-symmetrically distributed random variables.

Similar ideas led to Cauchy's sketch of the rigorous proof of a (if still rather specific) CLT in 1853, and also to Lyapunov's epochal proof of a very general form of the theorem in 1900/01. Cauchy had already begun in the 1820s to discuss "functions of great numbers," such that his work of 1853 was not only connected with error theory but was also produced in the broader context of his analytical studies. Dirichlet had, independent of Cauchy and actually even before him, also advanced similar ideas. He did not, however, publish his results, but only presented them in his lecture course of 1846.

2.4 Dirichlet's Proof of the Central Limit Theorem

Peter Gustav Lejeune Dirichlet (1805–1859) is renowned for his pioneering contributions to mathematical physics and number theory. In the field of probability theory, however, one can find only a few brief notices in Dirichlet's collected *Werke*. Actually, during his Berlin period (1828–1855), he quite frequently gave courses on probability and error theory presenting new and original ideas, as we can see from unpublished lecture notes (see [Fischer 1994]). In these lecture courses, Dirichlet's main concern was not the treatment of probabilistic fundamentals or applications, but rather the discussion of demanding analytical problems. He considered these problems as applications of the theory of definite integrals, and therefore plainly named several of the pertinent courses "Anwendungen der Integralrechnung" ("applications of integral calculus"). In one of these "Anwendungen"—dedicated to foundational issues of least squares that served as a 1-hr appendage to a 4-hr course on definite integrals in 1846[24]—one can find a very notable and innovative approach to a proof of the CLT.

Dirichlet's analytical style varied between an almost "epsilontic" presentation, as used in his publications, and a rather intuitive handling of problems, quite often connected with infinitistic methods. Evidence of this can be found in his lectures or unpublished drafts (see [Fischer 1994]). The style of Dirichlet's contribution to the CLT [1846] seems mainly of the second kind; yet, as we will see, all essential steps (only sketched out in the original source) can be taken using finitistic considerations which were within Dirichlet's scope.

2.4.1 Dirichlet's Modification of the Laplacian Method of Approximation

Dirichlet's main probabilistic interests lay in problems of approximating "functions of large numbers." Thus, he actually satisfied Laplace's hope that such questions would interest the "geometers" (see the introductional part of the present chapter). At the same time, one can see in Dirichlet's activities a shift from the typical objects of classical probability, concentrating on practical applicability, toward the discussion of the respective analytical methods.

In the 1830s, Dirichlet presented (e.g., [1838, 67 f.]) Laplace's original deduction of Stirling's formula in his lectures. He succeeded at least in deducing the law of Laplace's series (2.2), which Cauchy [1844, 68] would still consider to be unknown. As we have seen in Sect. 2.1.2, Laplace had set

$$\Gamma(s+1) = M \int_{-s}^{\infty} \mathrm{e}^{-z}(1+z/s)^s dz = M \int_{-\infty}^{\infty} \mathrm{e}^{-t^2} \frac{dz}{dt} dt,$$

[24] The corresponding lecture notes, written by an unknown author, are undated. From all we know about Dirichlet's teaching activities in probability theory, it seems evident, however, that the lecture notes pertain to Dirichlet's course in summer semester 1846 [Fischer 1994, 56, 60].

where z is a power series in t and $M = e^{-s}s^s$. Dirichlet differentiated the equality

$$e^{-z}(1 + z/s)^s = e^{-t^2}$$

by t to obtain

$$z\frac{dz}{dt} = 2t(s + z).$$

By employing the formula $z = k_1 t + k_2 t^2 + \cdots$ with unknown coefficients k_i ($z = 0$ if and only if $t = 0$) in the latter equation and by comparing the coefficients of powers of t, Dirichlet determined the first terms of $\sum_{n\geq 1} k_n t^i$. In essence, he developed the recursion formula

$$k_1 = \sqrt{2s}, \quad k_n = \frac{2k_{n-1}}{(n+1)k_1} - \frac{1}{2k_1}\sum_{i=2}^{n-1} k_i k_{n+1-i} \quad (n \geq 2).$$

From this, the series expansion

$$\Gamma(s+1) = s^{s+1/2}e^{-s}\sqrt{2\pi}\left(1 + \sum_{n\geq 1}\frac{1\cdot 3\cdot 5\cdots(2n+1)a_{2n+1}}{s^n}\right),$$

where

$$a_i = 2^{1-i}(\sqrt{2s})^{i-2}k_i$$

follows. (Dirichlet, however, made explicit only the first terms of the latter series expansion, which can also be deduced by different "modern" methods, see [Copson 1965, 53–57; Fischer 2006].)

In the 1840s, Dirichlet's interest in Stirling's formula no longer aimed at formal series expansions, but at a modification of the basic procedure concerning the Laplacian method of approximation, in exactly the sense which was described in Sect. 2.3.3 for the case of the CLT. Dirichlet [1841/42, 56–61] split the entire integral

$$\int_{-n}^{\infty} e^{-z}\left(1 + \frac{z}{n}\right)^n dz = \int_{-n}^{\infty} ydz = \Gamma(n+1)e^n n^{-n}$$

into the sum

$$\int_{-n}^{-n^m} ydz + \int_{-n^m}^{n^m} ydz + \int_{n^m}^{\infty} ydz = I_1 + I_2 + I_3,$$

where $\frac{1}{2} < m < \frac{2}{3}$. He set $y(z) = e^{-t^2(z)}$, and considering the convergent (!) series expansion of $\log y(z)$ around $z = 0$ (the abscissa of the maximum of y) he showed that I_1 and I_3 tend to 0 as n increases indefinitely, whereas

$$\frac{I_2}{\sqrt{2n}} \to \int_{-\infty}^{\infty} e^{-u^2}du = \sqrt{\pi}.$$

Thus, he obtained the expected result for $\Gamma(n+1)$ for "infinitely large" n (for more details see [Fischer 1994, 49 f.]).

2.4.2 The Application of the Discontinuity Factor

In order to adopt his reasoning from the case of Stirling's formula to the CLT, Dirichlet first needed an appropriate representation of the exact probabilities for sums or linear combinations of random variables. As one can see from the development of Dirichlet's ideas, as represented in his lectures of 1838 compared to his lectures of 1846, Poisson's discussion of the jump function (2.14) apparently led to Dirichlet's general method of calculating integrals over complicated domains with the aid of "discontinuity factors."

In his courses on Laplacian error theory as of 1838 and 1846, Dirichlet proposed the central problem of finding an approximate term for the probability P that the value of the linear combination $\alpha_1 x_1 + \alpha_2 x_2 + \cdots + \alpha_n x_n$ was within $-\lambda'$ and $+\lambda'$, where $\lambda' = \lambda\sqrt{n}$ and λ was a given positive constant. More precisely, $x_1, x_2, \ldots, x_n$ stood for independent observation errors (n being a large number), with expectations 0 and with (in general different) symmetric probability densities $f_1, f_2, \ldots, f_n$, vanishing beyond the finite interval $[-a; a]$.

Initially, Dirichlet [1838, 142–144] repeated Poisson's "combinatorial" procedure for the deduction of a formula for the probability that a linear combination of errors is within a given interval (see Sect. 2.2.2). But then, he presented—unlike Poisson also for the general case of arbitrary n and arbitrary λ—the application of "his" discontinuity factor

$$\frac{2}{\pi}\int_0^\infty \frac{\sin\varphi}{\varphi}\cos(k\varphi)d\varphi = \begin{cases} 0 \text{ for } |k| > 1 \\ 1 \text{ for } -1 < k < 1, \end{cases} \tag{2.25}$$

which he deduced from

$$\int_0^\infty \frac{\sin(kt)}{t}dt = \frac{\pi}{2} \quad (k > 0)$$

using trigonometric addition theorems. To this aim he calculated the probability

$$P = \int_G f_1(x_1)\cdots f_n(x_n)dx_1\cdots dx_n,$$

where

$$G = \{x \in \mathbb{R}^n | -\lambda' < \alpha_1 x_1 + \cdots + \alpha_n x_n < \lambda'\},$$

by use of the jump function (2.25), with the result

$$P = \frac{2}{\pi}\int_{[-a;a]^n}\int_0^\infty f_1(x_1)\cdots f_n(x_n)\frac{\sin\varphi}{\varphi}\cos\left[(\alpha_1 x_1 + \cdots \right.$$
$$\left. \cdots + \alpha_n x_n)\frac{\varphi}{\lambda'}\right] d\varphi dx_1\cdots dx_n. \tag{2.26}$$

Dirichlet [1839a;b;c] published three papers in which the jump function (2.25) was used for the calculation of specific multiple integrals that were important for the determination of space volumes and for potential theory, but he did not mention any applications in probability theory. In his 1846 course, he totally ignored Poisson's combinatorial approach, and by applying his discontinuity function he deduced (2.26) through a consideration of the analogies between probabilities and space volumes.[25] From (2.26) Dirichlet [1846, 27] deduced

$$
\begin{aligned}
P &= \frac{2}{\pi}\int_0^\infty \frac{\sin(\lambda'\varphi)}{\varphi}\int_{-a}^{a} f_1(x_1)\mathrm{e}^{\alpha_1 x_1\varphi\sqrt{-1}}dx_1\cdots\int_{-a}^{a} f_n(x_n)\mathrm{e}^{\alpha_n x_n\varphi\sqrt{-1}}dx_n\,d\varphi \\
&= \frac{2}{\pi}\int_0^\infty \frac{\sin(\lambda\sqrt{n}\varphi)}{\varphi}\int_{-a}^{a} f_1(x_1)\cos(\alpha_1 x_1\varphi)dx_1\cdots \\
&\qquad\cdots\int_{-a}^{a} f_n(x_n)\cos(\alpha_n x_n\varphi)dx_n\,d\varphi. \qquad (2.27)
\end{aligned}
$$

The interchange of the order of integration was not discussed. In his paper [1839c], however, Dirichlet—without referring to probabilistic applications—pointed out the need for a proof of such interchanges. He suggested multiplying the integrands with factors such as $\mathrm{e}^{-r\varphi}$. For $r > 0$ the absolute "convergence" of the modified integrals would be guaranteed (and, thus, the interchangeability of the order of integration). For both multiple integrals, the one before and the other after the interchange, one had finally to examine the limit $r \to 0$. Actually, this method is practical in the case of the probabilities of linear combinations of mutually independent random variables if one assumes for the densities of these variables certain—not very drastic—conditions, from which the absolute integrability of the function $\varphi \mapsto \frac{\sin\varphi}{\varphi}\int_{[-a;a]} f_1(x_1)\cdots f_n(x_n)\cos[(c_1x_1+\cdots+c_nx_n)\varphi]dx_1\cdots dx_n$ over $[0;\infty[$ follows for fixed $c_1,\ldots,c_n$. The hypotheses regarding the density functions, which Dirichlet supposed more or less tacitly, are in fact sufficient for this condition.[26]

2.4.3 Dirichlet's Proof

Dirichlet's discussion of the asymptotic distribution of linear combinations of observational errors can be reconstructed in the sense of a rigorous proof of the CLT, even from today's point of view.

---

[25] Glaisher [1872a, 195; 1872b, 98] was perhaps the first—of course without being directly influenced by Dirichlet—to publish the use of Dirichlet's factor in exactly the same way as Dirichlet had presented it in his 1846 lecture course. [Cauchy 1853d], as it seems without knowledge of Dirichlet's prior contributions, had already given a very similar consideration, see Sect. 2.5.2.

[26] For an account of post-Weierstrassian era on the problem of interchanging the order of integration in applying Dirichlet's factor, see [David 1909].

2.4.3.1 Tacit Assumptions and Proposition

As described above, Dirichlet discussed linear combinations $\alpha_1 x_1 + \cdots + \alpha_n x_n$ of random errors. The densities of these errors were not only considered to be symmetric and concentrated on a fixed interval, but also to be smooth (in the sense of the existence of continuous derivatives) and unimodal, as it appears from a picture in the lecture notes [1846, 21]. The latter assumption was, however, not absolutely necessary for Dirichlet's deductions. As we will see, Dirichlet tacitly presupposed that the sequence of the α_ν had a positive lower bound (named α by me) and a positive upper bound (A), and that all variances of the random errors should be uniformly bounded away from 0 (by a positive lower bound to which I refer as k). Such tacit assumptions were natural within error theory. For a rigorous completion of Dirichlet's line of proof in the case of non-identically distributed observation errors, one has to additionally assume a certain uniformity in the shape of all the density functions, such as, for example, the existence of an upper bound C such that $|f_\nu'(x)| < C$ for all $x \in [-a; a]$ and all ν. (From this condition one can already deduce the existence of the above-mentioned constant k.)

Expressed as a "modern" limit assertion, the main result of Dirichlet's lecture course on error theory in 1846 was

$$\left| P\left(-\lambda\sqrt{n} \leq \sum_{\nu=1}^{n} \alpha_\nu x_\nu \leq \lambda\sqrt{n}\right) - \frac{2}{\sqrt{\pi}} \int_0^{\lambda/r} \mathrm{e}^{-s^2} ds \right| \to 0 \quad (n \to \infty),$$

where

$$r = 2\sqrt{\frac{1}{n}\sum_{\nu=1}^{n} k_\nu \alpha_\nu^2}.$$

Even if the transcriber of the lecture notes did apparently not render all arguments entirely correctly, the basic ideas for a rigorous proof of this limit can be clearly discerned. At least in the special case of identically distributed errors a complete argumentation can be reached with such methods that Dirichlet himself used.[27]

2.4.3.2 Dirichlet's Discussion of the Limit

Analogous to his derivation of Stirling's formula, Dirichlet split the integral (2.27) with respect to φ into three parts

$$\frac{2}{\pi}\int_0^{\delta} \ldots d\varphi + \frac{2}{\pi}\int_{\delta}^{\Delta} \ldots d\varphi + \frac{2}{\pi}\int_{\Delta}^{\infty} \ldots d\varphi = p + q_1 + q_2,$$

where δ and Δ depend on n in such a way that

$$\delta\sqrt[4]{n} \to 0, \quad \delta\sqrt{n} \to \infty \quad (n \to \infty) \tag{2.28}$$

[27] For an edition of the original source see Appendix.

and

$$\Delta \propto n^{\gamma} \quad \text{with an arbitrary, but fixed } \gamma > 0. \tag{2.29}$$

Dirichlet represented the product $\Pi(\varphi)$ of the integrals

$$g_\nu(\varphi) := \int_{-a}^{a} f_\nu(x_\nu)\cos(\alpha_\nu x_\nu \varphi)dx_\nu$$

by

$$\Pi(\varphi) = \mathrm{e}^{-\sum k_\nu \alpha_\nu^2 \varphi^2} \mathrm{e}^{R(\varphi)}, \quad k_\nu := \frac{1}{2}\int_{-a}^{a} z^2 f_\nu(z)dz. \tag{2.30}$$

It was not explained in the lecture notes [Dirichlet 1846] that for general densities f_ν this representation with real $R(\varphi)$ is only valid for sufficiently small φ, and therefore only in the first of the three integrals for small δ. Since $g_\nu(\varphi) > 0$ for $0 \le \varphi \le \frac{\pi}{2Aa}$, $R(\varphi)$ exists for at least all $\varphi \in [0; \frac{\pi}{2Aa}]$. Dirichlet perhaps supposed unimodal densities f which diminish sufficiently fast with growing absolute values of the argument; then the term $\int_{-a}^{a} f(x)\cos(\alpha\varphi x)dx$ is positive for all α and all φ. As we will see below, however, it actually suffices that (2.30) holds for a small interval of φ-values.

In order to justify the asymptotic disappearance of

$$R(\varphi) = \sum_{\nu=1}^{n}\left(\log\left(\int_{-a}^{a} f_\nu(x_\nu)\cos(\alpha_\nu x_\nu\varphi)dx_\nu\right) + k_\nu\alpha_\nu^2\varphi^2\right)$$

in the first integral, Dirichlet expanded each logarithmic term into a power series of φ (in each case he explicitly took into account only the first nontrivial power of φ), and thus obtained for $0 \le \varphi \le \delta$ an estimate equivalent to the form

$$|R(\varphi)| < nL\delta^4 + nM\delta^6 + \cdots. \tag{2.31}$$

$L, M, \ldots$ designate the absolute values of the largest coefficients of $\varphi^4, \varphi^6, \ldots$ among all expansions of the individual logarithmic terms, and are therefore constants depending only on the functions f_ν and the multipliers α_ν. Dirichlet did not discuss the exact form of these constants. On the basis of (2.31) and (2.28) Dirichlet concluded that $R(\varphi)$ could be neglected in the first integral

$$p = \frac{2}{\pi}\int_0^{\delta} \frac{\sin(\lambda\sqrt{n}\varphi)}{\varphi}\Pi(\varphi)d\varphi = \frac{2}{\pi}\int_0^{\delta}\frac{\sin(\lambda\sqrt{n}\varphi)}{\varphi}\mathrm{e}^{-\sum k_\nu\alpha_\nu^2\varphi^2}\mathrm{e}^{R(\varphi)}d\varphi$$

as $n \to \infty$. For a complete justification (see [Fischer 2000, sect. 2.3.1]) of Dirichlet's hints, one can show, with the aid of the elementary inequalities

$$\cos z \ge 1 - \frac{z^2}{2},$$

$$\cos z \le 1 - \frac{z^2}{2} + \frac{z^4}{24},$$

$$\log(z) < z,$$

$$\log(1 - z) > -z - 2z^2 \quad (0 < z \leq \frac{1}{2}),$$

and by considering the above-mentioned "tacit presuppositions," that

$$|R(\varphi)| < nL\delta^4 \quad (L = \frac{a^4}{2}A^4). \tag{2.32}$$

In the integral p, Dirichlet now made the substitution of variables $\psi = \sqrt{n}\varphi$. The upper bound $\delta\sqrt{n}$ of the domain of integration of the new integral became equal to ∞ as $n \to \infty$ because of (2.28). Thus, for a "large" number of observations the relation

$$p \approx \frac{2}{\pi}\int_0^\infty \frac{\sin(\lambda\psi)}{\psi} e^{-\psi^2 \frac{\sum k_\nu \alpha_\nu^2}{n}} d\psi$$

followed. This relation can be rigorously deduced from the inequalities (which were not explicitly stated by Dirichlet):

$$\left| \int_0^{\delta\sqrt{n}} \frac{\sin(\lambda\psi)}{\psi} e^{-\psi^2 \frac{\sum k_\nu \alpha_\nu^2}{n}} d\psi - \int_0^{\delta\sqrt{n}} \frac{\sin(\lambda\psi)}{\psi} e^{-\psi^2 \frac{\sum k_\nu \alpha_\nu^2}{n}} e^{R(\psi/\sqrt{n})} d\psi \right|$$
$$< \max(e^{nL\delta^4} - 1; 1 - e^{-nL\delta^4})\lambda \int_0^\infty e^{-k\alpha^2\psi^2} d\psi =: \lambda C_1(n)$$

(based on (2.32)) and

$$\left| \int_{\delta\sqrt{n}}^\infty \frac{\sin(\lambda\psi)}{\psi} e^{-\psi^2 \frac{\sum k_\nu \alpha_\nu^2}{n}} d\psi \right| \leq \frac{1}{\delta\sqrt{n}} \int_0^\infty e^{-\psi^2 k\alpha^2} d\psi =: C_2(n).$$

From (2.28) one sees that the right sides of these inequalities tend to 0 as $n \to \infty$.

Finally, Dirichlet [1846, 30] concluded by use of well-known integral formulae that

$$p \approx \frac{2}{\sqrt{\pi}} \int_0^{\lambda/r} e^{-s^2} ds.$$

The justification that q_1 and q_2 tend to 0 as n increases, is only hinted at in the lecture notes [Dirichlet 1846, 30 f.], and seems to go as follows: $g_\nu(\varphi) = \int_{-a}^{a} f_\nu(x)\cos(\alpha_\nu\varphi x)dx$ is strictly monotonic decreasing in the interval—dependent on ν (!)—$[0; \varepsilon_\nu]$, and thus $g_\nu(\varphi_0) > |g_\nu(\varphi)|$ for all $\varphi_0 \in [0; \varepsilon_\nu]$ and all $\varphi > \varphi_0$.[28] From this, Dirichlet concluded (loosely translated) that there must exist a $\delta_n > 0$ such that for all $\varphi_0 \in [0; \delta_n]$ and all $\varphi > \varphi_0$ also $\Pi(\varphi_0) > |\Pi(\varphi)|$ holds. Apparently, the possible dependence of the ε_ν on ν, and thus of the δ_n on n, was not taken into consideration, and it was supposed that, for a sufficiently large n, the δ according to (2.28) would be smaller than δ_n, and therefore

[28] We have $g_\nu'(\varphi) < 0$ in a neighborhood of $\varphi = 0$ and $|g_\nu(\varphi)| < 1$ for $\varphi > 0$. Moreover, $g_\nu(\varphi) \to 0$ for $\varphi \to \infty$, as one can deduce after partial integration, see below. Finally, the asserted behavior of g_ν follows from its continuity with respect to all $\varphi \geq 0$.

$$|\Pi(\varphi)| < \Pi(\delta) \quad \forall \varphi > \delta \tag{2.33}$$

would apply. However, because δ_n might tend to 0 even faster than δ as $n \to \infty$, (2.33) only holds if a certain uniformity in the shape of the factors $g_\nu(\varphi)$ of $\Pi(\varphi)$ as functions of φ is presupposed. (Actually, this can be deduced from the "tacit assumptions," though, as it seems, only by methods which were not known to Dirichlet, see [Fischer 2000, Sect. 2.3.1].) From (2.33) one gets for sufficiently large n

$$|q_1| < \int_\delta^\Delta \left| \frac{\sin(\lambda\sqrt{n}\varphi)}{\varphi} \Pi(\varphi) d\varphi \right| < \lambda\sqrt{n}\Delta\Pi(\delta).$$

By definition $\Pi(\delta) = \mathrm{e}^{-\sum k_\nu \alpha_\nu^2 \delta^2} \mathrm{e}^{R(\delta)}$ and therefore, using the "tacit assumptions":

$$|q_1| < \lambda\sqrt{n}\Delta \mathrm{e}^{-nk\alpha^2\delta^2} \mathrm{e}^{R(\delta)} =: \lambda C_3(n).$$

If one sets $\delta = n^{-\frac{1}{3}}$, as suggested by Dirichlet [1846, 29] as an example of a possible δ in accordance with (2.28), the right side tends to 0 as n increases.

In order to justify that

$$q_2 = \frac{2}{\pi} \int_\Delta^\infty \frac{\sin(\lambda\sqrt{n}\varphi)}{\varphi} \Pi(\varphi) d\varphi$$

can also be neglected for "infinite" n, Dirichlet used the relation

$$\int_{-a}^{a} \cos(\alpha_\nu \varphi x) f_\nu(x) dx = \frac{2 f_\nu(a) \sin(\alpha_\nu \varphi a)}{\alpha_\nu \varphi} - \int_{-a}^{a} \frac{\sin(a\varphi x)}{\alpha_\nu \varphi} f_\nu'(x) dx,$$

which can be derived by partial integration. (For the existence of continuous derivatives of the densities see the "tacit assumptions.") From that, Dirichlet concluded that $|\Pi(\varphi)|$ must be smaller than $\left(\frac{c}{\varphi}\right)^n$ with a constant c independent of n, which is only true under the "tacit assumptions." Dirichlet's reasoning can be completed as follows: From the estimate $|\Pi(\varphi)| < \left(\frac{c}{\varphi}\right)^n$ one gets

$$|q_2| < \int_\Delta^\infty \frac{1}{\varphi} \left(\frac{c}{\varphi}\right)^n d\varphi = \left(\frac{c}{\Delta}\right)^n \frac{1}{n} =: C_4(n).$$

From the hypothesis (2.29) on the growth of Δ, the latter term tends to 0 as $n \to \infty$.

On the basis of the inequalities stated above, we can reconstruct Dirichlet's result by the inequality

$$\left| P\left(-\lambda\sqrt{n} \le \sum_{\nu=1}^{n} \alpha_\nu x_\nu \le \lambda\sqrt{n}\right) - \frac{2}{\sqrt{\pi}} \int_0^{\lambda/r} \mathrm{e}^{-s^2} ds \right|$$
$$\le \lambda C_1(n) + C_2(n) + \lambda C_3(n) + C_4(n),$$

which is valid for sufficiently large n. Presupposing

$$\delta = n^{-1/2+\varepsilon}, \quad 0 < \varepsilon < \frac{1}{4},$$

the bounds C_1, C_2, C_3, C_4 have the respective asymptotic orders

$$C_1(n) = \mathrm{O}(n^{-1+4\varepsilon}),\ C_2(n) = \mathrm{O}(n^{-\varepsilon}),\ C_3(n) = \mathrm{o}(n^{-\rho}),\ C_4(n) = \mathrm{o}(n^{-\rho}),$$

where ρ is an arbitrary positive constant. From this, we can see that Dirichlet's method gives an estimate for the error of approximation that is far from the optimal one as developed by modern methods. It was, however, not Dirichlet's intention at all to find a "very good" approximation error for the normal distribution. Apparently, he wanted to show that his modification of the Laplacian method of approximation could also be applied to the problem of probabilities of linear combinations of random errors. In this sense, the central CLT for Dirichlet served chiefly as an illustration of special analytical techniques and was less a problem which he treated in its own right.

2.5 Cauchy's Bound for the Error of Approximation

Augustin Louis Cauchy (1789–1857) provided fundamental contributions to a great number of mathematical subjects and essentially determined the development of mathematics during the 19th century. On probability theory in the narrow sense, Cauchy only published a few papers, in 1853, printed in the *Comptes rendus*, which referred to his dispute with Irénée Jules Bienaymé (1796–1878) over the Laplacian foundation of the method of least squares. In this scientific controversy, which occurred during the months of June, July, and August in the summer of 1853 at the Paris Academy, Bienaymé defended the Laplacian error theory, whose basic ideas were repeatedly criticized by Cauchy.[29] Cauchy's last article in a total of eight papers contains an interesting discussion of the approximate normal distribution of linear combinations of random errors. Basically, his line of analytical argumentation is similar to Dirichlet's and employs methods which are still being used in the modern treatment of the CLT. His (rather narrow) conditions are in essence the same as Dirichlet's.

2.5.1 The Cauchy–Bienaymé Dispute

From a historian's point of view, Cauchy's and Bienaymé's interest in treating stochastic problems in an almost purely mathematical manner, indicating a shift from classical toward mathematical probability, is especially important. However,

[29] For more details on this dispute see [Heyde & Seneta 1977] and [Fischer 2000, 76–97]. Bienaymé's contributions are, as listed in the Bibliography, [Bienaymé 1853a] to [Bienaymé 1853e], Cauchy's contributions are [Cauchy 1853a] to [Cauchy 1853h].

Cauchy's position of only accepting arguments within mathematics for a discussion of the error theoretic foundations (which became more and more adamant during the controversy), met with Bienaymé's opposition, who still demanded the critical "good sense" assessment of those problems.

The political and private connections of both opponents might have been especially important for the background of their scientific quarrel. As a consequence of the revolution of July 1830, which brought Louis-Philippe, the "king of the people," to power, Cauchy, being a supporter of the overthrown Charles X Bourbon, had to give up his positions in higher education and go into exile.[30] From 1833 to 1838 he was in charge of the education of Charles's eldest son in Prague. After the completion of his duties there, he went back to Paris and resumed work at the Academy. After the revolution of February 1848, which, for a brief period, reestablished the republic, he was able to return to teaching at the university. With the seizure of power by Napoleon III in 1851, Cauchy's official position remained unchanged. As a supporter of the house of Bourbon, however, he did not look on this political change especially enthusiasted.

In 1820 Bienaymé[31] set out on a brilliant career in government finance which remained entirely unscathed by the 1830 revolution. Whereas the revolution of 1848 had brought some advantages to Cauchy, Bienaymé had to resign from his positions. Consequently he delved into more scientific endeavors. Bienaymé, in contrast to Cauchy, sympathized with Napoleon III, and after his seizure of power regained a certain influence on the country's financial politics.

Apart from differences in their political views, Cauchy and Bienaymé seem to have had personal misgivings as well. As suggested by [Heyde & Seneta 1977, 13], these could have originated for one thing from different religious beliefs—Cauchy was a fanatic Catholic, and Bienaymé tended toward agnosticism. Furtherly, Bienaymé cultivated a close friendship with Antoine Auguste Cournot,[32] who was very influential in science back then, while Cournot and Cauchy were bitter enemies.

Bienaymé presented his essay on foundational problems of least squares (see Sect. 2.1.5.2) in 1852 at the Paris Academy. His good reception there contributed significantly to his election as an ordinary member of the Academy soon thereafter. It is only natural that Bienaymé would have been very interested in contributing to discussions on "his field," error calculus, at Academy conventions. He found a suitable opportunity when Cauchy once again presented his method of interpolation (introduced already in 1835); Cauchy suggested that this method be applied instead of least squares even in those cases which had not yet been taken into consideration when his procedure of interpolation was introduced.

In presenting his method in 1835, Cauchy began with the following problem: He assumed that a function $y(x)$ could be expanded into a convergent series of the form

---

[30] For biographical details on Cauchy see [Belhoste 1991].

[31] For biographical details see [Heyde & Seneta 1977].

[32] Regarding probability theory, Cournot became especially prominent by his elementary treatise [Cournot 1843], in which a clear distinction was made between the subjective and the objective notion of probability.

$$y(x) = au(x) + bv(x) + cw(x) + \cdots$$

with given functions $u(x)$, $v(x)$, $w(x)$, ..., but unknown coefficients $a, b, c, \ldots$. Assigned to the given abscissae $x_1, x_2, \ldots, x_n$ were observed function values $y_1, y_2, \ldots, y_n$, which were, however, subject to the observation errors $\epsilon_1, \epsilon_2, \ldots, \epsilon_n$. Cauchy searched for a method of "interpolation" with which one could jointly 1) assess, with regard to the order of magnitude of the observational errors, how many series terms had to be calculated to obtain a sufficiently exact approximation of the true function value for each arbitrary x, and 2) calculate those series terms in an easy way. Cauchy [1835/37, 8–16] presented a procedure by which the coefficients $a, b, c, \ldots$ could be approximated by a method that allowed one to calculate the coefficients with a simple correction from the ones already determined, if the number of the coefficients was increased by 1. From the error theoretic point of view, Cauchy's reasoning was based on the idea of minimizing the maximal possible error in each single stage of his procedure.

Cauchy's method of interpolation can be considered as a procedure for determining compromise solutions $\overline{a}_1, \overline{a}_2, \ldots$ of the overdetermined system

$$y_i = a_1 u_{i1} + a_2 u_{i2} + \cdots + a_r u_{ir} + \cdots \quad (i = 1, \ldots, n)$$

with the given u_{ih} (according to the function values $u(x_i), v(x_i), \ldots$) and y_i (the observations afflicted by errors), where, however, the number $r < n$ of the $\overline{a}_i$ needed is not known at the beginning. Yet it was obvious that Cauchy's procedure could also be applied to the case of overdetermined systems of linear equations with a fixed number r of variables.

Around 1840, Cauchy began to show increased interest in astronomy and especially in perturbation theory. Belhoste [1991, 205 f.] sees a connection with Cauchy's election to the "Bureau des Longitudes" in 1839, which had to be revoked because, being a royalist, Cauchy had refused to show any kind of allegiance to the "king of the people" Louis-Philippe. The works of astronomers Hervé Faye, Urbain Jean Joseph Leverrier (whose investigations in perturbation theory led to the discovery of Neptune in 1846), and Antoine François Yvon-Villarceau were influenced by Cauchy, and in turn stimulated some contributions by him. The problem of comparing observations and results obtained by perturbation theory kept Cauchy busy for most of the second half of 1847, when he issued a series of papers, and led him back to his own method of interpolation. Now, he [1847a] wanted to see this method also applied to overdetermined systems of linear equations with an a priori fixed number of unknowns. One can assume that this problem was being repeatedly discussed by the astronomy-prone members of the Academy. Cauchy [1847b] referred to a paper published by Villarceau in 1845 (this paper was not further specified) because approximation methods had apparently been used in it, analogous to his method of interpolation. Around 1849, Villarceau used Cauchy's method in extensive calculations of approximations of various orbit parameters [Heyde & Seneta 1977, 74]. Cauchy [1853a, 36] quoted a remark made by Faye on the usefulness of his interpolation procedure (the corresponding paper of Faye's cannot be bibliographically determined). So, when declaring himself to be partial to the method of least squares and against the method of interpolation, Bienaymé met not only with opposition from Cauchy, but from a whole group of astronomers.

2.5.2 Cauchy's Exceptional Laws of Error

Cauchy's initial line of argument was to minimize the maximum possible errors of approximation. Thus, he used a typical interpolation justification, which practically did not touch probability at all. Bienaymé [1853a, 5; 10], on the other hand, criticized this lack of probabilistic argumentation: Errors of observation are subject to chance. Thus, in order to fit the parameters to the observations, those methods should be preferred that can be analyzed and justified by stochastic considerations. In this way, Bienaymé emphasized the universal claim of classical probability being responsible for all fields in which complete knowledge of causes and laws could not be obtained. In response to this criticism, Cauchy began his probabilistic research. According to Schneider [1987a, 200 f.], Cauchy did not disapprove probability completely, but was only willing to accept probabilistic results which could be justified within mathematics. For Cauchy, the usual reasoning of classical probability, based on the unity of good sense and mathematics, had become obsolete. In the case of error theory, Laplace had claimed that the method of least squares should be preferred "in any case." Now, Cauchy set out to ridicule this claim by using Laplace's (and Bienaymé's) own probabilistic methods, although from a strictly mathematical point of view.

Like Laplace, Cauchy considered the system of n "approximative" equations

$$a_jx + b_jy + \cdots + g_jv + h_jw = k_j \quad (j = 1, \ldots, n)$$

with m "unknowns" $x, y, \ldots, v, w$ and n observed values $k_1, k_2, \ldots, k_n$. Cauchy approximated the "unknown" x by $\overline{x}=\sum_{j=1}^{n} \lambda_j k_j$, where the multipliers $\lambda_1, \ldots, \lambda_n$ had the additional property

$$\sum_{j=1}^{n} \lambda_j a_j = 1, \sum_{j=1}^{n} \lambda_j b_j = 0, \ldots, \sum_{j=1}^{n} \lambda_j h_j = 0. \tag{2.34}$$

From the "exact" equations

$$a_jx + b_jy + \cdots + g_jv + h_jw = k_j - \epsilon_j \quad (j = 1, \ldots, n),$$

where the ϵ_j represent the observational errors, it followed that

$$\sum_{j=1}^{n} \lambda_j a_j x + \sum_{j=1}^{n} \lambda_j b_j y + \cdots + \sum_{j=1}^{n} \lambda_j g_j v + \sum_{j=1}^{n} \lambda_j h_j w = \sum_{j=1}^{n} \lambda_j k_j - \sum_{j=1}^{n} \lambda_j \epsilon_j.$$

On account of (2.34) the estimate $\overline{x}$ was distorted by the "error" $\overline{x}-x=\sum_{j=1}^{n}\lambda_j\epsilon_j$. Cauchy restricted his discussion to the determination of $\overline{x}$ as being representative of all of the other variables. For the errors ϵ_j he presupposed a common symmetrical density $f(x)$, concentrated on the interval $[-\kappa;\kappa]$ with $\kappa \leq \infty$. For those densities Cauchy coined the term "indice de probabilité." Taking up the Laplacian characterization of the "most advantageous value," he demanded that

$$p = P\left(|x-\overline{x}| \leq \upsilon\right) = P\left(|\sum_{j=1}^{n}\lambda_j\epsilon_j| \leq \upsilon\right) = \max \qquad (2.35)$$

for all $\upsilon > 0$.

In his discussion of this condition, Cauchy made systematic use of the (now so-called) characteristic function, which he named "fonction auxiliaire." If $g(x)$ was the "indice de probabilité" of an error with values within $[\kappa_1;\kappa_2]$, then the "fonction auxiliaire" related to it was defined by[33]

$$\varphi(x) = \int_{\kappa_1}^{\kappa_2} \mathrm{e}^{-\mathrm{i}zx} g(z)dz \quad (\mathrm{i} = \sqrt{-1}).$$

Repeating arguments of his proof [1818; 1827, note VI] of the Fourier inversion formula,[34] he [1853d; 1853e] showed that for symmetrical densities f, defined as above, and their characteristic functions

$$\varphi(x) = 2\int_{0}^{\kappa} f(z)\cos(xz)dz$$

[33] The designation "indice de probabilité" is used, for example, in [Cauchy 1853f, 106], the designation "fonction auxiliaire" in [Cauchy 1853h, 125]. In a slightly different form compared with Cauchy's use, in modern probability theory the characteristic function of a random variable X is defined by $\mathrm{E}e^{+\mathrm{i}X\theta}$ instead of $\mathrm{E}e^{-\mathrm{i}X\theta}$. For symmetrically distributed random variables with zero means (which case was predominantly considered by Cauchy) both terms coincide.

[34] Fourier, Poisson, and Cauchy around 1820 (more or less independently) published very similar versions of the inversion formula [Laugwitz 1990, 30–34]. An early form, which remained, however, unpublished, had been presented by Fourier already in 1807 [Grattan-Guinness & Ravetz 1972]. The complex version of the formula

$$f(x) = \lim_{c\to\infty}\frac{1}{2\pi}\int_{-c}^{c}\int_{-\infty}^{\infty} f(t)\mathrm{e}^{\mathrm{i}u(t-x)}dtdu$$

for functions $f(x)$ continuous in x (precise properties of those functions were not explained for the time being) is essentially due to Cauchy. In the collection of Gauss's private papers ("Nachlass") an unpublished note (written presumably before 1813) on a complex version, with title "Schönes Theorem der Wahrscheinlichkeitsrechnung" ("beautiful theorem of probability calculus") [Gauss 1900, 88 f.] was found as well. From Gauss's remarks one can see that he derived his formulae on the basis of orthogonality relations like (2.3), by considering Fourier series with periods tending to infinity. One may suppose that Gauss was inspired to these observations by reading Laplace's *TAP*. For surveys of the history of the Fourier inversion formula during the 19th century, see [Burckhardt 1914, 1085–1097] (up to ca. 1850) and [Pringsheim 1907] (for the time 1850–1900). An outline of the entire development up to ca. 1940 is given by Cooke [2005].

the equation

$$f(x) = \frac{1}{\pi}\int_0^\infty \varphi(z)\cos(xz)dz \tag{2.36}$$

holds. For the characteristic function Φ of the linear combination $\lambda_1\epsilon_1 + \cdots + \lambda_n\epsilon_n$, where each of the mutually independent errors $\epsilon_1, \ldots, \epsilon_n$ obeys the law f, Cauchy derived:

$$\Phi(x) = \varphi(\lambda_1 x)\cdots\varphi(\lambda_n x). \tag{2.37}$$

From a modern point of view, Cauchy's proof [1853d, 86] for the latter identity was unnecessarily complicated, as it was not based on the now common conception of characteristic function as expectation. Instead, Cauchy used, in a rather intricate way, the jump function, very similar to Dirichlet's,

$$\frac{1}{2\pi}\int_{-\infty}^{\infty}\int_a^b e^{\theta(\tau-x)i}d\tau d\theta = \begin{cases} 1 & \text{for } x \in]a;b[\\ 0 & \text{for } x \notin [a;b], \end{cases}$$

which he derived by a (rather formal) use of the Fourier inversion formula.[35] Hinting at this jump function, Cauchy [1853e, 96] also stated

$$p = \frac{2}{\pi}\int_0^\infty \frac{\sin(\theta\upsilon)}{\theta}\Phi(\theta)d\theta, \tag{2.38}$$

where $\Phi(\theta)$ was defined as above (see (2.37)).

Cauchy [1853e, 98–101] gave a plausible justification that condition (2.35) is met if and only if $\kappa = \infty$ and

$$\varphi(x) = e^{-c|x|^N} \tag{2.39}$$

with positive constants c and N (see [Heyde & Seneta 1977, 82–85]). Cauchy's arguments for the "only if" were not sound.

From (2.37) to (2.39) it followed that

$$p = \frac{2}{\pi}\int_0^\infty e^{-c\theta^N\sum_{j=1}^n|\lambda_j|^N}\frac{\sin(\theta\upsilon)}{\theta}d\theta$$

[Cauchy 1853e, 102]. Independent of υ, p is maximized if $\sum_{j=1}^n|\lambda_j|^N$, under the constraint (2.34), is minimized. This implies, as Cauchy [1853e, 102 f.] showed, in the case for which a single element x is to be determined ($b_j = \cdots = g_j = h_j = 0$), that the condition

$$\lambda_j = \operatorname{sign}(a_j)|a_j|^{\frac{1}{N-1}}\left(\sum_{r=1}^n |a_r|^{\frac{N}{N-1}}\right)^{-1}$$

[35] Jump functions played an important role in Cauchy's analytical work. As a means for integration he used this device not until 1853, however. See [Burckhardt 1914, 963; 1320–1324] for a general account on the use of jump functions during the first half of the 19th century.

holds for p being maximal. Only for the case $N = 2$ are the λ_j the least square multipliers. Cauchy did not observe that only for exponents N with $N \leq 2$ the function $\varphi(x)$ in (2.39) was the characteristic function of a probability distribution. On the contrary, he assigned to the case $N = \infty$ an essential importance. As Cauchy [1853e, 103 f.] argued, this case corresponded to his own method of interpolation with multipliers $\lambda_j = \pm 1$.

With the aid of the inversion formula (2.36) Cauchy was able to determine the specific law of error corresponding to the constants c and N in two special cases: For $N = 2$ one gets the Gaussian law of error, and for $N = 1$ one gets the density

$$f(x) = \frac{k}{\pi} \frac{1}{1 + k^2 x^2} \quad \left(k = \frac{1}{2\sqrt{c}} \right).$$

Poisson (see Sect. 2.2.3.1) had already shown that the sum of independent identically distributed random variables with this density does not satisfy the CLT.

The main result of the article [Cauchy 1853e] was the fact that laws of error which are different from the Gaussian error law can lead to systems of multipliers entirely different from the least squares multipliers if Laplace's criterion for the "most advantageous value" is taken as a basis. Thus, from a purely mathematical point of view such as Cauchy's, the method of least squares was not distinguished from other fitting methods, but was in principle only one possible method among many equivalent methods.

Naturally, Cauchy knew that observation errors are bounded. Laplace had shown that linear combinations of identically distributed bounded errors were normally distributed in the asymptotic sense, and, on this basis, one could expect that the method of least squares would produce fitting values rather close to the "optimal" possible fitting values (assuming a large number of observations). But, what assertion concerning the method of least squares had been actually proven by Laplace? As Schneider [1998] has pointed out, it was Laplace's style to avoid formulations that permitted a refutation of his arguments. Phrases like "Preference should (!) thus be given [to the method of least squares]," or, "if we have a very great number of observations," without a closer specification of "how great," could hardly be disproved. For a mathematical refutation of Laplace's assertions, Cauchy had to transform Laplace's application-oriented propositions into mathematical claims. But this thrusted Cauchy into the dilemma that the presentation of some impractical counterexamples could hardly compromise Laplace's position, as Bienaymé immediately pointed out. Yet there was still Laplace's deduction of the approximate normal distribution, which no longer met the analytical standards of the mid-19th century, and did not produce an adequate and exact estimate of the deviation of the approximative distribution from the actual one. As we have seen (Sect. 2.1.5.2), the sore point in Laplace's argument was the assumption of a very substantial proximity (strictly speaking even of equality) of both distributions. Making precisely this point to the subject of discussion, Cauchy could argue that Laplace had not examined his approximations with sufficient scrutiny.

Cauchy [1853f] indeed gave a first discussion on the approximate normal distribution for linear combinations of errors, however without exactly discussing the quality of approximation. His account essentially endorsed Laplace's foundation of least squares. Still, Cauchy announced further critical examinations.

In [1853g] he actually presented several "candidates" for the failure of a sufficiently close proximity between approximate and exact distribution. One example referred to bounded errors, however with a density close to the above-mentioned "Cauchy-density" f_k. Another referred to cases in which large deviations concerning the order of magnitude among the least square multipliers λ_j occurred. From the point of view of common practice of observation and measurement, however, both examples seemed to be far-fetched, as Bienaymé would shortly point out.

2.5.3 Bienaymé's Arguments

Bienaymés reply to Cauchy's arguments is mainly contained in the "Considérations à l'appui de la découverte de Laplace sur la loi de probabilité dans la méthode des moindres carrés" [Bienaymé 1853e]. This article consists of four parts: In the first, one can find a general defense of the principles of Laplacian probability theory. The second part contains a discussion of the importance of the "mean of the squares of the differences of the errors and their mean value" which, in modern terminology, is simply the variance of observational errors. Through this discussion, Bienaymé confirmed his preference for least squares. In the third part, Bienaymé deduced the inequality which is now named after him and Chebyshev, however not aiming at the (now common) discussion of a weak law of large numbers, but for the sake of giving an additional intuitive argument for the superiority of least squares in the case of a large number of observations. Finally, the practical irrelevance of Cauchy's exceptional laws was discussed by Bienaymé in the fourth part.

The first part of Bienaymés considerations is well described by Heyde & Seneta [1977, 87 f.] and by Schneider [1987a, 208–210]. Here, Bienaymé pointed out the statistical importance of large samples, and, in the same context, the importance of Laplace's CLT. This exposition was connected with the refusal of small samples because of their insignificance, at least implicitly.

This refusal was discussed in more detail in the second part. Bienyaymé gave a plausibility consideration in order to show that for linear combinations $\sum_{j=1}^{n} h_j\epsilon_j$ of independent identically distributed observational errors (each with only a finite number of possible values) the asymptotic relation

$$\sqrt[m]{\mathrm{E}\left(\sum(h_j\epsilon_j - \mathrm{E}h_j\epsilon_j)\right)^{2m}} \approx \text{const.} \sum \mathrm{E}(h_j\epsilon_j - \mathrm{E}h_j\epsilon_j)^2 \tag{2.40}$$

is valid for each natural m (the constant "const." depending on m). In this context, Bienaymé criticized Gauss's remark [Gauss 1823, 6 f.] that the variance was not distinguished as a precision measure from other central moments of even order. Gauss had made this statement in the context of an arbitrary number of observational errors. Bienaymé, however, was apparently convinced that only the case of "large numbers" was worthy of consideration. Because in this case all central moments of even order of the deviation $\sum h_j\epsilon_j$ between the true and the estimated value could be reduced to the variance $\mathrm{Var}\epsilon_1 \sum h_j^2$ by virtue of (2.40), he maintained that "nothing is simpler, than to recognize that one has to render the sum of the squares of the factors h_j a minimum" [Bienaymé 1853e, 319].

Bienaymé's arguments in the second part were complemented by a discussion of Laplace's criterion (2.9) for the "most advantageous value." Bienaymé, applying a rather simple procedure (equivalent to the modern textbook proof of the Bienaymé–Chebyshev inequality), calculated the "form" of the probability of the deviation between the true and the estimated value in the case of identically distributed observational errors with zero means and the common variance σ^2:

$$P\left(\left|\sum h_j\epsilon_j\right| \leq t\sqrt{2\sigma^2}\right) = 1 - \frac{\theta f}{2t^2}\sum h_j^2,$$

where θ and f are positive "constants" less than 1, depending on the error law and the factors h_j. From this estimation, Bienaymé plausibly argued that Laplace's criterion is met if $\sum h_j^2$ becomes a minimum, which condition leads to the method of least squares. For a more exact discussion, however, Bienaymé, somewhat maliciously, referred to the article [Cauchy 1853f], in which a first reexamination of Laplace's normal approximation was given (still without suitable limits for the approximation error).

For a discussion of Cauchy's exceptional laws, Bienaymé confined himself to the examination of the now so-called "Cauchy distribution" with the density

$$f_k(\epsilon) = \frac{k}{\pi}\frac{1}{1+k^2\epsilon^2}.$$

This restriction was probably due to the fact that this density was the only one which could be given explicitly by an algebraic formula. However, it also seems that Bienaymé treated this density as representative of all exceptional laws. He argued first, with the aid of a table of $\int_{-a}^{a} f_1(x)dx$ for several values a, that, presupposing this error law, the probabilities of very large values were so high that no reasonable person would use a corresponding observation instrument. Second, Bienaymé advanced the argument that in the case of direct observations the probability of a certain deviation between the true value and the arithmetic mean would not depend on the number of observations, in contrast to all experiences of observational practice.[36] Bienaymé

[36] Bienaymé [1853e, 323] only noticed that it would be "very easy" to show this. The probably easiest way is the following: Let $y_j = x + \epsilon_j$ $(j = 1, \ldots, n)$, and let $\varphi(x) = \mathrm{e}^{-|x|}$ be the characteristic function of each single error ϵ_j. Then the characteristic function $\Phi(z)$ of the difference $\frac{\sum \epsilon_j}{n}$ between the arithmetic mean $\frac{\sum y_j}{n}$ and the real value x is

$$\Phi(z) = \left(\varphi\left(\frac{z}{n}\right)\right)^n = \left(\mathrm{e}^{-\frac{|z|}{n}}\right)^n = \varphi(z).$$

did not fail to remark that Poisson had already realized—in contrast to Cauchy, as it seemed—the practical irrelevance of the error laws f_k.

Bienaymé [1853e, 324] also discussed Cauchy's example of multipliers which considerably deviate in their respective orders of magnitude. He emphasized that such cases were far from any "well-planned and careful" application of the method of least squares.

Bienaymé's comments constitute a mix of purely mathematical arguments and reflection upon these arguments within the framework of the practice of observation. If Cauchy tried to transpose Laplace's statements into purely mathematical claims, then Bienaymé conversely tried to transform Cauchy's mathematical considerations into concrete situations of observation. In doing this, both mathematicians executed a separation of mathematics and its applications which had remained foreign to Laplace's classical probability. Bienaymé, however, did not share Cauchy's attitude of attributing the same value to any stochastic model which could be mathematically derived, but instead insisted on an assessment of any implication by "good sense."

2.5.4 Cauchy's Version of the Central Limit Theorem

In his last contribution to the scientific discussion with Bienaymé on least squares, Cauchy [1853h] established explicit upper bounds for the error of a normal approximation to the distribution of a linear combination $\sum_{j=1}^{n} \lambda_j \epsilon_j$ of identically distributed independent errors ϵ_j with a symmetric density f vanishing for arguments beyond the compact interval $[-\kappa; \kappa]$. He additionally required that the λ_j should have "the order of magnitude" of $\frac{1}{n}$ or less, and that $\sum \lambda_j^2 =: \Lambda$ should be of the order $\frac{1}{n}$. For a precise formulation of the first requirement, we have to assume that there exist positive constants α and β independent of n, such that for all $j = 1, \ldots, n$ there is a $\gamma(j) \geq 1$ with

$$\alpha \leq n^{\gamma(j)} |\lambda_j| \leq \beta. \tag{2.41}$$

Cauchy [1853h] only gave a sketch of proof (some details are discussed in the next section) that, for $\upsilon > 0$ with the notation $c := \int_0^{\kappa} x^2 f(x) dx$,

$$\left| P\left(-\upsilon \leq \sum_{j=1}^{n} \lambda_j \epsilon_j \leq \upsilon\right) - \frac{2}{\sqrt{\pi}} \int_0^{\frac{\upsilon}{2\sqrt{c\Lambda}}} e^{-\theta^2} d\theta \right| \leq C_1(n) + C_2(n, \upsilon) + C_3(n), \tag{2.42}$$

but at least he made the formulae for the bounds C_1, C_2, and C_3 explicit, which are valid for sufficiently large n and which tend to 0 as n increases.

In his doctoral thesis, Ivan Vladislavovich Sleshinskii [1892] gave detailed deductions for the constants C_1, C_2, and C_3, and he corrected apparent misprints in Cauchy's formulae (see [Heyde & Seneta 1977, 94–96] and [Seneta 1984, 48–50]), with the following result: Let $\Theta = n^{\frac{1}{2}+\delta}$ $(0 < \delta < \frac{1}{4})$; then

$$C_1 = \frac{1}{\pi \mathcal{N}} \mathrm{e}^{-\mathcal{N}}, \text{ with } \mathcal{N} = \frac{1}{2} \frac{r\Lambda\Theta^2}{1 + r\lambda^2\Theta^2}, \tag{2.43}$$

$$C_2(n, \upsilon) = \frac{2h\sqrt{3}}{\pi} \log\left(\frac{\Theta\upsilon}{\sqrt{3}} + \sqrt{1 + \frac{\Theta^2\upsilon^2}{3}}\right), \tag{2.44}$$

where

$$\lambda := \max(|\lambda_1|, \ldots, |\lambda_n|); \quad h := \max\left(\mathrm{e}^{\frac{1}{4}c\Lambda\lambda^2\Theta^4\kappa^2} - 1, 1 - \mathrm{e}^{-\frac{c^2\Lambda\lambda^2\Theta^4}{1 - c\lambda^2\Theta^2}}\right),$$

and

$$C_3(n) = \frac{\mathrm{e}^{-c\Lambda\Theta^2}}{\pi c\Lambda\Theta^2}. \tag{2.45}$$

There are minor differences with regard to C_2 and C_3 between Cauchy's original formulae and Sleshinskii's.

The quantity C_2 has the minor flaw that it is not independent of υ; it grows for fixed n together with υ. However, presupposing (2.41), and considering that $|\epsilon_j| \leq \kappa$, one can deduce

$$P\left(|\sum \lambda_j \epsilon_j| \leq \upsilon\right) = 1 \quad \text{if } \upsilon \geq \beta\kappa. \tag{2.46}$$

Since C_2 is monotonically increasing as a function of υ, one gets

$$\left| P\left(-\upsilon \leq \sum_{j=1}^{n} \lambda_j \epsilon_j \leq \upsilon\right) - \frac{2}{\sqrt{\pi}} \int_0^{\frac{\upsilon}{2\sqrt{c\Lambda}}} \mathrm{e}^{-\theta^2} d\theta \right| \leq C_1(n) + C_2(n, \beta\kappa) + C_3(n)$$

for $\upsilon \leq \beta\kappa$. On the other hand, for $\upsilon > \beta\kappa$, (2.46) yields

$$\left| P\left(-\upsilon \leq \sum_{j=1}^{n} \lambda_j \epsilon_j \leq \upsilon\right) - \frac{2}{\sqrt{\pi}} \int_0^{\frac{\upsilon}{2\sqrt{c\Lambda}}} \mathrm{e}^{-\theta^2} d\theta \right| \leq \frac{2}{\sqrt{\pi}} \int_{\frac{\beta\kappa}{2\sqrt{c\Lambda}}}^{\infty} \mathrm{e}^{-\theta^2} d\theta =: C_4(n).$$

Now, because Λ must be of the order of magnitude $\frac{1}{n}$, $C_4(n)$ tends to 0 independent of υ. Altogether, it follows that for any $\upsilon \in \mathbb{R}^+$,

$$\left| P\left(-\upsilon \leq \sum_{j=1}^{n} \lambda_j \epsilon_j \leq \upsilon\right) - \frac{2}{\sqrt{\pi}} \int_0^{\frac{\upsilon}{2\sqrt{c\Lambda}}} \mathrm{e}^{-\theta^2} d\theta \right|$$
$$\leq C_1(n) + C_2(n, \beta\kappa) + C_3(n) + C_4(n),$$

where the right side tends to 0 independent of υ.

Apparently, the convergence of C_2 to 0 as $n \to \infty$ (υ fixed) is the slowest among all of the "constants," because $C_2 = \mathrm{O}\left(\frac{\log n}{n^{1-4\delta}}\right)$. Thus, the order of magnitude of Cauchy's upper bounds was rather close to the optimal asymptotic order, which is, in the case at hand, and according to Harald Cramér [1928], equal to $\mathrm{O}(\frac{1}{n})$.

From today's point of view Cauchy's account can be interpreted as the more or less rigorous proof of the finite version of a CLT for linear combinations of independent identically distributed random variables. In fact, a "modern" CLT can be inferred from Cauchy's version by considering a sequence of independent random variables X_j, distributed like Cauchy's observational errors, and by setting $\lambda_j = \frac{1}{n}$, $\upsilon = \frac{a}{\sqrt{n}}$ $(a > 0)$, $c = \frac{1}{2}\mathrm{Var}X_1$. Then, by virtue of (2.42),

$$\left| P\left(-a\sqrt{n} \le \sum_{j=1}^{n} X_j \le a\sqrt{n}\right) - \frac{2}{\sqrt{\pi}}\int_0^{\frac{a}{2\sqrt{c}}} \mathrm{e}^{-x^2}dx \right|$$
$$\le C_1(n) + C_2(n, \frac{a}{\sqrt{n}}) + C_3(n) \to 0 \quad (n \to \infty).$$

Though Sleshinskii gave more precise explanations in comparison to Cauchy, he did not substantially go beyond the latter's ideas, and, in particular, he did not succeed in weakening Cauchy's still rather restrictive assumptions. Like Cauchy, Sleshinskii was primarily interested in solving an—although quite abstract—problem of error theory. Therefore, we may actually follow Freudenthal [1970–76, 142] in championing Cauchy for the "first rigorous proof" of the CLT, we must not forget, however, that his goals were quite different from those of modern probability theory.

2.5.5 Cauchy's Idea of Proof

There was a rule that only brief articles were accepted for publication in the *Comptes rendus*, and thus, Cauchy [1853h] had to restrict his presentation to a description of the major steps of his reasoning. The basic ideas, however, can be clearly discerned from his account. In particular, the deduction of (2.42) was based on Cauchy's use of characteristic functions and his modification of the Laplacian method of approximation, which he had already dealt with in several articles published in the 1840s [Cauchy 1844; 1845; 1849]. In [1849, 138–140], for example, Cauchy discussed the asymptotic behavior (as $n \to \infty$) of the integral

$$S = \frac{1}{2\pi}\int_{-\pi}^{\pi} \left(Z(r_0\mathrm{e}^{(p_0+\varphi)\sqrt{-1}})\right)^n d\varphi,$$

where $Z(z)$ is an analytic function whose derivative has the property $Z'(r_0\mathrm{e}^{p_0\sqrt{-1}}) = 0$, and whose modulus $|Z(r_0\mathrm{e}^{(p_0+\varphi)\sqrt{-1}})|$ attains its maximum at $\varphi = 0$. Cauchy split the entire integral S into a "major" part with a domain of integration close to the maximum and, in comparison with this part, very small remaining parts which vanish as $n \to \infty$. Thus, he developed asymptotic methods similar to those of Dirichlet, who, on the other hand, had not published his contributions.

Proving his version of the CLT, Cauchy [1853h, 125 f.] first summarized the most important properties of the "fonction auxiliaire" $\varphi(\theta) = 2\int_0^{\kappa} f(\epsilon)\cos(\theta\epsilon)d\epsilon$

for the error law $f(\epsilon)$, and in this context he repeated the fundamental relation

$$P\left(-\upsilon \leq \sum_{j=1}^{n} \lambda_j \epsilon_j \leq \upsilon\right) = \frac{2}{\pi} \int_0^{\infty} \Phi(\theta) \frac{\sin(\theta\upsilon)}{\theta} d\theta, \tag{2.47}$$

$$\Phi(\theta) = \varphi(\lambda_1\theta) \cdots \varphi(\lambda_n\theta).$$

Basically resuming his approach of [1853f], Cauchy from $\varphi(0) = 1$ and $|\varphi(\theta)| < 1$ for $\theta > 0$ concluded that

$$[\varphi(\theta)]^2 = \frac{1}{1 + \rho(\theta)\theta^2}$$

($\rho(\theta) > 0$ for $\theta > 0$). He briefly justified that $\rho(\theta)$ has a positive lower bound r, such that

$$[\varphi(\theta)]^2 \leq \frac{1}{1 + r\theta^2}. \tag{2.48}$$

For this justification he needed the estimate "$\rho(\infty) \geq \left[\frac{1}{2f(\kappa)}\right]^2$," which was, as we can see from a similar consideration in [1853f, 107], most likely obtained by partial integration under the tacit presupposition that f possessed a continuous derivative.[37]

Finally, he [1853h, 126] referred to a consideration in [Cauchy 1853f, 107 f.] (based on the mean value theorem of differential calculus as applied to $\sin(z)$ and $\log(1 - z)$) that for sufficiently small $\theta > 0$:

$$\varphi(\theta) = 1 - \int_0^{\kappa} \left(2 \sin \frac{\theta\epsilon}{2}\right)^2 f(\epsilon) d\epsilon = \mathrm{e}^{-\sigma\theta^2}$$

with

$$1 - \left(\frac{\theta\kappa}{2}\right)^2 < \frac{\sigma}{c} < \frac{1}{1 - c\theta^2} \quad \left(c = \int_0^{\kappa} x^2 f(x) dx\right).$$

By virtue of these estimates, Cauchy's further proceeding [1853h, 127–129] corresponded to his above-mentioned modification of the Laplacian method of approximation as applied to the integral (2.47). The integrand $\Phi(\theta)\frac{\sin(\upsilon\theta)}{\theta}$ of this integral attains its absolute maximum at $\theta = 0$. For Θ of an "order greater than $\sqrt{n}$ but smaller than $n^{\frac{3}{4}}$" (e.g., $\Theta = n^{\frac{1}{2}+\delta}, 0<\delta<\frac{1}{4}$), and for sufficiently large n, Cauchy established the following inequalities for the grade of accuracy regarding the approximation of the integrand by a bell-shaped function in the neighborhood of $\theta = 0$:

$$\left| \frac{2}{\pi} \int_0^{\Theta} \Phi(\theta) \frac{\sin(\theta\upsilon)}{\theta} d\theta - \frac{2}{\pi} \int_0^{\Theta} \mathrm{e}^{-cA\theta^2} \frac{\sin(\theta\upsilon)}{\theta} d\theta \right| < C_2(n, \upsilon),$$

[37] By partial integration one gets $\varphi(\theta) = 2\frac{f(\kappa)\sin(\theta\kappa) - \int_0^{\kappa} f'(x)\sin(\theta x)dx}{\theta}$. If we set $\eta(\theta) := |\int_0^{\kappa} f'(x)\sin(\theta x)dx|$ the inequality $\rho(\theta) > \frac{1}{4\left(f(\kappa)+\eta(\theta)\right)^2} - \frac{1}{\theta^2}$ ensues. By taking into account the relation $\lim_{\theta\to\infty} \eta(\theta) = 0$ (which for Cauchy most probably was a matter of course) the asserted estimate can be followed.

and

$$\left| \frac{2}{\pi} \int_0^{\frac{\upsilon}{2\sqrt{c\Lambda}}} \mathrm{e}^{-\theta^2} d\theta - \frac{2}{\pi} \int_0^{\Theta} \mathrm{e}^{-c\Lambda\theta^2} \frac{\sin(\theta\upsilon)}{\theta} d\theta \right| < C_3(n).$$

In order to estimate the "tail," Cauchy derived

$$\left| \frac{2}{\pi} \int_{\Theta}^{\infty} \Phi(\theta) \frac{\sin(\theta\upsilon)}{\theta} d\theta \right| < C_1(n).$$

The constants C_1, C_2, C_3 are already quoted in (2.43), (2.44), (2.45), respectively.

2.5.6 *The End of the Controversy*

Cauchy [1853h, 130] wrote that for "very large values" of n (the total number of errors) there would be "une grande approximation" between exact and approximate probability. He stated:

> The various formulae that we have just written down also permit us to assess, by reducing them to their true significance, the advantages of the employment of the one or the other system of factors, and consequently of the one or the other method.

Cauchy's "formulae," in particular those concerning the upper bounds C_1, C_2, C_3, were indeed appropriate, at least in cases of "large numbers" of observations, for confirming the closeness of the actual distribution of a linear combination of errors to the corresponding normal distribution, and therefore for confirming the superiority of the method of least squares. One could rarely use them for a rejection of least squares, however.

At the end of his article, Cauchy announced that he would return to the issue, but he did never resume his probabilistic studies. There does not exist any explicit evidence as to why he did not continue his discussion of the method of least squares. Beginning with Sleshinskii [1892], the common opinion has been established that Cauchy had come so close to Laplace's (and Bienaymé's) position with his asymptotic result that a continuation of the dispute did not appear advisible (see [Heyde & Seneta 1977, 96; Stigler 1974/1999]).

A closer examination, however, shows that Cauchy's result was not even properly suited—at least from the practical point of view of error theory—for a really sound justification of the Laplacian approach. As we have seen above, Cauchy's bounds for the difference of the actual and the normal probability distribution were quite appropriate in an asymptotic sense. In many cases of practical importance, however, his bounds were scarcely usable.

In the case of direct observations, for example, the equations of condition are

$$x = k_j - \epsilon_j \quad (j = 1, \ldots, n).$$

The least square multipliers are identical with $\frac{1}{n}$. If the errors obey a uniform density within the interval $[-1; 1]$, then for $\Theta \geq \sqrt{n}$ the constant r in (2.48) (which is only important for $x \geq \Theta$) can be assumed to be $r = 0.9$ if $n \geq 10$.[38] According to [Sleshinskii 1892, 255], Cauchy's estimates can be applied if

$$n > \max\left(8; \frac{4\beta^2}{\alpha^2}; \frac{8\kappa^2\beta^2}{r\alpha^2}\right)$$

and

$$\frac{2\sqrt{2n}}{\alpha\sqrt{r}} < \Theta < \frac{n}{\kappa\beta},$$

where $[-\kappa; \kappa]$ is the support of the error density f, and α, β are according to (2.41). In our case we can choose $\alpha = \beta = 1$, and the first of the latter conditions is satisfied for all $n \geq 9$. For $n = 10$, $\upsilon = 0.1$, and $r = 0.9$ the sum $C_1 + C_2 + C_3$ (dependent on a Θ which has still to meet Sleshinskii's second condition) is at its minimum (for $\Theta \approx 9.43$) approximately equal to 0.288. In the case at hand, the probability $P(-\upsilon \leq \sum \lambda_j \epsilon_j \leq \upsilon)$ with $\lambda_j = \frac{1}{n}$ can be directly calculated by use of the formula (2.1), which was already derived by Laplace in the 1770s. The exact value of this probability is (for $n = 10$, $\upsilon = 0.1$) equal to 0.41096, whereas the approximation by the normal distribution gives the value 0.4161. Similar calculations for other υ show that, already for $n = 10$, the difference between the exact and the approximate value is less than 1/100. If $n > 10$, for a comparison with the case of 10 observations we have to use values of υ which decrease in the ratio $\sqrt{10/n}$. For $n = 20$, $\upsilon = 0.1 \cdot \sqrt{0.5}$, and $r = 0.9$ the minimum sum $C_1 + C_2 + C_3$ is roughly 0.16 ($\Theta \approx 13.4$); for $n = 100$, $\upsilon = 0.1 \cdot \sqrt{0.1}$, and $r = 0.9$ the minimum sum is still about $6/100$ ($\Theta \approx 33.2$). A critical numerical discussion of this kind was certainly within Cauchy's reach, and his above-quoted reference to the "true significance" of his "formulae" might point in this direction.

Thus, within the framework of observational practice, by applying Cauchy's bounds one was able to confirm Laplace's point of view only if a really large number of observations appeared. Certainly, a great many observations were occasionally available in the context of astronomical problems. Bessel [1818, 18–21], in his comparison of the frequency distributions of the residuals of direct observations on the one hand, and normal distributions on the other, had used two series with 300 observations each, and one with 470.[39] Alexis Bouvard had considered approximately 130 equations of condition for Jupiter and another 130 for Saturn in his determination of the orbit elements of these planets. This work was described by Laplace [1812/20/86, 516] as an "immense travail."[40] In most cases, however, the number of observations was far below 100. Gauss [1811], for example, determined his "improvements" of elliptical elements of Pallas from only 11 equations of condition.

---

[38] In our special case the "fonction auxiliaire" is $\varphi(z) = \frac{\sin z}{z}$. For $z^2 \geq 10$ the estimate $\frac{1}{1+0.9z^2} \geq \frac{1}{z^2}$, and thus, $[\varphi(z)]^2 \leq \frac{1}{1+0.9z^2}$ is valid.

[39] See [Stigler 1986, 204; Hald 1998, 361–363].

[40] For a summary of Bouvard's work see [Bouvard 1821].

There exists a brief report [Cauchy 1853g'] in the *Comptes rendus* referring to Cauchy's remarks on Bienaymé's defense [1853e] (see Sect. 2.5.3) of Laplace's approach to least squares. Concerning Laplace's analytical methods, we read:

> The analysis by which he [Laplace] has established the properties of the method of least squares uses series expansions whose convergence is not proven. M. Cauchy has replaced this analysis by exact and rigorous formulae.

Thus, we can see that Cauchy clearly stressed his "new" analytical rigor as an exceptional merit as opposed to Laplace's style of reasoning. But, from the practical point of view of error theory, he neither succeeded in improving Laplace's analysis by establishing sufficiently close bounds for the error of approximation, nor did he succeed in giving convincing counterexamples concerning the method of least squares. We should not forget that Cauchy's main interest was originally to give an effective procedure for astronomical calculations (see Sect. 2.5.1). Thus, his turn toward an "abstract" point of view which scarcely considered questions like general applicability or computational simplicity was—in a certain sense—contrary to his original aims. From a purely mathematical point of view, however, Cauchy's contribution even enforced the Laplacian preference to least squares in the case of bounded errors. Naturally, Cauchy could not exclude the possibility of bounds more appropriate than his own (which in fact can be derived by modern methods). Bienaymé, however, whose analytical abilities were likewise at a respectable level, was unable to give an exact mathematical argument in favor of Laplace's position. On the contrary, by showing that (in modern terminology) the estimator obtained from *any* system of multipliers (if these are of an order of magnitude inversely proportional to the number of observations) converges stochastically to the true value, he showed at the same time, that the method of least squares could be, presupposing a "very large" number of observations, only slightly superior (according to Laplace's criterion) to other methods. Thus, the end of the scientific controversy was not so much determined by Cauchy's hypothetic fear of coming too close to Bienaymé's position, but rather by a situation in which neither of the two scientists was able to make further substantial contributions. In this sense, the dispute ended in a tie.

2.5.7 Conclusion: Steps Toward Modern Probability

Laplace's version of the CLT served mainly as a tool of "good sense" and therefore its importance was primarily determined by a field beyond mathematics. Around the mid-19th century, due to the contributions of Dirichlet and Cauchy, the CLT became part of mathematics in the narrow sense. In Dirichlet's work, it served as an illustration of special analytical techniques, whereas Cauchy used it for his approach to an error theory which was mainly determined by purely mathematical goals. In adjusting Laplacian approximation techniques to an analytical style different from algebraic analysis, they contributed to the development of new standards within analysis. Poisson, with his contributions to the CLT, however still according to the principles of classical probability, considerably influenced Dirichlet's and Cauchy's work through his innovative analytical techniques and through his discussion of the validity of normal approximations.

On the one hand, in Dirichlet's and Cauchy's contributions the CLT obtained a substantial intramathematical role. In Cauchy's work, it was connected with a rather abstract and therefore almost "modern" perspective of error theory. On the other hand, it had not yet reached an entirely independent status within mathematics. In particular, general statements independent of the original context of applications were still lacking. Full autonomy, according to Mehrtens [1990] an essential characteristic of the modernization of mathematics, was not reached for the CLT until Lyapunov published his epochal work on the "Theorem of Laplace" in 1900/1901.

Appendix: Original Text of Dirichlet's Proof of the Central Limit Theorem According to Lecture Notes from 1846

The following text is a transcription[41] of pages 25 to 31 of the lecture notes [Dirichlet 1846] (for closer bibliographic details see References and [Fischer 1994]). To the greatest extent possible, the original wording is reproduced to the letter, and the original punctuation is kept as well. As a rule, "mistakes" are therefore not due to misprints. The original page numbers are also referred to.

Seite 25
...„Es möge sich bei einer bestimmten Gattung von Beob. das Fehlergesetz von Beob. zur Beob. beliebig ändern, dabei aber doch, was ja immer erreicht werden kann, indem man nur das größte als Norm nimmt, sämtliche Fehlergesetze $f_1(x_1), f_2(x_2), f_3(x_3) \ldots f_n(x_n)$ zw. festen Grenzen $\pm a$ enthalten sein, man soll bestimmen wie groß die Wahrscheinlichk. ist, daß wenn man die Fehler der einzelnen Beobachtungen $x_1, x_2, x_3 \ldots$ mit den respectiven Constanten $\alpha_1, \alpha_2, \alpha_3 \ldots$ multiplizirt, die Productsumme zw. gegebenen Gränzen g und h liege; daß man also habe:"

$$g < \alpha_1 x_1 + \alpha_2 x_2 + \cdots + \alpha_n x_n < h$$

Zur Lösung dieser Aufgabe bemerken wir, daß in Folge des Vorhergegangenen die Probabilitäten, daß der erste Fehler zwischen

Seite 26
den Grenzen x_1 u. $x_1 + \partial x_1$, der zweite zw. den Grenzen x_2 u. $x_2 + \partial x_2$, der n^{te} zw. den Grenzen x_n u. $x_n + \partial x_n$ liege, ausgedrückt sind durch

$$f_1(x_1)\partial x_1, f_2(x_2)\partial x_2 \ldots f_n(x_n)\partial x_n$$

für die Größe der Probabilität, daß diese Fehler zw. den Grenzen g und h enthalten sind, hat man die Ausdrücke:

$$\int_g^h f_1(x_1)\partial x_1, \int_g^h f_2(x_2)\partial x_2 \ldots \int_g^h f_n(x_n)\partial x_n$$

und die zusammengesetzte Wahrscheinlichk., daß diese Grenzen bei allen gleichzeitig Statt finden, ist gleich dem Vielfachen Integrale:

$$\iint \cdots \iiint_g^h f_1(x_1) f_2(x_2) \cdots f_n(x_n)\partial x_1 \partial x_2 \cdots \partial x_n$$

Zur Discussion dieses Integrals, wollen wir fürs erste den Anfangsp. von dem aus man die Grenzen g und h zehlt in den P. $\frac{g+h}{2}$ verlegen. Es wird dann wenn man

[41] Courtesy of Institut für Geschichte der Naturwissenschaften, Universität München, Professor M. Folkerts.

$g = -\lambda$ setzt offenbar $h = +\lambda$, und unsere Ungleichung geht über in

$$-\lambda < \alpha_1 x_1 + \alpha_2 x_2 + \cdots + \alpha_n x_n < \lambda$$

$$\text{oder} \quad -1 < \frac{1}{\lambda}(\alpha_1 x_1 + \alpha_2 x_2 + \cdots + \alpha_n x_n) < +1$$

Wir wenden nun das bekannte Verfahren eines Multiplicators an. Man hat nemlich

$$\int_0^\infty \frac{\sin\varphi}{\varphi}\partial\varphi = \frac{\pi}{2}$$

oder wenn man $l\varphi$ st φ schreibt:

$$\frac{2}{\pi}\int_0^\infty \frac{\sin l\varphi}{\varphi}\partial\varphi = \pm 1$$

je nachdem l eine positive oder negative Constante vorstellt. Mittelst dieses Integrals kann man nun leicht sich den gewünschten Multiplicator verschaffen. Es ist nemlich:

$$\frac{2}{\pi}\int_0^\infty \frac{\sin\varphi}{\varphi}\cos k\varphi\partial\varphi = \frac{2}{\pi}\left\{\frac{1}{2}\int_0^\infty \frac{\sin(1+k)\varphi}{\varphi}\partial\varphi + \frac{1}{2}\int_0^\infty \frac{\sin(1-k)\varphi}{\varphi}\partial\varphi\right\}$$

woraus man mit Hilfe des vorhergehenden Integrales erhält:

Seite 27

$$\frac{2}{\pi}\int_0^\infty \frac{\sin\varphi}{\varphi}\cos k\varphi\partial\varphi = \begin{cases} 0 \text{ für } k > 1 \text{ absolut genommen} \\ 1 \ \text{,,} \quad -1 < k < 1. \end{cases}$$

In Folge unserer Ungleichheitsbedingung $-1 < \frac{1}{\lambda}(\alpha_1 x_1 + \alpha_2 x_2 + \cdots \alpha_n x_n) < 1$ kann man mit diesem Integral das zu untersuchende durch Multiplication verbinden, wodurch man erhält:

$$\frac{2}{\pi}\iint\cdots\iiint_{-a}^{a}\int_0^\infty f_1(x_1)f_2(x_2)\cdots f_n(x_n)\frac{\sin\varphi}{\varphi}\times$$
$$\times\cos(\alpha_1 x_1 + \alpha_2 x_2 + \cdots \alpha_n x_n)\frac{\varphi}{\lambda}\partial\varphi\partial x_1\partial x_2\partial x_3\cdots$$

Nun ist bekanntlich: $\sqrt{-1}\int_{-\mu}^{\mu}\sin\frac{\alpha_1 x_1+\alpha_2 x_2+\cdots\alpha_n x_n}{\lambda}\varphi\partial\varphi = 0$ und hieraus und dem obigen Ausdrucke wird durch Addition:

$$\frac{2}{\pi}\iint\cdots\iint\int_{-a}^{a}\int_0^\infty f_1(x_1)f_2(x_2)\cdots f_n(x_n)\frac{\sin\varphi}{\varphi}\times$$
$$\times e^{(\alpha_1 x_1+\alpha_2 x_2+\cdots\alpha_n x_n)\frac{\varphi}{\lambda}\sqrt{-1}}\partial x_1\partial x_2\partial x_3\cdots\partial\varphi$$

Setzt man $\lambda\varphi' = \varphi$, und läßt man nach geschehener Subst die Accente wieder weg, so erhält man:

$$\frac{2}{\pi}\iint\cdots\iiint_{-a}^{a}\int_{0}^{\infty} f_1(x_1)f_2(x_2)\cdots f_n(x_n)\frac{\sin\lambda\varphi}{\varphi}\times$$
$$\times e^{(\alpha_1x_1+\alpha_1x_1\cdots\alpha_nx_n)\varphi\sqrt{-1}}\partial x_1\partial x_2\cdots\partial x_n\partial\varphi$$

was man offenbar auch in folgender Form schreiben kann:

$$\frac{2}{\pi}\int_{0}^{\infty}\frac{\sin\lambda\varphi}{\varphi}\left\{\left(\int_{-a}^{a} f(x)e^{\alpha_1x_1\varphi\sqrt{-1}}\partial x\right)\left(\int_{-a}^{a} f_2(x_2)e^{\alpha_2x_2\varphi\sqrt{-1}}\partial x_2\right)\cdots\right.$$
$$\left.\cdots\left(\int_{-a}^{a} f_n(x_n)e^{\alpha_nx_n\varphi\sqrt{-1}}\partial x_n\right)\right\}\partial\varphi \quad (1)$$

Wir müssen uns nun fürs erste mit der Discussion des Integrales

$$\int_{-a}^{a} f(x)e^{\alpha x\varphi\sqrt{-1}}\partial x = \int_{-a}^{a} f(x)\cos(\alpha x\varphi)\partial x + \sqrt{-1}\int_{-a}^{a} f(x)\sin(\alpha x\varphi)\partial x$$
$$= \int_{-a}^{a} f(x)\cos(\alpha x\varphi)\partial x = \int_{-a}^{a} f(x)\cos\beta x\partial x \qquad \alpha\varphi = \beta$$

beschäftigen. Dasselbe erreicht für $\beta = 0$ sein Max, wo es dann, da $f(x)$ als Ausdruck einer Wahrscheinlichk., nie negativ werden kann, aus lauter positiven Elementen besteht. Die Reihe

$$\int_{-a}^{a} f(x)\cos\beta x\partial x = \int_{-a}^{a} f(x)\partial x - \frac{\alpha^2\varphi^2}{2}\int_{-a}^{a} x^2 f(x)\partial x + \cdots$$

Seite 28

in der α eine gegebene endliche Constante bezeichnet, convergirt für sehr kleine Werthe von φ, so schnell, daß fast der ganze Werth des Integrales in den beiden ersten Gliedern der Reihe enthalten ist, wodurch bewirkt wird, daß der ganze Werth unseres Integralproduktes sich im Anfange concentrirt. Setzt man zur Abkürzung

$$\frac{1}{2}\int_{-a}^{a} x_\nu^2 f(x_\nu)\partial x_\nu = k_\nu$$

wo k_ν eine Constante bezeichnet, die sich nur für die betreffenden Beobachtungen, von deren Fehlergesetz es abhängt, verändert, so erhält man wegen der Relation

$$\int_{-a}^{a} f(x)\partial x = 1$$

für einen Factor obigen Doppelintegrales den Ausdruck:

$$1 - k_\nu \alpha_\nu^2 \varphi^2 + \cdots$$

Nimmt man den Neper'schen Logarithmus, so bekommt man, wenn man dieselben in Reihen auflöst:

$$\log \operatorname{nat}(1 - k_\nu \alpha_\nu^2 \varphi^2 + \cdots) = -k_\nu \alpha_\nu^2 \varphi^2 + \cdots$$

wo man für ν die Zahlen $1, 2, 3, \ldots n$ zu setzen hat, und die so erhaltenen Ausdrücke sodann alle zu addiren. Geht man dann von den Logarithmen wieder zu den Zahlen über, so erhält man einen Ausdruck für das oben behandelte Produkt aus Integralfactoren. Man hat aber hiebei auch dafür zu sorgen, daß die weggelassenen Glieder der Reihe absolut klein seien, nicht blos klein im Verhältniße zum ersten Gliede. Denn in einer Exponentialgröße $e^{\alpha+\beta} = e^\alpha e^\beta$, wie sie hier auftritt, darf man offenbar nur dann den zweiten Theil β des Exponenten vernachlässigen, wenn es eine absolut verschwindende Größe ist.

Wenn man nun unsere n Gleichungen

$$\log \operatorname{nat}(1 - k_\nu \alpha_\nu^2 \varphi_\nu^2 + \cdots) = -k_\nu \alpha_\nu^2 \varphi_\nu^2 + l\varphi^4 + \cdots$$

Seite 29

summirt, so wird die Summe der Glieder der vierten Ordnung immer $\leq nl\varphi^4$ sein, wenn man nemlich mit l den größten vorhandenen Entwicklungscoeff bezeichnet, und also eine gewisse Constante sein wird in Beziehung auf φ, und n als der Index der einzelnen Beobachtungen als eine immer wachsende Größe gedacht werden muß. Nach dem oben Gesagten muß nun unser Bestreben immer dahin gehen, daß die Summe der höheren Entwicklungskoeff immer eine absolut kleine Zahl bleibt, und dieß bewerkstelligen wir dadurch daß wir in dem willkürlich eingeführten Integrale mit der Variablen φ diese letztere blos soweit wachsen lassen, daß das Produkt $n\delta^2$, wo δ einen Zustand von φ bezeichnet, immer wie groß auch n werden möge, sich der Grenze Null nähert. Dann reduzirt sich in vorstehender Formel das Product der Co... auf die Summe ihrer ersten Entwicklungsglieder, und man erhält statt unserer Formel (1) sogleich:

$$\frac{2}{\pi}\int_0^\delta e^{-\varphi^2 \sum\limits_{\nu=1}^{\nu=n} k_\nu \alpha_\nu^2} \frac{\sin \lambda\varphi}{\varphi} \partial\varphi + \frac{2}{\pi}\int_\delta^\infty e^{-\varphi^2 \sum\limits_{\nu=1}^{\nu=n} k_\nu \alpha_\nu^2} \frac{\sin \lambda\varphi}{\varphi} \partial\varphi = p + q$$

wo also δ durch die Bedingung bestimmt wird, daß $\delta \sqrt[4]{n}$ für ein zunehmendes n immer kleiner und kleiner wird. Beschäftigen wir uns zuerst mit dem ersten Integrale

$$p = \frac{2}{\pi}\int_0^\delta e^{-\varphi^2 \sum k_\nu \alpha_\nu^2} \frac{\sin \lambda\varphi}{\varphi} \partial\varphi$$

so wir für $\varphi = \frac{\psi}{\sqrt{n}}$ sogleich:

$$p = \frac{2}{\pi}\int_0^{\delta\sqrt{n}} e^{-\psi^2 \frac{\sum k_\nu \alpha_\nu^2}{n}} \frac{\sin \frac{\lambda}{\sqrt{n}}\psi}{\psi} \partial\psi$$

Nach unserer Bedingung soll nun $\delta\sqrt[4]{n}$ beständig abnehmen, womit aber durchaus nicht gesagt ist, daß dies auch mit $\delta\sqrt{n}$ der Fall sein müsse, welches im Gegentheil sogar immer größer werden kann, wie dieß z.B. für die Annahme $\delta = \frac{1}{\sqrt[3]{n}}$ statt findet. Bestimmen wir nun

Seite 30

das δ so, daß für ein zunehmendes n das Product $\delta\sqrt[4]{n}$ immer abnimmt, $\delta\sqrt{n}$ aber wächst, so erhalten wir für eine sehr große Anzahl von Beobachtungen offenbar:

$$p = \frac{2}{\pi}\int_0^\infty \mathrm{e}^{-\psi^2\frac{\sum k_\nu \alpha_\nu^2}{n}} \frac{\sin\lambda\psi}{\psi}\partial\psi$$

wobei noch zu bemerken ist, daß man hier, wie auch geschehen ist, auch das λ mit wachsendem n zunehmen muß, indem offenbar die Wahrscheinlichkeit, daß bei unendlich vielen Beobachtungen der Fehler zw. gegebenen festen Grenzen liege, Null ist. Wir setzen deshalb $\lambda\sqrt{n}$ statt λ. Nun ist aber zur Vereinfachung dieses Resultates bekanntlich:

$$\int_0^\infty \mathrm{e}^{-c^2\varphi^2}\cos\lambda\varphi\partial\varphi = \frac{\sqrt{\pi}}{2c}\mathrm{e}^{-\frac{\lambda^2}{4c^2}}$$

$$\int_0^\lambda \partial\lambda\int_0^\infty \mathrm{e}^{-c^2\varphi^2}\cos\lambda\varphi\partial\varphi = \int_0^\infty \mathrm{e}^{-c^2\varphi^2}\frac{\sin\lambda\varphi}{\varphi}\partial\varphi = \frac{\sqrt{\pi}}{2c}\int_0^\lambda \mathrm{e}^{-\frac{s^2}{4c^2}}\partial s$$

$$\text{oder}\quad \int_0^\infty \mathrm{e}^{-c^2\varphi^2}\frac{\sin\lambda\varphi}{\varphi}\partial\varphi = \sqrt{\pi}\int_0^{\frac{\lambda}{2c}} \mathrm{e}^{-s^2}\partial s$$

und hiemit wird:

$$p = \frac{2}{\sqrt{\pi}}\int_0^{\frac{\lambda}{2\sqrt{\frac{\sum k_\nu\alpha_\nu^2}{n}}}} \mathrm{e}^{-s^2}\partial s.$$

als Ausdruck der Wahrscheinlichk. daß die Größe $\sum\limits_{\nu=1}^{\nu=n}\alpha_\nu x_\nu$ zw. den Gränzen $\pm\lambda\sqrt{n}$ enthalten sei, oder daß sei

$$-\lambda\sqrt{n} < \sum_{\nu=1}^{\nu=n}\alpha_\nu x_\nu < \lambda\sqrt{n}$$

wenn n immer größer wird. Es ist hiedurch allein schon diese Wahrscheinlichkeit ausgedrückt, weil das zweite Integral

$$q = \int_\delta^\infty \mathrm{e}^{-\varphi^2\sum k_\nu\alpha_\nu^2}\frac{\sin\lambda\varphi}{\varphi}\partial\varphi$$

sich unaufhörlich der Null nähert. Diese Behauptung erweist man folgender maßen. Wir haben bei der Form (1) bemerkt, daß $2\int_0^a f(x)\cos(\alpha\varphi x)\partial x$ für $\alpha x = c$ ein absolutes Maximum und $= 1$ sei

Seite 31

und niemals mehr unter den später eintreten Maximis ein diesen an Größe gleich kommendes sich befinde. Man kann sich nun das Intervall so klein denken, daß die Funktion innerhalb deßselben nicht allein beständig abnimmt, sondern auch noch größer bleibt, als sie später irgendwo noch werden kann, *a fortiori* also das Product aller dieser analog gebildeten Functionen, und man kann diesen Zustand der Ungleichheit durch Verkleinerung des Intervalles von 0 bis δ soweit treiben, als verlangt wird, woraus unsere Behauptung folgt. Strenger läßt sie sich aber auf folgende Weise rechtfertigen. Man zerlege das Integral $\int_\delta^\infty$ in die Summe zweier andren $\int_\delta^\Delta + \int_\Delta^\infty$, so wird immer im ersteren die Function am Anfange größer sein, als irgendwo später. Es wird also das Integral kleiner sein, als die Differenz der Grenzen $\Delta - \delta$, also um so mehr kleiner als der Anfangswerth der Function, und es nähert sich deswegen, wenn man das Δ einer positiven Potenz von n proportional nimmt, der Werth des Productes der Grenze Null. Was nun noch das zweite Integral betrifft, so hat man durch partielle Integration:

$$\int \cos(\alpha\varphi x) f x \partial x = \frac{\sin(\alpha\varphi x)}{\alpha\varphi} f(x) - \int \frac{\sin(\alpha\varphi x)}{\alpha\varphi} f'(x)\partial x$$

$$\int_{-a}^{a} \cos(\alpha\varphi x) f(x)\partial x = \frac{2\sin(\alpha\varphi a)}{\alpha\varphi} f(a) - \int_{-a}^{a} \frac{\sin(\alpha\varphi x)}{\alpha\varphi} f'(x)\partial x$$

wo wir also jetzt auch noch annehmen müßen, daß $f(x)$ innerhalb der Integralgränzen endlich bleibt, welche Annahme durch die Natur der Fehlercurve vollkommen gerechtfertigt ist. Die Zähler beider Ausdrücke schwanken immer zw. gewissen Grenzen hin und her, und der Werth des Integrales wird daher kleiner als $\frac{c}{\varphi}$, und folglich der Werth des Productes n solcher Integrale $< \frac{c}{\varphi^n}$ welcher sich also mit wachsendem φ der Null nähert.

Chapter 3
The Hypothesis of Elementary Errors

In the framework of classical probability theory, the primary objective was to calculate probabilities of certain events, with the aim of making "rational" decisions based on these probabilities. Error or frequency functions[1] only played the role of auxiliary subjects. This paradigm, however, would change fundamentally during the course of the 19th century. In the field of biological statistics, for example, probability distributions became an independent object of research. In this context, it was the prevailing opinion for a long time that almost all quantities in nature obeyed normal distributions. For a justification of the apparently privileged role of normal distribution, a model was used in most cases which had originally arisen from error theory: the hypothesis of elementary errors. A random quantity obeying this hypothesis was assumed to be additively composed of a very large number of independent elements, each of them being insignificant compared with the total sum. In this case, the CLT guaranteed an approximate normal distribution of the random quantity under consideration.

The hypothesis of elementary errors was first stated by Hagen in 1837, and one year later considerably generalized by Bessel, who aimed at a deduction of normal distributions for errors of observation as indisputable as possible. The assumption of normally distributed errors of observation was originally needed for a simplified treatment of Gauss's "first" justification of the method of least squares. The use of error theoretic considerations for the examination of statistical quantities in biology, economy, and social sciences, particularly propagated by Quetelet, led to the application of the hypothesis of elementary errors even beyond errors of observation.

At the beginning, statistics of distributions was very closely related to error theory regarding its methods and concepts. However, growing awareness of the fact that the notion of biological "frequency laws" was entirely different from that of error laws was an important milestone in the development of statistical thinking. While errors of observation are deviations from a single real value, deviations from

[1] During the 19th century, "frequency function" was a term especially used to designate densities of random variables in biology, economy, and social sciences.

H. Fischer, *A History of the Central Limit Theorem*, Sources and Studies in the History of Mathematics and Physical Sciences, DOI 10.1007/978-0-387-87857-7_3,

a certain mean within biological samples have an entirely different quality, because that mean has a purely mathematical character.

At the beginning of the 20th century, statistics became more and more independent from error theoretic concepts, as there was a shift of its main objectives from the examination of empirical distributions toward hypothesis testing. Accordingly, elementary errors became more and more unimportant in statistics. Still in the 1920s, however, elementary errors were rather frequently discussed in connection with the CLT.

3.1 Gauss and "His" Error Law

The foundation of least squares according to Laplace was based on a CLT for linear combinations of errors of observation. This foundation competed against the two foundations presented by Gauss, of which in particular the first—including various modifications by later authors—was very popular during the 19th century, despite being based on the very special assumption of exclusively normally distributed errors.[2] Originally, Gauss [1809, 240–245] had deduced the error law, which would later be named after him, by virtue of the principle of the arithmetic mean being the "most probable" estimation for the true value in the case of direct observations $M_1, M_2, \ldots, M_\mu$, where each obeyed the same law of error $\varphi(\Delta)$, which was assumed to be symmetric and unimodal.[3] Gauss's arguments relied on inverse probabilities (see footnote 2, Chap. 2), and can be summarized, in a somewhat modernized form, as follows: If $dP(x|M_1, \ldots, M_\mu)$ designates the (infinitesimal) probability that x is the true value underlying the direct observations $M_1, \ldots, M_\mu$, and $dP(M_1, \ldots, M_\mu|x)$ the respective inverse probability, then, if one a priori presupposes all possible true values x uniformly distributed, both probabilities are proportional to each other. Therefore, the "most probable" estimation p for the true value is characterized by the condition that $dP(M_1, \ldots, M_\mu|x)$ and therefore also

$$\varphi(M_1 - x)\varphi(M_2 - x)\cdots\varphi(M_\mu - x) \tag{3.1}$$

is maximized for $x = p$. From this, under the assumption of the arithmetic mean being the "most probable" value, Gauss [1809, 244] derived the following condition for the error curve $\varphi(\Delta)$:[4]

$$\frac{\varphi'(M_1 - p)}{\varphi(M_1 - p)} + \frac{\varphi'(M_2 - p)}{\varphi(M_2 - p)} + \cdots + \frac{\varphi'(M_\mu - p)}{\varphi(M_\mu - p)} = 0 \tag{3.2}$$

[2] Comprehensive accounts on Gauss's "first" foundation of the method of least squares can be found in [Sheynin 1979] and [Hald 1998, 351–357].

[3] Strictly speaking, however, the latter properties are not required for Gauss's proofs if one only assumes—as Gauss naturally did—$\varphi(\Delta)$ being sufficiently smooth.

[4] The reader who compares the following equations with those in the original source should not be bewildered by different notations. Whereas Gauss used the abbreviation $\varphi'(\Delta)$ for $\frac{\mathrm{d}}{\mathrm{d}\Delta}\log(\varphi(\Delta))$, in the present book φ' simply stands for the derivative of φ.

for all natural μ if

$$p = \frac{M_1 + M_2 + \cdots + M_\mu}{\mu}.$$

In the special case $M_k = M_1 - \mu N$ ($k = 2, \ldots, \mu$, N an arbitrary constant) condition (3.2) yielded

$$\frac{\varphi'[(\mu-1)N]}{\varphi[(\mu-1)N]} = (1-\mu)\frac{\varphi'(-N)}{\varphi(-N)}.$$

From the latter equality, it could—in Gauss's own words—"easily" be deduced that

$$\frac{\varphi'(\Delta)}{\varphi(\Delta)\Delta} = \text{const.}$$

This differential equation finally led, on account of the constraints that (3.1) had to attain a "true" maximum and that $\int_{-\infty}^{\infty} \varphi(\Delta)d\Delta = 1$, to the "Gaussian" law of error

$$\varphi(\Delta) = \frac{h}{\sqrt{\pi}} e^{-h^2\Delta^2} \quad (h > 0). \tag{3.3}$$

Presupposing in the linear model for the observations d_i and the unknown true values x_j

$$d_i = \sum_{j=1}^{m} a_{ij}x_j + y_i \quad (i = 1, \ldots, n) \tag{3.4}$$

that all (mutually independent) errors y_i were identically distributed according to (3.3), the (infinitesimal) probability $dP(d_1, \ldots, d_n|x_1, \ldots, x_m)$ (the conditional probability that the observations d_i are made under the presupposition of the true values x_j) could be expressed by

$$dP(d_1, \ldots, d_n|x_1, \ldots, x_m) = \left(\frac{h}{\sqrt{\pi}}\right)^n e^{-h^2\sum_{i=1}^{n}(d_i - \sum_{j=1}^{m} a_{ij}x_j)^2} dy_1 \cdots dy_n. \tag{3.5}$$

$dP(d_1, \ldots, d_n|x_1, \ldots, x_m)$ attains its maximum value if $x_j = x'_j$ ($j = 1, \ldots, m$), where

$$\sum_{i=1}^{n}(d_i - \sum_{j=1}^{m} a_{ij}x'_j)^2 = \min. \tag{3.6}$$

The least squares condition therefore corresponded to the "most probable value system" x'_j [Gauss 1809, 245].

As already mentioned, justifications of least squares following the basic ideas of Gauss's "first" foundation remained very popular.[5] Gauss's arguments, however, were plagued by some serious conceptual problems.[6] Gauss himself no longer

[5] Merriman [1877c, 165] reports that the majority of books on error theory published up until that time preferred this method (occasionally modified) of introducing least squares.

[6] Knobloch [1992, 62–64] summarizes the contemporary criticism on Gauss's "first" foundation. Because in the case of direct observations (only one unknown x_1, $d_i = x_1 + \epsilon_i$) the method of

granted his justification too great a value in later times.[7] Primarily, he was bothered by the fact that the probabilities under consideration (see (3.5)) were only infinitesimal. What was the use if one could show that an infinitely small probability attains a maximum? Another problem was that Gauss's line of argument relied heavily on inverse probabilities, and was therefore quite intricate. Gauss designated the values of the least square estimators x'_j according to (3.6) as "the most probable." This notion, however, made perfect sense only if one had in mind that, in accord with the principles of inverse probabilities, $dP(x_1, \ldots, x_m | d_1, \ldots, d_n)$ was proportional to $dP(d_1, \ldots, d_n | x_1, \ldots, x_m)$, and therefore the most probable values for the x_j corresponded to the most probable observations d_i and errors y_i, respectively. In contrast to these rather complicated and in part—regarding the a priori equiprobability hypothesis for the true values—questionable considerations, many authors after Gauss simply equated the errors $y_i = d_i - \sum_{j=1}^{m} a_{ij} x_j$ with the residuals $y'_i = d_i - \sum_{j=1}^{m} a_{ij} x'_j$ (x'_j being the estimators for the x_j). The probability (3.5) was then conceived as a probability regarding residuals, and the most probable values of the residuals corresponded, naturally, to the most probable estimations [Czuber 1891, 48–52; Hald 2007, 106–109]. One of the most problematic aspects, however, was that, on the one hand, Gauss had hypothetically deduced "his" error law from the principle of the arithmetic mean as the "most probable" estimation (in [1809, 244] he called this principle "axiom"), and on the other hand, in so doing had also arrived at a statement of physical fact. The question which immediately arose from this apparent contrast was whether errors of observation actually obeyed an (approximate) Gaussian distribution or this kind of distribution was only a convenient hypothesis, possibly far from reality.

Apparently dissatisfied with his "first" foundation of least squares, Gauss [1823] showed that least squares conditions analogous to (3.6), for obtaining estimators x'_j for x_j in (3.4) of the form $x'_j = \sum_{i=1}^{n} k_{ji} d_i$, were equivalent to the requirement that

$$\mathrm{Var}(x'_j - x_j) = \mathrm{Var} \sum_{i=1}^{n} k_{ji} y_i = \min$$

under the constraint $\sum_{i=1}^{n} k_{ji} a_{il} = \delta_{jl}$, presupposing independent errors of observation with zero expectations.[8] Regarding the problem of a favored existence of special error functions he stated in the first part of his 1823 paper:

> Probably, in practice it will be as good as impossible to indicate this function a priori [Gauss 1823, 5].

From this quotation we can see that Gauss did not adhere to the idea of a "natural" predominance of normally distributed errors.

---

least squares leads, in turn, to the arithmetic mean ($x'_1 = \frac{1}{n} \sum_{i=1}^{n} d_i$), Stigler [1986, 141] criticizes a logical circle from the principle of arithmetic mean to Gaussian error law and least squares, and, finally, back to arithmetic mean. This aspect, however, apparently did not play a decisive role in the 19th-century discussions on Gauss's arguments.

[7] See [Gauss 1880, 523], letter to Bessel, 28 February 1839.

[8] For a detailed analysis of Gauss's arguments, which also included the more general case of possibly different error laws for different observations, see [Hald 1998, 472 f.].

So why did Gauss's "first" justification of least squares remain so favored until the beginning of the 20th century, despite its hypothetical character, its conceptual complexity, and the fundamental criticism of its originator? One reason may be that Gauss's arguments were already valid for a small number of observations in contrast to Laplace's CLT considerations, and that they were altogether easier to expound with respect to the analytical methods used. These advantages, however, applied even more to Gauss's "second" justification. The decisive advantage of Gauss's first justification was apparently that it was closely related to an explicit law of error. Knowledge of specific laws of errors was needed when the "best" among different competing estimators for a certain parameter was to be found. A typical example was given by Laplace [1812/20/86, 571–577][9] in his discussion of the accuracy of the arithmetic mean versus the median. Without the explicit knowledge of the law of error it is impossible to determine which of the two estimators has the larger probability for a certain maximum deviation from the true value.[10] A further example was the evaluation of different estimators for the "mean error" (the former name for the modern "standard deviation"), as can basically be traced back to Gauss [1816].[11] Slightly exaggerated (and un-historically expressed), the preference of Gauss's "first" justification—in many cases simplified by considering residuals instead of real errors, as seen above—was based on the advantages of parametric statistics in comparison to nonparametric statistics in the framework of error theory.

Therefore, Gauss's "first" justification in fact broached the "nature" of the law of error. Did this actually obey a Gaussian distribution in particular experimental settings, which still had to be described more exactly? This question could only be answered by comparing the relative frequencies of errors observed on the one hand, and Gaussian distributions on the other. This comparison had to be accompanied by discussing plausible models for the formation of errors of observation. In connection with such considerations, the hypothesis of elementary errors became particularly popular.

3.2 Hagen, Bessel, and "elementäre Fehler"

The entire work of the astronomer Friedrich Wilhelm Bessel (1784–1846) is distinguished by a comprehensive and detailed discussion of random as well as systematic errors of observation.[12] Concerning random errors depending on special instruments

[9] In the second supplement of the 3rd edn. of *TAP*, dated "February 1818." For a description of this supplement see [Hald 1998, 444–452].

[10] Dirichlet [1836; 1897b] later used this example for a general criticism of the method of least squares [Fischer 1994, 44–47].

[11] For comments on [Gauss 1816] and related work see [Czuber 1891, 128–145; 174–182; Hald 1998, 456–458; Sheynin 2005b, 144 f.].

[12] See [Lavrynovich 1995, 136–150] for a detailed account, however neglecting mathematical aspects.

of observation, he was particularly interested in the interplay between models of the emergence of errors, and actual error laws. His error theoretic work exhibits a see-saw between considering the Gaussian error law as merely plausible and useful for certain investigations on the one hand, and attributing a real character to it on the other (notwithstanding certain deviations from the mathematical model which occur in practice).[13] In 1818, in the framework of observations of right ascensions and declinations of stars, Bessel had already drawn a comparison between the relative frequencies of residuals of direct observations which fell into certain intervals and the corresponding probabilities calculated on the basis of Gaussian error laws. He found a good correspondence between the two sets of values [Bessel 1818, 18–21].[14] In his hitherto unpublished exchange of letters with Carl Gustav Jacobi[15] one can find an attempt dated from 1830 to prove the arithmetic mean being the most probable estimate (in Gauss's sense) of direct observations under very weak conditions and independent of a special law of error [Bessel 1830]. Jacobi [1830], however, immediately found serious errors in Bessel's arguments, which could not be eliminated. One can be quite sure that error theory played a significant role in the courses on astronomy given by Bessel at the University of Königsberg.

3.2.1 The Rediscovery of the Hypothesis of Elementary Errors by Gotthilf Hagen

With his courses, Bessel paved the way for the "invention" of the hypothesis of elementary errors by one of his favorite disciples, Gotthilf Hagen (1797–1884, Fig. 3.1). Indeed, as we can see from the unpublished exchange of letters between Bessel and Hagen (for closer details see below), the sole credit of an explicit formulation and justification of the hypothesis of elementary errors has to be assigned to the latter.

From a purely formal point of view, Hagen only "re-discovered" elementary errors. Daniel Bernoulli [1778] had already hinted at the idea of any observational error being the sum of a large number of very small errors. In his 1778 paper, he was still elaborating this idea in a rather qualitative manner, while in a subsequent article [1780] he thoroughly discussed the aberrations of pendulum clocks by means of a simple, yet quantitative binomial model. He assumed that the accumulation of equiprobable deviations which are equal in modulus, though not necessarily in sign, caused the accidental procedure or pursuing of the clock in each period [Sheynin 1972, 289–292; Hald 1990, 506].

---

[13] Already Gauss [1809, 244] had hinted at the property of normal distributions to allow arbitrarily large errors, at least in principle. On the other hand, he noticed that this "flaw" was unimportant in practice due to the rapid decrease of normal error laws.

[14] Bessel's tables are reprinted in [Schneider 1988, 278; Stigler 1986, 204; Hald 1998, 362].

[15] These are actually mathematical notes, which the two colleagues at Königsberg University exchanged between 1826 and 1846. They are now kept in the archive of the Academy of Sciences of Berlin.

Thomas Young [1819, 72–77] attempted to show by induction that the "probable error"[16] r is related to the expectation of the modulus of an observational error (which he called the "mean error" e) with $r \approx 0.85e$. Without any knowledge of Bernoulli's work, Young assumed the observational error to be the sum of a very large number n of partial errors, each taking the values $\frac{1}{\sqrt{n}}$ and $-\frac{1}{\sqrt{n}}$ with probability $\frac{1}{2}$. Sums of this kind are approximately normally distributed, and therefore Young, apparently without a clear insight into the error theoretic details, derived his result as being general although being only valid for normally distributed errors.[17]

Certain statements by Gauss suggest that he, too, had ideas which at least come close to elementary errors.[18] In §3 of his *Theoria Combinationis* [Gauss 1823, 4 f.] he distinguished between "partial errors" and "total errors," the latter being composed of "several simple errors" and having values which resulted "in infinitely many ways from the composition of the partial errors, which themselves are more or less probable." Gauss, however, did not explicitly state a hypothesis on the coaction of these "partial errors." Apparently, Gauss's remarks were not noticed by mathematicians who later worked on the hypothesis of elementary errors.[19] But due to the strong influence of his contributions to the development of error theory in general, one cannot exclude the possibility that his statements produced a certain impetus toward the idea of elementary errors. Gauss's distinction between "accidental" and "constant" errors was based on a discussion of error sources in the specific use of measuring instruments. Similar ideas can be found in the works of almost all proponents of elementary errors. D. Bernoulli [1780] had already made a distinction between "aberrationes chronicae" (systematic errors) and "aberrationes momentaneae" (random errors) in his work on pendulum clocks. However, Bernoulli's work remained unnoticed during the 19th century and was only appreciated by recent authors.[20]

It was left to Gotthilf Hagen—who apparently did not know of D. Bernoulli's contributions—to base error theory in a general way on a clearly formulated hypothesis about elementary errors, resorting to the principles of the use of measuring instruments. Hagen [1837, 34] made the following assumption:

> ... the error in the result of any measurement is the algebraic sum of an infinitely large number of elementary errors ["elementäre Fehler"], which are all equally large, and of which each single one can be just as positive as negative.

---

[16] For a probability density $f(x)$ which is symmetric with respect to $x = 0$ the probable error r is defined by the condition $\int_{-r}^{r} f(x)dx = \frac{1}{2}$.

[17] Young [1819, 78] boasts in a quite exaggerated way stating: "In other respects the results here obtained do not materially differ from those of LEGENDRE, BESSEL, GAUSS and LAPLACE: but the mode of investigation appears to be more simple and intelligible."

[18] I thank Ivo Schneider for drawing my attention to this fact.

[19] Neither in the exchange of letters between Bessel and Hagen (see footnote 22), nor in the discussion of elementary errors by the Gauss-adept Encke [1850, 334–352], nor in the surveys on error theory by Czuber [1891; 1899] or Pizzetti [1892] can one find any allusions to Gauss's remarks.

[20] See [Sheynin 1970; 1972, 286; Hald 1990, 500–504].

Fig. 3.1 Gotthilf Hagen

This hypothesis was related to a model of error causation, which was according to a very large number of drawings with replacements from an urn containing black and white balls corresponding to the positive and negative values of elementary errors.

After first studying mathematics and astronomy at Königsberg University (1816–1818), Hagen changed to civil engineering ("Baukunst"), because he felt more affiliated to the "practical side."[21] Mainly during his studies in Königsberg he had acquired his mathematical and astronomical skills, among them error theory. Those skills remained very useful in his new activities. He always kept close contact with Bessel.[22] From 1831, Hagen worked at the "Oberbaudeputation" in Berlin (the leading institution for civil engineering in Prussia at this time), and also became a teacher at the Berlin "Bauakademie" (the Prussian college for civil engineering). In his teaching as well as his scientific activities (primarily focusing on hydraulic engineering), he stressed in a quite unusual way compared to contemporary customs theoretical and mathematical aspects. He was particularly interested in applications of probability calculus to land survey ("lower geodesy") and was a quite isolated innovator in this respect [Hagen 1831].[23]

Despite resistance of his colleagues, Hagen's interest in probability calculus remained undiminished. In a letter dated 2 August 1836 Hagen [1836a] sent a draft of a little essay to Bessel, in which he presented and justified "his" hypothesis of

[21] For Hagen's professional and scientific career see [Ottmann 1934]. This biography is based on Hagen's extensive collection of private papers, which has been missing since the end of the Second World War. In Ottmann's book, one can also find many passages from letters and from Hagen's unpublished autobiography (which has also disappeared ever since).

[22] Parts of the unpublished exchange of letters between Bessel and Hagen can be found in the manuscript division of the Staatsbibliothek Preussischer Kulturbesitz in Berlin and the archive of the Academy of Sciences at Berlin.

[23] Quite frequently, there were controversial discussions in 19th-century Germany about the importance of mathematics for engineering. Hensel [1989] has given an account on the role of mathematics for engineering education during the second half of the 19th century in Germany. However, the corresponding development until about 1850 has been largely neglected by historians until now.

elementary errors and deduced a Gaussian distribution (Fig. 3.2) for observational errors. Hagen referred to the motivation for his research with the following words:

> I wanted to explain my conductors, who always speak about absolute precision in levelling, which consequences are caused by the accumulation of errors ...

In his reply, Bessel [1836] showed his above-mentioned ambivalent attitude of being willing to use appropriate models of error causation on the one hand, while on the other hand criticizing the lack of reality regarding hypotheses and particular laws of error. He expressed his doubts about the universal validity of Hagen's very simple assumptions and of the Gaussian distribution for errors deduced therefrom. At the same time, Bessel hinted at Young's 1819 contribution which had previously gone unnoticed by Hagen.

In another letter to Bessel (28 July 1836), Hagen [1836b] announced the publication of a book already planned a long time ago, titled *Grundzüge der Wahrscheinlichkeitsrechnung mit besonderer Anwendung auf die Operationen der Feldmeßkunst* ("Essential Features of Probability Calculus with Special Application to the Operations of Land Survey"). The book was published in 1837, a second edition in 1867, and a third in 1882. The contents of the book went far beyond the problems of surveying. Especially covered were further applications of probability to the examination of the strength of various materials or to the fitting of parameters in empirically determined formulae. The essential part of the book, however, was on the discussion of Hagen's hypothesis of elementary errors and on an analysis of least squares derived therefrom.[24] In its theoretical intentions, Hagen's book was entirely different from Christian Gerling's exposition on the method of least squares which was directed toward the same audience, appearing a little later in 1843. It was not Hagen's intention to give a collection of recipes; instead he aimed at the education of independently thinking technicians.

In the first edition, Hagen thoroughly discussed his hypothesis of elementary errors, whose value was not purely didactical from his point of view. In fact, in this hypothesis he saw substantial scientific progress regarding the foundations of the method of least squares. In the second and third edition, Hagen was considerably more reserved in this respect, perhaps because he had meanwhile realized the untenability of some aspects of the model, in particular the confinement to two-valued elementary errors.

In his book, Hagen chose exactly the same analytical method for his derivation of normal distributions by elementary errors that he had already communicated to Bessel in his letter [1836a]. This method was very similar to D. Bernoulli's procedure for approximating binomial distributions with success probabilities close to $\frac{1}{2}$ via differential equations. From this fact alone, however, we cannot conclude that Hagen had plagiarized Bernoulli's work, which remained (as mentioned above) virtually unknown during the 19th century.

Hagen [1837, 41–49] assumed that each of the $2n$ (for the sake of symmetry he only considered an even number) coacting elementary errors could only take one

---

[24] For a characterization of the book, in particular of Hagen's conception of "precision," see [Olesko 1995, 108 f.; 113–115].

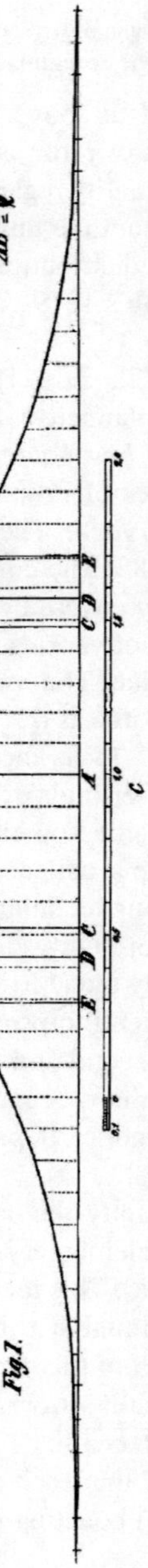

Fig. 3.2 One of the earliest graphical representations of the Gaussian error law in [Hagen 1837]

negative and one positive value, each with the same modulus $\frac{dx}{2}$, and each with probability $\frac{1}{2}$. If $y(m)$ ($m = -n, \ldots, +n$) designates the probability that the sum of these $2n$ elementary errors is equal to mdx (or $n + m$ of these elementary errors are positive and $n - m$ are negative), then, as a result of the properties of binomial distributions one gets[25]

$$y(m) - y(m-1) = -y(m-1)\frac{2m-1}{n+m}. \qquad (3.7)$$

Now, the number of elementary errors was assumed to be "infinitely large" and their unity $\frac{dx}{2}$ to be "infinitely small." According to Hagen, for the probability $y(x)$ that the sum of these elementary errors is $x = mdx$, an equation analogous to (3.7) could be obtained by substituting 1 by dx and m by x in (3.7). In this way, he concluded under the assumption of "infinitely large" n:

$$\frac{y(x) - y(x-dx)}{dx} = -y(x-dx)\frac{2x}{n},$$

wherefrom he inferred the differential equation

$$y'(x) = -y(x)\frac{2x}{n}. \qquad (3.8)$$

(3.8) has the solution

$$y(x) = y(0)\mathrm{e}^{-\frac{x^2}{n}}.$$

Because of the properties of the binomial distribution, and by use of Wallis's product,[26] Hagen was able to infer that, for "infinitely large" n,

$$y(0) = \frac{\sum y(x)}{\sqrt{\pi n}},$$

where the summation is with respect to all possible values of $x = -ndx$, $(-n+1)dx, \ldots, +ndx$. Finally, since $\sum y(x) = 1$:

$$y(x) = \frac{1}{\sqrt{\pi n}}\mathrm{e}^{-\frac{x^2}{n}}. \qquad (3.9)$$

Hagen's method of handling infinitely large and small quantities, as just described, was not only difficult to understand, it was even flawed since the quantities x and dx were applied in a rather inconsistent manner.[27] Probably, the reader was even more puzzled by Hagen's further line of argument [1837, 49–51], in which the term $\frac{1}{\sqrt{n}}$ in

---

[25] Hagen [1837, 43] gave a slightly erroneous version of this formula, "$\frac{2m+1}{n+m}$" instead of $\frac{2m-1}{n+m}$. This mistake did, however, not influence his following arguments.

[26]

$$\frac{\pi}{2} = \frac{2}{1}\cdot\frac{2}{3}\cdot\frac{4}{3}\cdot\frac{4}{5}\cdot\frac{6}{5}\cdot\frac{6}{7}\cdot\frac{8}{7}\cdot\frac{8}{9}\cdots$$

[27] In a logically consistent way, (3.7) has to be replaced by

(3.9) was substituted, by means of a nebulous "rescaling," by the "precision" h (the reciprocal value of the standard deviation of the total error times $\sqrt{2}$).[28] The obvious difficulties for Hagen's readers are highlighted by a controversial discussion between Charles Kummell and Mansfield Merriman, in the 1877 edition of the *Journal of the Franklin Institute*, in which contemporary problems using the "infinite" are exemplified.[29]

After the publication of Hagen's book, several articles appeared with the intention of eliminating the flaws of the original deduction.[30] Hagen himself, however, kept his version of the proof without any substantial modifications in the second and third editions of his book. The basic idea common to all of the "improved" deductions—even if each single author claimed to have derived an entirely different approach—was to complement the hypothesis of elementary errors by the (in fact unjustified) assumption that the two values of each elementary error were $\pm\frac{\Delta x}{2}$ with $\Delta = \frac{1}{h\sqrt{n}}$, where h was a positive constant. For n "infinitely large," the equation for the (infinitely small) probability $y(x)$ that the sum of the $2n$ elementary errors is equal to x was:

$$\begin{aligned} y(x) - y(x - \Delta x) = -y(x - \Delta x)\frac{2x - \Delta x}{n\Delta x + x} &= -y(x - \Delta x)\frac{2x - \Delta x}{n(\Delta x)^2 + x\Delta x}\Delta x \\ &= -y(x - \Delta x)\frac{2h^2x - h^2\Delta x}{1 + h^2x\Delta x}\Delta x, \end{aligned}$$

from which, since Δx is "infinitely small," the differential equation

$$y' = -2h^2xy$$

could be deduced. The solution of this differential equation is $y(x) = y(0)e^{-h^2x^2}$. With

$$\sum y(x) = 1,$$

$$y(x) - y(x - dx) = -y(x - dx)\frac{2x - dx}{ndx + x}.$$

From this equation, however, it is not possible to obtain (3.8). Incidentally, similar "inaccuracies" can be found in D. Bernoulli's contributions mentioned above.

28 Hald [1998, 367] has given an exposition of Hagen's procedure from the perspective of modern analysis, in which the inconsistencies just described are not discussed.

29 The discussion, which comprises [Merriman 1877a], [Kummell 1877], and [Merriman 1877b] in temporal order, started with critical remarks by Merriman on Kummell's [1876] modified version of Hagen's "proof."

30 Merriman [1877b, 330; 1877c, 182] cites Quetelet [1846, 384–387] (an exposition substantially following Hagen's original version), Wittstein [1849, 348–354], Encke [1850, 330–350] (rather incorrectly, because Encke relied on Stirling's formula without making any use of the differential equation (3.8), see also [Czuber 1891, 81–83]), [Dienger 1852, 149–155], Price [1865, 376–379], Tait [1865] (with a method similar to [Encke 1850]), Natani [1866, 16–33], Faà-di-Bruno [1869, 44–45], Meyer [1874, 215], and Kummell [1876, 133–135].

it followed[31]

$$y(0) = \frac{1}{\sum e^{-h^2x^2}} = \frac{\Delta x}{\sum e^{-h^2x^2}\Delta x} = \frac{\Delta x}{\int_{-\infty}^{\infty} e^{-h^2x^2}dx},$$

and therefore

$$y(x) = \Delta x \frac{h}{\sqrt{\pi}} e^{-h^2x^2}.$$

This result corresponded to the Gaussian error law.

One of the main goals of Hagen's book, as expressed in the preface [1837, v f.], was to present the basic ideas of probability theory in a simple and clear manner. In his particular deduction of the error law he did not base his arguments on inverse probabilities, in contrast to Gauss. As a consequence, Hagen [1837, 66–70] gave a justification of least squares which did not rely on inverse probabilities either: he simply equated errors of observation with residuals (see Sect. 3.1), and he was apparently one of the first authors to do so. Hagen did not comment on this—actually rather obscure—puzzling of two different mathematical subjects.[32] Later in the text, discussing the problem of estimating the probability of a given deviation between the actual and estimated value, Hagen [1837, 76–84] carefully differentiated between "real" errors and residuals, which latter he designated "difference between the result of the observation and that of the calculation."

3.2.2 *Bessel's Generalization of the Hypothesis of Elementary Errors*

As mentioned above, Bessel was rather reserved regarding the overall validity of Hagen's hypothesis of elementary errors. He [1836] also questioned the universality of the Gaussian error law:

> Laplace, and after him Gauss[33] have stated the probability law of an error υ to be arbitrary, in general. I think Laplace is right there.

On the other hand, in his 1838 article "Untersuchungen über die Wahrscheinlichkeit der Beobachtungsfehler" ("Analyses of the Probability of Observational Errors"), where he considerably generalized the hypothesis of elementary errors compared to Hagen's, Bessel took a rather positive attitude toward the possible predominance of

---

[31] Dienger and Natani (footnote 30), for example, chose this way. Instead of this method it was also possible to follow Hagen's original arguments, and to infer, by use of Wallis's product, that $y(0) = \frac{1}{\sqrt{\pi n}}$, and then to substitute $\frac{1}{\sqrt{n}}$ in the latter term by $h\Delta x$, with an "infinitely small" Δx (see [Kummell 1876, 135] for a particularly clear exposition).

[32] Hald [2007, 107 f.] interprets these considerations in the sense of early maximum likelihood arguments, however.

[33] Apparently, Bessel was alluding to Gauss's remark in the context of the latter's "second" justification of least squares, as quoted in Sect. 3.1.

Gauss's error law. This article was certainly motivated by the intention to demonstrate his disciple who the true master was. Bessel's mathematical ambitions and skills are shown by the quality of his purely mathematical contributions, and, in various cases, by his exchange of letters with Gauss [1880] and Jacobi (see footnote 15). However, the greater motivation for Bessel might have been his constant interest in specific properties of error laws as an independent subject of research. Hagen's model did not fit reality in Bessel's opinion. Therefore, he tried to establish a model corresponding better to the practice of measurement than Hagen's.

Bessel [1838a] (Fig. 3.3) communicated the success of his effort to his mathematical "advisor" Jacobi with the words:

> You can henceforth take for granted, dearest!, that many coacting causes of error always yield a probability of the entire error which is close to the exponential law.

In the very next sentence, however, Bessel expressed his skeptical attitude:

> Whether the "always" is carrying things too far has to be investigated though; therefore rather "in general."

Bessel assumed that an observational error was additively composed of a very large number of independent elementary errors $x, y, z, \ldots$, with ranges of values $[-a; a]$, $[-b; b]$, $[-c; c]$, respectively, and densities $\varphi, \varphi_1, \varphi_2, \ldots$, each symmetric with respect to 0.[34] At the beginning of his paper, Bessel [1838c, 377–390] tried to give an exact formula for the density function of a sum of elementary errors of this kind using convolutions of the single densities. This procedure, however, led to formidable difficulties, so that Bessel had to restrict his considerations to a maximum total number of 4.

In the course of these explanations, Bessel also developed a general formula by use of a trigonometric jump function. He [1838c, 377 f.] discretized the errors by subdividing the number line in pieces of length $\frac{1}{i}$, where i was an "infinitely large" natural number; he only considered those values $x, y, z, \ldots$ of elementary errors which could be represented by $x = \overline{x} \cdot \frac{1}{i}$, $y = \overline{y} \cdot \frac{1}{i}$, $z = \overline{z} \cdot \frac{1}{i}, \ldots$, where $\overline{x}$, $\overline{y}$, $\overline{z}, \ldots$ were—in general "infinitely large"—integers. Bessel [1838c, 377] noticed that the probability for the value x was $\frac{1}{i}\varphi(x)$, where φ was the density of the associated elementary error. Apparently, it was a matter of course for Bessel, which did not need any further explanation, that after the discretization one had to imagine the entire probability mass of the interval between x and $x + \frac{1}{i}$ concentrated in x. The probability that the sum of $\mu + 1$ elementary errors $x + y + z + \cdots$ was equal to n,[35] Bessel initially represented by

$$\frac{1}{i}\psi(n) = \frac{1}{i^{\mu+1}} \sum_{x+y+z+\cdots=n} \varphi(x)\varphi_1(y)\varphi_2(z)\cdots, \tag{3.10}$$

[34] As usual for the 19th century, Bessel did not make any terminological difference between errors (in the sense of random variables), and the respective values of errors. In accordance with the contemporary usage, too, he [1838c, 374] noticed that $\phi(x)dx$ was the probability that an error with density $\phi(x)$ "falls between x and $x + dx$."

[35] The use of "n" for the sum of elementary errors by Bessel was slightly unfavorable, because, in general, n was not a natural number.

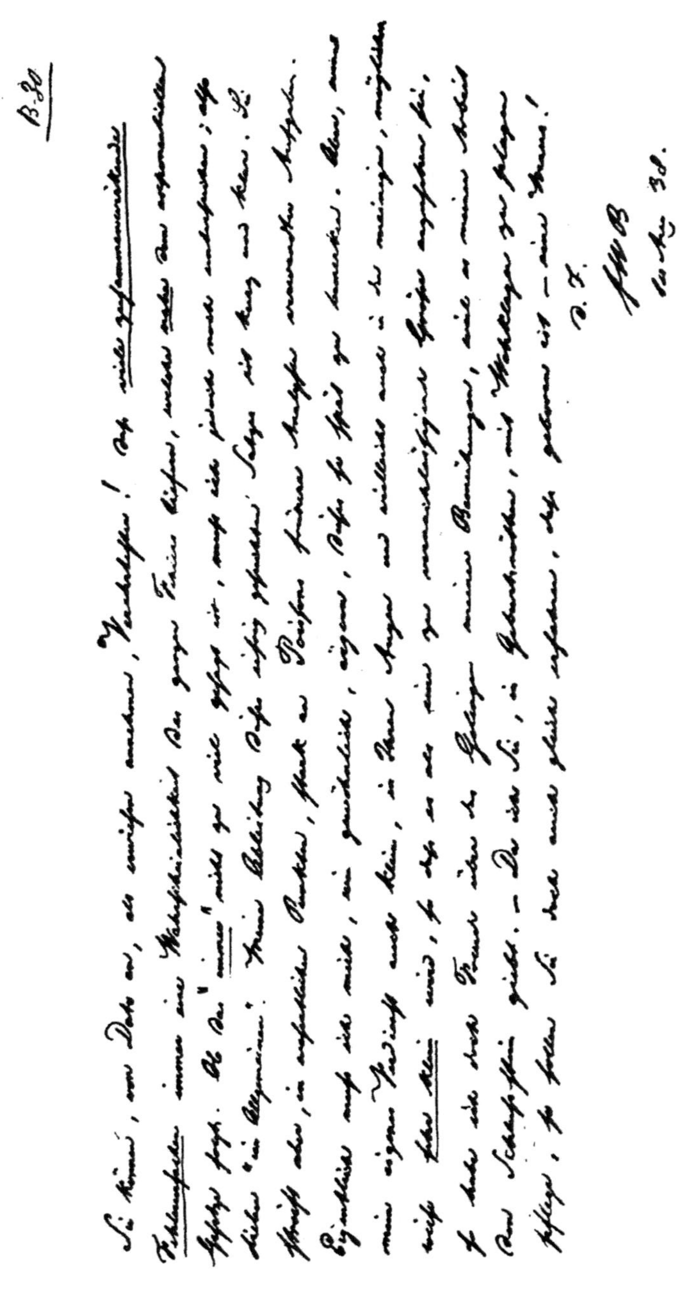

Fig. 3.3 Letter from Bessel to Jacobi, 14 August 1838. For a transcription and an English translation of the full text, see Appendix

where ψ was the density function of the sum.

For the further evaluation of this formula, Bessel used—apparently inspired by his colleague Jacobi[36]—the jump function

$$\frac{1}{2i\pi}\int_{-i\pi}^{i\pi} e^{(x+y+z+\cdots-n)u\sqrt{-1}}du = \begin{cases} 1 \text{ if } & x+y+z+\cdots-n=0 \\ 0 & \text{else,} \end{cases}$$

which is valid for such $x, y, z, \ldots, n$ for which $(x+y+z+\cdots-n)i$ is a (possibly "infinitely large") integer. By use of this jump function, Bessel derived from (3.10):

$$\begin{aligned}\psi(n) &= \frac{1}{i^{\mu}}\sum_{x,y,z,\ldots}\frac{1}{2i\pi}\int_{-i\pi}^{i\pi} e^{(x+y+z+\cdots-n)u\sqrt{-1}}du\varphi(x)\varphi_1(y)\varphi_2(z)\cdots \\ &= \frac{1}{2\pi}\int_{-i\pi}^{i\pi}\left(\sum_x \frac{1}{i}\varphi(x)e^{ux\sqrt{-1}}\right)\left(\sum_y \frac{1}{i}\varphi_1(y)e^{uy\sqrt{-1}}\right)\cdots e^{-un\sqrt{-1}}du.\end{aligned}$$

The summations are with respect to all $x \in [-a;a]$, $y \in [-b;b]$, ... for which xi, yi, ... are integers. Taking into account that i was "infinitely large," and under the assumption that all density functions were symmetric with respect to 0, Bessel concluded

$$\psi(n) = \frac{1}{2\pi}\int_{-\infty}^{\infty}\int_{-a}^{a}\varphi(x)\cos(ux)dx\int_{-b}^{b}\varphi_1(x)\cos(uy)dy\cdots\cos(un)du.$$

Following the basic ideas of the Laplacian method of approximation, Bessel set

$$\int_{-a}^{a}\varphi(x)\cos(ux)dx\int_{-b}^{b}\varphi_1(y)\cos(uy)dy\times \\ \times\int_{-c}^{c}\varphi_2(z)\cos(uz)dz\cdots = e^{-U(u)}, \quad (3.11)$$

where $U(u)$ was a power series which had still to be determined. By expanding the left side of (3.11) into a power series in u, Bessel was able to calculate the first terms of $U(u)$, with the result

$$\begin{aligned}e^{-U(u)} &= \exp\left(-\frac{[\mu_2]}{2}u^2 - \frac{3[\mu_2^2]-[\mu_4]}{24}u^4 - \frac{30[\mu_2^3]-15[\mu_2\mu_4]+[\mu_6]}{720}u^6 - \cdots\right) \\ &= \exp\left(-\frac{[\mu_2]}{2}u^2\right)\left(1 - \frac{3[\mu_2^2]-[\mu_4]}{24}u^4 - \frac{30[\mu_2^3]-15[\mu_2\mu_4]+[\mu_6]}{720}u^6 - \cdots\right),\end{aligned}$$

where[37]

---

[36] This can be inferred from a letter (24 August 1838), which Bessel [1838b] wrote to Jacobi; see [Fischer 2000, 107] for closer details.

[37] The present notation is slightly deviating from Bessel's, it preserves the "Gaussian brackets," however, which most authors of 19th-century error theory used for designating sums.

$$[\mu_j^r \mu_k^s] = \left(\int_{-a}^{a} \varphi(x)x^j\,dx\right)^r \left(\int_{-a}^{a} \varphi(x)x^k\,dx\right)^s + \\ + \left(\int_{-b}^{b} \varphi_1(x)x^j\,dx\right)^r \left(\int_{-b}^{b} \varphi_1(x)x^k\,dx\right)^s + \\ + \left(\int_{-c}^{c} \varphi_2(x)x^j\,dx\right)^r \left(\int_{-c}^{c} \varphi_2(x)x^k\,dx\right)^s + \cdots .$$

By use of the formula

$$\int_0^\infty x^{2j} \cos rx \mathrm{e}^{-a^2x^2}\,dx = (-1)^j \frac{\sqrt{\pi}}{2a} \frac{\mathrm{d}^{2j}}{\mathrm{d}r^{2j}} \mathrm{e}^{-\frac{r^2}{4a^2}} \quad (j \in \mathbb{N}_0,\ a, r > 0), \quad (3.12)$$

which had been already established by Laplace [1812/20/86, 98], the first terms of a series expansion for $\psi(n)$ resulted in the form

$$\psi(n) = \frac{\mathrm{e}^{-\frac{n^2}{2[\mu_2]}}}{\sqrt{[\mu_2]2\pi}} \left(1 - \frac{3[\mu_2^2] - [\mu_4]}{24[\mu_2]^2} \left(3 - \frac{6n^2}{[\mu_2]} + \frac{n^4}{[\mu_2]^2}\right) - \right. \\ \left. - \frac{30[\mu_2^3] - 15[\mu_2\mu_4] + [\mu_6]}{720[\mu_2]^3} \left(15 - \frac{45n^2}{[\mu_2]} + \frac{15n^4}{[\mu_2]^2} - \frac{n^6}{[\mu_2]^3}\right) - \cdots \right) \quad (3.13)$$

$$= \frac{\mathrm{e}^{-\frac{n^2}{2[\mu_2]}}}{\sqrt{[\mu_2]2\pi}} (1 - a_1 - a_2 - \cdots). \quad (3.14)$$

With this series expansion, Bessel achieved a remarkable result, which was significantly beyond Laplace's and Poisson's corrections of the normal density function by additional terms. It would have been within Bessel's reach (see formula (3.12)) to represent his result by

$$\psi(n) = \frac{1}{\sqrt{[\mu_2]2\pi}} \left(\mathrm{e}^{-\frac{n^2}{2[\mu_2]}} + \frac{[\mu_4] - 3[\mu_2^2]}{4!} \frac{\mathrm{d}^4}{\mathrm{d}n^4} \mathrm{e}^{-\frac{n^2}{2[\mu_2]}} + \right. \\ \left. + \frac{15[\mu_2\mu_4] - 30[\mu_2^3] - [\mu_6]}{6!} \frac{\mathrm{d}^6}{\mathrm{d}n^6} \mathrm{e}^{-\frac{n^2}{2[\mu_2]}} + \cdots \right).$$

This is the Charlier A series (see Sect. 3.4.2.2) of a sum of random variables in the special case of symmetric densities.[38] Priorists may therefore conceive Bessel's result as an anticipation of those series expansions. However, Bessel was not interested in a discussion of any systematics which concerned expansions in—later so called—Hermite polynomials. Only around the last third of the 19th century, motivated by problems different from the CLT, a growing interest emerged in series of this kind. Only then, in turn, was there a beginning of systematic research on series expansions for densities or distribution functions of sums of independent random

[38] See [Hald 1998, 327–329] for a more detailed discussion of Bessel's result in the light of Charlier expansions.

variables. First results in the latter context are due to Chebyshev [1887/90], who, however, did not refer to Bessel (see Sect. 3.4.1).

Under the assumption that among all densities the ν-th moments are of the same "order of magnitude" k^ν (k a "fixed" quantity), Bessel [1838c, 388] inferred that the term a_1 in the expansion (3.14) was "of the order of $\frac{1}{\mu+1}$," and the subsequent terms $a_2, \ldots$ were "of the order of $\frac{1}{(\mu+1)^2}$, etc." The "etc." gives reason for the suspicion that Bessel supposed a_j in (3.14) being of the order $(\mu + 1)^{-j}$ generally. This conjecture, however, would have been wrong. a_3 (corresponding to a Hermite polynomial of degree 8) is of the same order of magnitude as a_2.[39] From the "etc." Bessel followed:

> The expression [the density of the normal distribution] can be conceived as an approximation to $\psi(n)$ with the greater right, the larger the number of the coacting causes of error is.

He was convinced, however, that the series expansion (3.13) was divergent. The reason he stated for this assertion was that, in the case of a finite number $\mu + 1$ of elementary errors, the expression $\psi(n)$ was "discontinuous," and "discontinuous" expressions could not be represented by convergent series in Bessel's opinion. Moreover, the Laplacian method of approximation, which Bessel had used for the derivation of his series, led, in his own words, "in general" to divergent series expansions. These arguments show that Bessel in 1838, despite his mathematical talents and skills, was no longer acquainted with the contemporary state of the art of analysis. In the first part of his article he had shown that, in the case of a finite number of elementary errors, explicit algebraic formulae for $\psi(n)$ could in general only be given by case differentiations. Therefore, in 18th century mathematical language, which was different from the terminology that was established during the 19th century and is in common use now, ψ was "discontinuous." Moreover, from the point of view of 18th century analysis, algebraic formulae had a "general" validity [Jahnke 2003b, 131; Lützen 2003, 161 f.]. As it seems, in the special case of (3.13) the consequence of this perception for Bessel was that the left side $\psi(n)$ had to possess the same algebraic properties as the right side. Thus, it had to be possible to represent $\psi(n)$ by the same algebraic expression for all $n \in \mathbb{R}$. Bessel tried to resolve this apparent inconsistency by assuming the series expansion to be divergent.

The full significance of his series expansion, even for a minor number of elementary errors, was not recognized by Bessel. He was interested in the deduction of an—at least approximate—Gaussian distribution for observational errors, and he did not discuss possible deviations from this particular distribution. Bessel summarized the preconditions for the validity of a normal distribution with the following hypotheses:

> The first of these assumptions is that *many* causes coact in the generation of the observational errors; the second that, among the mean errors generated from the single causes,[40] no one considerably surpasses the others [Bessel 1838c, 389].

[39] For a determination of the order of magnitude of the respective terms in Charlier A series, see, e.g., [Cramér 1946, 226].

[40] With "mean errors" Bessel probably designated the standard deviations of the single elementary errors.

Bessel did not comment on the consequence that, presupposed a large number of elementary errors, the respective variance of each elementary error had to be very small.

As he deplored in a letter to Gauss (edited in [Gauss 1880, 522]), Bessel was unsatisfied with his analytic methods, because in his opinion they led to divergent series expansions. As he mentioned in the letter to Jacobi [Bessel 1838a] (depicted above), he was also angry about having realized too late the affinity of his arguments to "Poisson's previous analyses of similar problems." Therefore, from Bessel's point of view, the concluding part of his article where he again tried to propagate his basic idea of a general hypothesis of elementary errors was especially important. In the special case of Reichenbach's meridian circle, he altogether specified 13 sources of elementary errors showing that they met his two basic assumptions.[41] With these considerations, Bessel apparently wanted to convince his readers that his hypothesis of elementary errors was not only suitable for a computational model, but was even close to reality, at least in certain cases.

3.3 The Reception of Hagen's and Bessel's Ideas

After the publication of Hagen's and Bessel's contributions it took some time until the concept of elementary errors was accepted and further developed within error theory. This quite tepid reception was presumably due to the very limited number of scientists interested in theoretical questions of error theory. The first edition of Hagen's book did not sell very well. In fact, the further development of elementary errors was also strongly motivated by problems beyond error theory.

3.3.1 Normal Distributions in Statistics of Biological and Social Phenomena

From about 1840, statistical distributions became, in particular through Adolphe Quetelet's research program (see [Stigler 1986, 203–214]), more and more important. The longlasting dogma about the primacy of normal distributions in the statistics of biological and social phenomena (Fig. 3.4), from the point of view of the 20th century designated as "Queteletism," was usually justified by hypotheses similar to Hagen's. Quetelet, in his discussion of additively coacting "random causes," from which a Gaussian distribution could be assumed to be generated, referred to a simple "black-and-white" urn model: The probability of a certain value of a (suitably discretized) statistical quantity was identified with the probability that, out of a large number of drawings with replacement from an urn, a certain number of black (or white) balls was obtained. Whether Quetelet, who developed these ideas

---

[41] For closer details see [Lavrynovich 1995, 149 f.].

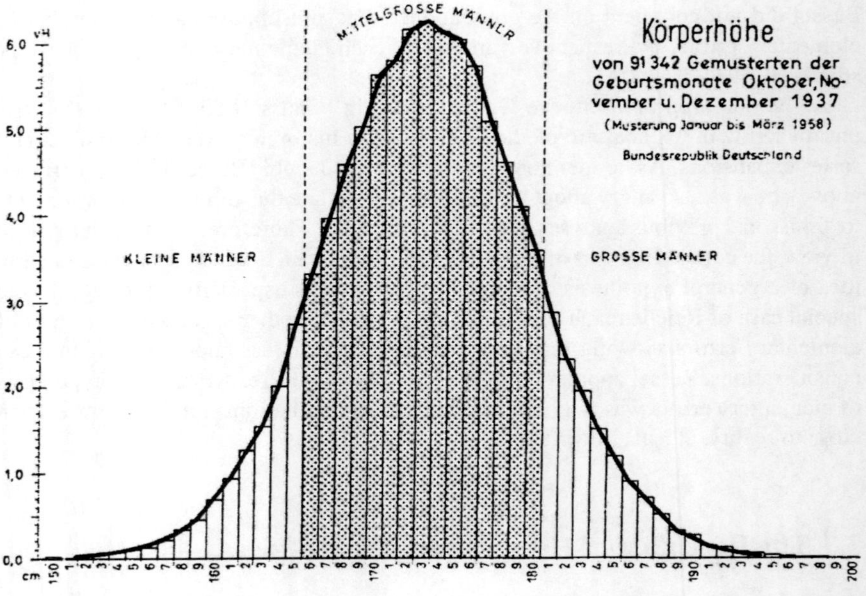

Fig. 3.4 A modern example of Queteletism? Frequency distribution of body heights of German recruits [Barth & Haller 1994, 276]. May it be possible to explain the normal distribution by an appropriate model related to the CLT?

between 1835 and 1844, was decisively inspired by reading Hagen's book, cannot be decided. However, as we see from quotations in some of his works ([Quetelet 1846, 385–387], for example), he was familiar with Hagen's book; moreover, his methods for fitting normal distributions to empirically obtained frequency curves were closely related to Hagen's. Quetelet's analyses were straightforward based on the urn model, whereas Hagen, for whom the concept of elementary errors was preeminent, used the urn model only as an illustration. Until ca. 1880, concerning statistical issues beyond error theory neither Quetelet nor any of his successors made the attempt to take Bessel's more generalized hypothesis of elementary errors as a basis for research. One reason for this might have been the rather poor mathematical knowledge of many 19th-century biological or social statisticians. Perhaps it was more decisive that any hypothesis on the causation of deviations could only be a very rough model due to the complexity of biological and social issues. Convenience therefore suggested resorting to the most simple model. Whereas Bessel analyzed 13 distinct types of elementary errors in the context of a particular measuring instrument, a similarly precise approach was impossible in the biological or social context, because there was insufficient knowledge of the microstructure of the phenomena considered. Moreover, the "black-and-white urn model" yielded a concrete and easily comprehensible mechanism for the emergence of random deviations, which more complicated models, as Bessel's, could not provide.

The adoption of only the simplest model to general statistics, however, did not obstruct its advancement, as can be seen by the example of Francis Galton. In accordance with his rather poor mathematical skills and apparently without any knowledge of more general concepts, Galton in his 1875 article "Statistics by Intercomparison with Remarks on the Law of Frequency of Error" expressed his amazement that in nature one could find so many normally distributed quantities, in spite of the fact that their existence allegedly depended on the very restrictive conditions of Hagen's hypothesis. However, this misunderstanding concerning the universality of the binomial scheme motivated him to a discussion of fundamentally different models which did not resort to error theoretic analogies. According to Stigler [1986, 272–281], Galton with these ideas achieved a "breakthrough," in abolishing the paradigm of the conceptual identity between statistical variation and errors of observation, despite his adherence to the primacy of the normal distribution.[42] It was precisely this insight of the fundamental difference between deviations within a population and deviations caused by errors, which, by the end of the 19th century, paved the way for an increasing research on nonnormal distributions. In turn, only in connection with this research can one find considerably generalized hypotheses of elementary "errors" in biological and social statistics, in Edgeworth's work (see Sect. 3.4.2.3), and, especially in the so-called "Scandinavian school" of Gram, Thiele, Charlier, and others (see Sects. 3.4.2.2, 3.4.3.2). Hald [1981, 6] has argued that the discussion of Bessel's version of elementary errors in the monograph of the Danish geodesist Karl Christian Zachariae [1871] prepared the ground for the formation of this "school."

3.3.2 *Advancement Within Error Theory*

Many authors, who contributed to error theory, contented themselves with Hagen's simple hypothesis. This might have been motivated by a didactic intention in some cases, as Czuber [1891, 80] has observed. The circumstance, however, that the variances of all elementary errors had to be very small in Bessel's model suggested a reduction to Hagen's model as well, and that in particular if one adopted an "atomistic" interpretation of elementary errors, as Johann Franz Encke [1850, 334–352] did. He compared elementary errors with "elements," "oscillating" in a certain sense and defended the apparently exaggerated simplicity of Hagen's model arguing that quantitative differences could be neglected because of the "subtlety" ("Feinheit") of those oscillations [Encke 1850, 350].[43] From Encke's point of view, however, Hagen's hypothesis had only the character of an illustration, without a direct relation to reality.

[42] For a comprehensive discussion of Galton's ideas in this context, see also [Porter 1986, 128–146].

[43] Encke's support of Hagen's hypothesis, and his simultaneous neglect of Bessel's, may have also been caused by Encke's temporary aversion toward Bessel (see [Bruhns 1869, 267–287; Wattenberg 1976, 13; 31–39; Lavrynovich 1995, 82 f.]).

Yet several authors felt uneasy with the apparent discrepancy between hypothesis and reality in Hagen's model. Bessel was not the only one who wanted to "prove" the "real" existence of the Gaussian law of error, at least under certain conditions, by generalizing Hagen's version. Some authors were satisfied with only a slight generalization toward rectangularly distributed elementary errors. On the other hand, even Bessel's model was further generalized by William Crofton and Paolo Pizzetti.[44]

3.3.2.1 Rectangularly Distributed Elementary Errors

George Biddell Airy [1861, 7] proposed the hypothesis that each error of observation had to be conceived as "produced by the algebraic combination [i.e., summation] of a great many independent causes of error" in his textbook on error theory. Without citing Hagen and Bessel, he referred to Laplace for this basic idea. However, he admitted that—strictly speaking—elementary errors did not exist "in the language of Laplace." With certainty, Laplace's and Poisson's "causation systems" (see Sects. 2.1.5.1 and 2.2.3.2) have advanced the later hypothesis of elementary errors in a conceptual sense; the hypothesis itself, however, is explicitly stated neither in Laplace's nor in Poisson's work, even if some authors express a contrary opinion.[45] Airy [1861, 8–15] precisely described Laplace's method of approximating the distribution of a sum of independent random variables with identical rectangular distributions, as it can be found in the latter's discussion of the comet problem (see Sect. 2.1.5.1). He [1861, 15] stated that one could hardly question "the accordance of the result with our general ideas on the frequency of errors." In the third edition of his book, however, Airy replaced the—as he called it [1879, iv]—"Laplacian" derivation by that of "Thomson and Tait"[46] without stating any reason for this change.

Charles H. Kummell [1882, 177], in his attempt to obtain an exact formula for the density of a linear combination

$$a_1\Delta_1 + a_2\Delta_2 + \cdots + a_n\Delta_n,$$

Δ_i being independent errors, each rectangularly distributed within the interval $[-\alpha_i;\alpha_i]$ $(i = 1,\ldots,n)$, directly referred to Bessel's [1838c] article. He was able to establish the formula sought for by use of Dirichlet's discontinuity factor. The asymptotic treatment of this formula led to a result analogous to Bessel's.

Arnold Sommerfeld, who now is chiefly known for his work on quantum mechanics, has also contributed to the summation of independent elementary errors

[44] Surveys of the different hypotheses of elementary errors during the 19th century can be found in [Czuber 1891, 61–99], and, based on Czuber's account, in [Eisenhart 1983, 554–557].

[45] Stigler [1986, 202] as well as Hald [1990, 507] have erroneously stated that the hypothesis of elementary errors played an important role in Laplace's treatment of the CLT.

[46] Peter Guthrie Tait and William Thomson (Lord Kelvin) had essentially described the deduction of the Gaussian error law by John Herschel [1850] (see [Merriman 1877c, 211]) in the third chapter of their *Treatise of Natural Philosophy,* Vol. 1 [1867]. For Herschel's account see also [Czuber 1891, 103–108].

with identical rectangular distributions. Sommerfeld [1904] gave a very intuitive "geometrical" derivation of the (already well known) exact formula (2.1) for the density of the sum, and thereby he anticipated some elementary ideas concerning central B-splines [Butzer, Schmidt, & Stark 1988, 143–147]. Regarding the limit case of "infinitely many" elementary errors, Sommerfeld referred to an article of Ludwig Maurer [1896] on repeated arithmetic means: If f denotes a bounded function such that $\int_{-h}^{h} f(x+\xi)d\xi$ exists for some $h>0$ and all $x \in \mathbb{R}$, the n-fold arithmetic mean of f with respect to the interval $[x-h; x+h]$ is defined by

$$f_n(x) = \left(\frac{1}{2h}\right)^n \int_{-h}^{h} \cdots \int_{-h}^{h} f(x+\xi_1+\cdots+\xi_n)d\xi_1\cdots d\xi_n.$$

Maurer [1896, 265–270] represented the latter integral by use of Dirichlet's factor, and he showed that

$$f_n(x) = \frac{1}{\pi h}\int_{-nh}^{nh} f(x+u)P(u)du, \tag{3.15}$$

where

$$P(u) = \frac{h\sqrt{\pi}}{k}\mathrm{e}^{-\frac{u^2}{k^2}} + \mathrm{O}\left(\frac{1}{n\sqrt{n}}\right), \quad k = h\sqrt{\frac{2}{3}n}.$$

Without referring in this respect to other authors, Maurer used techniques similar to those applied by Dirichlet or Cauchy in connection with the CLT. However, the situation which Maurer had to master was considerably simpler than that of a CLT under reasonably general assumptions. As Sommerfeld briefly noticed, Maurer's result comprised (if only implicity) a local CLT for independent random variables, if each of them had the same rectangular distribution. In fact, Sommerfeld's assertion can be verified by setting

$$f(x) = \begin{cases} \frac{1}{2h} & -h \le x \le h \\ 0 & \text{else,} \end{cases}$$

and by taking into account that in this case $f_n(x)$ is equal to the $n+1$-fold convolution of $f(x)$ with itself. By representing f_n according to (3.15), one can show that

$$\sqrt{n}\, f_n(x\sqrt{n}) \to \frac{1}{\sigma\sqrt{2\pi}}\mathrm{e}^{\frac{-x^2}{2\sigma^2}} \quad (\sigma^2 = \frac{h^2}{3}).$$

As it seems, Maurer's work was not influenced by, and in turn did not have any influence on the development of the CLT. Maurer's contribution shows, however, that around the end of 19th century the development of analytic methods had advanced so far that greater success also in the realm of the CLT became possible.

3.3.2.2 Crofton's Hypothesis

Contributions based on rather simple assumptions on the distributions of elementary errors were contrasted by others, in which Bessel's hypothesis was even more generalized. In 1870, a very general hypothesis on additively coacting independent elementary errors was presented by Morgan William Crofton (1826–1915, Fig. 3.5), who, additionally, expounded an analytical method entirely different from that of Laplace and Poisson. One of his tricks anticipated an essential idea, which played an important role in the modern proofs of the CLT by Lyapunov, Lévy, and Lindeberg. In Lyapunov's case, a direct reference to Crofton's artifice is even probable (see Sect. 5.1.3).

Fig. 3.5 Morgan William Crofton

Crofton [1870, 175] described his main objective with the words:

> (...) to give the mathematical proof, in its most general form, of the law of single errors of observation [i.e., the Gaussian], on the hypothesis that an error in practice arises from the joint operation of a large number of independent sources of error, each of which, did it exist alone, would produce errors of extremely small amount as compared generally with those arising from all the other sources combined.

For the elementary errors he allowed laws[47] of "utmost generality" [1870, 179]. He supposed [1870, 183 f.], however, that all moments of each of them should be "infinitesimal," such that the moments of an order greater than the second could be "neglected" in calculations regarding the "compound" error. More precisely, he assumed each "component error" to be the "diminutive" of some error "of finite importance," such that, for an infinitesimal η, the "mean value" E_1, "mean square"

[47] Strictly speaking, Crofton did not represent laws of error by densities of probability distributions, but he used idealized absolute frequencies, which corresponded to discrete or continuous distributions; also a mixture of both types was possible. In his deductions, Crofton referred only to the second case. He stated, however, that his results were valid for all types of distributions. A deeper reason for Crofton's preference to absolute frequencies does not appear in his exposition. For simplicity and consistency of presentation, in the following Crofton's arguments are described in the language of probabilities.

E_2, "mean cube" E_3, etc. of the latter yielded the expectations ηE_1, $\eta^2 E_2$, $\eta^3 E_3$, respectively, of the former. Crofton [1870, 177 f.] conceded that a proof for the "truth of this hypothesis" seemed impossible to him, he tried, however, to show the "reasonableness" of the hypothesis by a recourse to the practice of measuring "in the case of refined and delicate observations."

The essence of Crofton's mathematical analysis (including an introductory plausibility consideration [1870, 184 f.]) involved the relation

$$y(x) = \mathrm{e}^{\frac{1}{2}(h-i)\mathrm{D}^2} f(x-m), \tag{3.16}$$

which should be valid for the law of error y of any error consisting of an "infinitely large" number of elementary errors, if f was the density of an arbitrary, not necessarily bounded, elementary error. In this formula, D denoted the derivative operator $\frac{\mathrm{d}}{\mathrm{d}x}$, and m, h, i the sum of the expectations, of the mean squares, and of the squares of the expectations of the elementary errors. As it appears, Crofton tacitly assumed m, h, i to be finite quantities.[48] Crofton's deduction of (3.16) rested upon an iteration of convolutions. If the compound error was made up of two elementary errors only, with the respective laws f (as above) and φ (positive only within the interval $[-b; a]$[49]), then, according to Crofton, for the law $y(\xi)$ of the sum the following formula had to valid:

$$y(\xi) = \int_{-b}^{a} f(\xi - x)\varphi(x)dx.$$

Crofton expanded $f(\xi - x)$ into a power series in x, and thus obtained

$$y(\xi) = f(\xi) - \alpha f'(\xi) + \frac{\lambda}{2} f''(\xi) - \frac{\sigma}{2\cdot 3} f'''(\xi) + \cdots,$$

where

$$\alpha = \int_{-b}^{a} x\varphi(x)dx, \quad \lambda = \int_{-b}^{a} x^2\varphi(x)dx, \quad \sigma = \int_{-b}^{a} x^3\varphi(x)dx.$$

Because of the assumption that the mean powers of orders greater than two could be neglected, the equation

$$y(\xi) = (1 - \alpha\mathrm{D} + \frac{\lambda}{2}\mathrm{D}^2) f(\xi)$$

followed. After the addition of another elementary error, it ensued

$$y(\xi) = (1 - \beta\mathrm{D} + \frac{\mu}{2}\mathrm{D}^2)(1 - \alpha\mathrm{D} + \frac{\lambda}{2}\mathrm{D}^2) f(\xi),$$

---

48 This assumption actually damaged the "utmost generality" of Crofton's model of elementary errors as "diminutives" of finite errors. If, for example, the elementary errors of an "infinitely large" number n are identically distributed just like the random variable ηX, where $\mathrm{E}X \neq 0$, then we have $m = \eta n \mathrm{E}X$ and $h = \eta^2 n \mathrm{E}X^2$. It is impossible, however, that η and η^2 are of the same order of magnitude $\sim \frac{1}{n}$.

49 The cases $a = \infty$ or $b = \infty$ not excluded.

and so on, such that, finally, in the case of a composition of "all" elementary errors:

$$y(\xi) = (1 - \alpha\mathrm{D} + \frac{\lambda}{2}\mathrm{D}^2)(1 - \beta\mathrm{D} + \frac{\mu}{2}\mathrm{D}^2)(1 - \gamma\mathrm{D} + \frac{\nu}{2}\mathrm{D}^2)\cdots f(\xi).$$

$\alpha, \lambda, \beta, \mu, \gamma, \nu$, etc. had to be considered as "infinitely small," and therefore Crofton was able to convert the latter equation into the form

$$y(\xi) = \mathrm{e}^{-(\alpha+\beta+\gamma+\cdots)\mathrm{D}+\frac{1}{2}(\lambda-\alpha^2+\mu-\beta^2+\cdots)\mathrm{D}^2} f(\xi),$$

from which he inferred that

$$y(x) = \mathrm{e}^{\frac{1}{2}(h-i)\mathrm{D}^2}\mathrm{e}^{-m\mathrm{D}} f(x) = \mathrm{e}^{\frac{1}{2}(h-i)\mathrm{D}^2} f(x-m).$$

Crofton now generally showed that, for a and k being positive

$$\mathrm{e}^{a\mathrm{D}^2}\mathrm{e}^{-kx^2} = \frac{1}{\sqrt{1+4ak}}\mathrm{e}^{-\frac{kx^2}{1+4ak}}. \tag{3.17}$$

For the proof he derived a partial differential equation of first order in the variables a and k for the left side in (3.17), and then he determined the general solution of this differential equation. Now Crofton assumed—and this was his decisive trick—that

$$f(x) = \frac{1}{\theta\sqrt{\pi}}\mathrm{e}^{-\frac{x^2}{\theta^2}}.$$

By use of this particular auxiliary density, and on account of (3.16) and (3.17), the law of the error summed up from all elementary errors including the one with density f resulted in

$$y(x) = \frac{1}{\sqrt{\pi(2(h-i)+\theta^2)}}\mathrm{e}^{-\frac{(x-m)^2}{2(h-i)+\theta^2}}.$$

As Crofton noticed, the influence of the particular f could be abandoned after setting $\theta = 0$, and therefore the law of the compound error, which was composed of the elementary errors with arbitrary densities, could be deduced as being

$$y = \frac{1}{\sqrt{2\pi(h-i)}}\mathrm{e}^{-\frac{(x-m)^2}{2(h-i)}}.$$

In his article "Probability" in the 1885 edition of the *Encyclopædia Brittanica*, Crofton [1885, 781] presented a modification of the just-described procedure for deducing the Gaussian law of error from the hypothesis of elementary errors. He now assumed that $f(x)$ was the law of an error which was already composed of a very large number of elementary errors. If one further elementary error with an infinitesimal expectation α and an infinitesimal mean square λ was added, then the new compound error obeyed the error law

$$y(x) = (1 - \alpha \mathrm{D} + \frac{\lambda}{2}\mathrm{D}^2) f(x). \tag{3.18}$$

From this equation, Crofton concluded that each elementary error contributed to the law of the compound error only through its expectation and its mean square. If still another elementary error was superposed, then a law of the compound error resulted according to

$$y(x) = \left\{1 - (\alpha + \alpha_1)\mathrm{D} + \frac{\lambda + \lambda_1 - \alpha^2 - \alpha_1^2 + (\alpha + \alpha_1)^2}{2}\mathrm{D}^2\right\} f(x).$$

From this, Crofton inferred that the totality of all elementary errors contributed to the law f of the compound error only by algebraic terms in $x - m$ and $h - i$, and therefore the relation

$$z = f(x) = F(x - m, h - i) \tag{3.19}$$

was valid. On account of (3.18) Crofton concluded that each single elementary error with expectation 0 and variance δh influenced the law of error z of the rest of the elementary errors with the increment

$$\delta z = \frac{\delta h}{2} \frac{\mathrm{d}^2}{\mathrm{d}x^2} z.$$

From this consideration the differential equation

$$\frac{\partial^2 z}{\partial x^2} = 2\frac{\partial z}{\partial h}$$

resulted, and therefrom

$$\frac{\partial^2 z}{\partial \xi^2} = 2\frac{\partial z}{\partial \eta}, \tag{3.20}$$

where

$$\xi = x - m \text{ and } \eta = h - i.$$

A second differential equation

$$\xi\frac{\partial z}{\partial \xi} + 2\eta\frac{\partial z}{\partial \eta} + z = 0 \tag{3.21}$$

was obtained by Crofton through a consideration of the change of (3.19) if all elementary errors were substituted by their $(1 + \omega)$-fold values, ω being an infinitesimal quantity. The integration of the differential equations (3.20) and (3.21) led to a Gaussian error law z. Edgeworth later adopted this method for his deduction of asymptotic series expansions for densities of sums of elementary errors (see Sect. 3.4.2.3).

The procedure as described in the *Encyclopædia Brittanica* was also propagated by Czuber in the first volume of his very popular textbook *Wahrscheinlichkeitsrechnung*, which appeared in many editions between 1901 and 1938. However, Crofton's deduction, which did not include any precise discussions of the conditions needed for its single steps, could lead the reader to the impression that in the case of sums of a great number of independent random variables an approximating Gaussian law of error was a matter of course. Crofton's "proof," therefore, has to be connected with a point of view, very common up to the first decades of the 20th century, which associated probability calculus with "natural science" rather than with mathematics proper. Some fifty years after Crofton, however, a similar approach via differential equations was rigorously established in the context of random walks, through contributions by Kolmogorov [1931b; 1933c], Khinchin [1933], and Petrovskii [1934].

3.3.2.3 Pizzetti's Account on the Hypothesis of Elementary Errors

In 1892, Paolo Pizzetti (1860–1918) published a survey of error theory, which regarding its fundamental goals, if not entirely its extent, can be compared with Czuber's already frequently cited monograph [1891]. Whereas Czuber preferred an impartial description rather close to the original sources of 19th-century error theory, Pizzetti assessed the individual contributions and modified them if this seemed appropriate. Right at the beginning of Pizzetti's account one can find a discussion of Crofton's version of the hypothesis of elementary errors. Later in his work, however, Pizzetti [1892, 224] designated the analytical method of Crofton's (first) account as not rigorous. Consequently, he derived his own analytic approach, which basically followed Poisson's contributions to the CLT (see Sects. 2.2.4, 2.2.5), and substantially employed infinitesimal considerations. In this framework Pizzetti rather carefully observed whether the single asymptotic arguments he used could actually be justified, and he tried to give a clear description of the assumptions needed.

Pizzetti [1892, 123–133] assumed an "infinitely large" number s of independent elementary errors with values within $[-a_k; b_k]$, and for them he considered arbitrary discrete or continuous laws, or even distributions mixed from the two types. He further supposed $s = s' + s''$, where s'' was a finite number, such that, for s' elementary errors, the absolute values of the error bounds $-a_k$ and b_k were of the common infinitesimal order $\frac{1}{s'}$, and for the remaining s'' elementary errors these error bounds were of an infinitesimal order as well, although of an order greater than $\frac{1}{s'}$. From this condition, Pizzetti inferred that, except for a finite number of elementary errors, the respective i-th moments were of the order of magnitude $\frac{1}{s'^i}$. This property, however, necessarily led to an infinitely small variance of the sum of elementary errors, and therefore to an absurd condition, not noticed by Pizzetti. A more appropriate assumption would have been that the error bounds of the s' elementary errors were of an order $\frac{1}{\sqrt{s'}}$ and, simultaneously, the sum of the expectations of all elementary errors was not "infinitely large."

By aid of the discontinuity factor

$$\frac{1}{\pi}\int_{-\infty}^{\infty} \mathrm{e}^{(\sigma-t)u\sqrt{-1}}\,\frac{\sin ut}{u}\,du = \begin{cases} 1 \text{ for } 0 < \sigma < 2t \\ 0 \text{ for } \sigma < 0 \text{ or } \sigma > 2t \end{cases}$$

Pizzetti calculated the probability P_{2t} that the sum of elementary errors was between 0 and $2t$ in the form

$$P_{2t} = \frac{1}{\pi} \sum_{z_1=-a_1}^{b_1} \sum_{z_2=-a_2}^{b_2} \cdots \sum_{z_s=-a_s}^{b_s} \int_{-\infty}^{\infty} e^{(\sigma-t)u\sqrt{-1}} Z_1 Z_2 \cdots Z_s \frac{\sin ut}{u} du.$$

In this formula, Z_k denoted the (possibly "infinitely small") probability for a single value z_k of the elementary error with index k, and $\sigma = z_1 + z_2 + \cdots + z_s$. By interchanging the summation and integration (not explicitly discussed by Pizzetti, compare the similar approach of Dirichlet, Sect. 2.4.2) it followed

$$P_{2t} = \frac{1}{\pi} \int_{-\infty}^{\infty} A_1 A_2 \cdots A_s e^{-tu\sqrt{-1}} \frac{\sin ut}{u} du,$$

where

$$A_k = \sum_{z_k=-a_k}^{b_k} Z_k e^{uz_k\sqrt{-1}}.$$

In the same way as Poisson (see Sect. 2.2.2) had already done, Pizzetti used the abbreviations

$$A_k = \rho_k e^{\theta_k\sqrt{-1}}, \quad R = \rho_1 \cdots \rho_s, \quad \psi = \theta_1 + \cdots + \theta_s$$

and obtained

$$P_{2t} = \frac{2}{\pi} \int_0^{\infty} R \cos(\psi - ut) \frac{\sin ut}{u} du. \tag{3.22}$$

Along the lines of Poisson, Pizzetti expanded $\log R$ and ψ in series of powers of u. The coefficients depended on the moments of the elementary errors in increasing order, and, thus, Pizzetti was able to make use of his assumption about the smallness of these moments. He split the integral in (3.22) into a sum of two integrals from 0 to υ and from υ to ∞, where $\upsilon = Hs'^{\frac{3}{5}}$ with a positive H.[50] From a discussion of the order of smallness of the single series terms of $\log R$ he concluded that, for s "infinitely large," in the first integral:

$$R \cos(\psi - ut) = e^{-u^2\alpha^2} \cos\left(u(\sum g_r - t)\right) + M,$$

where M between 0 and υ was an evanescent quantity. In this equation g_r designates the expectation of the r-th elementary error, and α^2 the sum of the variances of the single elementary errors divided by 2. Pizzetti was rather vague about why the second integral was evanescent with $s = \infty$, whereas the first became equal to

$$\frac{2}{\pi} \int_0^{\infty} e^{-u^2\alpha^2} \cos\left(u(\sum g_r - t)\right) \frac{\sin ut}{u} du.$$

---

50 Regarding the assumptions modified with respect to Pizzetti's original (see above), we would have to set $\upsilon = Hs'^{\frac{1}{7}}$.

After the substitutions $2t = x$, $\sum g_r = a$, and $h = \frac{1}{2\alpha}$, and by use of some well-known integral formulae, he was able to show

$$P_x = \frac{h}{\sqrt{\pi}} \int_0^x e^{-h^2(a-t)^2} dt,$$

where P_x was the probability that the sum of "infinitely many" elementary errors was between 0 and x.

Pizzetti's contribution exemplifies a growing tendency toward analytical rigor in deducing approximating normal distributions by Poisson's method at the end of the 19th century. The splitting of the integral representing the probability of the sum into one "main-" and several "lateral-" integrals, an idea which could already be found in the works of Dirichlet and Cauchy, was analogously used by Czuber [1891, 267–270], in his account on asymptotic probabilities of linear combinations of errors, albeit with manifest gaps in the line of arguments. Czuber as well as Pizzetti might have been brought to the idea of splitting by reading of [Poisson 1824], in which this device occurs at least implicitly (see Sect. 2.2.4).[51] Only Lyapunov succeeded in a proof of the CLT under very general conditions, in which the "idea of splitting" was used strictly according to the analytic standards of the post-Weierstrassian era.

3.3.2.4 Schols, and Elementary Errors in Plane and Space

There was a growing interest in planar and spatial errors beginning in the middle of the 19th century, although, in most cases, this field was still perceived as marginal compared with problems of "linear" error theory. The discussions on errors in several dimensions had two essential starting points: firstly, the idea already indicated by Laplace and further examined by Bienaymé that a joint consideration of all elements was necessary (see Sect. 2.1.5.2) in applications of least squares to the estimation of more than one element; secondly, research on shooting errors, a field which directly led to problems in two-dimensional errors and which was treated rather frequently [Stigler 1986, 317]. By the end of the 19th century, a third aspect of multidimensional errors came along: Their coordinates served as a paradigm of correlated quantities. The theory of errors in a plane and space thus became a starting point of correlation theory within mathematical statistics.[52]

The early history of the CLT for sums of independent random vectors was connected in a natural way with multidimensional error theory. Priorists, however, may let this history begin with Lagrange's approximation of the multinomial distribution,

---

[51] Neither Pizzetti [1892] nor Czuber [1891] gave a reference to Cauchy's work on the CLT; therefore, it is improbable that they were influenced by Cauchy's modifications of the Laplace–Poisson method. The possibility that Czuber's exposition may have influenced Pizzetti's can be excluded, too, because the latter scarcely had any knowledge of Czuber's survey (published in 1891) when writing his own, rather extensive, article. In fact, Pizzetti does not cite [Czuber 1891] in his very comprehensive bibliography.

[52] For Edgeworth's contributions to these problems see [Stigler 1986, 315–325], on K. Pearson's see [Lancaster 1971].

based on a method analogous to de Moivre's approximation of the binomial distribution [Lagrange 177?, 204–209].[53] It is more than questionable, however, whether the respective numbers of successes in the case of several disjoint events were conceived as sums of the coordinates with possible values 0 and 1 of random vectors. It is also questionable whether the probabilities for the joint occurrence of linear combinations $\alpha_1\epsilon_1+\alpha_2\epsilon_2+\cdots$, $\beta_1\epsilon_1+\beta_2\epsilon_2+\cdots$, etc. ($\epsilon_k$ being one-dimensional errors), as discussed in context with Laplace's and Bienaymé's accounts on the method of least squares in the case of several elements (see Sect. 2.1.5.2), were actually conceived as probabilities of sums of random vectors. Such an interpretation was certainly used in those applications in which the parameters to be estimated were coordinates of points [Czuber 1899, 189]. The history of the CLT for random vectors in a stricter sense did therefore not start before sums of palpably multidimensional entities were considered.

As it seems, the Dutch geodesist Christian Schols (1849–1897) was the first[54] to derive approximations of densities of sums of two- and three-dimensional errors, which can, from the point of view of priority, be interpreted as "archetypes" of local CLTs. Schols [1875/86][55] tried, apparently motivated by work on shooting statistics and least squares applied to planar surveying, to give a "theory" of errors in plane and space.

Schols [1875/86, 125–132] started his article with an extension of Gauss's inequalities for one-dimensional densities[56] to probability laws $\overline{F}(\rho)$, associated with two- and tree-dimensional errors, which are "independent of the direction." In the case of three dimensions, for example, $\overline{F}(\rho)$ is connected with an ordinary probability density $F(x, y, z)$ by the equation

$$\overline{F}(\rho) = \int_0^{2\pi} \int_{-\pi/2}^{\pi/2} F(\rho\cos\varphi\cos\vartheta, \rho\sin\varphi\cos\vartheta, \rho\sin\vartheta)\rho^2\cos\vartheta\, d\vartheta\, d\varphi.$$

For an interpretation of planar and spatial probabilities Schols quite frequently used mechanical notions. The range of values of an error, endorsed with the law of error as the "mass density," he named "probability solid."[57] Justifying his inequalities, Schols imagined that the "polar moment of inertia" $M^2 := \int_0^\infty \rho^2\overline{F}(\rho)d\rho$ could be "minimized" by assuming the partial masses of the probability solid as only existent at discrete points. Similar ideas were used by Winckler [1866] for his

[53] For closer details see [K. Pearson 1978, 598–603; Dale 1991, 81–86; Hald 1998, 44 f.].

[54] At least according to Czuber [1891, 363; 1899, 221]. For a biographical sketch on Schols see [Ramaer 1924].

[55] Schols's 1875 article, written in Dutch, was translated in 1886 without any modifications into French.

[56] Let $f(x)$ be a unimodal and smooth density which attains its maximum at $x = 0$, and let $m := \int_{-\infty}^{\infty} x^2 f(x)dx$ and $\mu := \int_{-\lambda m}^{\lambda m} f(x)dx$ for $\lambda > 0$. In 1823 Gauss proved that $\lambda \le \mu\sqrt{3}$ if $\mu \le \frac{2}{3}$, and $\lambda \le \frac{2}{3\sqrt{1-\mu}}$ if $\mu > \frac{2}{3}$. See Sect. 4.1 for closer details.

[57] In general, Schols assumed the probability solids to be bounded. In all integrations, however, he used the limits $-\infty$ and ∞, presupposing that the considered error functions vanished beyond finite domains.

generalizations of Gauss's inequalities, and later also by Markov [1884a;b; 1886] and Stieltjes [1884d] in connection with further moment problems (see Sect. 4.3). Apparently without any knowledge of Winckler's work, Schols [1875/86, 127] showed that, for

$$\mu = \int_0^{\lambda M} \overline{F}(\rho) d\rho,$$

the estimate—which corresponds to the Bienaymé–Chebyshev inequality[58]—

$$\mu \geq 1 - \frac{1}{\lambda^2}$$

was valid. In the three-dimensional case, Schols's version [1875/86, 130 f.] of the "Gauss inequality" was as follows:
If $F(x, y, z) \geq F(x', y', z')$ for $x^2 + y^2 + z^2 \leq x'^2 + y'^2 + z'^2$, then

$$\lambda \leq \begin{cases} \sqrt{\frac{5}{3}}\sqrt[3]{\mu} & \text{for } \mu \leq \frac{2}{5} \\ \sqrt[3]{\frac{2}{5}}\frac{1}{\sqrt{1-\mu}} & \text{for } \mu > \frac{2}{5}. \end{cases}$$

The main focus of Schols's article [1875/86, 136–152] was on sums of independent planar and spatial errors. Schols determined the moments of the error sum as depending on the respective moments of the individual errors. He further derived exact formulae for the density of the sum by use of convolutions of the individual densities, and to this aim he developed some remarkable tricks, which he discussed in more detail in an additional paper [1887a], and which were also used by Czuber [1891, 67–76] in his account on sums of independent "linear" (i.e., one-dimensional) errors.

By a rather simple argument Schols [1875/86, 147–149] reduced the discussion of the "limit law" of a sum of independent planar and spatial errors to the discussion of sums of linear errors. He referred to Laplace's *TAP* and to Bessel's [1838c] for the assertion that the density of an error composed of linear elementary errors "converges" to a function of the form $\frac{1}{M\sqrt{2\pi}}e^{-\frac{u^2}{2M^2}}$, presupposing all elementary errors having zero means and none of them having a variance (denoted quite confusingly "valeur moyenne" by Schols) which is "considerably larger than any of the others." Based on an analogous condition for planar and spatial elementary errors, Schols argued that the projection of the compound error onto an arbitrary straight line, enclosing the angles α, β, γ with the coordinate axes, was equal to the sum of the projections of the respective elementary errors, and, therefore, followed a Gaussian "limit law." Schols assumed the chosen axes to be the principal axes of the probability solid related to the multidimensional law of the compound error. Under this assumption he had already shown at a previous place in his article the following equation for the variance $M^2_{\alpha\beta\gamma}$ of the projection onto the line:

58 Schols, however, did not mention the contributions of Bienaymé [1853e] and Chebyshev [1867].

$$M_{\alpha\beta\gamma}^2 = M_x^2 \cos^2\alpha + M_y^2 \cos^2\beta + M_z^2 \cos^2\gamma,$$

where

$$M_x^2 = \int_{-\infty}^{\infty}\int_{-\infty}^{\infty}\int_{-\infty}^{\infty} x^2 F(x,y,z)dxdydz \quad \text{etc.}$$

By use of this relation he concluded that the law of the projection of the compound error onto the line was asymptotically equal to

$$\frac{\mathrm{e}^{-\frac{u^2}{2(M_x^2\cos^2\alpha+M_y^2\cos^2\beta+M_z^2\cos^2\gamma)}}}{\sqrt{2\pi(M_x^2\cos^2\alpha + M_y^2\cos^2\beta + M_z^2\cos^2\gamma)}}.$$

This law, however, could be conceived, as Schols had shown previously in his paper, as the density of the sum of three independent normally distributed errors with variances $M_x^2\cos^2\alpha$, $M_y^2\cos^2\beta$, and $M_z^2\cos^2\gamma$, respectively. Therefore, the projection of the compound error onto an arbitrary straight line containing the origin of the coordinate system which coincides with the system of principal axes had to obey the same probability law as the sum of the projections of three independent linear errors, placed on the x,y, and z-coordinate axis, and normally distributed with variance M_x^2, M_y^2, and M_z^2, respectively. Schols [1875/86, 149] inferred from this fact that the sum of a very large number of elementary errors "obeys the same law as the resultant of its three projections onto the principal axes, if these [projections] are assumed independent." Thus, the law of error of a sum of a very large number of elementary errors was, with respect to a coordinate system which was identical with the system of the principal axes of the compound error, approximately equal to

$$\frac{\mathrm{e}^{-\frac{1}{2}\left(\frac{x^2}{M_x^2}+\frac{y^2}{M_y^2}+\frac{z^2}{M_z^2}\right)}}{M_x\sqrt{2\pi}M_y\sqrt{2\pi}M_z\sqrt{2\pi}}.$$

In a subsequently published article, Schols [1887b] attempted a "direct proof" of the "limit law" for sums of planar and spatial elementary errors. He used the (multidimensional) Fourier integral for representing the density of the sum, and then adapted Bessel's procedure to the case of two and three dimensions. However, also with this method, Schols considered only one particular coordinate system, which coincided with the principal axes of the compound error.

3.4 Nonnormal Distributions, Series Expansions, and Modifications of the Hypothesis of Elementary Errors

From Quetelet on, one of the predominant problems of 19th-century statistics was the search for distribution laws in biological and social populations. Whereas the priority of the normal distribution had been propagated and justified by assuming the coaction of many insignificant "causes" in the works following Quetelet, non-normal distributions became more and more prominent at the end of the century. There was a growing tendency to regard such distributions not as accidental deviations from a regular case but to grant them the same right as normal distributions.

It seems that conceptual differences between observational errors and other statistical quantities were only taken seriously from the last third of the 19th century on. In error calculus, the main object was to give an "optimal" estimate of the true value of a physical quantity and to minimize random deviations from this value as far as possible. In biological and social statistics, however, the variations among the single sample elements were the main objects of investigation. Whereas normal, or at least unimodal and symmetric distributions, and (connected with them) characteristic least square estimators such as the arithmetic mean were especially important within error theory, additional tools were needed for more general statistical investigations. Gustav Fechner [1897, 16] wrote in his posthumously published *Kollektivmaßlehre* (something like "doctrine of measurement of collectives"):

> For *Kollektivmaßlehre* the aspect [the Gaussian law of error] which implies the privilege of arithmetic mean in the theory of physical and astronomical measurement is without any importance. All exemplars from one *Kollektivgegenstand* [collective object], even if they deviate in any order from the arithmetic mean or from any other principal value, are equally real and true, and a privileged consideration of the one before the other . . . does not really make any sense.

In fact, some important methods of general statistics could be adapted from fields less prominent within error theory. This especially concerned a nascent discussion of observational outliers, and, in this context, of so-called "robust" estimators, such as the median [Stigler 1973; Harter 1988].

As it was characteristic of 19th-century science, mechanistic ideas also influenced biological and social statistics, with the result that universal laws for the distributions of important characteristics and fundamental stochastic mechanisms producing these laws were sought. Regarding such problems, error theory was able to provide useful ideas for general statistics well into the first decades of the 20th century even if there was a decline in the paradigm of the normal distribution. There was actually a certain success in generalizing the "exponential" or "Gaussian" law toward distributions which could be interpreted as generated according to modifications of the hypothesis of elementary errors, and, thus, appeared to be especially "natural." In particular, this concerned methods of expanding density or distribution functions by series of derivatives of a normal density or distribution, respectively. If an expansion of this kind provided a sufficiently good approximation with a minor number of terms, it appeared reasonable to assume that the distribution considered was caused by a certain—possibly moderate—number of elementary errors.

3.4.1 Approximations of "Arbitrary" Probability Functions by Series in Hermite Polynomials

While discussing the probability density of a sum of elementary errors, Bessel carefully observed, in addition to the normal density, correction terms corresponding to a series expansion (see Sect. 3.2.2). He only used this expansion, however, to demonstrate the asymptotic character of the Gaussian error law. Bessel's result can be summarized in modern form by the following expansion for the density ψ of a sum of independent random variables with zero expectations and symmetric densities:

$$\psi(x) = \frac{1}{\sigma\sqrt{2\pi}} \mathrm{e}^{-\frac{x^2}{2\sigma^2}} \left[1 + a_4 H_4(x) + a_6 H_6(x) + \cdots\right],$$

where

$$\sigma^2 := \sum \mathrm{Var} X_i \text{ and } H_k(x) = \mathrm{e}^{\frac{x^2}{2\sigma^2}} \frac{\mathrm{d}^k}{\mathrm{d}x^k} \mathrm{e}^{-\frac{x^2}{2\sigma^2}}.$$

After Bessel, some other authors, like Bienaymé [1852] (see Sect. 2.1.5.2), gave similar series expansions related to the CLT. Chebyshev [1887/90] eventually derived a result for the probability that a sum of random variables was within a given interval, which was analogous to Bessel's, but explicitly used derivatives of the normal density and specifically aimed at an already small number of summands.

Only recently has the history of series expansions in statistics been thoroughly examined. The following survey owes very much to Anders Hald's contributions in this field, in particular to his monograph [2002], which the reader may consult for further details.

According to Czuber [1899, 201], the first person to tackle the problem of approximating arbitrary functions y by linear combinations $\sum_{k=0}^{m} A_k X_k$ of normed polynomials X_k of degree k via the method of least squares was Gustave Plarr [1857]. Plarr showed that the condition

$$\int_{-1}^{1} (y - \sum_{k=0}^{m} A_k X_k)^2 dx = \min$$

led to what are now known as "Legendre polynomials" X_k.[59]

A little later, Chebyshev [1859], based on his earlier work [Chebyshev 1855/58] on discrete least squares approximation, treated problems of this kind from a far more general point of view. Given the function $F(x)$ (not further specified) and any weight function $\Theta^2(x)$, he looked for a polynomial $g(x)$ of maximum degree r such that

$$\int_a^b (F(x) - g(x))^2 \,\Theta^2(x) dx \leq \int_a^b (F(x) - f(x))^2 \,\Theta^2(x) dx \quad \forall f \in \mathbb{P}_r.$$

[59] For closer historical details on these polynomials see Sect. 4.2.1.

For certain particular weight functions $\Theta^2(x)$ and limits of integration a, b Chebyshev represented the solution $g(x)$ by linear combinations of orthogonal polynomials, acquired from the partial denominators of certain continued fractions (see Sect. 4.2.3 for details). In his discussion of the approximation problem $a = -b = \infty$ and

$$\Theta^2(x) = \sqrt{\frac{k}{\pi}} \mathrm{e}^{-kx^2},$$

he dealt with (what are now known as) "Hermite polynomials" $\psi_j^{(k)}(x)$ in particular detail,[60] including the fundamental relations

$$\psi_j^{(k)}(x) = \mathrm{e}^{kx^2} \frac{\mathrm{d}^j}{\mathrm{d}x^j} \mathrm{e}^{-kx^2}$$

and

$$\int_{-\infty}^{\infty} \psi_j^{(k)}(x) \psi_{j'}^{(k)}(x) \mathrm{e}^{-kx^2} dx = 0 \quad (j \neq j'). \tag{3.23}$$

In the article [Chebyshev 1887/90], already mentioned above, Hermite polynomials were finally used in a series expansion of the form

$$P\left(t \leq \frac{\sum X_i}{\sqrt{2 \sum \mathrm{Var} X_i}} \leq t'\right) = \frac{1}{\sqrt{\pi}} \int_t^{t'} \left[1 + A_3 \psi_3^{(1)}(x) + A_4 \psi_4^{(1)}(x) + \cdots\right] \mathrm{e}^{-x^2} dx, \tag{3.24}$$

where X_i were independent, not necessarily symmetrically distributed random variables with zero expectations. Chebyshev did not refer, however, to the problem of the convergence (or divergence) of the series.

Priorists may champion Laplace for the discovery of Hermite polynomials, however. In [1811, 375–387; 1812/20/86, 294–300], he discussed a parabolic differential equation which modeled the mixing of black and white balls when alternating drawings were made from one urn to another. The solution of the differential equation Laplace represented by means of a series of polynomials proportional to Hermite polynomials, and he calculated the unknown coefficients by use of orthogonality relations analogous to (3.23) (see [Molina 1930, 384 f.; Hald 1998, 337–343]). In contrast to Chebyshev, however, Laplace's chief concern was not a problem of approximating arbitrary functions by linear combinations of polynomials, but of finding the general solution of a specific differential equation. Consequently,

[60] In modern statistics, Hermite polynomials h_ν are usually defined as connected with the density ϕ of the standard normal distribution by

$$h_\nu(x) = (-1)^\nu \frac{\phi^{(\nu)}(x)}{\phi(x)}.$$

Hermite was always a little later than Chebyshev in publishing important results regarding "his" polynomials, see [Hald 2000, 239 f.].

the significance of Laplace's contribution regarding series expansions for arbitrary functions was not observed during the 19th century.

Apparently with a poor knowledge of prior work on approximation theory, Jørgen Gram (1850–1916) in his 1879 doctoral dissertation[61] gave a comprehensive discussion on the least squares approximation of a given function, with respect to an arbitrary weight function, by linear combinations of orthogonal functions (not necessarily polynomials). The systems of orthogonal functions considered were represented by Gram via determinants which now bear his name. In contrast to Chebyshev, Gram also tackled the problem of the pointwise convergence of his series, if extended to an infinite number of terms, to the given function. Among other topics, Gram posed the problem of finding for a given function $f(x)$ the function

$$g_{a_0,\ldots,a_r}(x) = \mathrm{e}^{-x^2}(a_0 + a_1x + \cdots + a_rx^r)$$

for which

$$\int_{-\infty}^{\infty} \mathrm{e}^{x^2}\left(g_{a_0,\ldots,a_r}(x) - f(x)\right)^2 dx \leq \int_{-\infty}^{\infty} \mathrm{e}^{x^2}\left(g_{b_0,\ldots,b_r}(x) - f(x)\right)^2 dx \quad \forall b_0,\ldots,b_r \in \mathbb{R}.$$

Regarding this particular problem, Gram was apparently influenced by the earlier work of his teacher Ludvig Oppermann, which remained, however, unpublished.[62] Gram's solution was:

$$g_{a_0,\ldots,a_r}(x) = \sum_{j=0}^{r} \frac{\int_{-\infty}^{\infty} \mathrm{e}^{x^2}\varphi^{(j)}(x) f(x)dx}{\int_{-\infty}^{\infty} \mathrm{e}^{x^2}\left[\varphi^{(j)}(x)\right]^2 dx}\varphi^{(j)}(x), \quad \varphi^{(j)}(x) = \frac{\mathrm{d}^j}{\mathrm{d}x^j}\mathrm{e}^{-x^2}.$$

At first glance, the strongly increasing weight function e^{x^2}, used by Gram, seems to have been introduced mainly for algebraic reasons, because the approximation by the transcendent function $g_{a_0,\ldots,a_r}(x)$ could then be reduced to an approximation by a polynomial, as one can immediately see from the equation

$$\int_{-\infty}^{\infty} \mathrm{e}^{x^2}\left(g_{a_0,\ldots,a_r}(x) - f(x)\right)^2 dx = \int_{-\infty}^{\infty} \mathrm{e}^{-x^2}\left((a_0 + a_1x + \cdots + a_rx^r) - \mathrm{e}^{x^2/2} f(x)\right)^2 dx.$$

Weight functions of this kind, however, might have been also introduced by virtue of manifest statistical reasons. Let us assume that a frequency function $f(x)$ is determined by a large number s of independent random experiments, counting the respective numbers of cases in which a certain numerical characteristic falls within the intervals $[a; a+h[, [a+h; a+2h[, \ldots, [a+(m-1)h; b]$ (where h is very small). If $f(x)$ can already be roughly approximated by a normal distribution—for the sake of simplicity with zero expectation and variance $\frac{1}{2}$, which can always can be achieved by an appropriate scaling—then representing $f(x)$ by

[61] The essential parts of his dissertation are summarized in the article [Gram 1883], see [Hoem 1983, 217; Hald 1998, 540–550].

[62] See [Charlier 1905c, 12; Hald 1981, 6].

$$f(x) \approx \sqrt{\frac{1}{\pi}} \mathrm{e}^{-x^2} \left(a_0 + a_1 x + a_2 x^2 + \cdots + a_r x^r\right) \tag{3.25}$$

suggests itself. The absolute frequency of the hits within the j-th interval obeys a binomial distribution with a probability of success

$$p_j \approx h\sqrt{\frac{1}{\pi}} \mathrm{e}^{-x_j^2},$$

where x_j is the center of the interval. Because h is very small, the variance of the relative frequency of hits within the j-th interval is approximately equal to

$$\frac{1}{s} h\sqrt{\frac{1}{\pi}} \mathrm{e}^{-x_j^2} \cdot \left(1 - h\sqrt{\frac{1}{\pi}} \mathrm{e}^{-x_j^2}\right) \approx \frac{1}{s} h\sqrt{\frac{1}{\pi}} \mathrm{e}^{-x_j^2}.$$

If one attempts a least squares fitting of $f(x)$ by (3.25), the approximative weight $\left(\frac{1}{s} h\sqrt{\frac{1}{\pi}} \mathrm{e}^{-x_j^2}\right)^{-1}$ of the relative frequency r_j of hits within the j-th interval has to be considered.[63] In this way, one obtains the least squares condition:

$$\sum_{j=1}^{m} \left(\frac{1}{\sqrt{\pi}} \mathrm{e}^{-x_j^2} (a_0 + a_1 x_j + a_2 x_j^2 + \cdots + a_r x_j^r) - \frac{r_j}{h}\right)^2 hs\sqrt{\pi} \mathrm{e}^{x_j^2} = \min.$$

In the case of "infinitely small" h and $a = -b = \infty$, from the latter relation the condition

$$\int_{-\infty}^{\infty} \mathrm{e}^{x^2} \left(\mathrm{e}^{-x^2} (c_0 + c_1 x + c_2 x^2 + \cdots + c_r x^r) - f(x)\right)^2 dx = \min,$$

used by Gram, follows.

Thorvald Nicolai Thiele (1838–1910), who had dealt, already in [1873], with estimations of the coefficients in the series expansions of the type (3.25) by moment methods, and had also introduced an equivalent form for those series by employing derivatives of normal densities [Hald 2000, 243; 2002, 14], took up and developed further Gram's methods in a comprehensive account published in 1889. Instead of resorting to least squares, Thiele [1889/2002, 74 f.] motivated the use of series expansions in Hermite polynomials by a plausibility consideration on approximating both densities $f(x)$ and definite integrals $\int_{x-p/2}^{x+p/2} f(z)dz$ $(p > 0)$ by the same series type: Because the application of Taylor's theorem to $f(z)$ (with the abbreviation D for the derivative) yielded the representation

[63] The reciprocal variance as a weight for fitting by means of least squares in case of observations of different accuracy was introduced by Gauss. In the context of approximation of empiric frequency functions via least squares, Seth Carlo Chandler [1872]—in the framework of an approach entirely different from (3.25)—had already used weights analogous to $\frac{s}{p_j(1-p_j)}$, see [Seal 1979, 238].

$$\int_{x-p/2}^{x+p/2} f(z)dz = pf(x) + p^3\frac{\mathrm{D}^2 f(x)}{3!4} + p^5\frac{\mathrm{D}^4 f(x)}{5!16} + p^7\frac{\mathrm{D}^6 f(x)}{7!64} + \cdots,$$

Thiele proposed to expand $f(x)$ in a series

$$f(x) = \sum_{r\geq 0}(-1)^r \frac{k_r}{r!}\mathrm{D}^r \xi(x) \tag{3.26}$$

of derivatives of any function ξ. Consequently, $\int_{x-p/2}^{x+p/2} f(z)dz$ had the "same form [as (3.26)], just with other constants in the series expansion." Thiele recommended using the "simple exponential error function $\xi = \mathrm{e}^{-x^2/2}$," and finally [1889/2002, 92; 178 f.] arrived at a representation analogous to

$$f(x) = k_0\phi_{\mu,\sigma^2}(x) - k_1\phi'_{\mu,\sigma^2}(x) + \frac{k_2}{2!}\phi''_{\mu,\sigma^2}(x) - \frac{k_3}{3!}\phi'''_{\mu,\sigma^2}(x) + \cdots. \tag{3.27}$$

The emergence of Thiele's new device of "half-invariants" was apparently motivated by the problem of determining the coefficients of these or analogous series expansions as simply as possible. In order to calculate the coefficients k_j by means of the orthogonality relations (3.23) one has to perform the integrations

$$\int_{-\infty}^{\infty} f(x)H_j(x)dx, \quad \text{where } H_j(x) = (-1)^j \left(\phi_{\mu,\sigma^2}(x)\right)^{-1}\frac{\mathrm{d}^j}{\mathrm{d}x^j}\phi_{\mu,\sigma^2}(x).$$

This procedure virtually consists of determining the central moments of f up to the order j.[64] Thiele [1889/2002, 84 f.] saw the drawback that, especially regarding the effort in calculating empirical moments, the numerical values of moments of even order were considerably increasing with their order. Hald [2000, 242] suggests that this circumstance may have led to Thiele's definition of what he called the "half-invariants" κ_i,[65] by the recursion formula

$$m_{r+1} = \sum_{i=0}^{r}\binom{r}{i} m_{r-i}\kappa_{i+1} \quad (r = 0, 1, \ldots),$$

m_r denoting the empirical or theoretical moments of the order r of the distribution under consideration. This recursion formula is similar to another recursion formula of Thiele [1889/2002, 81] which relates ordinary moments and central moments to each other. Thiele [1899/2002, 227] introduced only in a subsequent paper the

[64] In the general case, the derivatives $\phi^{(j)}_{\mu,\sigma^2}$ satisfy the orthogonality relations

$$\int_{-\infty}^{\infty}\frac{1}{\phi_{\mu,\sigma^2}(x)}\phi^{(j)}_{\mu,\sigma^2}(x)\phi^{(j')}_{\mu,\sigma^2}(x)dx = \frac{1}{\sigma^{j+j'}}\delta_{jj'} \quad (j, j' \in \mathbb{N}_0).$$

The Hermite polynomials H_j related to ϕ_{μ,σ^2} are connected to the Hermite polynomials h_j related to the standard normal distribution by $H_j(x) = \frac{1}{\sigma^j}h_j\left(\frac{x-\mu}{\sigma}\right)$.

[65] The modern designations are "cumulants" or "semi-invariants," see [Hald 2000, 241].

now common definition of half-invariants (see Sect. 3.4.2.1). Yet another aspect was very decisive for the introduction of half-invariants by Thiele: In the case where f is the density of a sum of independent random variables—important in numerous applications, such as, for example, the discussion of the distribution of the arithmetic mean [Thiele 1889/2002, 129 f.]—the moments of an order greater than 1 cannot be obtained from the sums of the respective moments of the single variables. Even for the central moments this is only possible up to the third order. As Thiele [1889/2002, 103] showed, for the half-invariants $\kappa_j(X)$ of order j (associated with the random variable X) the following property (in modern notation) is valid:

$$\kappa_j\left(\sum a_r X_r\right) = \sum a_r^j \kappa_j(X_r), \tag{3.28}$$

if X_r are independent random variables, and the a_r are arbitrary real numbers. Altogether, there was a good deal of advantages in determining, on the basis of orthogonality relations, the coefficients k_j of (3.27) in terms of half-invariants instead of moments, and Thiele [1889/2002, 91 f.] gave a list of formulae for the coefficients up to k_8 in terms of the half-invariants $\kappa_1, \ldots, \kappa_8$.

Compared with Gram's and Thiele's approaches, Ernst Heinrich Bruns [1897] gave a very different derivation of the integral version

$$\int_a^b f(x)dx = \sum_{j\geq 0}(-1)^j k_j \left(\Phi^{(j)}_{\mu,\sigma^2}(b) - \Phi^{(j)}_{\mu,\sigma^2}(a)\right) \tag{3.29}$$

of the series (3.27). His basic idea was to represent $H(a,b) := \int_a^b f(x)dx$ by

$$H(a,b) = \int_{-\infty}^{\infty} f(x)E(x),$$

where $E(x)$ was a step function defined by

$$2E(x) = \operatorname{sign}(b-x) - \operatorname{sign}(a-x).$$

By means of an approximate representation of the function $E(x)$ via Fourier integrals (the error of approximation could be made arbitrarily small) and subsequent expansions of the (to some extent arbitrary) integrands in power series, Bruns finally derived (3.29) as a special case. In contrast to Thiele, Bruns tried to justify his arguments from the point of view of contemporary analytical rigor, in particular with regard to the problem of whether or not $H(a,b)$ could actually be represented by the respective series. Concerning his final step toward (3.29), however, he was only able to give plausibility arguments in support of the use of the series in practice, where its calculation could be restricted to a small number of terms.

Friedrich Lipps [1897; 1901, 171–175], whom we will come across again in context with further generalizations of elementary errors (see Sect. 3.4.3.1), gave additional derivations for "Bruns's series." The second [1901, 174 f.] of them reminds one to a certain extent of Thiele's plausibility consideration on the joint representation of integrals and integrands described above. The impact of Lipps's contributions on statistical development was apparently almost nil, however. Among prominent statisticians, who used series analogous to (3.27) or (3.29) around the turn of the 19th and 20th centuries, only Czuber [1902, 357]—in a footnote—referred to his work.

For Thiele, as well as for Bruns and Lipps, the primary motivation for discussing series such as (3.27) or (3.29), respectively, was the fitting of frequency functions to sampling data as advantageously as possible regarding the calculating effort. The coefficients of the series were to be determined by empirical moments. Thiele and Bruns, however, also used "their" series for "theoretical" considerations, in particular in the context of sums of independent random variables. Thiele [1889/2002, 129 f.] represented the density of the arithmetic mean of observations by means of (3.27), presupposing the same distribution for each of the observations. He showed that all coefficients except the first of the series tended to 0, and therefore he "proved" a CLT for the arithmetic mean of independent identically distributed random variables. Bruns [1897, 339 f.] discussed the series (3.29) in the case of sums of independent but not necessarily identically or symmetrically distributed "errors" up to the 6th term, and thereby derived a result which generalized that of Bessel (who had only considered symmetrically distributed elementary errors). However, from the point of view of contemporary analytical rigor, neither Thiele nor Bruns was willing or able to give sound arguments for the fact that the distributions of the considered sums were actually represented by the series employed, or that there was actually a convergence to the normal distribution. Moreover, there was scarcely any hint at similar "classical" methods established by Laplace or Poisson. At best Chebyshev's reference to the series (3.24) in connection with his "proof" of the CLT might have related to this theme.[66]

3.4.2 The "Natural" Role of the Normal Distribution and Its Derivatives

Thiele as well as Bruns and Lipps had stressed the fact that their series could be made up in derivatives of *any* arbitrary function, at least in principle. The use of the normal distribution had only a more or less conventional character. Therefore, the reason for the exceptional role of the Gaussian error law in connection with sums of independent random variables was not clarified by the principles that were made explicit in connection with those derivations of the series expansions.

[66] There are controversial speculations on Chebyshev's presumable analytical basis for setting up the series (3.24), see [Hald 2002, 10–12].

3.4.2.1 Hausdorff's "Kanonische Parameter"

Apparently, Felix Hausdorff [1901, 169–178] was the first to give a thoroughly clear explanation of the connections between deductions of the CLT in succession of Laplace on the one hand, and series expansions in derivatives of the normal distribution on the other. In a survey of problems of probability theory, Hausdorff [1901, 169] introduced generating functions associated with a random variable with density φ by the equation

$$\Phi(u) = \int_{-\infty}^{\infty} \varphi(x)\mathrm{e}^{xu}\,dx = \mathrm{De}^{xu} \quad \text{("D" from the German "Durchschnitt")},$$

corresponding to the modern notation $\Phi(u) = \mathrm{Ee}^{uX}$, X being a random variable. For representing $\varphi(x)$ through $\Phi(u)$, Hausdorff used Fourier's inversion formula

$$2\pi\varphi(x) = \int_{-\infty}^{\infty} \Phi(\mathrm{i}u)\mathrm{e}^{-\mathrm{i}xu}\,du, \tag{3.30}$$

which he ascribed to Gauss, who had already written down the formula around 1813 in a posthumously published note [Gauss 1900, 88 f.] (see footnote 34, Chap. 2). In this context, however, Hausdorff explicitly stated that he did not aim at a closer examination of the validity of the formulae employed. He expanded $\log \Phi(u)$ into a (formal) power series

$$\log \Phi(u) = M_1 u + \frac{M_2}{2!}u^2 + \frac{M_3}{3!}u^3 + \cdots, \tag{3.31}$$

and named the coefficients M_α "kanonische Parameter" ("canonical parameters") of the error law $\varphi(x)$. Hausdorff gave an easy proof for

$$\mathrm{Ee}^{u(X_1+X_2)} = \mathrm{Ee}^{uX_1}\mathrm{Ee}^{uX_2} \quad (X_1, X_2 \text{ independent})$$

by referring to the basic property of expectations $\mathrm{E}(Y_1Y_2) = \mathrm{E}Y_1\mathrm{E}Y_2$ (Y_1, Y_2 independent random variables). Using this fundamental property of generating functions he showed that for the "kanonische Parameter" a relation analogous to (3.28) is valid, thus rediscovering Thiele's half-invariants. In terms of the "kanonische Parameter" Hausdorff [1901, 173 f.] also derived a sufficient condition for the CLT: By P_α he designated the arithmetic means of the cumulants of order α of the independent random variables $X_1, \ldots, X_n$, each of which he assumed to be continuously distributed with a zero expectation. His condition for the convergence of the density of $\frac{X_1 + \cdots + X_n}{\sqrt{2nP_2}}$ to the normal density $\varphi_{0,\frac{1}{2}}$ was that, for $\alpha \geq 3$, the terms $P_\alpha : \sqrt{n^{\alpha-2}P_2^\alpha}$ vanished as $n \to \infty$. In fact, taking into account the property (3.28), this condition allows each series term of (3.31) to vanish asymptotically, except for the second, which is equal to $\frac{1}{4}$. The function $\Phi(u) = \exp(\frac{1}{4}u^2)$ is the generating function of a normal distribution with zero expectation and variance $\frac{1}{2}$. In this "proof" of the CLT two substantial gaps existed: Firstly, the convergence of the generating function $\Phi(u)$ could not be rigorously inferred alone from the asymptotic properties of each single term within the series expansion (3.31). Secondly, it was not clear whether the convergence of the generating function of the normed sum to the generating function of the normal density actually implied the convergence of the density of the sum to the normal density. Apparently, it was not Hausdorff's intention to give a rigorous proof of the CLT, but he at least presented an example of a sequence of elementary errors, in which the order of magnitude of the single elements was decreasing with the index, such that the limiting law

$$\varphi(x) = 1 : \left(e^{\frac{\pi}{2}x} + e^{-\frac{\pi}{2}x}\right)$$

of the sum of these errors was different from the normal distribution.[67] This counterexample would play an important role about 30 years later (see Sect. 6.1.1). Despite the lack of an exact proof, Hausdorff [1901, 173] concluded his considerations with the remark:

> On exactly the same two causes, the additive character and the vanishing of the canonical parameters in the case of the Gaussian law, also the approximate validity of the Gaussian law for a total error that results from the numerous partial errors of slightly different orders of magnitude is based.

From (3.30) and (3.31) Hausdorff [1901, 174 f.] formally inferred that

$$2\pi\varphi(x) = \int_{-\infty}^{\infty} du \exp(-\frac{u^2}{4} + ixu) \exp\left(\sum_{\alpha=3}^{\infty} \frac{(-1)^\alpha (iu)^\alpha}{\alpha!} M_\alpha\right).$$

He expanded the "second exponential function" into the series

$$1 - \frac{M_3}{3!}(iu)^3 + \frac{M_4}{4!}(iu)^4 - \frac{M_5}{5!}(iu)^5 + \frac{M_6 + 10M_3^2}{6!}(iu)^6 + \cdots.$$

By integrating term by term and using the relations

$$\int_{-\infty}^{\infty} du (iu)^\alpha \exp\left(-\frac{u^2}{4} + ixu\right) = \frac{d^\alpha}{dx^\alpha} \int_{-\infty}^{\infty} du \exp\left(-\frac{u^2}{4} + ixu\right)$$
$$= 2\sqrt{\pi}\frac{d^\alpha}{dx^\alpha} e^{-x^2} =: 2\sqrt{\pi} e^{-x^2} s_\alpha,$$

he obtained the following series:

[67] Hausdorff's counterexample was "dual" to the one of Poisson (see Sect. 2.2.3.1) in which the characteristic function of the limiting law was $2 : \left(e^{\frac{\pi}{4}x} + e^{-\frac{\pi}{4}x}\right)$.

$$\varphi(x) = \frac{e^{-x^2}}{\sqrt{\pi}}\left[1 - \frac{M_3}{3!}s_3(x) + \frac{M_4}{4!}s_4(x) - \frac{M_5}{5!}s_5(x) + \right.$$
$$\left. + \frac{M_6 + 10M_3^2}{6!}s_6(x) - \cdots\right]. \quad (3.32)$$

Regarding the application of this series, Hausdorff [1901, 174] only hinted at the condition of M_α for $\alpha \geq 3$ being "small compared with $M_2 = \frac{1}{2}$." He did not consider the order of magnitude of the coefficients of the series in the case of a sum of elementary errors explicitly, but he only discussed the possible "semi-convergence" of the series [1901, 177] in a rather vague manner.

Already in 1899, Thiele had deduced series expansions in derivatives of the normal density by considering the generating function of a given probability law. Compared with his first derivation of 1889, Thiele [1899/2002] employed an alternative way, basing himself on a new and "direct" definition of half-invariants κ_k, which was equivalent to (3.31) for $\kappa_k = M_k$. Instead of Fourier's inversion formula, Thiele used symbolic calculations associated with the operator

$$\exp\left(\sum_{k\geq 1}(-1)^k\frac{\kappa_k D^k}{k!}\right) \quad (D = \frac{d}{dx}),$$

thus arriving at the relation between two error densities $\phi^1(x)$ and $\phi(x)$ with half-invariants κ_k^1 and κ_k, respectively:

$$\phi^1(x) = e^{\frac{\kappa_3 - \kappa_3^1}{3!}D^3 - \frac{\kappa_4 - \kappa_4^1}{4!}D^4 + \cdots}\phi(x). \quad (3.33)$$

In the special case of

$$\phi(x) = \frac{1}{\sqrt{2\pi}}e^{\frac{-(x-\kappa_1)^2}{2\kappa_2^2}}$$

a series analogous to (3.32) could be achieved.[68] In contrast to Hausdorff, Thiele did not explain any connections between his derivations and proof methods for the CLT, and, in his account, the prominent role of the normal density was primarily due to the fact that in the case of this error law the calculating effort in connection with (3.33) was significantly reduced, because all cumulants of order greater than 2 vanished.

Neither Thiele's 1899 article nor Hausdorff's 1901 paper had a significant impact on other authors. Only by contributions of Charlier and Edgeworth in 1905 and later years, in which the derivation of series expansions $\sum_{i\geq 0} c_i \phi^{(i)}_{\mu,\sigma^2}$ was based on hypotheses of elementary errors, was the connection between analytic devices in the realm of the CLT on the one hand, and series expansions on the other, perceived by a broader audience.

[68] For closer details see [Hald 1998, 345–347; 2002, 20 f.].

3.4.2.2 Charlier's A Series

Carl Vilhelm Ludvig Charlier (1862–1934)[69] was in a more substantial manner occupied with probability and statistics only in a relatively late period of his work. Initially, his scientific activities were in perturbation theory, photometry, and scientific photography. It may be that he became interested in the problem of approximating empirical frequency curves by linear combinations of special functions in context with his research on stellar statistics. This conjecture is supported by the content of a letter he wrote to Chebyshev. In this letter Charlier [1888], referring to the latter's work on series in Hermite polynomials (including [Chebyshev 1855/58; 1859; 1887/90]), asked for an assessment of the applicability of series in orthogonal polynomials (especially Hermite polynomials) for representing arbitrary frequency functions.

In 1905, Charlier published the article "Über das Fehlergesetz" ("On the Law of Error"), in which he tried to demonstrate that the representation of "arbitrary frequency curves" by series expansions in derivatives of a normal density function "could be followed from the Laplacian theory of errors in an un-coerced way" [Charlier 1905a, 9]. The phrasing "*the* law of error" shows that Charlier intended to derive a general representation for a large class of different frequency functions. His aim was to enforce the universal validity of this representation by its deduction from a fundamental stochastic concept, the hypothesis of elementary errors. In so doing, Charlier tried to advance the methods of Laplace and Poisson for calculating probabilities of sums of independent random variables. Apparently, however, Charlier was not familiar with more recent work on the CLT (by Cauchy, Sleshinskii, or Lyapunov, for example) in which the original methods were further advanced in accordance with the contemporary standards of analytical rigor. He was unable to do justice to his own demand [1905a, 2] for a rigorous performance of the "necessary considerations on convergence."

Charlier assumed a number s of (tacitly independent) elementary errors, and, like Laplace, started by discretizing the single error values, aiming at a representation of the density $f(z)$ of the sum by means of characteristic functions. Unlike Laplace, however, he performed the transition from discrete to continuous variables already in the exact representation of f. In this context, Charlier [1905a, 4] erroneously assumed the orthogonality relation

$$\frac{1}{2\pi}\int_{-\pi}^{\pi} \mathrm{e}^{mx\sqrt{-1}}\mathrm{e}^{m'x\sqrt{-1}}dx = \delta_{mm'},$$

to be true even for noninteger m, m'. Consequently he arrived at

$$f(z) = \frac{1}{2\pi}\int_{-\pi}^{\pi} P_1(\omega)P_2(\omega)\cdots P_s(\omega)\mathrm{e}^{-z\omega\sqrt{-1}}d\omega,$$

where

[69] For biographical details see [Malmquist 1960].

$$P_i(\omega) = \int_{-\infty}^{\infty} f_i(x) e^{x\omega\sqrt{-1}} dx,$$

instead of the correct version

$$f(z) = \frac{1}{2\pi} \int_{-\infty}^{\infty} P_1(\omega) P_2(\omega) \cdots P_s(\omega) e^{-z\omega\sqrt{-1}} d\omega.$$

Essentially following Poisson's line of argument, for the characteristic function

$$\varphi_s(\omega) = P_1(\omega) P_2(\omega) \cdots P_s(\omega)$$

he derived a series expansion corresponding to

$$\varphi_s(\omega) = e^{\mu\omega\sqrt{-1} - \frac{\sigma^2\omega^2}{2}} \left(1 + \sum_{n \geq 3} \gamma_n (-\omega\sqrt{-1})^n \right) \tag{3.34}$$

(where μ and σ^2 designate the expectation and the variance of the sum, respectively). Charlier [1905a, 4 f.] maintained—this was a second mistake—that his deduction of (3.34) was only valid in the case where $|\varphi_s(\omega)|$ attains "considerable values" in a small neighborhood of $\omega = 0$ exclusively. He did, however, need this assumption for the conclusion

$$\int_{-\pi}^{\pi} \varphi_s(\omega) e^{-z\omega\sqrt{-1}} d\omega \approx \int_{-\infty}^{\infty} \varphi_s(\omega) e^{-z\omega\sqrt{-1}} d\omega$$
$$= \int_{-\infty}^{\infty} e^{-\frac{\sigma^2\omega^2}{2}} \left(1 + \sum_{n \geq 3} \gamma_n (-\omega\sqrt{-1})^n \right) e^{(\mu - z)\omega\sqrt{-1}} d\omega.$$

Via term-by-term integration he finally deduced the series expansion corresponding to

$$f(z) = \phi_{\mu,\sigma^2}(z) + \sum_{n=3}^{\infty} (-1)^n \gamma_n \phi_{\mu,\sigma^2}^{(n)}(z). \tag{3.35}$$

Charlier explicitly calculated the coefficients γ_3 and γ_4 as dependent on the moments of the individual elementary errors up to the third and fourth order, respectively. In a further paper he [1905c] gave a plausibility consideration on the determination of the coefficients of general series in derivatives of a given function, which in the particular case of the series (3.35) amounted to Gram's and Thiele's methods. Charlier, however, failed to estimate the respective orders of magnitude of the coefficients in (3.35) as depending on the number of elementary errors.

According to Charlier's arguments with regard to $|\varphi_s(\omega)|$, the validity of the series expansion (3.35) was indisputable only for a very large number of elementary errors. In this case, however, the deviation between the exact distribution of the compound error and a Gaussian error law could only be small. In the case of "arbitrary" laws of error $f(z)$, Charlier, in contrast to his claim, did not succeed in establishing the equation (3.35) from the "Laplacian theory of errors."

A further article "Die zweite Form des Fehlergesetzes" ("The Second Form of the Law of Error"), which appeared soon after, might have been motivated by the insight of its author in the inconsistencies of the former paper. Certainly, the fact that the series (3.35) could not be cut off after a small number of terms if it was to be used for approximating considerably asymmetric laws caused a new search for alternative series expansions. At first only dealing with lattice distributions, Charlier [1905b, 2] discussed a model of elementary errors, each of which had the values 0 and α exclusively, with probabilities $p_i \approx 1$ and $q_i = 1 - p_i$, respectively. In a slightly erroneous and rather obscure manner Charlier [1905b, 4–7] deduced the representation for the probability A_r of the value $r\alpha$ ($r \in \mathbb{N}_0$) of the compound error

$$A_r = \sum_{m=0}^{\infty} \mu_m \Delta^m \psi_\lambda(r), \tag{3.36}$$

where $\lambda = \sum_{i=1}^{s} q_i$, $\Delta^{m+1}\psi_\lambda(z) = \Delta^m \psi_\lambda(z) - \Delta^m \psi_\lambda(z-1)$, and

$$\psi_\lambda(z) = \frac{\mathrm{e}^{-\lambda}}{\pi} \int_0^\pi \mathrm{e}^{\lambda \cos\omega} \cos[\lambda \sin\omega - z\omega]d\omega.$$

In the particular case $z \in \mathbb{N}_0$ the identity $\psi_\lambda(z) = \lambda^z \frac{\mathrm{e}^{-\lambda}}{z!}$ holds, and that is why [Charlier 1905b, 7] considered the expansion (3.36) as an improvement of the Poisson approximation to the binomial distribution.

Charlier named (3.35) "form A," and (3.36) "form B" of the law of error, thus coining designations which are still in use. From the purely mathematical point of view, the discovery of the B-series was his greater achievement. However, others had anticipated him also in this respect. Thiele in 1889, and especially Lipps in 1897 had already discussed the use of the B-series for approximating empirical distributions. In contrast to other authors, however, Charlier, who dedicated a good deal of his papers to the discussion of series expansions and related topics[70]—a third type, the C-series, was adjoined in 1928—reached a broader audience. The influence of Charlier's work is also witnessed by Särndal, who [1971, 375] even sees Charlier as the real founder of the "Scandinavian School."

In Särndal's interpretation, this "Scandinavian School" was a rival of Karl Pearson's "school." Aiming at approximating probability functions of hypergeometric distributions by differentiable functions $y(x)$, Karl Pearson had deduced the differential equation

$$\frac{\mathrm{d}y}{\mathrm{d}x} = -y\frac{x-a}{b_0 + b_1x + b_2x^2}$$

in 1895. The solutions of this differential equation, which coincides in the particular case $b_1 = b_2 = 0$ with Hagen's (see Sect. 3.2.1), yield the four-parameter "Pearson system of curves." This system, however, in contrast to Charlier series, could not be deduced "genetically" through a plausible stochastic model, like the hypothesis of elementary errors [Särndal 1971, 379].[71] Members of the Scandinavian school

[70] See Hald [2002, 49–62] for a comprehensive survey.

[71] The genesis of Pearson's system of curves between 1893 and 1895 was originally motivated, if in a more general respect, by the hypothesis of elementary errors. Already around 1895, however,

were convinced that models of elementary errors—which had to be generalized even further if necessary—could provide helpful ideas regarding the preselection of hypothetical probability distributions, not only in the context of fitting them to empirical frequencies, but also in a variety of testing and estimating problems.

3.4.2.3 Edgeworth and "The" Law of Error

Francis Ysidro Edgeworth (1845–1926)[72] was perhaps the statistician with the greatest mathematical abilities at the end of the 19th century. Later on, however, he became prominent due to his work on economics rather than his statistical contributions. Bowley [1928, 2] explains this circumstance by stating the relatively low practical application of Edgeworth's statistical work, which in most cases was dedicated to questions of predominantly theoretical interest. Edgeworth, who had originally studied ancient languages and law, acquired his mathematical skills as an autodidact mainly by reading the works of "classical" authors like Laplace, Poisson, and Fourier [Stigler 1978, 290]. He did not become familiar with the analytical development of the beginning modern era. Regarding his stochastic concepts and analytic methods, Edgeworth was especially influenced by apparently very carefully reading Laplace's *TAP*. The latter's remarks on possible alternatives to the method of least squares for parameter estimation were adopted by Edgeworth and further advanced, in particular in context with his frequent discussions of nonnormal distributions. As a consequence of his autodidactic education, Edgeworth cultivated an analytic style which reminds of the 18th rather than the early 20th century. His presentation of mathematical issues was often sketchy and not always straightforward. Altogether, these circumstances make his statistical work quite difficult to read. On the other hand, his permanent readiness to advance the discussion of statistical models far beyond their momentary practical applicability give his contributions a very modern touch.

Within Edgeworth's statistical work, the problem of the characteristic properties of *The* "Law of Error" was especially prominent. The designation "Law of Error" was mainly used by Edgeworth for "frequency laws" which expressed the probability distribution of a random variable resulting from the coaction of several "elements." Those "elements" were essentially independent elementary errors whose sum obeyed an approximate normal distribution, at least roughly. According to Edgeworth [1917, 412], an error law of this kind had "the advantage of being based on a *vera causa*, perhaps the most universal law of nature." And therefore, as Edgeworth pointed out again and again (see [1898, 672 f.], for example), these error laws could be assessed regarding their use for representing empirical frequencies

Pearson took the standpoint that reality was too complex to be explained by a particular stochastic model [Stigler 1986, 335 f.; 339 f.]. The term "genetic" was coined by Wicksell [1917], see [Särndal 1971, 378].

[72] Comprehensive accounts on Edgeworth's life and work on mathematical statistics can be found in [Bowley 1928], [Stigler 1978; 1999, sect. I.5], and [Stigler 1986, 300–325]. For details on Edgeworth expansions see also [Hald 2002, 42–48].

in two ways: Firstly by an "*a priori* consideration" of the plausibility of a possible mechanism based on elementary errors, by which the statistical quantity considered could be produced; secondly by testing whether the law of error (after an appropriate fitting of its parameters) could actually provide a sufficiently exact representation of the distribution of the sampling data. The applicability of a frequency curve of the "Pearson family," however, could only be checked by the second procedure.

What was the nature of the most general error law that could be based on the hypothesis of essentially independent and additively coacting "elements"? In the case of an "infinite" number of elementary errors one could expect a Gaussian distribution, apart from "pathological" exceptions. Edgeworth, however, intended to substantiate the frequent occurrences of moderate deviations from normal distributions by an appropriate hypothesis of elementary errors as well. The assumption of only a modest number of elementary errors yielded modifications of the Gaussian distribution through series, which could represent "The Law of Error" quite accurately. In the context of strongly asymmetric empirical distributions, the usual model of elementary errors could no longer be used. But even in such cases, Edgeworth tried to maintain the "natural" character of his approximations to probability curves by means of his "method of translation" (explained in more detail in the subsequent section).

In his first statistical paper, Edgeworth [1883] had already thoroughly discussed the use of correctional terms in addition to a normal density for representing general "facility-curves." In 1894 he submitted an article, in which he further elaborated the idea of modifying normal densities by series expansions, including a discussion on the adjustment of these expansions to statistical material [Stigler 1986, 338]. With the exception of a summary [Edgeworth 1894], this article remained unpublished, probably due to the rather clumsy presentation of the contents in the paper [Stigler 1986, 341]. A revised version was only printed in 1905. The core of this essay, which had apparently been written without knowledge about similar contributions by other authors, was the expansion of "frequency-loci"—Edgeworth's general designation for graphic representations of probability functions—assigned to sums of independent random variables by the now so-called "Edgeworth series." Edgeworth explicitly considered only lattice distributed random variables with a common lattice distance Δx, without giving reasons for this restriction. As "frequency functions" for variables X of this kind he perceived functions f with the property

$$P(X = x_k) = f(x_k)\Delta x \quad (x_k = -m\Delta x, (-m+1)\Delta x, \ldots, (n-1)\Delta x, n\Delta x),$$

m and n being natural numbers. In most cases Edgeworth—more or less tacitly—also carried over his results to continuous random variables and their densities.

Let us consider a sum of—for the sake of simplicity—identically distributed independent random variables X_i, each having the same density with zero expectation, variance 1, and cumulants κ_r; then the Charlier A series for the density of the (normed) sum is given by

$$f_n(x) = \frac{\mathrm{d}}{\mathrm{d}x} P\left(\frac{X_1 + \cdots + X_n}{\sqrt{n}} \leq x\right) = \phi(x)(1 - \frac{c_3}{3!}h_3(x) + \frac{c_4}{4!}h_4(x) - \cdots),$$

where the h_i designate the Hermite polynomials associated with the standard normal density ϕ. The coefficients c_j depend on the cumulants κ_r and the number n of the random variables by

$$c_3 = -\frac{\kappa_3}{\sqrt{n}},\ c_4 = \frac{\kappa_4}{n},\ c_5 = -\frac{\kappa_5}{\sqrt{n}^3},\ c_6 = \frac{\kappa_6}{n^2} + \frac{10\kappa_3^2}{n}, \ldots .$$

c_3 is therefore of the order of magnitude $n^{-\frac{1}{2}}$, c_4 is of the order n^{-1}, and c_5 is of the order $n^{-\frac{3}{2}}$. The illusive law of the coefficients c_j being of order $n^{-\frac{j-2}{2}}$ is violated from $j = 6$ on. c_6 is of the same order as c_4, and similar "irregularities" are found with higher coefficients. This flaw can be removed, however, by reordering the series according to

$$f_n(x) = \phi(x)(1 + \frac{q_1(x)}{\sqrt{n}} + \frac{q_2(x)}{n} + \cdots), \tag{3.37}$$

where

$$q_1(x) = \frac{\kappa_3}{3!}h_3(x),\ q_2(x) = \frac{\kappa_4}{4!}h_4(x) + \frac{1}{2}\left(\frac{\kappa_3}{3!}\right)^2 h_6(x),\ \text{etc.}$$

In 1905, Edgeworth, apparently without knowledge of prior work on Charlier A series, showed that a series expansion (3.37) in terms of the order of magnitude $(\sqrt{n})^{-r}$, $r = 1, 2, \ldots$, could be generally established up to an arbitrary number of terms. In so doing he also made it plausible that the difference between $f_n(x)$ and the series cut off after a certain number of terms vanishes as $n \to \infty$.

The 1905 article "The Law of Error" crowns Edgeworth's two-decade-long efforts concerning sums of elementary errors and simultaneously summarizes his previously achieved results on this and related topics. This voluminous paper is quite difficult to read, though by no means long-winded, and full of interesting details. "Edgeworth expansions" are introduced and derived by three different methods. There is also a discussion of—in Edgeworth's own words—"reproductive" distributions (stable distributions in modern terminology), and their significance as limit laws for sums of identically distributed random variables is demonstrated.[73] Edgeworth illustrated his series expansions by discussing particular cases, such as sums of two-valued or rectangularly distributed random variables. He also tried to generalize his results toward multidimensional errors, certain functions (not only sums) of elementary errors, or weakly dependent elementary errors. Edgeworth explicitly stated and discussed his assumptions on the properties of the elementary errors, even if in a verbose and not always entirely precise form. In his analytic methods, he followed Laplace and Poisson regarding the use of Fourier analysis, he was influenced by Crofton's idea of partial differential equations, and he applied Karl Pearson's method of moments. Apparently, Edgeworth neither knew modern contributions to the CLT or the analytic theory of moments, nor was he familiar with the analytic style of post-Weierstrassian era. His 1905 article, of which only a few aspects can be discussed here, nevertheless is impressive because of the analytic and stochastic intuition of its author, as well as due to its quest for far-reaching generalizations.

[73] The complete systematics, however, was reconsidered only by Lévy, see Sect. 5.2.6.

The first $t+1$ ($t \in \mathbb{N}$) terms of "his" series in the case of the density of a sum of m "elements," each with zero expectation, Edgeworth represented in the form

$$y_t = \mathrm{e}^{-\frac{k_1}{3!}\left(\frac{\mathrm{d}}{\mathrm{d}x}\right)^3+\frac{k_2}{4!}\left(\frac{\mathrm{d}}{\mathrm{d}x}\right)^4-\mathrm{etc}+(-1)^t\frac{k_t}{(t+2)!}\left(\frac{\mathrm{d}}{\mathrm{d}x}\right)^{t+2}}\frac{1}{\sqrt{2\pi k_0}}\mathrm{e}^{-\frac{x^2}{2k_0}}. \tag{3.38}$$

Edgeworth [1905, 36] named this expression the "$(t+1)$st approximation" to the "actual locus." As the "first" approximation he designated the Gaussian error law $y_0 = \frac{1}{\sqrt{2\pi k_0}}\mathrm{e}^{-\frac{x^2}{2k_0}}$, where k_0 was the sum of the variances of the single elementary errors (see Fig. 3.6 for a practical example). The exponential expression of the differential operators in (3.38) has to be conceived as this part of the series

$$\sum_{i=0}^{\infty}\frac{1}{i!}\left(\sum_{r=1}^{t}(-1)^r\frac{k_r}{(r+2)!}\left(\frac{\mathrm{d}}{\mathrm{d}x}\right)^{r+2}\right)^i$$

which, after expanding and rearranging the multinomials $(\dots)^i$ in groups with the same value $\nu+\mu$ for k_ν^μ, contains multiples of k_ν^μ with $\nu+\mu \le t+1$. The coefficients k_r in (3.38) are, as Edgeworth [1905, 44] explained, equal to the difference between the moment of the order $r+2$ of the actual error law y of the sum of elementary errors and the moment of the same order of y_{r-1} ($y_{-1} := 0$). In this way Edgeworth obtained

$$y = y_0 + (y_1 - y_0) + (y_2 - y_1) + \cdots, \tag{3.39}$$

where

$$\begin{aligned} y_0 &= \frac{1}{\sqrt{2\pi k_0}}\mathrm{e}^{-\frac{x^2}{2k_0}},\\ y_1 - y_0 &= -\frac{k_1}{3!}\frac{\mathrm{d}^3 y_0}{\mathrm{d}x^3},\\ y_2 - y_1 &= \frac{k_2}{4!}\frac{\mathrm{d}^4 y_0}{\mathrm{d}x^4} + \frac{k_1^2}{2!3!3!}\frac{\mathrm{d}^6 y_0}{\mathrm{d}x^6}, \qquad \text{etc.} \end{aligned}$$

A first justification for the approximative and asymptotic character of his series expansion was given by Edgeworth [1905, 40–45] by means of the method of moments.[74] It was shown that in the case of a large number of elementary errors the central moments of the sum of the elementary errors up to a certain order were only slightly different from those of the approximating law y_t. Regarding this criterion on moments, Edgeworth [1905, 41] did not refer to the results of the analytic theory of moments, as produced by Chebyshev, Markov, or Stieltjes (see Sects. 4.3, 4.4), but cited Karl Pearson, who himself apparently did not possess any knowledge of the recent analytic development of moments. Estimating parameters of error laws from empirical moments, however, had become a commonly used device in error theory

[74] Closer details are discussed in [Bowley 1928, 39–45].

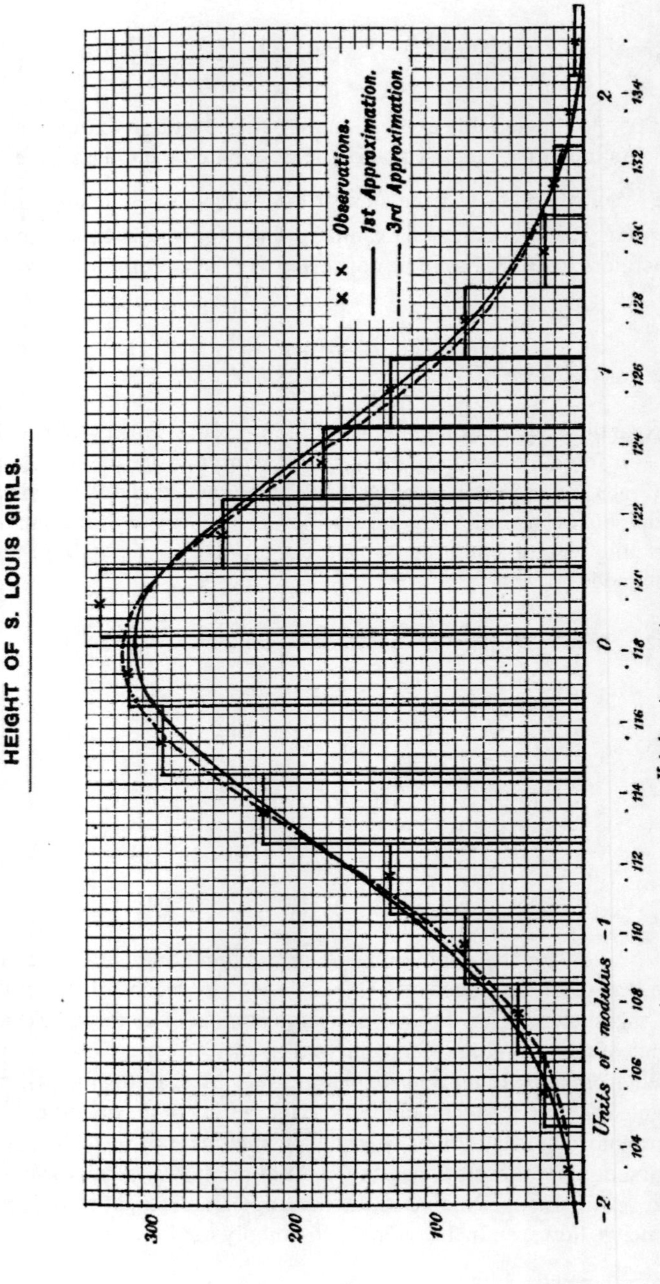

Fig. 3.6 Fitting a normal curve and an Edgeworth expansion up to the coefficient k_2 to data [Edgeworth 1906]

and statistics since Gauss (see Sect. 4.1). As Edgeworth [1905, 41] noticed, Pearson had discussed some practical examples, in which—in modern terminology—the L^1-norm between empirical histograms and approximating curves became smaller the greater the order was up to which identical moments within both systems existed (see [Stigler 1986, 334 f.]).

Edgeworth [1905, 39] assumed independent elementary errors $\xi_1, \ldots, \xi_m$, each with zero expectation, however having different densities, which vanish beyond given finite intervals. For each t from $t = 1$ up to a "considerable" t he assumed the moments of order t to attain approximately the same magnitude among all elementary errors. For a closer discussion of the moments of the sum $X = \sum \xi_i$, Edgeworth [1905, 41 f.] started with the equation

$$\mathrm{E}e^{\theta X} = \mathrm{E}e^{\theta(\xi_1 + \cdots + \xi_m)} = \mathrm{E}e^{\theta\xi_1} \cdots \mathrm{E}e^{\theta\xi_m}.$$

From this, with the designation $x^{(i)} = \mathrm{E}X^i$ and analogous notations for the moments of the ξ_q, he inferred

$$\sum_{i=0}^{\infty} \frac{1}{i!}\theta^i x^{(i)} = \sum_{i_1=0}^{\infty} \frac{1}{i_1!}\theta^{i_1}\xi_1^{(i_1)} \cdots \sum_{i_m=0}^{\infty} \frac{1}{i_m!}\theta^{i_m}\xi_m^{(i_m)}$$
$$= e^{\log\left(\sum_{i_1=0}^{\infty} \frac{1}{i_1!}\theta^{i_1}\xi_1^{(i_1)}\right) + \cdots + \log\left(\sum_{i_m=0}^{\infty} \frac{1}{i_m!}\theta^{i_m}\xi_m^{(i_m)}\right)}$$
$$= e^{\sum_{q=1}^{m}\left[\frac{\theta^2}{2!}\xi_q^{(2)} + \frac{\theta^3}{3!}\xi_q^{(3)} + \theta^4\left(\frac{\xi_q^{(4)}}{4!} - \frac{1}{2^3}(\xi_q^{(2)})^2\right) + \cdots\right]}.$$

Altogether he derived

$$\mathrm{E}e^{\theta X} = e^{\frac{k_0}{2!}\theta^2 + \frac{k_1}{3!}\theta^3 + \frac{k_2}{4!}\theta^4 + \cdots + \frac{k_t}{(t+2)!}\theta^{t+2} + \cdots}, \tag{3.40}$$

where

$$k_0 = \sum_{q=1}^{m} \xi_q^{(2)}, \quad k_1 = \sum_{q=1}^{m} \xi_q^{(3)}, \quad k_2 = \sum_{q=1}^{m} \left(\xi_q^{(4)} - \frac{4!}{2^3}(\xi_q^{(2)})^2\right), \quad \text{etc.}$$

On the basis of his conditions on the magnitude of the moments and the boundedness of the single elementary errors, Edgeworth showed that, after redefining $x^{(2)}$ as a "unity," the quantities k_r, as well as the products $k_{s_1} \cdot k_{s_2} \cdots k_{s_n}$ with $s_1 + \cdots + s_n = r$, had an order of magnitude $\frac{1}{(\sqrt{m})^r}$.[75] Because of the fact that $x^{(s)}/s!$ was equal to the coefficient of θ^s in the expansion of (3.40) as a power series of θ, and using his considerations on the orders of magnitude of the k_r, Edgeworth was able to prove that

[75] This means, in modern terminology, that these orders of magnitude are valid for the standardized sum of the elementary errors.

1) $\frac{x^{(2p)}}{(\sqrt{x^{(2)}})^{2p}} = \frac{2p!}{p!2^p} + r_{2p}(m)$, where $r_{2p}(m)$ is a finite sum of terms of order of magnitude $\frac{1}{m}$ and below,

2) $\frac{x^{(2p+1)}}{(\sqrt{x^{(2)}})^{2p+1}} = \frac{(2p+1)!}{3!(p-1)!2^{p-1}} k_1' + R_{2p+1}(m)$, where $k_1' = \frac{k_1}{\sqrt{x^{(2)}}^3}$ is of the order of magnitude $\frac{1}{\sqrt{m}}$ and $R_{2p+1}(m)$ is a finite sum of terms of order of magnitude $\frac{1}{m^{\frac{3}{2}}}$ and below.

From these two relations one could immediately see that, presupposing a considerable m, the difference between the moments of the sum X of the elementary errors and those of the corresponding normal distribution y_0 was "small," relative to the respective power of $\sqrt{x^{(2)}}$. From

$$\int_{-\infty}^{\infty} x^t \frac{\mathrm{d}^{t-s}}{\mathrm{d}x^{t-s}} y_0 dx = \frac{t!}{s!} \int_{-\infty}^{\infty} x^s y_0 dx$$

Edgeworth [1905, 44] also deduced that the partial sums with terms of a common order of magnitude in $r_{2p}(m)$ and $R_{2p+1}(m)$ are successively equal to the quotients of the moments of order $2p$ and $2p+1$, respectively, of $y_1 - y_0$, $y_2 - y_1$, etc., and the corresponding power of $\sqrt{x^{(2)}}$, if the approximations y_1, y_2, etc. are according to (3.39). Thus, there was a good accordance between the moments of the sum of the elementary errors X and the moments of the higher approximations up to a certain order, if the number m of the elementary errors was considerable. Edgeworth [1905, 44 f., 132 f.], however, also hinted at the possibility of an increasing discrepancy between the corresponding moments, if they were of an order too high, because then the number of the terms in $r_{2p}(m)$ and $R_{2p+1}(m)$, respectively, which were of the same order of magnitude corresponding to a power of $\frac{1}{\sqrt{m}}$, was growing together with p.

It is not surprising that, at a subsequent place in his article, Edgeworth [1905, 51–54], discussing Laplace's methods of deriving approximations to probabilities of sums, achieved a result for $\mathrm{E}e^{\theta X\sqrt{-1}}$ analogous to (3.40), by use of which he could show that for the density $y(x)$ of the sum of the elementary errors:

$$y(x) = \frac{1}{2\pi} \int_{-\infty}^{\infty} \mathrm{E}e^{\alpha X\sqrt{-1}} e^{-\alpha x\sqrt{-1}} d\alpha$$
$$= \frac{1}{2\pi} \int_{-\infty}^{\infty} e^{\sum_{t=2}^{\infty} (\sqrt{-1})^t \frac{\alpha^t k_{t-2}}{t!}} e^{-\alpha x\sqrt{-1}} d\alpha. \quad (3.41)$$

This relation yielded, by cutting off the series at $t = s$, substituting the exponential term $e^{\sum_{t=2}^{s} (\sqrt{-1})^t \frac{\alpha^t k_{t-2}}{t!}}$ by a power series, reordering the series according to such k_ν^μ with common $\nu + \mu$, and finally integrating term by term, approximations y_{s-2} corresponding to Edgeworth expansions. Edgeworth's arguments concerned the algebraic form only; he did not derive upper bounds for $|y(x) - y_r(x)|$ or discuss the asymptotic behavior of this difference as $m \to \infty$. Expressions equivalent to (3.41) had also been deduced by Hausdorff and Charlier, who, however, had not systematically considered the order of magnitude of the respective series terms.

Edgeworth presented a further method for establishing his series, in adapting Crofton's idea of describing the coaction of elementary errors by partial differential equations. Whereas Crofton had only considered the contributions of the expectations and the mean squares of the single errors to the compound law, Edgeworth, aiming at more precise approximations, tried to advance this procedure by including mean powers of higher order. It is probable that Edgeworth originally reached his higher approximations $y_1, y_2, \ldots$ by Crofton's method, because, in his first paper on elementary errors, he [1883, 304] had already indicated how a "second approximation" (i.e., y_1) could be obtained by aid of the "theorem given by Mr. Crofton." The general discussion in [Edgeworth 1905, 45–51] is, however, difficult to comprehend at several places, and rather gives the impression of a plausibility consideration.

If another elementary error with expectation 0, variance ∂k_0, and a (no more negligible) mean third power ∂k_1 is added to a sum of elementary errors with the law y, then in analogy to Crofton's (3.18) for the new law $y + \delta y$ we have

$$y + \delta y \approx (1 + \frac{\partial k_0}{2} D^2 - \frac{\partial k_1}{3!} D^3) y. \tag{3.42}$$

Edgeworth [1905, 47] considered k_0 and k_1 as independent variables, and therefore he concluded

$$\frac{\partial y}{\partial k_1} \approx -\frac{1}{3!}\frac{\partial^3 y}{\partial x^3}.$$

If one also takes into account the dependence of the error law y on k_1, then y_1 with

$$\frac{\partial y_1}{\partial k_1} = -\frac{1}{3!}\frac{\partial^3 y_1}{\partial x^3} \tag{3.43}$$

can be conceived as a more precise approximation to the law of error than the normal density y_0, for which "Crofton's differential equation"

$$\frac{\partial y}{\partial k_0} = \frac{1}{2}\frac{\partial^2 y}{\partial x^2}$$

is valid. Egeworth now assumed that y_1 could be represented by

$$y_1 = y_0 + k_1 \left[\frac{\partial y_1}{\partial k_1}\right]_{k_1=0} + \cdots.$$

Apparently, he interpreted y_0 as $(y_1)_{k_1=0}$, because with reference to (3.43) he inferred

$$y_1 = y_0 + k_1 \left[-\frac{1}{3!}\frac{\partial^3 y_0}{\partial x^3}\right].$$

A further continuation of this procedure with consideration of even higher moments led to certain difficulties, because, as Edgeworth explained, the corresponding moments of the additional elementary error could no longer be conceived as differentials of variables which are independent of k_0 and k_1. The moment of

4th order k_2 of the compound error, for example, also depends on the moments of 2nd order of the single elementary errors. Nevertheless, Edgeworth made it plausible that, in his approach, $k_0, k_1, k_2, \ldots$ could be treated as if being independent, and he generalized (3.42)—even if in a somewhat obscure manner—toward a representation of δy by a linear combination of (also higher) differentials of $k_0, k_1, k_2, \ldots$. Taking into account the different orders of magnitude of the k_i, he finally reached, in a manner similar to his derivation of y_1, the verification of his general approximation y_t.

Edgeworth's comprehensive account also comprised the discussion of a further, in his own words [1905, 54] "fresh," condition on the "sought" law of error, to be "reproductive" (i.e., "stable" in modern terminology). According to Edgeworth, these probability densities are called "reproductive" which belong to a certain "family" of functions with the following property:

> (...) if two or more independently fluctuating quantities [i.e., random variables] $A, B, \ldots$ assume different values with a frequency designated by a member of the family represented by the sought function, then Q a quantity formed by adding together each pair (triplet, etc.) of concurrent values presented by $A, B, \ldots$ will also assume different values with a frequency designated by a member of the sought family.

This portion of text, which is characteristic of Edgeworth's idiosyncratic style, was far from being a precise definition of "reproductive." Only by the following analytical explanations was it clarified that a "family" consisted of all "frequency functions" of the form $\frac{1}{c} f\left(\frac{x}{c}\right)$, where c was any positive number and f a given density function.

Edgeworth [1905, 54 f.] also hinted at the fact that "frequency-curves" of random variables which are composed of a "great number" of independent identically distributed elementary errors are necessarily reproductive: If A and B are independent random variables, and both variables may be assumed to be additively composed of "great numbers" m_1 and m_2 of elementary errors of the "typical sort," then the densities of A and B belong to the same family. $A + B$ is a fortiori composed of a "great number" $m_1 + m_2$ of independent identically distributed elementary errors of the "typical sort," and therefore has a frequency function from the family under consideration as well.

Edgeworth now assumed that random variables $A, B, \ldots$ of the number m had the same reproductive frequency function $f(x)$ with the property $f(x) = f(-x)$. Consequently $Q = A + B + \cdots$ had a frequency function $y = \frac{1}{c} f(\frac{x}{c})$, where c was a constant depending on m. Using "Laplace's analysis," that is, using characteristic functions $\mathrm{E}e^{\alpha A\sqrt{-1}}$, which he represented without any justification by $e^{\psi(\alpha)}$, Edgeworth expressed the latter condition through $\psi(c\beta) = m\psi(\beta)$. The solution of this functional equation could be determined from well-known—though not rigorously proven—rules. For $\beta \in \mathbb{R}_0^+$ this solution is of the general form $\psi(\beta) = \beta^t \cdot a$, wherefrom Edgeworth inferred that

$$f(x) = \frac{1}{\pi} \int_0^\infty e^{a\alpha^t} \cos \alpha x d\alpha.$$

He did not discuss the nontrivial question of whether such an $f(x)$ (as depending on a and t) was actually a probability density.

Edgeworth [1905, 56 f.] even broached the problem of asymmetric "reproductive" laws. He showed that reproductive laws were symmetric (with respect to 0) if they possessed a finite variance. Symmetric and reproductive laws y with a finite variance and zero expectation, however, could be assumed as being made up of a very large number of elementary errors, and therefore conformed to Crofton's equation $y(x) = F(x, k_0)$ (see (3.19)). On the basis of this equation Edgeworth deduced $y(x) = \frac{1}{\sqrt{2\pi k_0}} e^{-\frac{x^2}{2k_0}}$. He called this deduction of the Gaussian error law a "variant" of Crofton's method. This variant, however, served only as a first approximation, and could not be used for the derivation of a series expansion.

In a first publication on error laws which meet certain conditions, such as the hypothesis of elementary errors or the property of being reproductive, Edgeworth [1883, 305–307] had already discussed densities like

$$f(x) = \frac{1}{2\pi} \int_{-\infty}^{\infty} e^{-|\alpha|^t} \cos \alpha x d\alpha,$$

where t was not further specified. He was only able to give explicit algebraic expressions in the case $t = 1$ by the function $f(x) = \frac{1}{\pi(1+x^2)}$, "noticed by Poisson," and for $t = 2$ by the Gaussian law. He judged the existence of frequency functions which were reproductive, but different from the Gaussian density, as a piece of evidence

> ... that the "ancient solitary reign" of the exponential law of error should come to an end ... [Edgeworth 1883, 306].

Against possible objections that "common sense" would already eliminate the possibility of such error laws, Edgeworth [1883, 308] objected:

> But in Chance, as in other provinces of speculation which have been invaded by mathematics, common sense must yield to symbol.

Even if Edgeworth later won more practical experience in dealing with statistical data, and therefore did no longer assign any arbitrary hypothesis on errors the same right, still his opinion on the priority of mathematics remained basically unchanged. He always did his best to discuss statistical problems with regard to all aspects, not only the practical ones. This attitude reminds one of Cauchy's approach in his dispute over the method of least squares with Bienaymé, in which stable laws likewise played an important role. Most probably Edgeworth, who usually carefully referred to the results of other authors, was without any knowledge on this work by Cauchy. We may therefore ascribe to Edgeworth the merit of a "rediscovery" of stable distributions—at least the symmetric ones—and especially the discovery of their importance as limiting laws for sums of independent identically distributed random variables.

In his 1905 article Edgeworth also considered several generalizations of the results described above. Edgeworth [1905, 115 f.] established, for example, his series expansion for sums of two-dimensional elementary errors, using the method of moments. Concerning his assumption of bounded elementary errors he mentioned that for the existence of a normal limiting distribution it would only be important that the moments of the single errors were finite and approximate among each other. Edgeworth recommended to substitute in the proofs unbounded elementary errors by bounded ones, such that the latter had almost the same moments as the first. With these—rather qualitative and only verbal remarks—he came close to the idea of truncated random variables as it was developed by Markov at exactly the same time (see Sect. 5.1.5).

Very noteworthy, but quite difficult to understand due to their rather sketchy presentation, are Edgeworth's ideas [1905, 121–126, 139–141] on generalizing additivity of elementary errors. To this aim the compound error was assumed to be a polynomial in elementary errors of a certain degree, where it was presupposed that in this polynomial the coefficients of higher order were small compared with the first ones. In modern terminology, Edgeworth probably meant the following: Let (X_j) be a sequence of independent elementary errors, which meet the usual conditions (regarding zero expectations, orders of magnitude of the single moments, etc.). Additionally, let k be a fixed natural number and

$$Y_n = \sum_{\{\alpha \in \mathbb{N}_0^n | 1 \leq |\alpha| \leq k\}} a_\alpha^{(n)} X_1^{\alpha_1} \cdots X_n^{\alpha_n},$$

where the coefficients $a_\alpha^{(n)}$ have the order of magnitude $\sqrt{n}^{1-|\alpha|}$.[76] Then the approximation of the density of $Y_n/\sqrt{\mathrm{Var} Y_n}$ by the corresponding Edgeworth expansion, cut off after a few terms, becomes more and more exact the greater the number of elementary errors n is. The characteristic feature of Edgeworth's polynomial was that a sum of independent random variables (corresponding to the linear part of the polynomial) was augmented by additional quantities depending on these random variables, whose influence, however, was small compared with the elements of the linear part. In turn, Edgeworth [1905, 126 f.] tried to cover possible stochastic dependencies among elementary errors by polynomial models of this kind. In fact, von Mises [1935; 1936], without any reference to Edgeworth, proved a CLT for sample statistics, which concerned a similar situation (see also [Cramér 1946, 218 f., 352–367]).

3.4.3 The Method of Translation

Both Edgeworth and Charlier A expansions (in practice cut off after a few terms) could only serve for fitting such frequency curves to empirical material that did not deviate too much from a Gaussian bell-shaped curve. The practical application of

---

[76] More exactly:

$$\exists r, s > 0\, \forall n \in \mathbb{N}\, \forall \alpha \in \mathbb{N}_0^n : 1 \leq |\alpha| \leq k \Rightarrow r \leq |a_\alpha^{(n)}| \sqrt{n}^{|\alpha|-1} \leq s.$$

Charlier B-series was restricted to those cases in which the frequency had a relatively sharp maximum near the upper or lower bound of the statistical quantity under consideration. In all other cases one had to resort to Pearson's system of curves (see Sect. 3.4.2.2) or to apply the method of translation, where the latter—at least from the point of view of its propagators—could preserve, in contrast to Pearson's approach, a certain "genetic" character.

The basic idea of the method of translation is to discuss the random variable $f(X)$ rather than the random variable X, where f is an appropriate function.[77] Särndal [1971, 382] ascribes this principle to Galton [1879] in the particular case $f(x) = \log(x)$, and in its general version to Edgeworth [1898]. Jacobus Cornelius Kapteyn [1903] discussed translations $f(x) = (x + k)^q$ $(x > -k, k, q \in \mathbb{R})$, and he tried to justify this special choice of f. Finally, Sven Dag Wicksell [1917], who relied on Kapteyn's ideas, presented a very general model of elementary errors, by which the choice of a favorable translation function could be facilitated.

3.4.3.1 The Log-Normal Distribution

Johann Heinrich Lambert in his 1760 *Photometria* had discussed, in context with photometric measuring, the geometric mean as a possible alternative to the arithmetic mean. In 1863 Philipp Ludwig Seidel argued that it would not make sense to consider, in the error theoretic analysis of photometric data, the deviations of the light intensities measured from their true value; it was rather appropriate to take the differences between the logarithms of the values observed and the logarithm of the true value.[78] In 1834, Ernst Heinrich Weber had already stated the later so-called Weber–Fechner law (proportionality between sensation and logarithm of stimulus), which became the main topic of Gustav Fechner's 1860 *Psychophysik*.[79]

Galton directly referred in the first part of the 1879 article "The Geometric Mean"—the second, mathematical, part was due to Donald McAlister—to Fechner. Galton discussed a photometric test arrangement, in which the geometric mean apparently had to be preferred as opposed to the arithmetic mean. Galton [1879, 367] alluded to Gauss's deduction of the normal law from the principle of the arithmetic mean being the most probable value, and he pointed out that the assumption of the geometric mean being the most probable value would lead to another type of error law. Moreover, Galton made it plausible that in numerous cases the frequency function of a statistical characteristic was inevitably asymmetric, and therefore deviated from the "usual" Gaussian error law.

McAlister, in his part of the article, considered the situation of observations $x_1, \ldots, x_n$ (each tacitly assumed to be positive), from which the true value a was

---

[77] If, for example, the range of values of X is equal to the interval $]a;\infty[$, if $f\ :]a;\infty[\rightarrow \mathbb{R}$ is onto and strictly monotonic increasing, and if $f(X)$ obeys a normal distribution Φ_{μ,σ^2}, then $P(X \leq x) = \Phi_{\mu,\sigma^2}(f(x))$ for $x > a$, and $P(X \leq x) = 0$ for $x \leq a$.

[78] For closer details concerning Lambert's and Seidel's contributions see [Knobloch 1992, 273–275].

[79] For Fechner's work referring to this matter see [Stigler 1986, 243–254].

to be estimated. He imposed the condition that the geometric mean $\sqrt[n]{x_1 \cdots x_n}$ was the "most probable value" for a, and, accordingly, the arithmetic mean $\frac{1}{n}[\log(x_1) + \cdots + \log(x_n)]$ was the "most probable value" for $\log a$. In analogy to Gauss's line of arguments (see Sect. 3.1), for the logarithm $y = \log x$ of a measurement x the frequency law

$$\frac{h}{\sqrt{\pi}} e^{-h^2(y-\log a)^2}$$

resulted, and from that, because $dy = \frac{1}{x}dx$, the frequency law

$$\frac{h}{\sqrt{\pi}x} e^{-h^2(\log \frac{x}{a})^2}$$

for the quantity x was deduced. Distributions with densities of this type are now called "log-normal distributions." Under the heading "Quetelet's Method" McAlister [1879, 372 f.] discussed how the new frequency law could be deduced from a hypothesis of elementary errors which was along the lines of Hagen's. McAlister assumed the ratio of an observation x and the true value a as being equal to the product of $2n$ elementary errors, each with the equiprobable values $\alpha \approx 1$ and α^{-1}. The exponents $-2n, \ldots, 2n$ of the possible values α^{-2n}, α^{2-2n}, ..., α^{-2}, 1, α^2, ..., α^{2n-2}, α^{2n} of x/a were given according to a binomial distribution, and therefrom, for a large number n, an approximate normal distribution of the exponents, and a log-normal distribution of the observations x, resulted. As already noticed above (see Sect. 3.3.1), around the mid-1870s Galton had abandoned the belief that each normal distribution was a result of the existence of elementary errors. As a consequence, McAlister's discussion of elementary errors was rather brief.

The comprehensive discussion of the log-normal distribution as an alternative to the Gaussian law, however, did not imply that Galton abandoned his fundamental preference for normal distributions [Porter 1986, 139]. On the contrary, log-normal distributions could in most cases be accurately fitted to asymmetric empiric distributions, and since log-normal distributions could be considered as descendants of normal distributions, Galton's positive attitude toward normal laws was confirmed. In turn, he criticized Karl Pearson's approach as not being capable of a "rational" justification [Stigler 1986, 336], although only in private letters.

Fechner's ideas inspired Galton to present his account on the log-normal distribution, but Fechner did not publish anything on this issue during his lifetime. In his posthumous book *Kollektivmaßlehre* [1897], which had been extensively reworked and supplemented by Friedrich Lipps, an entire chapter was dedicated to this distribution, the concept of which was at least in principle due to Fechner. In the book, possible asymmetries of frequency distributions of biological or social data were discussed at considerable length. Apparently, for Fechner the quite universal applicability of the log-normal distribution for representing empirical data was an indicator for the failure of the normal distribution in general. Lipps, who at the time of the publication of Fechner's book was an ardent worshiper of the hypothesis of elementary errors, also gave a derivation, based on elementary considerations, of the log-normal distribution, which was similar to McAlister's. Later on, Lipps [1901,

163–166] fundamentally changed his attitude. Now, he spoke up for a thorough rejection of any hypotheses on elementary errors, because the assumption of mutual independence of elementary errors became untenable in his opinion.

3.4.3.2 Wicksell's General Model of Elementary Errors

Edgeworth [1898, 674 f.] presented a general "method of translations," but he only vaguely discussed the advantages of this new idea. Corresponding to the "prevalence" of the hypothesis of elementary errors, those "formulae" should be preferred which were related to the normal distribution to some extent. According to Edgeworth it was "probable" that modifications of the usual conditions on elementary errors (independence, additivity, relative smallness, large number) would lead to certain deviations from normal frequency laws. A precise explanation of specific relations between modified models of elementary errors and particular translation functions cannot be found in Edgeworth's contribution, however.

Wicksell expounded a general concept of elementary errors in 1917, from which the method of translation could be deduced and in which specific assumptions on elementary errors were connected with particular types of translation functions. He referred to ideas, developed by Kapteyn [1903], concerning the stochastic dependence of different increments on the current value of the compound error to whose accumulation they are contributing. Wicksell's theory, however, was far more general than Kapteyn's.

[Wicksell 1917, 7] assumed "sources" $Q_1, Q_2, \ldots, Q_s$ of elementary errors acting in succession and thus producing a compound error z. The respective "error impulses" $x_1, x_2, \ldots, x_s$ (presupposed as being independent) had probability densities with the property that the frequency function of $\sum_{i=1}^{s} x_i$ could be approximated as precisely as necessary by the first terms of a Charlier A or B series. Wicksell assumed that z had already reached the size z_{i-1}, directly before the action of Q_i; the increment of z caused by the action of Q_i should be $x_i \cdot \Theta(z_{i-1})$, where Θ was a function, whose properties were still unspecified. Expressed by formulae, the equation

$$z_i = z_{i-1} + x_i \Theta(z_{i-1}), \tag{3.44}$$

and therefore also

$$z = z_0 + \sum_{i=1}^{s} x_i \Theta(z_{i-1})$$

should be valid. [Wicksell 1917, 8 f.] now introduced the transformation function $A(z)$ by

$$A(z) = \int_{z_0}^{z} \frac{du}{\Theta(u)}.$$

Assuming each of the differences $z_i - z_{i-1}$ to be very small, he followed by virtue of (3.44) that

$$A(z) \approx \sum_{i=1}^{s} \frac{z_i - z_{i-1}}{\Theta(z_{i-1})} = \sum_{i=1}^{s} \frac{x_i \Theta(z_{i-1})}{\Theta(z_{i-1})},$$

and therefore

$$A(z) \approx x_1 + x_2 + \cdots + x_s. \tag{3.45}$$

On account of (3.45), the frequency function of the random variable $A(z)$ could be represented by the Charlier series related to the sum of independent random variables $x_1, \ldots, x_s$.

Wicksel's hypothesis of elementary errors had the advantage of providing a plausible stochastic model which not only implied the principle of transformation function, but also led to the specific form of this function, if substantiated assumptions could be made on the particular shape of the function Θ. A prominent role among all the special functions Θ Wicksell [1917, 16–18] discussed played $\Theta(u) = u$ and, consequently, $A(z) = \log\left(\frac{z}{z_0}\right)$. Presupposing a great number of—approximately equally small—error impulses, the result was a log-normal distribution.

3.4.3.3 The Further Fate of the Hypothesis of Elementary Errors

A controversy arose between the defenders of the hypothesis of elementary errors on the one hand, and of Pearson's system of curves on the other. However, it was never disputed in a particularly sharp manner. The main issues of the controversial discussion were computational and statistical advantages and disadvantages of the procedures applied. However, the search for an appropriate method of representing empirically obtained frequency distributions which could be connected with the inner nature of the considered characteristics was even more important. Regarding universal applicability, Pearson's system, which also covered strongly asymmetric probability laws, had the advantage. But it was based on an arbitrary stochastic model, for which a connection to natural processes could hardly be established. The hypothesis of elementary errors was questionable—even in Wicksell's general model—since it required independence of the single errors (or error impulses), a condition which was hardly in line with reality, especially in the biological realm. Edgeworth's poorly presented model of 1905 for weakening independence (see Sect. 3.4.2.3) remained unnoticed, even though it was taken up again in an additional paper [Edgeworth 1906]. Studies in which sums of chained random variables were discussed, following the work of Markov, remained without any influence on the hypothesis of elementary errors, because of their predominantly theoretical orientation.

In statistics, elementary errors became ever more unimportant after ca. 1920 since this issue remained only of secondary interest in the "new" field of hypothesis testing. On the other hand, mathematicians, like Lévy, Lindeberg, or Cramér, took up the hypothesis of elementary errors as an "applied" motivation for their research on the CLT. Whereas in Lévy's and Lindeberg's accounts elementary errors were treated as examples (and counterexamples) for the occurrence of the CLT in the physical world, Cramér's direct concern was an important "practical" question: How exactly could the profit of an insurance company, resulting from all single contracts (which played the role of elementary errors), be approximated by a Charlier or Edgeworth series cut off after its first terms? Around the turn of the century, this problem was behind several relevant works on insurance risk, for example by Hausdorff [1897] or Bohlmann [1901]. Yet it would only finally be solved on the basis of fresh results in analytic probability theory, mainly due to Lyapunov (whom Bohlmann already had referred to), and von Mises (see Sects. 5.1.3 and 5.2.2).

Appendix: Letter from Bessel to Jacobi, 14 August 1834

German Transcription

Sie können, von Dato an, als erwiesen annehmen, Verehrtester! daß viele zusammenwirkende Fehlerursachen immer eine Wahrscheinlichkeit des ganzen Fehlers liefern, welche nahe dem exponentiellen Gesetze folgt. Ob das "immer" nicht zu viel gesagt ist, muß ich jedoch noch untersuchen; also lieber "im Allgemeinen". Meine Ableitung dieses eifrig gesuchten Satzes ist kurz und klar. Sie streift aber, in wesentlichen Punkten, stark an Poissons frühere Analysen verwandter Aufgaben. Eigentlich muß ich mich, wie gewöhnlich, ärgern, dieses so spät zu bemerken. Aber, wenn mein eigenes Verdienst auch klein, in Ihren Augen und vielleicht auch in den meinigen, möglicherweise sehr klein wird, so daß es als eine zu vernachlässigende Größe anzusehen sei, so habe ich doch Freude über das Gelingen meiner Bemühungen, weil es meiner Arbeit den Schlußstein gibt. — Da ich Sie, in Geburtsnöthen, mit Wehklagen zu plagen pflege, so sollen Sie doch gleich erfahren, daß geboren ist — eine Maus!

d. I.
FWB
14 Aug 38.

English Translation

You can henceforth take for granted, dearest!, that many coacting causes of error always yield a probability of the entire error which is close to the exponential law. Whether the "always" is carrying things too far has to be investigated though; therefore rather "in general." The derivation of this theorem, which I was eagerly seeking for, is clear and concise. Yet there is an essential relation to Poisson's previous analyses of similar problems. As usual, I actually get angry about realizing this so late. Although the credit I deserve is small, maybe very small in your eyes and possibly in mine as well, not more than a negligible quantity in the end, I'm still happy about my successful efforts which put the capstone on my work. Usually bothering you about my laments when suffering from birth pain, I shall come right to the point—a mouse is born!

Chapter 4
Chebyshev's and Markov's Contributions

From an historical point of view, the treatment of the CLT by the method of moments is strongly connected with the contributions of Chebyshev and Markov. According to the standard historical interpretation, both mathematicians with their contributions began to apply that mathematical rigor to probability theory which would become the norm during the first decades of the 20th century.

As far as the mathematical merits of Pafnutii Lvovich Chebyshev (1821–1894) are concerned, beginning with the accounts of Lyapunov [1895] and Vasilev [1898/ 1900] a hagiographic valuation of his work was established and has remained basically unchanged. A really exultant esteem of Chebyshev's contributions was maintained during the Soviet regime, as many essential features of his work could be interpreted in complete accordance with the official materialistic point of view, apparently without any problems.[1] The main characteristics of the "standard" perception of Chebyshev's life and (stochastic) work can briefly be summarized as follows:

— Chebyshev, the founder of "St. Petersburg school," with Markov and Lyapunov being his most prominent disciples, was an excellent and highly motivating teacher.

— Chebyshev was especially interested in mathematical problems which could be applied to practical issues, according to his "realistic" (in the interpretation of Soviet historians "materialistic") perception of mathematics.

— The characteristics of Chebyshev's mathematical methods are: Reduction of solutions of problems to elementary procedures and operations, approximations with a precise description of the errors committed, and, as a consequence of this approach, analytic rigor in a finitary sense, despite an apparent lack of interest in the foundations of analysis.

With regard to Chebyshev's probabilistic activities, these statements seem to be particularly appropriate. His favored methods were decisive especially for limit

[1] For Chebyshev's contributions to probability theory see [Bernshtein 1945/2004a; Maistrov 1974, 188–208; Gnedenko & Sheynin 1992, 251–262; Sheynin 1994; Sheynin 2005b, 214–226]. The latter two sources by Sheynin take a more balanced position. For Chebyshev's life and life's work see [Prudnikov 1964; Yushkevich 1970–76a; Bernshtein 1947/2001]. For an interesting collection of quotations from different authors (including Chebyshev himself) on Chebyshev's "realistic" attitude toward mathematics, see [Steffens 2006, 69–75].

H. Fischer, *A History of the Central Limit Theorem*, Sources and Studies in the History of Mathematics and Physical Sciences, DOI 10.1007/978-0-387-87857-7_4,

theorems, due to his demand for explicit estimates of the deviations between the exact formula in the case of a finite number of trials and the respective limit term. Only with knowledge of those estimates was a complete and rigorous discussion of a limit problem possible, according to Chebyshev (see Sect. 4.6.1). This principle suggested a research program which not only entailed analytic rigor but also focused on practical applicability. As far as the CLT is concerned, Chebyshev's demand was only fully satisfied by the work of Berry and Esseen during the 1940s (see Sect. 5.2.8.2).[2]

Chebyshev's disciples Lyapunov and Markov played a decisive role in the development of early modern probability theory after 1900, in pursuing Chebyshev's research program, at least in part. Concerning Chebyshev's influence on modern probability, Khinchin [1937/2005, 41] presumed to claim that the process of "Russian theory of probability" reaching its "exceptional standing" (see also [Maistrov 1974, 208]) was "completely" thanks to Chebyshev. In a similar way, Gnedenko and Sheynin [1992, 281] have observed that "the theory of probability was shaped as a general mathematical discipline" by Chebyshev.

After a closer examination of Chebyshev's work on moments and on probability, certain distinctions arise in each of the flat observations just described. There are indications that the cooperation between teacher and disciples in Chebyshev's "St. Petersburg school" did not have the quality commonly associated with the framework of a scientific school (see Sect. 4.3.1). Chebyshev's proof of the CLT did not meet any of the criteria regarding limit theorems that Chebyshev himself had established. It even turns out that Chebyshev's work on probability cannot be seen in the light of an independent discipline. His discussion of the CLT served mainly as an illustration of certain analytic methods, in particular his method of moments. One does not observe a really substantial effort to deal with the CLT in its own right, neither within mathematics, by seeking conditions as weak as possible for the CLT, nor beyond pure analysis, with respect to stochastic applications.

In the first articles by Andrei Andreevich Markov (1856–1922)[3] on the CLT, moment theoretic intentions were clearly prominent as well. Contrary to these articles, in his courses Markov's exposition of probability theory and especially of the CLT focused mainly on applications. There, he quite frequently settled for intuitive considerations or proofs which were not entirely rigorous. From about 1905 on, Markov considerably intensified his research in probability, now apparently mainly motivated by the specifically stochastic character of its problems. This was possibly due to his rivalry with Lyapunov, who, on the basis of Poisson's methods, had presented rigorous proofs of the CLT under very weak assumptions around 1900.

[2] Kolmogorov [1947/2005, 72] has expressed the opinion that Chebyshev's main contribution in connection with the CLT was the explicit formulation of the problem of finding error bounds in the case of approximating the exact formula for a finite number of summands by the normal distribution (see also [Maistrov 1974, 207]).

[3] For scientific biographies see [Yushkevich 1970–76b] and [Grodzenskii 1987].

4.1 Chebyshev's Moment Problem

The method of moments for proving the (integral version) of the CLT plays a marginal role in modern expositions of probability theory (for example, in connection with particular cases concerning Markov chains). The reason may be that rather "technical" and complicated considerations are needed, which effort seems hardly worthwhile for the goal of proving the CLT alone. In fact, in Chebyshev's work as well as in Markov's early contributions, the CLT served primarily as an illustration of general theorems on moments and continued fractions.

Roughly speaking, the theory of moments deals with the problem of finding out about as many properties as possible of a monotonically increasing function $\mu \geq 0$, defined on the interval $[a; b]$, from the knowledge of its moments

$$M_0 := \int_{x\in[a;b]} d\mu(x),\ M_1 := \int_{x\in[a;b]} x\,d\mu(x), \ldots,\ M_n := \int_{x\in[a;b]} x^n\,d\mu(x)$$

up to a certain order n. If one additionally assumes that the function μ is continuous from the right, then $\mu(x)$ can be interpreted as the mass of the segment $[a; x]$ of a rod.

In their most general form, moment problems for mass distributions $\mu(x)$ are due to Stieltjes [1894/95], who in this context also developed the notion of the (now so-called) "Stieltjes integral." In contrast to Stieltjes's approach, Chebyshev and Markov considered functions $\mu(x) = \int_a^x f(t)dt$, where f was nonnegative. In order to be able to cope with the situation of discrete mass points as well, they allowed case-specific functions f, which today would be designated as δ-like.[4] Less frequently they also considered, instead of integrals, sums over a continuous index range, where the summands could attain nonnegative finite values as well as "infinitely small" ones.

In the summary of a lecture held at a congress of the "Association française pour l'avancement des Sciences" in Lyon, Chebyshev [1874a, 157] maintained that he had encountered those specifically new problems on moments by his reading of the article [Bienaymé 1853e]. In this article, which was written during the Cauchy–Bienaymé controversy (see Sect. 2.5.3), Bienaymé dealt specifically with (central) moments of second order as measures of precision for laws of error. Bienaymé's paper also contained the derivation (with a silly mistake) of what is now known as the Bienaymé–Chebyshev inequality in the special case of linear combinations $\sum_{i=1}^n h_i\epsilon_i$ of identically distributed observational errors ϵ_i, each having a finite number of discrete values. "His" (amended) inequality was equivalent to the following relation:

$$P\left(\left|\sum_{i=1}^n h_i\epsilon_i - \sum_{i=1}^n h_i \mathrm{E}\epsilon_1\right| \leq t\sqrt{2\sigma^2}\right) = 1 - \frac{\theta f}{2t^2}\sum_{i=1}^n h_i^2, \tag{4.1}$$

[4] Chebyshev considered discontinuous (!) integral functions with δ-like integrands [Chebyshev 1887/89, 300 f., 315], for example. (For a closer inspection of the second reference see below, Sect. 4.4.)

where θ and f designate positive values less than 1 (depending on the specific properties of the errors under consideration), and σ^2 is the abbreviation for the variance of each error of observation.

In 1867, Bienaymé's 1853 article was reprinted in the *Journal de Liouville* in immediate proximity to the French version of Chebyshev's paper "Des valeurs moyennes." In that contribution, Chebyshev had proved the weak law of large numbers for arithmetic means using the following inequality for sums of discrete and (tacitly assumed) independent random variables X_i:

$$P\left(\sum \mathrm{E}X_i - \alpha\sqrt{\sum \mathrm{E}X_i^2 - \sum(\mathrm{E}X_i)^2} \le \sum X_i \right.$$
$$\left. \le \sum \mathrm{E}X_i + \alpha\sqrt{\sum \mathrm{E}X_i^2 - \sum(\mathrm{E}X_i)^2}\right) > 1 - \frac{1}{\alpha^2}. \quad (4.2)$$

This inequality is analogous to (4.1).[5] Bienaymé's and Chebyshev's proofs for (4.1) and (4.2), respectively, were based on the same idea, which is still used for the common textbook proof of the Chebyshev–Bienaymé inequality. The crux of the proof rests on the inequalities

$$\int_{x>r} x^n dV(x) > r^n \int_{x>r} dV(x) \quad (r, n > 0) \quad \text{if } V(r) < 1,$$

where V is the distribution function of a random variable.

If Chebyshev's reference to Bienaymé in his 1874 paper is to be taken seriously, then Chebyshev must already have studied the article [Bienaymé 1853e] thoroughly before completing his 1867 paper. In an article published even later, Chebyshev [1887/89, 305] maintained that his 1867 paper had been written in the framework of more extensive research on moments (which assertion, however, can hardly be approved by the content of his publications during that time). It may be that Chebyshev, when reading Bienaymé's article, realized that Bienymé's method could also be applied to individual errors with probability density f, with the result

$$\int_{\mathrm{E}X-r}^{\mathrm{E}X+r} f(x)dx > 1 - \frac{\int_{-\infty}^{\infty} x^2 f(x)dx - \left(\int_{-\infty}^{\infty} x f(x)dx\right)^2}{r^2},$$

in accordance with (4.1). It was an obvious step from inequalities of this kind to the more general moment problem of seeking accurate upper and lower bounds for integrals $\int_a^b f(x)dx$ if $a < b$ are arbitrarily chosen from the domain of definition $[A; B]$ of the nonnegative function f and if the moments

$$M_0 := \int_A^B f(x)dx, \quad M_1 := \int_A^B x f(x)dx, \quad M_2 := \int_A^B x^2 f(x)dx, \ldots$$

[5] No one exactly knows why Bienaymé's and Chebyshev's articles were printed so close together [Heyde & Seneta 1977, 13–15, 122 f.]. The proximity of the articles to each other may indicate that the editors were aware of their common core.

$$\ldots M_m := \int_A^B x^m f(x)dx$$

are given up to a certain order m.

In contrast to these conjectures, one can also reasonably assume that Chebyshev's reference to the "authorship" of Bienaymé was primarily intended as a gesture of homage to a French colleague in a lecture before a predominantly French audience. Gauss [1823, 10–12] had already—as also noted by Pizzetti [1892, 183] and Czuber [1899, 153]—contributed to inequalities for integrals $\int_a^b f(x)dx$ (f being a probability density). Aiming at the discussion of common features among unimodal laws of error $f(x)$ with the peak at $x = 0$, Gauss proved the following inequality[6] for $m := \sqrt{\int_{-\infty}^{\infty} x^2 f(x)dx}$ and $\mu := \int_{-\lambda m}^{\lambda m} f(x)dx$ ($\lambda > 0$):

$$\lambda \leq \begin{cases} \mu\sqrt{3} & \text{if } \mu \leq \frac{2}{3} \\ \frac{2}{3\sqrt{1-\mu}} & \text{if } \mu > \frac{2}{3}. \end{cases} \tag{4.3}$$

In 1866, Anton Winckler succeeded in generalizing Gauss's result. Winckler assumed a probability density $f(x)$, where $f(x) = f(-x)$, and $f(x) = 0$ for x beyond the compact interval $[-a; a]$. He further presupposed $f(x)$ monotonically decreasing for positive x, continuous over the interval $[-a; a]$, and $f(x)$ positive for $x \in]-a; a[$. According to Winckler [1866, 19–21], with the abbreviations $k_n := \sqrt[n]{2\int_0^a |t|^n f(t)dt}$ and $y := 2\int_0^x f(t)dt$ we have

$$x \leq \begin{cases} yk_n \sqrt[n]{n+1} & \text{if } y \leq \frac{n}{n+1} \\ \frac{nk_n}{(n+1)\sqrt[n]{1-y}} & \text{if } y > \frac{n}{n+1}. \end{cases}$$

The idea that a law of error was—at least within certain limits—determined by its moments up to a certain order, may also have been suggested by the usual treatment of the asymptotic behavior of sums of independent random variables according to Laplace and Poisson. The product of integrals $\int_{-a}^{a} f_j(x)e^{x\varphi\sqrt{-1}}dx$ (f_j being the density of the jth random variable with values in $[-a; a]$), which occurred in pertinent formulae, like (2.12)), was expanded into a series of powers of φ. The coefficients of this series are proportional to the moments of the sum, which in turn are decisive for the distribution of the sum.

In 1816 already, Gauss introduced a method of estimating the parameters of a law of error from sample moments, and he also used moments of an order higher than the second. He comprehensively discussed the precision of the estimations of

[6] For proof details see [Hald 1998, 462–464]. Both expressions on the right side of (4.3) are less than $\frac{1}{\sqrt{1-\mu}}$. The inequality $\lambda < \frac{1}{\sqrt{1-\mu}}$ implies $\mu > 1 - \frac{1}{\lambda^2}$, and therefore, in the case of zero expectations, the Bienaymé inequality. At the same time, this consideration shows how rough the Bienaymé inequality is compared with Gauss's, the latter, however, valid only under additional conditions.

the parameter h in the error law $\frac{h}{\sqrt{\pi}}e^{-h^2x^2}$, assuming these estimations were based on the arithmetic mean of powers of absolute values of known errors or residuals.[7]

Around the middle of the 19th century, error theory provided several techniques for estimating parameters of probability laws from observed frequencies, and the method of moments became the most commonly used among them. Therefore, it is rather improbable that Chebyshev in his moment theoretic research was solely motivated by Bienaymé with respect to probabilistic sources. The obvious conjecture that Chebyshev came upon working on moments by his activities in mechanics cannot be confirmed by the reading of his contributions to this field [Chebyshev 1948b]. In mechanics even moments of third and fourth order occur, if as auxiliary variables only. Such moments emerge from spatial moments of inertia, that is, moments of second order, in the computational transition to systems of a lower dimension.[8] At least for Markov and Stieltjes, the mechanical interpretation of moment problems was very productive.

As we will see below, numerical integration was a further strand of development, which was very important, perhaps even decisive, for the emergence of moment theory.

In his first paper on moments, Chebyshev [1874a, 158 f.] only gave the solution of the moment problem for specific limits of integration a, b, and without proof: Let f be a positive function and $\frac{\varphi(z)}{\psi(z)}$ one of the partial fractions (strictly speaking the mth partial fraction) of the continued fraction

$$\frac{1}{\alpha_1 z + \beta_1 + \frac{1}{\alpha_2 z + \beta_2 + \frac{1}{\alpha_3 z + \beta_3 + \cdots}}},$$

assigned to the integral

$$\int_A^B \frac{f(x)}{z - x} dx.$$

If $z_1 < z_2 < \cdots < z_l < \cdots < z_n < \cdots < z_m$ are the roots of the equation $\psi(z) = 0$, we have

$$\sum_{i=l+1}^{n-1} \frac{\varphi(z_i)}{\psi'(z_i)} < \int_{z_l}^{z_n} f(x)dx < \sum_{i=l}^{n} \frac{\varphi(z_i)}{\psi'(z_i)}. \tag{4.4}$$

In this inequality moments do not occur explicitly. As we will see, however, the mth partial fraction, obtained by cutting off the continued fraction after $\frac{1}{\alpha_m z + \beta_m}$, is

---

[7] Gauss's 1816 paper has already been referred to in Sect. 3.1.

[8] For example, let the mass distribution of a disk with radius R be given by a two-dimensional density depending only on the distance from the center. Then there exists a nonnegative function f such that the moment of inertia M of the disk with respect to a perpendicular axis through the center is given by

$$M = \int_{|x| \le R} |x|^2 dm = 2\pi \int_0^R r^3 f(r) dr.$$

The moment of inertia of the disk therefore corresponds to the third-order moment of a rod with length R and a linear mass density $2\pi f(r)$.

uniquely determined by the moments M_0 to M_{2m-2} of f. Chebyshev [1874a] did not mention that the z_i are always simple roots and lie within $]A; B[$. He only made this fact explicit and proved it in a later article [Chebyshev 1887/89, 294, 299].[9]

Chebyshev also described a second moment problem: Consider a rod of which the length, mass, position of the center of gravity, and the moment of inertia with respect to an axis through the center of gravity and perpendicular to the rod are given; find estimates for the mass of any part of this rod. Chebyshev [1874a, 159 f.] presented a solution of this problem without any discussion of its derivation. If one assumes that the left endpoint of the rod of length l is the origin of a one-dimensional coordinate system, and if one designates by f the linear mass density along the rod under the condition that its total mass is equal to 1, then the center of gravity has the coordinate

$$d = \int_0^l xf(x)dx.$$

The moment of inertia with respect to an axis perpendicular to the rod and going through the center of gravity is

$$k = \int_0^l (x-d)^2 f(x)dx.$$

Chebyshev gave the following case distinction:
Case 1: $0 \leq x < d - \frac{k}{l-d}$; then it is

$$0 \leq \int_0^x f(z)dz \leq \frac{k}{(d-x)^2+k}.$$

Case 2: $d - \frac{k}{l-d} \leq x \leq d + \frac{k}{d}$;[10] then it is

$$\frac{(x-d)(l-d)+k}{lx} \leq \int_0^x f(z)dz \leq \frac{(l+d-x)(l-d)-k}{l(l-x)}.$$

Case 3: $d + \frac{k}{d} < x \leq l$; then it is

[9] To avoid confusion, possibly emerging from the fact that partial fractions of a continued fraction may exist in different equivalent forms, in the present study "mth partial fraction of the continued fraction $\frac{a_1}{b_1+\frac{a_2}{b_2+\dots}}$" means only the algebraic term $\frac{A_m}{B_m}$ which is generated by cutting off the continued fraction after b_m and by converting the cut-off term to a fraction with only one slash, without any further simplification. This procedure corresponds to the recursion formula

$$A_m = b_m A_{m-1} + a_m A_{m-2}, \quad A_0 = 0,\ A_1 = a_1$$

and

$$B_m = b_m B_{m-1} + a_m B_{m-2}, \quad B_0 = 1,\ B_1 = b_1.$$

The term A_m (B_m) resulting from the procedure just described designates the mth partial numerator (denominator) of the continued fraction. Two continued fractions are called "equivalent" if there exists a real sequence (c_m) ($c_m \neq 0$), such that for any $m \in \mathbb{N}$ for the mth partial numerator (denominator) A_m (B_m) of the one, and A'_m (B'_m) of the other, the following equations are valid: $A'_m = c_m A_m$ and $B'_m = c_m B_m$.

[10] In [Chebyshev 1874a, 160] the right limit for x is misprinted.

$$\frac{(d-x)^2}{(d-x)^2+k} \le \int_0^x f(z)dz \le 1.$$

Dealing with his moment problem, Chebyshev had apparently seen a connection to some of his articles on continued fractions. This concerns in particular the articles [Chebyshev 1854; 1855/58; 1859].[11] The original motivation for these papers was the problem of approximating a function by polynomials according to the method of least squares. In the discrete case one looks for a polynomial f with maximum degree m, such that for given real numbers $x_1, \ldots, x_n$ and $y_1, \ldots, y_n$ $(n > m+1)$ the expression $\sum_{i=1}^n [y_i - f(x_i)]^2 \Theta^2(x_i)$, where Θ^2 is a given weight function, attains its minimum. In the continuous case one has to find, for a function $y(x)$, $x \in [a;b]$, a polynomial f of a certain maximum degree, such that $\int_a^b \Theta^2(x)\,(y(x)-f(x))^2\,dx$ attains its minimum. Chebyshev succeeded in expressing the approximating polynomials by linear combinations of the partial denominators of continued fractions representing

$$\sum_{i=1}^{n} \frac{\Theta^2(x_i)}{x-x_i} \quad \text{or} \quad \int_a^b \frac{\Theta^2(x)}{z-x}dx. \tag{4.5}$$

Expansions of definite integrals by continued fractions had been applied at least since Euler's work "De fractionibus continuis observationes," which had already been read before the St. Petersburg Academy in 1739, but was not published until 1750. In this article, Euler had given, for example, a continued fraction representing the integral $\int_0^1 \frac{dx}{1+x}$. Except for the sign, this corresponds to the particular case $\Theta(x) = 1$ and $z = -1$.

A common procedure for obtaining a continued fraction representing the sum or the integral (4.5), which in principle can be found in Euler's work as well, is as follows: If one expands $\frac{1}{z-x}$ into powers of x and then integrates this series term by term—which is allowed for sufficiently large |z| if $[a, b]$ is finite—then one gets

$$\int_a^b \frac{\Theta^2(x)}{z-x}dx = \int_a^b \Theta^2(x)\sum_{i=0}^{\infty}\frac{x^i}{z^{i+1}}dx = \sum_{i=0}^{\infty}\frac{1}{z^{i+1}}\int_a^b \Theta^2(x)x^i\,dx. \tag{4.6}$$

By successively equating the partial sums and the—initially unknown—partial fractions of the continued fraction

$$\cfrac{\alpha}{b+\cfrac{\beta}{c+\frac{\gamma}{d+\ldots}}},$$

a series of powers of $\frac{1}{z}$ can be converted into a continued fraction, which is—according to modern terminology—"equivalent" to the series.[12] This had already been demonstrated by Euler [1748, 362–390] in the 18th chapter of the first volume

[11] A summary description of these papers can be found in [Vasilev 1898/1900, 17–24] and [Akhiezer 1998, 38–49].

[12] Following the standard monograph on analytic continued fractions [Perron 1913, 205], a series and a continued fraction are called "equivalent" if the nth partial sum and the nth partial fraction are identical, respectively. According to Perron, the designation "equivalent" is due to Ludwig Seidel [1855].

of his textbook *Introductio in analysin infinitorum* and illustrated by numerous examples.

Continued fractions of the form

$$\frac{1}{\alpha_1 z + \beta_1 + \frac{1}{\alpha_2 z + \beta_2 + \frac{1}{\alpha_3 z + \beta_3 + \cdots}}}, \tag{4.7}$$

which correspond, in Oskar Perron's terminology [1913, 376], to a continued fraction "associated" with (4.6), do not exist in Euler's work. Chebyshev [1855/58; 1859], however, chose exactly this type of continued fraction,[13] and he found orthogonality relations for its partial denominators (see Sect. 4.2.3), from which the coefficients α_i and β_i could be determined. In the case of a continued fraction (4.7) associated with the integral in (4.5) the nth partial denominators $\psi_n(x)$ ($n \in \mathbb{N}_0$, $\psi_0(x) = 1$) have the following property:

$$\int_a^b \psi_n(x)\psi_m(x)\Theta^2(x)dx = \frac{(-1)^n}{\alpha_{n+1}}\delta_{nm} \quad (n, m \in \mathbb{N}_0). \tag{4.8}$$

By means of this orthogonality relation and the recursion formula

$$\psi_0(x) = 1,\ \psi_1(x) = \alpha_1 x + \beta_1,\ \psi_n(x) = (\alpha_n x + \beta_n)\psi_{n-1}(x) + \psi_{n-2}(x)$$

the coefficients α_k and β_k can be determined. Regarding the latter coefficients one can additionally show that

$$\beta_n = (-1)^n \alpha_n^2 \int_a^b x\psi_{n-1}^2(x)\Theta^2(x)dx. \tag{4.9}$$

Since ψ_n is a polynomial of degree n, it follows from the formulae (4.8) and (4.9) that α_n and β_n depend on the moments $M_i = \int_a^b \Theta^2(x)x^i dx$ up to the order $2n-1$. Those formulae allow a comfortable treatment of the continued fraction (4.7), which can be determined (independent of the convergence or divergence of the related series (4.6)) if only the moments M_i of an arbitrarily high order exist.

From (4.8) and (4.9) a close relationship between continued fractions, systems of orthogonal polynomials, and moment problems becomes evident. The first and the second field were, since the fundamental work by Gauss [1814] and Jacobi [1826],

---

[13] Perron [1913, 376] termed "associated" only those continued fractions which have the particular form

$$\frac{k_1}{z + l_1 - \frac{k_2}{z + l_2 - \frac{k_3}{z + l_3 - \cdots}}}.$$

Due to the different forms used by different authors it is advisable to also consider continued fractions equivalent to the latter, in particular those of the form (4.7) with the property $\alpha_i \neq 0$, as associated with the series (4.6), if they coincide up to the term of the power $\frac{1}{z^{2n}}$ with the expansion of the nth partial fraction in powers of $\frac{1}{z}$. In exactly this sense Heine [1878, 291] expressed the relation between the continued fraction (4.7) and the power series (4.6), a relation which had already been hinted at, if less clearly, by [Chebyshev 1855/58].

closely connected with numerical integration. It may be the case that Chebyshev found the inequality (4.4) within this framework of related topics.

4.2 Quadrature Formulae, Continued Fractions, Orthogonal Polynomials, Moments

An explicit and rigorous proof of the inequality (4.4) was published by Chebyshev himself only in 1891 in connection with a more general problem (see [Vasilev 1898/1900, 35 f.; Akhiezer 1998, 57 f.]).[14] Yet by this time, mainly owing to the activities of Markov and Stieltjes, who in 1884 had almost simultaneously published their first articles containing proofs of Chebyshev's inequality, the development of moment theory had passed by Chebyshev.

Whereas Markov [1884a] directly focused on the need for a proof of Chebyshev's inequality, Stieltjes [1884a] focused on a discussion of the generalized Gaussian method of quadrature, in which he also delivered a proof of Chebyshev's inequality practically identical to Markov's. At the insistence of Markov, Stieltjes [1885b] published a notice in which the priority of the former was appreciated. Markov's paper had in fact appeared shortly before Stieltjes's. However, the latter made credible that it was impossible for him to have had any knowledge about Markov's article when submitting his own [1884a], and admitted that he had overlooked Chebyshev's original paper [1874a].

Chebyshev's inequality established a relationship between integrals and continued fractions. In the remarkably similar papers by Markov and Stieltjes of 1884, mathematical disciplines which seemed heretofore far away from each other, such as numerical integration, continued fractions, or systems of orthogonal polynomials, were combined. In fact, extensive results concerning the interconnectivity among the above-mentioned fields were achieved around 1880, as shown by the pertinent passages in the two volumes of the second edition of Eduard Heine's *Handbuch der Kugelfunctionen* [1878, 286–297; 1881, 1–31].

4.2.1 The Gaussian Procedure of Quadrature

Let us start with the Gaussian method of numerical integration. It was Gauss's main objective to find, for any given $n \in \mathbb{N}$, pairwise different nodes x_i, such that the general approximation formula

[14] As in his earlier work, Chebyshev [1891/1907] argued within a network consisting of continued fractions, orthogonal polynomials, and moments. He did not directly refer to numerical integration. In his discussion of equation systems, however, he applied methods based on continued fractions, which were analogous to those used in the determination of nodes and weights for numerical integration (see Sect. 4.2.3). Whether Chebyshev's (rather complicated) considerations correspond to his original method of finding the inequality (4.4) cannot be said.

$$\int_a^b f(x)dx \approx \sum_{i=1}^{n} A_i f(x_i), \tag{4.10}$$

$$A_i = \frac{1}{U'(x_i)} \int_a^b \frac{U(t)dt}{t - x_i}, \quad U(t) = (t - x_1)\cdots(t - x_n) \tag{4.11}$$

was exact for all $f \in \mathbb{P}_{2n-1}$.

Gauss [1814, 165] referred to work by Isaac Newton (1642–1727) and Roger Cotes (1652–1716), in which the idea of an approximate quadrature by integration of an interpolating polynomial was hinted at. Gauss [1814, 168] noted in particular the selection *Harmonia Mensurarum* of Cotes's posthumous writings [1722], whose chapter "De Methode Differentiali Newtoniana" contains quadrature formulae for up to 10 equidistant nodes [Cotes 1722, 25]. Cotes [1722, 24] in turn cited Newton's "Methodus Differentialis" [Newton 1711, 11],[15] in which the latter had given, in addition to interpolation formulae, an approximation formula with four nodes equivalent to

$$\int_a^b f(x)dx \approx (b-a)\frac{f(a) + f(b) + 3\left(f(a + \frac{b-a}{3}) + f(a + 2\frac{b-a}{3})\right)}{8}.$$

Neither Newton nor Cotes discussed numerical integration by means of interpolation from a more general point of view, and they did not consider the attainable precision. They apparently gained their formulae by the rather cumbersome integration of Newtonian interpolation polynomials. Gauss, in contrast, aimed at a theory as general as possible; he used the Lagrangian representation of interpolation polynomials, however without referring to the latter. Lagrange [1795, 286] had established a polynomial $\tilde{f}(x)$ of maximum degree $n-1$ interpolating the n points $(x_1; f(x_1))$, $(x_2; f(x_2))$, ..., $(x_n; f(x_n))$, which had the form

$$\tilde{f}(x) = f(x_1)\frac{(x-x_2)(x-x_3)\cdots(x-x_n)}{(x_1-x_2)(x_1-x_3)\cdots(x_1-x_n)} + f(x_2)\frac{(x-x_1)(x-x_3)\cdots(x-x_n)}{(x_2-x_1)(x_2-x_3)\cdots(x_2-x_n)} + \cdots + f(x_n)\frac{(x-x_1)(x-x_2)\cdots(x-x_{n-1})}{(x_n-x_1)(x_n-x_2)\cdots(x_n-x_{n-1})}.$$

This formula was also used by Gauss. The equivalent notation

$$\tilde{f}(x) = \sum_{i=1}^{n} \frac{f(x_i)U(x)}{(x-x_i)U'(x_i)}, \quad U(x) = (x-x_1)\cdots(x-x_n)$$

was, strictly speaking, only applied by Jacobi [1826, 5]. By setting $\int_a^b f(x)dx \approx \int_a^b \tilde{f}(x)dx$ one immediately obtains the relations (4.10) and (4.11). Gauss

[15] Newton's activities in interpolation can be traced back to the year 1676 [Goldstine 1977, 71]. Also in his main work, the 1687 *Philosophiae Naturalis Principia Mathematica*, Newton gave a concise survey of numerical integration (third book, lemma v, annotated German translation of this passage in [Kowalewski 1917, 79–83]).

assumed that the integrand $f(x)$ could be represented by the power series $f(x) = \sum_{j=0}^{\infty} K_j x^j$, and that the nodes $x_1, \ldots, x_n$ lie within the domain of integration. The difference

$$R_n = \int_a^b f(x)dx - \sum_{i=1}^{n} A_i f(x_i)$$

between the integral and the approximation formula, where A_i is as in (4.11), has the form

$$R_n = \sum_{l=0}^{\infty} k_l K_l,$$

where

$$k_l = \int_a^b x^l dx - \sum_{i=1}^{n} A_i x_i^l.$$

Independent of the particular choice of x_i we have $k_0 = k_1 = \cdots = k_{n-1} = 0$, due to the fact that the approximation formula (4.10) is exact for all polynomials with a maximum degree of $n-1$, especially for all monomials x^l $(l = 0, 1, \ldots, n-1)$ [Gauss 1814, 169].[16] If one additionally demands that $k_n = \cdots = k_{2n-1} = 0$, then one gains n equations

$$\int_a^b x^n dx = \sum_{i=1}^{n} A_i x_i^n, \ldots, \int_a^b x^{2n-1} dx = \sum_{i=1}^{n} A_i x_i^{2n-1} \tag{4.12}$$

for the unknowns $x_1, \ldots, x_n$. As Gauss [1814, 183] argued, it was "easy to infer" that nodes $x_1, \ldots, x_n \in [a; b]$ could "always" be found such that (4.12) was valid, and therefore the approximation formula (4.10) was exact for all polynomials f with $\deg(f) \leq 2n-1$.

In his discussion of the procedure of determining the nodes $x_1, \ldots, x_n$, Gauss [1814] restricted himself to the special case $[a; b] = [-1; 1]$. By rather indirect arguments[17] he showed that, for this particular domain of integration, the nodes could be obtained as the roots of the nth partial denominator W_n of the continued fraction associated with

$$\varphi(t) := t^{-1} + \frac{1}{3}t^{-3} + \frac{1}{5}t^{-5} + \cdots = \frac{1}{2}\int_{-1}^{1} \frac{dx}{t-x}.$$

He also proved that

$$W_n(t) = t^n \mathrm{F}(-\frac{1}{2}(n-1), -\frac{1}{2}n, -(n-\frac{1}{2}), t^{-2}),$$

where F was "his" hypergeometric function.[18]

[16] If f is a polynomial of degree $\leq n-1$, then $\tilde{f} = f$.

[17] For descriptions of Gauss's 1814 paper see [Fuchs 1973; Goldstine 1977, 224–232].

[18] The hypergeometric function (see [Gauss 1813]) is defined by

Gauss in his 1814 paper did not comment on the fact that the partial denominators W_n were identical with the Legendre polynomials of degree n related to the interval $[-1; 1]$. The strong connection between the nodes in Gauss's procedure of numerical integration and Legendre functions was explained by Carl Gustav Jacobi [1826; 1827] only later.[19]

In [1826], Jacobi considered polynomials f of degree $n+p$ and the associated interpolation polynomial $\tilde{f}$ of degree $n-1$, such that, for the given interpolation points $x_1, \dots, x_n \in [0; 1]$:

$$\tilde{f}(x) = \sum_{i=1}^{n} \frac{f(x_i)U(x)}{U'(x_i)(x-x_i)}, \quad U \text{ as above.}$$

We have $\int_0^1 \tilde{f}(x)dx = \sum_{i=1}^n A_i \tilde{f}(x_i) = \sum_{i=1}^n A_i f(x_i)$, where A_i is determined according to (4.11). Basing himself on Gauss's fundamental ideas, Jacobi sought the maximal number p such that, presupposing an optimal choice of the nodes $x_1, \dots, x_n$:

$$\int_0^1 f(x)dx - \int_0^1 \tilde{f}(x)dx = 0 \quad \forall f \in \mathbb{P}_{n+p}.$$

Jacobi showed that this condition was met if and only if for

$$U(x) = (x-x_1)\cdots(x-x_n): \quad \int_0^1 x^k U(x)dx = 0 \quad (k = 0, 1, \dots, p). \tag{4.13}$$

The maximum number p which is according to this condition for an appropriate $U = \tilde{U}$ is $p = n-1$. Jacobi [1826, 8] succeeded in showing, on the basis of (4.13), that

$$\tilde{U}(x) = \frac{1}{2n(2n-1)(2n-2)\cdots(n+1)} \frac{\mathrm{d}^n x^n (x-1)^n}{\mathrm{d}x^n}. \tag{4.14}$$

From this result one could, as Jacobi [1826, 8] noticed, infer "by the doctrine of equations" that $\tilde{U}$ had n real roots between 0 and 1 (and that these roots were different from each other, which fact, however, was not made explicit by Jacobi).

In the subsequent volume of Crelle's Journal, Jacobi [1827] published a note on Legendre polynomials $X_{(i)}$ belonging to the interval $[-1; 1]$. These polynomials had been introduced by Legendre in 1784 as "coefficients" in the expansion

$$\frac{1}{\sqrt{1-2xz+z^2}} = 1 + X_{(1)}(x)z + X_{(2)}(x)z^2 + \cdots.$$

Legendre had subsequently proven several properties of these polynomials, which he summarized in the second volume of his *Exercises du calcul intégral* [1817]. In particular, he succeeded in showing that all roots of the polynomials named after him are single roots and lie within the interval $]-1; 1[$. He further showed that these polynomials are orthogonal to each other, in the sense of

---

$$\mathrm{F}(\alpha, \beta, \gamma, z) := 1 + \frac{\alpha\beta}{1\cdot\gamma}z + \frac{\alpha(\alpha+1)\beta(\beta+1)}{1\cdot 2\gamma(\gamma+1)}z^2 + \cdots.$$

[19] For Jacobi's respective work see [Goldstine 1977, 261–264], [Gautschi 1981, 78 f.], and [Heine 1881, 13 f.].

$$\int_{-1}^{1} X_{(m)} X_{(n)} dx = \frac{2}{2n+1} \delta_{mn}.$$

From this latter relation (which is analogous to (4.13)) Jacobi [1827] deduced the formula

$$X_{(n)}(x) = \frac{1}{2^n} \frac{1}{1 \cdot 2 \cdot 3 \cdot n} \frac{\mathrm{d}^n (x^2 - 1)^n}{\mathrm{d}x^n},^{20}$$

which is analogous to (4.14). Thus, Jacobi had shown that Gauss's polynomials W_n were identical to the Legendre polynomials.

4.2.2 *Generalizations of Gauss's Quadrature Formula, Systems of Orthogonal Polynomials*

Jacobi died in 1851. An extensive article was found among his private papers, apparently ready for press. This article was published under the title "Untersuchungen über die Differentialgleichung der hypergeometrischen Reihe" ("Studies on the Differential Equation of the Hypergeometric Series") in 1859. Jacobi generalized results obtained through his studies of Gauss's procedure of numerical integration. In particular, Jacobi aimed at the characterization of finite hypergeometric series (see footnote 18), Legendre polynomials being special cases of them, by their derivatives, and at a representation of these series by partial denominators of certain continued fractions. For $X_n := \mathrm{F}(-n, \alpha + n, \gamma, x)$, where $n \in \mathbb{N}_0$, $\gamma > 0$, and $\alpha > \gamma - 1$, Jacobi [1859, 194 f.] established the orthogonality relation

$$\int_0^1 X_m X_n x^{\gamma-1} (1-x)^{\alpha-\gamma} dx = 0 \quad (m \neq n).$$

His main result [1859, 196 f.] was that the $2n$th and $(2n+1)$th denominators of the continued fraction

$$\cfrac{A}{x - \cfrac{a}{1 - \cfrac{b}{x - \cfrac{c}{1 - \mathrm{etc}}}}}, \tag{4.15}$$

representing the integral

$$\int_0^1 \frac{t^{\gamma-1}(1-t)^{\alpha-\gamma}}{t-x} dt \quad (\gamma > 0, \alpha > \gamma - 1),^{21}$$

are proportional to $\mathrm{F}(-n, \alpha+n, \gamma, x)$ and $x\mathrm{F}(-n, \alpha+n+1, \gamma+1, x)$, respectively.

---

[20] As Heine [1878, 20 f.] pointed out, a certain Rodrigues (most likely Olinde Rodrigues) has to be credited for this formula, in an article from 1816 (not further specified by Heine, probably the article which in the bibliography is cited as [Rodrigues 1816]).

[21] The continued fraction (4.15) is connected with the integral in the sense that the series $\sum_{i=0}^{\infty} \frac{M_k}{x^{k+1}}$ (representing the integral) coincides with the expansion of the nth partial fraction up to the term $\frac{M_{n-1}}{x^n}$. In Perron's terminology [1913, 375], continued fractions of this type are called "regular" and "corresponding" to the integral.

Jacobi [1859] did not make it explicit that, because of the perfect analogy between the more general situation considered by himself, and the special situation $\alpha = \gamma = 1$ studied by Gauss [1814], the zeros of X_n could serve as nodes for the quadrature formulae

$$\int_0^1 g(x)x^{\gamma-1}(1-x)^{\alpha-\gamma}dx = \sum_{i=1}^{n} A_i g(x_i) \quad \forall g \in \mathbb{P}_{2n-1}.$$

This fact, however, was noted shortly after, and thus Jacobi's 1859 paper became a starting point for the discussion of generalized "Gaussian" quadrature formulae

$$\int_a^b f(x)g(x)dx \approx \sum_{i=1}^{n} A_i g(x_i), \tag{4.16}$$

where f was an arbitrary nonnegative "weight function." This discussion also included continued fractions which were associated with, or regular and corresponding to, the integrals $\int_a^b \frac{f(t)}{x-t}dt$.[22] Chebyshev [1859] was likewise an initiator of the general discussion of continued fractions of this type, but he never broached the subject of their relationship to Gaussian quadrature (see Sect. 4.2.3). This relationship was a theme of Ferdinand Gustav Mehler [1864], who, in a manner similar to Jacobi, considered integrals of the form $\int_{-1}^{1} f(x)g(x)dx$, where $f(x) = (1-x)^{\lambda}(1+x)^{\mu}$ $(\lambda, \mu > -1)$, but, unlike Jacobi, also discussed specific problems of numerical integration, such as the determination of the weights A_i or the approximation error connected with the use of (4.16).

A variety of further works followed, among which [Heine 1866], [Possé 1875], and [Christoffel 1877] were particularly important and influential.[23] The major results of these works, which were also summarized in the two volumes of Heine's *Handbuch der Kugelfunctionen* (*Compendium of Spherical Functions*) [Heine 1878, 286–297; 1881, 1–31] are:

— If $\frac{V_n(x)}{W_n(x)}$ designates the nth partial fraction of a continued fraction associated with $\int_a^b \frac{f(t)}{x-t}dt$, where f is nonnegative and $\int_a^b f(x)dx$ is positive, then for all polynomials s with $\deg(s) \leq 2n-1$ the following generalized Gaussian quadrature formula is valid:

$$\int_a^b f(x)s(x)dx = \sum_{i=1}^{n} s(x_i)A_i,$$

$$\text{where } A_i = \int_a^b f(x)\frac{W_n(x)}{(x-x_i)W_n'(x_i)}dx. \tag{4.17}$$

---

[22] The $2n$th partial fraction of a continued fraction which is regular and corresponding to an integral of this type is equal to the nth partial fraction of that continued fraction which is associated with this integral.

[23] The present author follows [Gautschi 1981, 79–84] and [Brezinski 1991, 214–217] in this assessment.

— W_n is proportional to that polynomial of degree n which corresponds to the orthonormal system of polynomials with respect to the inner product $\int_a^b (\cdot)(\cdot) f(x)dx$,[24] and the nodes x_i are the zeros of W_n, all real and different from each other, and located within $]a;b[$.
— V_n can be expressed in terms of W_n by

$$V_n(x) = \int_a^b \frac{W_n(x) - W_n(t)}{x - t} f(t)dt. \tag{4.18}$$

Because $W_n(x_i) = 0$, one can directly infer from (4.17) and (4.18):

$$A_i = \frac{V_n(x_i)}{W_n'(x_i)}. \tag{4.19}$$

On account of the (in cases only formal) relation

$$\int_a^b \frac{f(t)dt}{x - t} = \sum_{i=0}^{\infty} \frac{M_i}{x^{i+1}},$$

the moments $M_i = \int_a^b f(x)x^i dx$ play the role of auxiliary quantities, from which the coefficients of the continued fractions associated with $\int_a^b \frac{f(t)dt}{x-t}$ and (due to (4.19)) the weights A_i in the (generalized) Gaussian quadrature formula can be derived. It was Chebyshev's merit to formulate a research program in which moments should play a leading part in the discussion of properties of the function f. As it seems, Chebyshev's motivation for this research at least partly arose from probability theory; it is not surprising, however, that in his discussion of moment problems analytic devices were adopted from the theory of continued fractions, to which he had significantly contributed. In these contributions he had anticipated—but not completely justified—some of the results just described, in particular with regard to systems of orthogonal polynomials. Chebyshev's work on continued fractions was motivated by least squares approximation rather than by numerical integration, however.

4.2.3 Chebyshev's Contributions

Chebyshev is commonly considered one of the most important contributors to 19th-century approximation theory (see, for example, [Steffens 2006, Chapt. 2]). One of his most influential works in this field appeared in 1855, and it was also published in a French translation (which had been made by Bienaymé) in 1858 under the title "Sur les fractions continues." This article was especially distinguished by its innovative combination of approximation problems and continued fractions, and by the significance of its results regarding orthogonal polynomials.

---

[24] These polynomials can be considered as generalized Legendre polynomials with respect to the interval $[a;b]$ and the weight function f.

Chebyshev considered the values $y_0, \dots, y_n$ of a polynomial $F(x) = \sum_{k=0}^{m} a_k x^k$ as being obtained by observation such that, for given arguments $x_0, \dots, x_n$, one has

$$y_i = F(x_i) + \epsilon_i \quad (i = 1, \dots, n),$$

where ϵ_i are the observation errors. He posed the problem to find, for any natural $m < n$, a polynomial $\overline{F(X)}$ of a degree less than or equal to m such that

$$\sum_{i=0}^{n} (\overline{F(x_i)} - y_i)^2 \Theta^2(x_i) \le \sum_{i=0}^{n} (F(x_i) - y_i)^2 \Theta^2(x_i) \quad \forall F \in \mathbb{P}_m, \tag{4.20}$$

where Θ^2 was an arbitrary weight function. This latter function Chebyshev interpreted in the sense that $k_i^2 := \frac{1}{\Theta^2(x_i)}$ was the variance of the error ϵ_i. He showed that[25]

$$\overline{F(X)} = \sum_{k=0}^{m} (-1)^k A_{k+1} \psi_k(X) \sum_{i=0}^{n} \psi_k(x_i) \Theta^2(x_i) y_i. \tag{4.21}$$

In this formula ψ_k ($\psi_0 = 1$) designates the kth partial denominator of the continued fraction

$$\cfrac{1}{q_1 + \cfrac{1}{q_2 + \cdots}}$$

associated with $\sum_{i=0}^{n} \dfrac{\Theta^2(x_i)}{x - x_i}$, where $q_1, q_2, \dots$ are linear functions of the form $q_k = A_k x + B_k$. Setting $y_i := \psi_m(x_i)$ $(i = 1, \dots, n)$—in this case we have $\overline{F(X)} = \psi_m(X)$—Chebyshev [1855/58, 222] from (4.21) deduced the orthogonality relations

$$\sum_{i=0}^{n} \psi_m(x_i) \psi_l(x_i) \Theta^2(x_i) = \frac{(-1)^m}{A_{m+1}} \delta_{ml} \quad (l, m = 0, 1, \dots, n). \tag{4.22}$$

The basic ideas of Chebyshev's proof were as follows:[26] $\overline{F(X)}$ was represented by the general ansatz

$$\overline{F(X)} = \lambda_0(X) y_0 + \lambda_1(X) y_1 + \cdots + \lambda_n(X) y_n$$

with

$$\sum_{i=0}^{n} \lambda_i(X) x_i^k = X^k \quad (k = 0, \dots, m). \tag{4.23}$$

The least squares condition (4.20) implied

$$k_0^2 \lambda_0(X)^2 + k_1^2 \lambda_1(X)^2 + \cdots + k_n^2 \lambda_n(X)^2 = \min. \tag{4.24}$$

Chebyshev [1855/58, 207] succeeded in expressing the conditions (4.23) and (4.24) on λ_i by a system of equations

---

[25] Chebyshev had already communicated this result for the special case $\Theta^2 \equiv 1$ in a note [1854].

[26] For details of the proof see [Hald 1998, 528–531; Akhiezer 1998, 38–45; Steffens 2006, 52–54].

$$X^k = \sum_{i=0}^{n} \Theta^2(x_i)\varphi_X(x_i)x_i^k, \quad k = 0, \dots, m, \tag{4.25}$$

where $\varphi_X(x)$ denoted a polynomial of degree m, such that

$$\lambda_i(X) = \Theta^2(x_i)\varphi_X(x_i).$$

With respect to the form, the equation system (4.25) was analogous to the system

$$\int_a^b x^l dx = \sum_{i=1}^{n} A_i x_i^l \quad (l = 0, 1, \dots, 2n-1)$$

connecting the nodes x_i and weights A_i in Gauss's method of numerical integration. Actually, in discussing equation systems of this type by means of continued fractions, Chebyshev chose an approach similar to that of Gauss's. The basic idea [Chebyshev 1855/58, 208] was that φ_X solves the system (4.25) if and only if in the expansion of

$$\sum_{i=0}^{n} \frac{\Theta^2(x_i)\varphi_X(x_i)}{x - x_i} - \frac{1}{x - X}$$

into a series of powers of $\frac{1}{x}$ the terms $\frac{1}{x}, \dots, \frac{1}{x^{m+1}}$ do not occur. This latter condition in turn implied further conditions on the continued fraction associated with $\sum_{i=0}^{n} \frac{\Theta^2(x_i)}{x - x_i}$. In this way, Chebyshev reduced the approximation problem (4.20) to an algebraic problem which also referred to continued fractions. That mode of reasoning was characteristic of a considerable part of Chebyshev's work.

Around 1859 Chebyshev extended the discussion of partial denominators of continued fractions from those associated with sums to those associated with integrals, in particular with regard to orthogonality. In his publication "Sur le développement des fonctions à une seule variable," Chebyshev [1859], in a manner characteristic of his methods for solving problems of this kind, simply used the analogies between the properties of "discrete" sums $\sum_{i=1}^{n} \frac{\Theta^2(x_i)}{x - x_i}$—as treated in his 1855/58 paper—and integrals $\int_a^b \frac{f(z)}{x-z} dz$ without giving any specific arguments. According to Chebyshev, all results which had been derived for continued fractions associated with $\sum_{i=1}^{n} \frac{\Theta^2(x_i)}{x - x_i}$, had likewise to be valid in connection with $\int_a^b \frac{\Theta^2(z)}{x-z} dz$, even for $a = -\infty$ or $b = \infty$. In this way, Chebyshev discussed the cases $\Theta^2(z) = \frac{1}{\sqrt{1-z^2}}$, $a = -b = 1$ (the partial denominators are proportional to polynomials $\cos n \arccos x$, the now so-called "Chebyshev polynomials"), $\Theta^2(z) = 1$, $a = -b = 1$ (the partial denominators are proportional to the Legendre polynomials).[27] He comprehensively discussed $\Theta^2(x) = k\mathrm{e}^{-x}$, $a = 0$, $b = \infty$ (the partial denominators are proportional to the now so-called "Laguerre polynomials" [Laguerre 1879]), and, especially important for later probabilistic applications, $\Theta^2(x) = \sqrt{\frac{k}{\pi}}\mathrm{e}^{-kx^2}$, $a = -b = \infty$ (the partial denominators are proportional to

[27] Chebyshev [1854, 702] had already hinted at this fact.

the now so-called "Hermite polynomials" [Hermite 1864]). As Chebyshev [1859, 504] argued, it was "easy to assure oneself" that in the latter case the continued fraction was equivalent to one whose partial denominators were given by

$$\psi_l(x) = \mathrm{e}^{kx^2} \frac{\mathrm{d}^l \mathrm{e}^{-kx^2}}{\mathrm{d}x^l}.$$

In analogy to the discrete case (in particular with respect to (4.21) and (4.22)), Chebyshev [1859, 503] also inferred that

$$g(x) = \frac{\int_{-\infty}^{\infty} \sqrt{\frac{k}{\pi}} \mathrm{e}^{-kx^2} \psi_0(x) F(x) dx}{\int_{-\infty}^{\infty} \sqrt{\frac{k}{\pi}} \mathrm{e}^{-kx^2} \psi_0^2(x) dx} + \frac{\int_{-\infty}^{\infty} \sqrt{\frac{k}{\pi}} \mathrm{e}^{-kx^2} \psi_1(x) F(x) dx}{\int_{-\infty}^{\infty} \sqrt{\frac{k}{\pi}} \mathrm{e}^{-kx^2} \psi_1^2(x) dx} + \cdots \quad (4.26)$$

provided a polynomial approximation to the given function $F(x)$ such that the "error to be expected" would become minimal if the abscissae x obeyed the error law $n(x) = \sqrt{\frac{k}{\pi}} \mathrm{e}^{-kx^2}$. This somewhat obscure statement is likely to be interpreted (in modern terminology) as follows: Let X be a normally distributed random variable, F a sufficiently smooth function defined in $\mathbb{R}$. Then we have for a polynomial $g \in \mathbb{P}_r$ defined in accordance with (4.26):

$$\mathrm{E}(g(X) - F(X))^2 \leq \mathrm{E}(f(X) - F(X))^2 \quad \forall f \in \mathbb{P}_r.$$

In contrast to other authors who worked on orthogonal polynomials, Chebyshev did not start with Gaussian quadrature but with least squares approximation in its probabilistic interpretation. However, with his analytic methods, chiefly regarding the use of continued fractions, he also came close to numerical integration, the more so as he [1874b] also published on that topic.

4.3 Moment Problems Around 1884: Markov and Stieltjes

4.3.1 Markov's Early Work on Moments

The first published proof of Chebyshev's inequality (4.4) is due to Markov, who in 1884 derived the two inequalities for $m, n, l \in \mathbb{N}$, $2 \leq l \leq m$, $1 \leq n < m$:

$$\int_A^{z_{l-1}} f(z) dz < \sum_{i=1}^{l-1} \frac{\varphi(z_i)}{\psi'(z_i)} \quad (4.27)$$

and

$$\int_{z_{n+1}}^{B} f(z) dz < \sum_{i=n+1}^{m} \frac{\varphi(z_i)}{\psi'(z_i)}, \quad (4.28)$$

under the assumptions which have already been specified in Sect. 4.1.[28] As Markov [1884a, 174] showed, from these relations Chebyshev's inequality can be easily derived. To prove the inequality (4.27), Markov [1884a, 175–177] on the basis of well-known properties of the partial fractions of the continued fraction associated with $\int_A^B \frac{f(x)}{z-x}dx$, constructed a polynomial Φ of degree $2m-2$, such that

$$\int_A^B \Phi(x)f(x)dx = \sum_{i=1}^{m} \Phi(z_i)\frac{\varphi(z_i)}{\psi'(z_i)}, \tag{4.29}$$

$$\begin{aligned} \Phi(z_1) = \Phi(z_2) = \cdots = \Phi(z_{l-1}) = 1; \\ \Phi(z_l) = \Phi(z_{l+1}) = \cdots = \Phi(z_m) = 0, \end{aligned} \tag{4.30}$$

$$\forall z \le z_{l-1} : \Phi(z) \ge 1; \qquad \forall z \in [A;B] : \Phi(z) \ge 0. \tag{4.31}$$

From (4.31) it follows that

$$\int_A^{z_{l-1}} f(x)dx < \int_A^{z_{l-1}} f(x)\Phi(x)dx < \int_A^B f(x)\Phi(x)dx.$$

Taking into account (4.29) and (4.30), (4.27) can be derived from the latter relation. The second inequality (4.28) was analogously proven by Markov [1884a, 177 f.], through the use of an appropriate auxiliary polynomial.[29]

According to the principles of Gaussian quadrature, in particular (4.19), the fractions $\frac{\varphi(z_i)}{\psi'(z_i)}$ are equal to the weights A_i in $\int_A^B p(x)f(x)dx = \sum_{i=1}^m p(z_i)A_i$. This equation is exact for all polynomials p of degree $\le 2m-1$. Thus, (4.29) is "automatically" valid because $\Phi(x)$ is a polynomial of degree $2m-2$. $\Phi(x)$ interpolates the function

$$g(x) := \begin{cases} 1 \text{ if } x \in [A; z_{l-1}] \\ 0 \text{ else} \end{cases}$$

in $x = z_1, \ldots, z_m$. Therefore the (approximate) equalities

$$\int_A^{z_{l-1}} f(x)dx = \int_A^B g(x)f(x)dx \approx \int_A^B \Phi(x)f(x)dx = \sum_{i=1}^{l-1} A_i$$

suggest themselves. From this consideration the idea of estimating the partial integral $\int_A^{z_{l-1}} f(x)dx$ by a partial sum of the weights arises. Therefore, most probably, Markov's idea of proof was based on a consequent interpretation of Chebyshev's inequality in the sense of Gaussian quadrature.

Markov [1884a, 179 f.] also proved Chebyshev's estimates concerning the problem of mass distribution along a rod. Basically, he used the idea of obtaining an

---

[28] f is assumed to be positive on $]A;B[$, $\frac{\varphi(z)}{\psi(z)}$ designates the mth partial fraction of the continued fraction (4.7) associated with the integral $\int_A^B \frac{f(x)}{z-x}dx$, and the z_i designate the roots of the equation $\psi(z) = 0$.

[29] For further details of the proof see [Akhiezer 1998, 55 f.].

upper and lower bound, respectively, for the "partial mass" $\int_0^x f(s)ds$ by sums of discrete masses concentrated at certain points on the rod, such that the moments of zeroth, first, and second order of this discrete mass distribution coincide with the corresponding moments related to the density f.

In 1884, Markov held an instructor position at St. Petersburg University as subordinate to Chebyshev. Therefore, it seems somewhat strange that Markov [1884a, 178] explicitly thanked Konstantin Aleksandrovich Possé (a former disciple of Chebyshev) for suggestions, but did not consider Chebyshev in his acknowledgment. Actually, one would assume that Chebyshev, who at least had developed basic ideas for a proof of his inequality, would have discussed this issue with his disciples. As a fragmentary manuscript written by Markov in 1921 and recently edited by Sheynin [2004a, 111–122] reveals, the cooperation between Chebyshev and his disciples of the "St. Petersburg school" apparently did not correspond to today's understanding of this term. Markov reports on Chebyshev's 1874 note, in which the inequality (4.4) had been presented:

> This note had gone unnoticed until the beginning of the 1880s when it caused a lively exchange of opinions among Petersburg scientists, and mostly between me and my respected teacher, K.A. Possé (...) Professor Korkin even doubted that the Chebyshev inequality was valid. Finally, in 1883, I was able not only to prove it, but to solve his problem [the problem of estimating the mass distribution of a rod from its moments up to the second order].

Markov's words clearly indicate that Chebyshev did not maintain any significant scientific contact with his disciples at least after 1882, when he had retired from lecturing. As an apparent consequence, Markov designates Possé his teacher, not Chebyshev.[30] He also mentions Aleksandr Nikolaevich Korkin, who lectured on partial differential equations and variational calculus at St. Petersburg University and organized the so-called "Korkin Saturdays," when members of the St. Petersburg school would meet at his home and discuss current mathematical questions [Steffens 2006, 83]. So, we can assume that the members of the school actually had an active scientific exchange; Chebyshev, as the master, however lived in his own world being quite isolated from the rest of his school.

Markov remained very interested in the theory of moments during the last decades of the 19th century, especially in regards to further generalizations of inequalities like Chebyshev's (4.4). In his doctoral thesis [1884b] (of which the paper [1884a] just described represented only a small part) Markov had already succeeded in deriving formulae—again by means of continued fractions—for the "maximum and minimum value" of integrals $\int_0^x f(y)\Omega(y)dy$,[31] if $f : [0;l] \to \mathbb{R}$ was a nonnegative function with positive moments $\int_0^l x^k f(x)dx$ (from $k = 0$ up to a certain order $k = n$), and Ω was in $C^{n+1}([0;l])$, such that

$$\Omega(x) \geq 0, \Omega'(x) \geq 0, \Omega''(x) \geq 0, \ldots, \Omega^{(n+1)}(x) \geq 0 \quad \forall x \in [0;l].$$

30 Chebyshev is commonly considered as Markov's doctoral advisor, see Mathematics Genealogy Project http://www.genealogy.math.ndsu.nodak.edu/, for example.

31 For further details see (aside from the Russian original of Markov's thesis): [Markov 1886; Possé 1886, 90–136; Akhiezer 1998, 64–69].

Markov even considerably generalized these problems, by introducing an extension of the notion of moment.[32]

4.3.2 Stieltjes's Early Work on Moments

Thomas Jan Stieltjes (1856–1894)[33] was a versatile mathematician, who contributed to almost all branches of analysis. He became best known through his work on the analytic theory of continued fractions, and, in this context, on moment problems. He started exploring these topics with a discussion of Gaussian quadrature by means of continued fractions.

In 1883 he had already shown that, given an arbitrary positive weight function f, the Gaussian quadrature formula for a certain class of functions g converges to the integral $\int_a^b g(x)f(x)dx$ if the number of nodes grows [Stieltjes 1883, 314–316]. The particular functions g Stieltjes considered were uniformly convergent series in Legendre polynomials assigned to the finite interval $[a, b]$.[34] In close connection to this proof was the analysis of the properties of the coefficients α_i and λ_k of the continued fraction

$$\cfrac{M_0}{x-\alpha_0-\cfrac{\lambda_1}{x-\alpha_1-\cfrac{\lambda_2}{x-\alpha_2-\cfrac{\lambda_3}{x-\alpha_3-\cdots}}}}$$

associated with

$$\int_a^b \frac{f(z)}{x-z}dz = \frac{M_0}{x} + \frac{M_1}{x^2} + \frac{M_2}{x^3} + \cdots.$$

Under the assumption that $f(x) > 0$ for all $x \in]a; b[$, Stieltjes proved that $\lambda_k > 0$ for all $k \in \mathbb{N}$ and $\alpha_i \in]a; b[$ for all $i \in \mathbb{N}_0$.

After a short break, Stieltjes resumed his studies on "mechanical quadrature" in 1884. Regarding the main results of numerical integration, he [1884a, 378] referred to Heine's [1878; 1881] monograph, and, by means of the same auxiliary polynomial that Markov had used, he proved inequalities which were—as reclaimed by Markov shortly after—equivalent to (4.27) and (4.28).[35] Let P_n be the generalized Legendre polynomial of degree n related to the weight function f and to the interval $[a; b]$, let $x_1, \dots, x_n$ be the zeros of this polynomial, and let $A_1, \dots, A_n$ be weights such that $A_i = \int_a^b f(x)\frac{P_n(x)}{(x-x_i)P_n'(x_i)}dx$. Then, as Stieltjes [1884a, 384–388] proved, the inequalities

[32] See [Krein 1951/59] for a concise survey of this work by Markov.

[33] For biographical details see [Bernkopf 1970–76].

[34] Legendre polynomials $\overline{X}_{(m)}$ assigned to an interval $[a; b]$ are connected with the "usual" Legendre polynomials $X_{(m)}$ assigned to the interval $[-1; 1]$ by the formula $\overline{X}_{(m)}(x) = X_{(m)}\left(\frac{2x-a-b}{b-a}\right)$.

[35] Stieltjes assumed a nonnegative weight function f being above a certain positive lower bound on an arbitrarily small interval. Thus, his condition on the weight function was more general than Chebyshev's and Markov's.

$$\sum_{i=1}^{k} A_i > \int_a^{x_k} f(x)dx \quad (1 \leq k \leq n) \tag{4.32}$$

and

$$\sum_{i=1}^{k} A_i < \int_a^{x_{k+1}} f(x)dx \quad (1 \leq k \leq n-1) \tag{4.33}$$

hold. If one takes into consideration that $A_i = \frac{V_n(x_i)}{W_n'(x_i)}$, where $\frac{V_n(x)}{W_n(x)}$ is the nth partial fraction of a continued fraction associated with $\int_a^b \frac{f(z)}{x-z} dz$, then the equivalence of Stieltjes's inequalities (4.32) and (4.33) to the Chebyshev–Markov inequalities (4.27) and (4.28) is an immediate consequence. This equivalence would have certainly been a matter of course for Stieltjes if he had known the respective works by Chebyshev and Markov.

Stieltjes [1884a, 392–394] used "his" inequalities for proving that the (generalized) Gaussian quadrature formula approximating the integral $\int_a^b f(x)F(x)dx$ (f being the weight function) actually converges to this integral as $n \to \infty$,[36] if F is bounded and integrable, and f additionally meets the condition

$$\forall \alpha, \beta \in [a; b],\ \alpha < \beta : \quad \int_\alpha^\beta f(x)dx > 0.$$

In the proof, Stieltjes [1884a, 389–392] made essential use of the following property of generalized Legendre polynomials P_k related to the interval $[a; b]$ and to the weight function f: If $[\alpha; \beta]$ is an arbitrarily small partial interval of $[a; b]$, such that $\int_\alpha^\beta f(x)dx > 0$, then, for all sufficiently large natural n, the polynomials P_n have at least one zero within $[\alpha; \beta]$. An analogous property of the zeros of Hermite polynomials (with range of integration $]-\infty; \infty[$ and weight function $\frac{1}{\sqrt{\pi}} e^{-x^2}$) would, together with the Chebyshev–Markov inequalities, form the basis for Markov's rigorous proof of the CLT by means of moment methods in [1898].

As we have already seen, Stieltjes was interested from the beginning of his work in relations between numerical integration and continued fractions. In his article [1884a], such relations were not made explicit. In a subsequent note, however, Stieltjes [1884b] removed this deficit. In his proof that a continued fraction associated with $\int_a^b \frac{f(x)}{z-x} dx$ (f as in [1884a], $[a; b]$ finite) converges to this integral for all $z \in \mathbb{C} \setminus [a; b]$, he applied his inequalities (4.32), (4.33) for discussing the position of the zeros of the partial denominators.

Up to this point, Stieltjes had not explicitly dealt with moment problems. In his subsequent papers, however, he steered attention toward this topic. In the note [1884d] he communicated (without proof) inequalities like

$$A_1 + \cdots + A_n \leq \int_0^1 f(x)dx,$$

[36] The question of convergence of the Gaussian quadrature formula to the integral had already been treated by Heine [1881, 16–19]. He only dealt with the particular case that $f(x) = 1$ and F is represented by a power series.

f being a nonnegative function, where the A_k are solutions—together with x_k—of systems

$$\int_0^1 x^{\lambda_i} f(x)dx = \sum_{k=1}^{n} A_k x_k^{\lambda_i} \quad (i = 0, \ldots, 2n-1), \tag{4.34}$$

λ_i designating different nonnegative (integer?) numbers. In the particular case $\lambda_i = i$ this equation system is, as Stieltjes hinted at, identical with the $2n$ equations for determining nodes and weights for Gauss's method of numerical integration. He also interpreted the system (4.34) in the sense that the moments $\int_0^1 x^{\lambda_i} f(x)dx$ of the density f are equal to the moments $\sum_{k=1}^{n} A_k x_k^{\lambda_i}$ of that discrete mass distribution, for which the masses A_k are placed at the points x_k. The same interpretation was employed by Markov [1886] (see Sect. 4.3.1) in an only slightly different context. Nobody, it seems, used this manifestly physical interpretation before Markov and Stieltjes, not even in the case of Gaussian quadrature ($\lambda_i = i$). The designation "weight" for A_k became commonly used only around the end of the 19th century, and was apparently due to a conception of the integral as the weighted sum of certain single values of the integrand. The phrase "mechanical integration" for numerical integration, which was very familiar during the 18th and 19th centuries, did not refer to an interpretation in terms of moments. In fact, according to a tradition which can be traced back at least to the late Renaissance,[37] all methods of approximation were considered as "mechanical."

According to his moment-theoretic interpretation of the equation system (4.34), Stieltjes, seemingly without any knowledge of either Chebyshev's or Markov's works, advanced problems on moments of density functions which were similar to those of Chebyshev and Markov. This conclusion also applies to the content of the papers [Stieltjes 1884c; 1885a], in which a practical problem of geophysics (properties of the mass density of the earth) was solved by moment methods.

4.4 Chebyshev's Further Work on Moments

By the end of the year 1884, which can be designated as an "annus mirabilis" of moment theory, numerous results were achieved, which surpassed by far the original version of Chebyshev's inequality. Chebyshev in his own papers of the time after 1884 did not hint at Markov's and Stieltjes's contributions. He even presented several results due to these mathematicians without giving any reference.

[37] Apparently, a shift occurs in the conception of "mechanical" between the times of ancient Greeks and early modern times. In Greek geometry, "mechanical constructions" were characterized by the use of certain instruments in addition to compass and straightedge. These constructions were not considered as being imprecise, however (see [Hoppe 1920; Steele 1936]). On the other hand, Albrecht Dürer [1525], in his book on practical geometry, designated approximate constructions as "mechanical." As it seems, Dürer was one of the first among the authors of early modern times who, dealing with practical problems, differentiated between approximate and exact constructions.

This case also applies to a paper by Chebyshev, whose Russian version appeared in 1885 in the communications of St. Petersburg Academy, and which was published in French in 1887 under the title "Sur la représentation des valeurs limites des intégrales par des résidus intégraux." In this article, Chebyshev communicated, again without proof,[38] generalizations of his inequality (4.4), as they had also been treated in a similar form in Markov's dissertation. Possé [1886] even showed the equivalence of Chebyshev's and Markov's version of these general inequalities. Chebyshev in his assertions used Cauchy's concept of residues, and this circumstance explains the particular phrasing of his paper's title.

Let φ be a rational function, which—considered as a function of a complex variable—has poles only on the real axis. Together with Chebyshev we use the symbol (basically due to Cauchy [1826], see [Smithies 1997, 113–146])

$$\mathscr{E}_a^b \varphi(z)dz,$$

where $a < b$ are real numbers, for the sum of residues within the rectangle with the lower left vertex $a - \eta\sqrt{-1}$ and the upper right vertex $b + \eta\sqrt{-1}$, η being an arbitrary positive number. If a pole coincides with one side of the rectangle, the corresponding residue contributes to the sum with its half. Cauchy's residue theorem [1825, 54] yields

$$\mathscr{E}_a^b \varphi(z)dz = ⨍_a^b \varphi(x - \eta\sqrt{-1})dx + \sqrt{-1} ⨍_{-\eta}^{\eta} \varphi(b + y\sqrt{-1})dy -$$
$$- ⨍_a^b \varphi(x + \eta\sqrt{-1})dx - \sqrt{-1} ⨍_{-\eta}^{\eta} \varphi(a + y\sqrt{-1})dy,$$

if the principal values—denoted by $⨍$ at this place—of all integrals exist. Let $\frac{\varphi_m(z)}{\psi_m(z)}$ be the mth partial fraction of a continued fraction associated with $\int_A^B \frac{f(x)}{z-x}dx$ (f being "positive") and let $\psi_m(z_i) = 0$ ($i = 1, \ldots, m$).[39] Then for natural α, β, $1 \leq \alpha < \beta \leq m$ we have, for example,

$$\int_{z_\alpha}^{z_\beta} f(x)dx < \sum_{i=\alpha}^{\beta} \frac{\varphi_m(z_i)}{\psi'_m(z_i)} = \mathscr{E}_{z_\alpha-\omega}^{z_\beta+\omega} \frac{\varphi_m(z)}{\psi_m(z)}dz,$$

where ω is an—in Chebyshev's own words—"infinitely small" quantity.

Generalizing such relations, Chebyshev [1885/87, 39] even derived the following inequalities for all $\upsilon \in [A; B]$ (the interval was not necessarily assumed to be finite[40]):

$$\mathscr{E}_{A-\omega}^{\upsilon-\omega} F(z)dz \leq \int_A^{\upsilon} f(x)dx \leq \mathscr{E}_{A-\omega}^{\upsilon+\omega} F(z)dz, \tag{4.35}$$

---

[38] For the case of discrete distributions this proof is contained in [Chebyshev 1891/1907], see also footnote 14.

[39] Note that all roots of $\psi_m = 0$ are simple and real.

[40] Chebyshev did not explicitly discuss the cases $A = -\infty$ or $\upsilon = \infty$. In these cases, however, the following inequalities remain valid if one sets $-\infty - \omega = -\infty$ and $\infty + \omega = \infty$.

where

$$F(z) = \cfrac{1}{\alpha_1 z + \beta_1 - \cfrac{1}{\alpha_2 z + \beta_2 - \cfrac{1}{\ddots - \cfrac{1}{\alpha_m z + \beta_m - \cfrac{1}{Z}}}}},$$

and ω is "infinitely small." The continued fraction

$$\cfrac{1}{\alpha_1 z + \beta_1 - \cfrac{1}{\alpha_2 z + \beta_2 - \cfrac{1}{\alpha_3 z + \beta_3 - \cdots}}} \tag{4.36}$$

was considered to be associated with the integral $\int_A^B \frac{f(x)}{z-x}dx$. With regard to the function Z one has to differentiate the two cases of an even or an odd number of given moments $M_0, \ldots, M_{2m-1}$ or $M_0, \ldots, M_{2m}$ of the function f. If $\frac{\varphi_i(z)}{\psi_i(z)}$ denotes the ith partial fraction of (4.36), then, according to Chebyshev [1885/87, 41 f.], in the case of an even number $2m$ of given moments, the function Z can be expressed by

$$Z = \gamma(z - \upsilon) + \frac{\psi_{m-1}(\upsilon)}{\psi_m(\upsilon)},$$

where

$$\gamma := \max\left(\frac{1}{A-\upsilon}\left[\frac{\psi_{m-1}(A)}{\psi_m(A)} - \frac{\psi_{m-1}(\upsilon)}{\psi_m(\upsilon)}\right], \frac{1}{B-\upsilon}\left[\frac{\psi_{m-1}(B)}{\psi_m(B)} - \frac{\psi_{m-1}(\upsilon)}{\psi_m(\upsilon)}\right]\right).$$

In the case of an odd number of given moments, Z is even more complicated [Chebyshev 1885/87, 41 f.].

The awkwardness of Chebyshev's estimates is also documented by the following lengthy section, which was dedicated to an application to the problem of mass distribution along a rod [Chebyshev 1885/87, 45–55].

In 1887, Chebyshev published an article, which appeared two years later in French under the title "Sur les résidus intégraux qui donnent des valeurs approchées des intégrales," and in which the already obtained estimates (4.35) were considerably simplified for the case of an even number of given moments.[41] Chebyshev's main result [1887/89, 308] was the inequality

$$\left|\int_A^\upsilon f(x)dx - \mathscr{E}_{A-\omega}^{\upsilon} F(z)dz\right| \leq \frac{1}{2}\frac{1}{\sum_{i=0}^{m-1}\frac{\psi_i^2(\upsilon)}{\int_A^B \psi_i^2(x)f(x)dx}}, \tag{4.37}$$

[41] For a description of the chief arguments and results of this article, see [Vasilev 1898/1900, 31–34].

where ω is "infinitely small" again, and the ψ_i are the partial denominators of the continued fraction (4.36).

The right-hand side of (4.37) is dependent only on the moments of the "positive" function f up to the order $2m-1$. Thus, Chebyshev [1887/89, 309 f.] was able to conclude that for two "positive" functions f and f_1 with the properties

$$\int_A^B x^i f(x)dx = \int_A^B x^i f_1(x)dx \quad \text{for all } i = 0, \ldots, 2m-1$$

the following inequality is valid for $-\infty \le A < B \le \infty$:

$$\left| \int_A^\upsilon f(x)dx - \int_A^\upsilon f_1(x)dx \right| \le \frac{1}{\sum_{i=0}^{m-1} \frac{\psi_i^2(\upsilon)}{\int_A^B \psi_i^2(x) f(x)dx}} \quad \text{for all } \upsilon \in [A; B]. \quad (4.38)$$

In principle, it would have been possible to base the discussion of moment problems even with an infinite number of given moments on this inequality. It is not true[42] that questions of this kind were beyond Chebyshev's scope. There is a remark in [Chebyshev 1887/89, 310] that the right-hand side of (4.38) decreases as m increases, and tends to 0 if the series in the denominator is divergent. Chebyshev and other members of the St. Petersburg school did not succeed in a general discussion of the divergence of this series, however.

This lack of success becomes plausible, if one considers the difficulties which Chebyshev [1887/89, 311–322] had to overcome in obtaining an estimate of the right-hand side of (4.38) for the special case $A = -\infty$, $B = \infty$, $f(x) = \frac{q}{\sqrt{2\pi}} \mathrm{e}^{-\frac{q^2}{2}x^2}$. Resorting to his 1859 article "Sur le développement des fonctions à une seule variable," Chebyshev followed that in this particular case the polynomials ψ_i were Hermite polynomials with the following properties:

$$\psi_i(z) = \mathrm{e}^{\frac{q^2}{2}} \frac{\mathrm{d}^i \mathrm{e}^{-\frac{q^2}{2}x^2}}{\mathrm{d}x^i},$$

$$\psi_i(z) = -q^2 z \psi_{i-1}(z) - (i-1)q^2 \psi_{i-2}(z) \quad (i \ge 2),$$

and

$$\int_{-\infty}^{\infty} \psi_i^2(x) f(x)dx = i!q^{2i}.$$

In order to determine the asymptotic properties of the right-hand side of (4.38), Chebyshev examined the sum

$$\sum_{i=0}^{m-1} T_i, \quad \text{where } T_i := \frac{\psi_i^2(\upsilon)}{i!q^{2i}}.$$

[42] The author's opinion differs from Kjeldsen's [1993, 21] in this respect.

He considered the function $\Theta(t) = \sum_{i=0}^{\infty} T_i t^i$, for which, due to the recursion formula of Hermite polynomials, a certain first-order differential equation was valid. Taking into consideration that $\Theta(0) = \psi_0^2(\upsilon) = 1$, he obtained

$$\Theta(t) = \frac{e^{\frac{q^2\upsilon^2 t}{1+t}}}{\sqrt{1-t^2}}.$$

The subsequent arguments show Chebyshev's pragmatic use of infinitely large and small quantities. He introduced a nonnegative function Y which attained positive values only in a one-sided "infinitely small neighborhood" of the integers 0, 1, 2, 3, ..., "such that the integrals

$$\int_0^{\omega} Y\,dx, \int_{1-\omega}^{1} Y dx, \int_{2-\omega}^{2} Y dx, \ldots$$

tend to the quantities $T_0, T_1, T_2, \ldots$ if ω tends to 0" [Chebyshev 1887/89, 315]. According to Chebyshev, for such a δ-like function the equation

$$\int_0^{\infty} Y t^x dx = \sum_{i=0}^{\infty} T_i t^i = \Theta(t)$$

was valid. To the "mass density" Yt^x he now applied the results of the rod problem with given moments

$$\int_0^{\infty} Y t^x dx = \Theta(t), \int_0^{\infty} x Y t^x dx = t\Theta'(t), \int_0^{\infty} x^2 Y t^x dx = t\Theta'(t) + t^2\Theta''(t).$$

For a certain $\tau \in\,]0; 1[$ (depending on m) satisfying the condition

$$\frac{\tau\Theta''(\tau)}{\Theta'(\tau)} + 1 = m + 1 \tag{4.39}$$

(the issue of existence of τ was not broached) Chebyshev concluded that

$$\int_0^{m-1} Y\tau^x dx = \sum_{i=0}^{m-1} T_i \tau^i \geq \Theta(\tau) - \frac{\tau(\Theta'(\tau))^2}{\Theta'(\tau) + \tau\Theta''(\tau)}.$$

Because of $\tau < 1$ the inequality

$$\sum_{i=0}^{m-1} T_i > \Theta(\tau) - \frac{\tau(\Theta'(\tau))^2}{\Theta'(\tau) + \tau\Theta''(\tau)} \tag{4.40}$$

was an immediate consequence. Finally Chebyshev [1887/89, 317–320], taking into account the particular form of Θ, the relation (4.39), and the fact that $\tau \in\,]0; 1[$, derived from (4.40) the inequality

$$\sum_{i=0}^{m-1} T_i > \frac{2(m-3)^2\sqrt{m-1}}{3\sqrt{3}(m^2-2m+3)^{\frac{3}{2}}(q^2\upsilon^2+1)^3}.$$

By substituting this latter relation into the right side of (4.38), he achieved the result: If a "positive" function f_1, defined in $\mathbb{R}$, has, up to the order $2m - 1$, the same moments as the normal density $n(x) = \frac{q}{\sqrt{2\pi}} e^{-\frac{q^2}{2}x^2}$, then

$$\left| \int_{-\infty}^{v} f_1(x)dx - \frac{1}{\sqrt{\pi}} \int_{-\infty}^{\frac{qv}{\sqrt{2}}} e^{-x^2} dx \right| < \frac{3\sqrt{3}(m^2 - 2m + 3)^{\frac{3}{2}}(q^2v^2 + 1)^3}{2(m-3)^2\sqrt{m-1}} \tag{4.41}$$

for all real v. With this inequality Chebyshev solved the now so-called "Hamburger moment problem"[43] in a particular case. As he [1887/89, 311] noted, this case was very important for probability theory. Because the right-hand side of (4.41)[44] tends to 0 as $m \to \infty$, the immediate consequence is: If a "positive" function, defined in $\mathbb{R}$, coincides in all of its moments with those of a normal density n, then this function is identical with n.

4.5 The Stieltjes Moment Problem

From today's point of view one connects the relationship between moments and continued fractions, and the notion of "moment problem," chiefly with Stieltjes's name.[45] As we have already seen, Stieltjes had in the 1880s started research on the relations between moments and continued fractions. He had given first estimates for the mass distribution, if the moments were known up to a certain order. Discussing continued fractions which correspond to $\int_0^\infty \frac{f(u)}{z+u} du$ (f nonnegative), Stieltjes around 1892 encountered the problem that there might exist nonnegative "functions" f and f_1[46] with the property

$$\int_0^\infty x^k (f(x) - f_1(x))dx = 0 \quad \forall k \in \mathbb{N}_0,$$

despite these functions being different from each other. This seemingly paradoxical result[47] motivated Stieltjes to examine the dependence of mass distributions on their moments anew. He considered moments of indefinitely high order, and he posed the following problem:

---

[43] Hans Ludwig Hamburger [1920–21] solved the problem of existence and uniqueness of a mass distribution defined in $\mathbb{R}$ if its moments are given in each order [Kjeldsen 1993, 37–40].

[44] In 1892, Nikolai Yakovlevich Sonin succeeded in simplifying the right-hand side of (4.41) considerably, replacing it by $\sqrt{\frac{\pi}{2m-1}}$ (see [Vasilev 1898/1900, 33]).

[45] The following survey of Stieltjes's major contributions partially relies on [Kjeldsen 1993] and Bernkopf [1970–76].

[46] These "functions" had, as Stieltjes only intuitively conceived, a δ-like character, corresponding to the densities of discrete mass distributions.

[47] Kjeldsen [1993, 26, 33] erroneously argued that only this surprising case caused Stieltjes's activities in moment problems.

Given a sequence of positive numbers (c_k), then, for each positive x find the mass $F(x)$ assigned to the segment $[0; x]$, such that $\int_0^\infty x^k dF(x) = c_k$ for all $k \in \mathbb{N}_0$!

Stieltjes [1894/95] succeeded in finding necessary and sufficient conditions which have to be imposed on a moment sequence (c_k), such that for $x \in \mathbb{R}_0^+$ a mass distribution $F(x)$ associated with this sequence exists and is uniquely determined.

In order to simultaneously treat discrete and continuous mass distributions, Stieltjes used "his" integral, which, however, had not emerged in the direct context of moment problems. Originally, in a 1892 letter to Hermite, he had used this device for jointly representing the limiting functions of partial fractions of odd and even order [Kjeldsen 1993, 32].

Whereas Chebyshev's moment problem, which had been discussed to a certain extent also by Stieltjes and Markov during the 1880s, was on estimates of the distribution function in the case of a finite number of given moments, Stieltjes since 1892 put the center of his activities on other questions. Now, he was chiefly interested in questions on the convergence of analytic continued fractions. His moment problem mainly served to contribute to the solution of these convergence questions, and was basically less general than Chebyshev's. Chebyshev, on the other hand, was only able to solve a problem on infinite moment sequences in the particular case of normal distributions.

4.6 Moment Theory and Central Limit Theorem

4.6.1 Chebyshev's Probabilistic Work

Chebyshev only published four works of essentially probabilistic content. The main object of his master thesis, which appeared in printed form in 1845 but remained without greater influence, was to treat important problems of classical probability by methods more elementary than those of Laplace's *TAP*. In this work he already discussed the quality of approximations by estimates. In the case of Stirling's formula he [1845, 40–43] found

$$\mathrm{e}^{-x+\frac{1}{12x}} > \frac{x!}{\sqrt{2\pi}x^{x+\frac{1}{2}}} > \mathrm{e}^{-x+\frac{1}{12x}-\frac{1}{36x^3}}.$$

This quest for explicit estimates of approximation errors was certainly in part motivated by numerical necessities. But it also seems that Chebyshev for the sake of rigor regarded it as imperative to discuss explicit error bounds if approximations to "functions of large numbers" were employed. This program comprised finitary aspects and constructive ideas, in the sense that the deviation between an exact and a limiting function had to be precisely specified regarding its dependence on the essential parameters.

In an 1846 paper, Chebyshev in the context of Poisson's law of large numbers clearly expressed the fundamental ideas of his program. In modern notation the law of large numbers is equivalent to

$$\forall \varepsilon > 0: \quad P\left(\left| f_\mu - \frac{\sum_{i=1}^{\mu} p_i}{\mu} \right| \le \varepsilon\right) \to 1 \quad (\mu \to \infty), \tag{4.42}$$

f_μ being the relative frequency of the number of successes in a sequence of μ Bernoulli trials with—possibly different—success probabilities p_i. Poisson in his 1837 *Recherches* had deduced (4.42) as a corollary of the approximation (again in modern notation, H_μ designates the number of successes among μ trials, u a positive real number):

$$P\left(\sum p_i - u\sqrt{2\sum p_i(1-p_i)} \le H_\mu \le \sum p_i + u\sqrt{2\sum p_i(1-p_i)}\right)$$
$$\approx 1 - \frac{2}{\sqrt{\pi}} \int_u^\infty e^{-t^2}\, dt + \frac{1}{\sqrt{2\pi \sum p_i(1-p_i)}} e^{-u^2}. \tag{4.43}$$

Starting with a representation of $P(H_\mu = m) =: U$ according to

$$2\pi U = \int_{-\pi}^{\pi} \prod_{i=1}^{\mu} (p_i e^{x\sqrt{-1}} + q_i e^{-x\sqrt{-1}}) e^{-(m-n)x\sqrt{-1}} dx, \; (m+n = \mu,\ q_i = 1 - p_i),$$

Poisson [1837, 246–252] first expanded the logarithm of the characteristic function

$$\prod_{i=1}^{\mu} (p_i e^{x\sqrt{-1}} + q_i e^{-x\sqrt{-1}})$$

into a series of powers of x. After substituting $x = \frac{z}{\sqrt{\mu}}$ he stated for the case of "very large" μ that all series terms except for the first and the second could be neglected if the existence of a lower bound for all products $p_i q_i$ independent of i was presupposed. By use of some well-known integration formulae, Poisson [1837, 252 f.] obtained an (approximate!) expression for U. (4.43) was finally derived by use of Euler's summation formula, where, again, terms of an order of magnitude less than $\frac{1}{\sqrt{\mu}}$ were neglected. Taking additionally into account that the limit relation

$$\frac{u\sqrt{2\sum p_i q_i}}{\mu} \to 0 \quad (\mu \to \infty)$$

was valid for any arbitrarily large (but fixed) positive u, Poisson [1837, 254] from (4.43) inferred a statement, only expressed in words by him, which corresponds in modern mathematical notation to (4.42).

Chebyshev heavily criticized Poisson's deduction of (4.42) with the following words:

> This fundamental theorem of probability calculus, which covers the law of Jacob Bernoulli as a particular case, is deduced by Mr. Poisson by means of a formula [i.e., (4.43)], which he obtains by approximately calculating the value of a definite, however too complicated

integral (see the Recherches sur les probabilités des jugements, chapter IV). Even if the method applied by the famous geometer is very skillful, it still does not provide the bound of the error which yields his approximate analysis, and, due to this uncertainty about the value of the error, the proof of the theorem lacks rigor [Chebyshev 1846, 17].

In the main part of his 1846 paper, Chebyshev for $s = \sum_{i=1}^{\mu} p_i$ and natural m, n $(n < s-1, m > s+1)$ deduced the inequality

$$P(n < H_\mu < m) > 1 - \frac{1}{2(m-s)}\sqrt{\frac{m(\mu-m)}{\mu}}\left(\frac{s}{m}\right)^m\left(\frac{\mu-s}{\mu-m}\right)^{\mu-m+1} - \\ - \frac{1}{2(s-n)}\sqrt{\frac{n(\mu-n)}{\mu}}\left(\frac{s}{n}\right)^{n+1}\left(\frac{\mu-s}{\mu-n}\right)^{\mu-n}, \quad (4.44)$$

which Poisson's weak law of large numbers was a consequence of. As we have already seen in Sect. 2.1.4, around 1850 a still rather vague feeling had emerged that the method of deriving CLTs according to Laplace and Poisson by cutting off series expansions was not rigorous. Chebyshev's demand for explicit error bounds of the respective approximations guaranteed analytic rigor, but, for the sake of rigorous proofs of limit theorems alone, this demand was overstated. There exists a good deal of bequeathed sayings of Chebyshev, in which he pointed out that, in his opinion, mathematical research was only reasonable if the results could be applied for practical purposes. In this respect, in the case of a *finite* number of trials, one had to give advantage to numerically usable and sufficiently sharp estimates opposite mere limit theorems. Markov, Chebyshev's disciple, repeated this request in the preface of his popular book on probability (see Sect. 4.7.3). But again, this aspect does not touch the real question of analytical rigor either. The inequality (4.44) gives a rather sharp lower bound for $P(n<H_\mu<m)$,[48] but the use of this inequality for numerical considerations was not discussed in Chebyshev's article.

In an 1867 article, Chebyshev established "his" (and Bienaymé's) famous inequality, and thus continued his research on weak laws of large numbers, but again without discussing applications. Chebyshev's somewhat longwinded style was consistent with his intention of using methods as elementary as possible. In a way similar to many of his contributions to continued fractions, Chebyshev [1867] only considered discrete distributions with finite numbers of mass points; the transition to more general distributions was seen as a matter of course, apparently.

In contrast to the small number of published articles, Chebyshev quite frequently gave courses on probability calculus between 1860 and 1882 [Sheynin 1994, 322]. As we can see from lecture notes written by Lyapunov in 1879/80, Chebyshev

[48] If we take a Bernoulli process as a test case with $\mu = 10000$ trials and the success probability $\frac{1}{2}$, then we obtain $P(4950 \leq H_\mu \leq 5050) > 0.41$ according to Chebyshev's (4.44), $P(4950 \leq H_\mu \leq 5050) \geq 0(!)$ according to the Bienaymé–Chebyshev inequality, and $P(4950 \leq H_\mu \leq 5050) > 0.012(!)$ according to Jakob Bernoulli's estimates (see Sect. 1.6). The exact value of the probability under consideration is 0.683. The inequality (4.44) has the disadvantage, however, that for given $a > 0$ and p $(0 < p < 1)$ the minimum number μ, such that $P(|H_\mu - s| < \mu a) > p$, can be found by a rather tedious numerical procedure only.

[1936/2004] conceived probability theory as imbedded into the theories of definite integrals and of finite differences. In this way he pointed out the analytic relevance of probabilistic contents and especially of limit theorems, at least implicitly.

With regard to the CLT, Chebyshev in the just-mentioned lectures recapitulated Laplace's method of establishing an approximate normal distribution without any significant modifications, and he did not formulate a proper limit theorem [Chebyshev 1936/2004, 198–202]. At least he noticed [1936/2004, 202] that the presented line of proof was not rigorous. He maintained that for the time being (around 1880) analysis was not capable of deriving error bounds for the deviation of the exact probabilities from the corresponding limit expressions (see also [Sheynin 1994, 334]). However, this assessment did not entirely apply. Cauchy [1853h] had already given adequate estimates, although under rather restrictive assumptions during his dispute with Bienaymé. Due to the fact that Chebyshev did not care too much about the work of others (see [Steffens 2006, 71]), it appears plausible that he did not know Cauchy's contribution.

4.6.2 *Chebyshev's Uncomplete Proof of the Central Limit Theorem from 1887*

The original Russian version of Chebyshev's article "Sur deux théorèmes relatifs aux probabilités" was published in 1887 in *Zapiski akademii nauk*; the French translation was issued in 1890 in *Acta mathematica*. In this work, Chebyshev used a method somewhat different from that applied by Laplace, Poisson, and also Cauchy for proving the CLT. Instead of representing the considered probabilities through Fourier transforms like those authors, he used Laplace transforms instead, initiating by this modification an approach by means of moments.

At the beginning of the article, Chebyshev recapitulated the result of his 1867 paper "Des valeurs moyennes." He argued that this had already been obtained within the framework of his research program on moments. He then repeated the inequality (4.41) and announced his intention to show

> how this theorem on integrals leads to a theorem on probabilities, by which the most precise determination of the unknowns can be reduced to the method of least squares, if one has a large number of equations with more or less considerable random errors [Chebyshev 1887/90, 307].

Chebyshev therefore tried to imbed both the weak law of large numbers and the CLT into moment theory, and with the remark on the application of the CLT he primarily referred to the foundation of least squares according to Laplace.

Chebyshev [1887/90, 307] introduced the CLT in the following version: He considered a sequence of (implicitly assumed) independent random variables ("quantités") u_i, each with zero expectation. For these random variables Chebyshev presupposed nonnegative densities φ_i with moments of arbitrarily high order. He assumed that, for each order, an upper and lower bound of the moments existed, uniformly for all random variables. These bounds depended possibly on the order, however.[49] Under these assumptions, Chebyshev stated that for any $t < t' \in \mathbb{R}$:

$$\lim_{n\to\infty} P\left(t \leq \frac{\sum_{i=1}^{n} u_i}{\sqrt{2\sum_{i=0}^{n} \mathrm{E}u_i^2}} \leq t'\right) = \frac{1}{\sqrt{\pi}}\int_t^{t'} \mathrm{e}^{-x^2}dx. \tag{4.45}$$

The beginning of the history of the CLT in the strict sense of limit theorem is connected to [Chebyshev 1887/90]. Even if, in the context of approximate normal distributions for sums of independent random variables, some authors before Chebyshev, such as Poisson [1824], hinted at the fact that the difference between exact and approximating probability gets "infinitely small" with an "infinitely large" number of summands, such an assertion did not clearly express the convergence of a sequence of probabilities as it was demanded by the common analytic standards at the end of the 19th century. In order to translate Poisson's statements on the approximation of probabilities of sums of a "very large number" of random variables into the assertion of a limit theorem, Chebyshev chose the most simple way in formulating a limit theorem for normed partial sums assigned to a given sequence of random variables. Only during the 1920s, beginning with Bernshtein (see Sect. 5.2.7.1), was the more general situation of row sums in triangular arrays taken into consideration.

In his proof of the CLT, Chebyshev focused on the (today so-called) Laplace transform $\int_{-\infty}^{\infty} \mathrm{e}^{sx} f(x)dx$ of the the density f of the normed sum $\frac{\sum_{i=1}^{n} u_i}{\sqrt{n}}$, "$s$ a given constant."[50] By elementary considerations he derived the following formula:

$$\int_{-\infty}^{\infty} \mathrm{e}^{sx} f(x)dx = \prod_{i=1}^{n} \int_{-\infty}^{\infty} \mathrm{e}^{\frac{su_i}{\sqrt{n}}} \varphi_i(u_i)du_i. \tag{4.46}$$

Chebyshev now expanded the natural logarithm of the right-hand side of (4.46) into a series of powers of s. To this aim he [1887/90, 310] used the "approximate expression which is exact up to the term of order s^{2m-1} inclusively," obtained by a term-by-term integration:

49 Chebyshev's phrasing is a little obscure concerning this issue. It remains unclear whether there exists a common upper and a common lower bound, respectively, for all moments of all random variables or these bounds are possibly different for each order. An interpretation in the former sense would imply, however, that Chebyshev's theorem would not even apply to normally distributed random variables.

50 Chebyshev did not use a special name for the Laplace transform, and he did not discuss any properties of this entity. In particular, he did not consider the fact that the existence of the Laplace transform for a real s is secured only if certain conditions in addition to the mere existence of moments of arbitrary order are valid.

$$\int_{-\infty}^{\infty} \mathrm{e}^{\frac{su_i}{\sqrt{n}}} \varphi_i(u_i) du_i$$
$$= \int_{-\infty}^{\infty} \varphi_i(u_i) du_i + \frac{s}{1 \cdot \sqrt{n}} \int_{-\infty}^{\infty} u_i \varphi_i(u_i) du_i + \frac{s^2}{1 \cdot 2 \cdot (\sqrt{n})^2} \int_{-\infty}^{\infty} u_i^2 \varphi_i(u_i) du_i + \cdots$$
$$\cdots + \frac{s^{2m-1}}{1 \cdot 2 \cdots (2m-1)(\sqrt{n})^{2m-1}} \int_{-\infty}^{\infty} u_i^{2m-1} \varphi_i(u_i) du_i.$$

By this procedure Chebyshev reached an expansion of the form

$$\int_{-\infty}^{\infty} \mathrm{e}^{sx} f(x) dx = \mathrm{e}^{\frac{s^2}{2q^2} + \frac{M^{(3)}}{\sqrt{n}} s^3 + \dots + \frac{M^{(2m-1)}}{(\sqrt{n})^{2m-3}} s^{2m-1} + \dots}. \tag{4.47}$$

In this latter formula $\frac{1}{q^2}$ designated the arithmetic mean of the second-order moments. For $n \to \infty$ each single fraction $\frac{M^{(k)}}{(\sqrt{n})^{k-2}}$ tended to 0 on account of the common boundedness of the moments of each order.

Without any further discussions Chebyshev [1887/90, 312] concluded that "in the special case $n = \infty$" equation (4.47) is simplified toward

$$\int_{-\infty}^{\infty} \mathrm{e}^{sx} f(x) dx = \mathrm{e}^{\frac{s^2}{2q^2}}. \tag{4.48}$$

$\frac{1}{q^2}$ now designates the limit of the arithmetic mean of the second-order moments as $n \to \infty$. Apparently, Chebyshev tacitly assumed, besides the—by no means guaranteed—existence of this limit, the existence of a density f such that

$$\frac{\mathrm{d}}{\mathrm{d}x} P\left(\sum_{i=1}^{n} \frac{u_i}{\sqrt{n}} \leq x\right) \to f(x) \quad (n \to \infty).$$

The latter circumstance shows that he basically did not differentiate between local and integral versions of the CLT.

Chebyshev's line of argument as a matter of course presupposed that $\frac{1}{q^2}$ remains positive as $n \to \infty$, an assumption which was, however, not explicitly stated among the assumptions at the beginning of the paper. By expanding both sides of (4.48) into a series of powers of s, and by equating terms of common order, he obtained the result that the limit function f and the function $\frac{q}{\sqrt{2\pi}} \mathrm{e}^{-\frac{q^2}{2}x^2}$ have the same moments in each order. Chebyshev finally resorted to his estimate (4.41), which implied the limit theorem (4.45).

By these considerations Chebyshev had "proven," as he [1887/90, 315] also observed, a limit theorem "merely." He did not reach the main goal he had put for himself, to obtain inequalities for the deviation of the exact probability from the limit expression in the case of a finite number of random variables. At least, he hinted at the fact, without discussing any details, that, corresponding to the results of his article [Chebyshev 1859], the exact probability for a finite number of random variables could be expressed by a series expansion in polynomials $e^{x^2}\frac{d^i e^{-x^2}}{dx^i}$ (see Sect. 3.4.1).

Chebyshev basically pursued a program which committed to avoiding "approximate" arguments, in particular those employing cut-off series expansions (Sect. 4.6.1). Yet, when proving the CLT, he used exactly arguments of this kind. In this respect there was no considerable difference between his and Laplace's or Poisson's approach. In contrast to the latter authors, he passed over from the limit of the generating functions to the limit distribution itself not by means of Fourier methods (which procedure would have been easier), but by a quite complicated consideration on moments. Using this device, however, his mode of reasoning was again no way more rigorous in comparison with Laplace's and Poisson's accounts. Altogether, the reader reaches the impression that, in Chebyshev's article, the "proof" of the CLT was chiefly to serve for demonstrating important methods and results of moment theory. Apparently, the rigorous deduction of a mathematical theorem which should be studied in its own right came second.

4.6.3 Poincaré: Moments and Hypothesis of Elementary Errors

Even before Markov succeeded in giving rigorous proofs of the CLT by means of moments, Poincaré in his *Calcul des probabilités* [1896] had already taken on Chebyshev's idea of deriving the CLT by moment methods. Henri Poincaré (1854–1912) was one of the leading figures of mathematics and physics of the late 19th and early 20th centuries. With regard to probability theory, which discipline was not in the center of his scientific activities, he is still well known today by his discussion of the notion of "random,"[51] and by establishing the later so-called "method of arbitrary functions."[52] Poincaré's *Calcul des probabilités* (the first edition appeared in 1896, the second, somewhat modified edition in 1912) was one of the most influential textbooks in the period of transition from classical to modern probability outside of Russia.[53] The style of this book was not in accord, however, with that of contemporary textbooks on analysis. There is, in many places, a lack of precise conditions and entirely rigorous deductions. This circumstance may be taken as evidence that Poincaré did not conceive of probability theory as a discipline of mathematics proper.

With regard to the hypothesis of elementary errors, Poincaré [1896, 169–187; 1912, 189–206] tried to prove that the sum of a large number of independent errors with (in general) different but symmetric densities approximately obeys a Gaussian law. Somewhat vaguely, Poincaré presupposed the elementary errors to have "approximately the same order of magnitude" and to contribute only a "small part" to the total error. As already noted, Poincaré based his proof on moment methods.

[51] See, in particular, the chapter *Le Hasard* of his *Science et Méthode* [1908].

[52] See [von Plato 1983].

[53] For a résumé of Poincaré's work on probability see [Sheynin 1990].

He did not cite work by Chebyshev, Markov, or Stieltjes. Taking into consideration Poincaré's interest in analysis it is improbable that he was not acquainted with the contemporary development of moment theory, however.

Poincaré gave an amazingly simple, albeit erroneous,[54] proof that a density is equal to the Gaussian error function if it coincides in all of its moments with the latter. Then he proved that the moments of the expression $\frac{\sum_{i=1}^{n} y_i}{\sqrt{n}}$, y_i being identically distributed errors with symmetric densities and moments of arbitrarily high order, tend to the corresponding moments of a normal distribution with zero expectation and with variance $\mathrm{Var}(y_1)$. He also sketched, under not completely clear conditions, the proof for the analogous property in the case of non-identically distributed symmetric errors. Poincaré in this way anticipated, if in a not entirely general and rigorous way, Markov's proof of the convergence of the moments of normed sums of random variables to the moments of a Gaussian distribution. That proof was published two years later.

From the approximate equality of the respective moments of a sum of numerous elementary errors and a normal distribution, Poincaré [1896, 187; 1912, 206] concluded without any further explanations:

> In this case the resulting error very precisely obeys Gauss's law. This is, as it seems to me, the best reasoning which can be given in favor of Gauss's law.

4.6.4 Markov's Rigorous Proof

After Chebyshev had retired from giving lecture courses in 1882, Markov became his successor in teaching probability theory [Maistrov 1974, 218]. For the time being, Markov scarcely did any active research in this field, however. Only around 1898 did he develop an intensified interest in the moment theoretic proof of the CLT, as we can see from his exchange of letters with Vasilev.[55] Markov [1899/2004, 130] clearly expressed the flaws of Chebyshev's proof with the words

> ... Chebyshev's demonstration is very involved since it is based on preliminary investigations.[56] ... However, Chebyshev's derivation is expounded in such a way that its rigor may be doubted. A question therefore arises, whether Chebyshev's proof is distinguished from the previous one [by Laplace and Poisson] not in essence, but only by needless complexity, or can it be made rigorous.

As Markov (same place) stated, there had already been a "long-standing desire to simplify, and, at the same time, to make his [Chebyshev's] analysis quite rigorous" for him. His motivation for dealing with this problem had even been "strengthened" by his reading of the summarizing description of Chebyshev's deduction of the CLT in Vasilev's biography on Chebyshev, whose Russian version appeared in 1898.

---

[54] For an analysis of Poincaré's mistake see [Fischer 2000, 166].

[55] These letters from September/October 1898 were in part published as [Markov 1899] (English translation [Markov 1899/2004]). See also [Maistrov 1974, 209 f; Yushkevich 1970–76b, 127 f.].

[56] Markov most probably refers to Chebyshev's estimate (4.41).

An important reason why Markov began serious studies on the CLT with a certain delay only, might be that Markov, as we have already seen (Sect. 4.6.1), did not regard probabilistic limit theorems as essentially important. He rather gave preference to approximate formulae, for which he demanded estimates of the precision, however. As a possible consequence, Markov's 1898 article primarily focused on results on moments and continued fractions, which the CLT could be followed from as a corollary.

In Sect. 4.3.2 we have already briefly described Stieltjes's proof [1884a] for the assertion that, in any arbitrarily given subinterval of $[a;b]$, the partial denominators of the continued fraction associated with $\int_a^b \frac{f(x)dx}{z-x}$ have at least one zero from a certain order on. By virtue of this fact Stieltjes showed that the Gaussian quadrature formula assigned to the weight function f converges under very general conditions to the corresponding integral. Stieltjes had restricted his consideration to finite ranges of integration $[a;b]$.

Markov now extended Stieltjes's basic ideas[57] to the domain of integration $]-\infty;\infty[$ and the particular weight function $f(x) = \mathrm{e}^{-x^2}$. Whereas the formulae (4.17), (4.18), and (4.19), which are essential for the Gaussian procedure of integration, were more or less tacitly applied also for infinite domains of integration, if the weight function had appropriate properties,[58] a zero point property analogous to that of Stieltjes could not be asserted with the same self-evidence. Markov [1898/1912, 259–264] therefore had to show first that the partial denominators of a continued fraction associated with $\int_{-\infty}^{\infty} \frac{\mathrm{e}^{-t^2}}{x-t} dt$ (which are, as is generally known, proportional to the Hermite polynomials) from a certain order on possess at least one zero in any given interval.

The next step in Markov's proof was to show, by aid of this zero point property and "his" inequalities (4.27) and (4.28), that for each finite real interval $[\alpha;\beta]$ the following assertion is valid: Let $\psi_m(x)$ and $\varphi_m(x)$ be the denominators and numerators, respectively, of the mth partial fraction of a continued fraction associated with $\int_{-\infty}^{\infty} \frac{\mathrm{e}^{-t^2}}{x-t} dt$, and let $x_i^{(m)}$ be the roots of ψ_m in ascending order (i.e., $x_1^{(m)} < x_2^{(m)} < \cdots$). If $\xi_1^{(m)} < \alpha$, $\xi_2^{(m)} > \alpha$, $\eta_2^{(m)} < \beta$, and $\eta_1^{(m)} > \beta$ designate those zeros of ψ_m which are next to α and β, respectively, then one obtains (for $m \to \infty$)

$$\sum_{\xi_k^{(m)} \le x_i^{(m)} \le \eta_k^{(m)}} \frac{\varphi_m(x_i^{(m)})}{\psi_m'(x_i^{(m)})} \to \int_\alpha^\beta \mathrm{e}^{-t^2} dt \quad (k = 1, 2). \tag{4.49}$$

[57] Markov's esteem for Stieltjes and his knowledge of Stieltjes's work on continued fractions up to [1894/95] is evidenced by quotations in *Difference Calculus* [Markov 1896, 97], a textbook, in which—the title considerably differs from the content—mainly problems concerning Gaussian quadrature and continued fractions were discussed.

[58] Radau [1883] and Gourier [1883] explicitly discussed the validity of the Gaussian procedure with weight functions decreasing exponentially for infinite arguments (see [Gautschi 1981, 83]). The usual deductions of (4.17) and (4.18) are actually not dependent on the circumstance whether the range of integration is finite or not, if the weight function has moments of arbitrarily high order.

This means, in the language of Gaussian quadrature, that, for the weight function e^{-t^2} and the constant integrand identical to 1, the quadrature formula converges to the corresponding integral even with regard to arbitrary subintervals.

Markov was now ready to prove a theorem, which differed, according to his own words [Markov 1898/1912, 266], "only in minor details from a theorem of Chebyshev":

If all functions $f_n(\overline{x})$ of the sequence

$$f_1(\overline{x}), f_2(\overline{x}), f_3(\overline{x}) \ldots$$

obey the inequality

$$f_n(\overline{x}) \geq 0,$$

and if the sums

$$\sum_{-\infty}^{+\infty} f_n(\overline{x}), \sum_{-\infty}^{+\infty} \overline{x} f_n(\overline{x}), \sum_{-\infty}^{+\infty} \overline{x}^2 f_n(\overline{x}), \ldots$$

one after another approach the limits

$$\int_{-\infty}^{\infty} e^{-x^2} dx, \int_{-\infty}^{\infty} x e^{-x^2} dx, \int_{-\infty}^{\infty} x^2 e^{-x^2} dx, \ldots,$$

as soon as n grows indefinitely, then the sum

$$\sum_{\alpha}^{\beta} f_n(\overline{x}),$$

taken for all values of $\overline{x}$ which lie within a given interval (α, β), approaches the limit

$$\int_{\alpha}^{\beta} e^{-x^2} dx,$$

if n grows beyond all limits.

With these words Markov articulated a theorem which Chebyshev had neither stated nor justified in any way. The reference to Chebyshev might have been an act of courtesy toward Markov's correspondent and Chebyshev's hagiographer Vasilev. The mode of using sums in Markov's theorem is noticeable. Markov apparently tried to treat discrete and continuous distributions jointly. He did not use the device of Stieltjes integral, which was certainly known to him but did not correspond to St. Petersburg customs.

From (4.49) Markov, with the denotations used in this relation, was able to infer that for a given $\varepsilon > 0$

$$\left| \sum_{\xi_k^{(m)} \leq x_i^{(m)} \leq \eta_k^{(m)}} \frac{\varphi_m(x_i^{(m)})}{\psi_m'(x_i^{(m)})} - \int_{\alpha}^{\beta} e^{-t^2} dt \right| < \frac{\varepsilon}{2} \quad (k = 1, 2), \tag{4.50}$$

if m was sufficiently large.

Markov considered next the partial fraction of mth order $\frac{\overline{\varphi}_m(x)}{\overline{\psi}_m(x)}$ of a continued fraction associated with $\sum_{-\infty}^{\infty} \frac{f_n(z)}{x-z}$. He proved that the roots $\overline{x}_i^{(m)}$ and $x_i^{(m)}$ of the equations $\overline{\psi}_m(x) = 0$ and $\psi_m(x) = 0$, which correspond to each other in ascending order, for sufficiently large n differ only by arbitrarily small amounts from each other. This is true because the roots $\overline{x}_i^{(m)}$ are continuously dependent on the coefficients of the polynomial $\overline{\psi}_m$, and these coefficients continuously on

$$\sum_{-\infty}^{+\infty} f_n(\overline{x}), \sum_{-\infty}^{+\infty} \overline{x} f_n(\overline{x}), \sum_{-\infty}^{+\infty} \overline{x}^2 f_n(\overline{x}), \ldots, \sum_{-\infty}^{+\infty} \overline{x}^{2m-1} f_n(\overline{x}),$$

which expressions converge, according to the presuppositions, to

$$\int_{-\infty}^{\infty} \mathrm{e}^{-x^2} dx, \int_{-\infty}^{\infty} x\mathrm{e}^{-x^2} dx, \int_{-\infty}^{\infty} x^2\mathrm{e}^{-x^2} dx, \ldots, \int_{-\infty}^{\infty} x^{2m-1}\mathrm{e}^{-x^2} dx.$$

The roots $x_i^{(m)}$ are dependent on the latter quantities in an analogous way. Since, by the reasons just mentioned, not only the coefficients of $\overline{\psi}_m$ and ψ_m, but also of $\overline{\varphi}_m$ and φ_m get arbitrarily close to each other for a sufficiently large n, Markov [1898/1912, 268] was able to conclude that

$$\frac{\overline{\varphi}_m(\overline{x}_i^{(m)})}{\overline{\psi}'_m(\overline{x}_i^{(m)})} - \frac{\varphi_m(x_i^{(m)})}{\psi'_m(x_i^{(m)})} \to 0 \quad (n \to \infty)$$

for pairs of roots $\overline{x}_i^{(m)}$ and $x_i^{(m)}$ which correspond to each other.

For those roots $\overline{\xi}_k^{(m)}$ and $\overline{\eta}_k^{(m)}$ of $\overline{\psi}_m(x) = 0$ $(k = 1,2)$ which are analogous to the already defined roots $\xi_k^{(m)}$ and $\eta_k^{(m)}$ of $\psi_m(x) = 0$, respectively, one obtains, for sufficiently large n and $k = 1, 2$:

$$\left| \sum_{\xi_k^{(m)} \le x_i^{(m)} \le \eta_k^{(m)}} \frac{\varphi_m(x_i^{(m)})}{\psi'_m(x_i^{(m)})} - \sum_{\overline{\xi}_k^{(m)} \le \overline{x}_i^{(m)} \le \overline{\eta}_k^{(m)}} \frac{\overline{\varphi}_m(\overline{x}_i^{(m)})}{\overline{\psi}'_m(\overline{x}_i^{(m)})} \right| < \frac{\varepsilon}{2}. \tag{4.51}$$

By virtue of Chebyshev's inequalities, applied to the function f_n, the inequalities

$$\sum_{\alpha}^{\beta} f_n(x) > \sum_{\overline{\xi}_2^{(m)} \le \overline{x}_i^{(m)} \le \overline{\eta}_2^{(m)}} \frac{\overline{\varphi}_m(\overline{x}_i^{(m)})}{\overline{\psi}'_m(\overline{x}_i^{(m)})}$$

and

$$\sum_{\alpha}^{\beta} f_n(x) < \sum_{\overline{\xi}_1^{(m)} \le \overline{x}_i^{(m)} \le \overline{\eta}_1^{(m)}} \frac{\overline{\varphi}_m(\overline{x}_i^{(m)})}{\overline{\psi}'_m(\overline{x}_i^{(m)})}$$

hold.

Altogether, Markov showed that, for an arbitrarily small $\varepsilon > 0$, one can always find a natural number $m(\varepsilon)$ such that (4.50) is valid. Moreover, for all natural n above a certain bound $N(m(\varepsilon))$, (4.51) had to be valid. Taking into consideration the inequalities just quoted for $\sum_{\alpha}^{\beta} f_n(x)$, it could be inferred that for all $n > N(m(\varepsilon))$

$$\int_{\alpha}^{\beta} \mathrm{e}^{-x^2} dx - \frac{\varepsilon}{2} - \frac{\varepsilon}{2} < \sum_{\alpha}^{\beta} f_n(x) < \int_{\alpha}^{\beta} \mathrm{e}^{-x^2} dx + \frac{\varepsilon}{2} + \frac{\varepsilon}{2},$$

which was the chief assertion of his article.

Markov [1898/1912, 269 f.] argued that "almost immediately, as Chebyshev has noticed," the CLT could be followed:
Let $u_1, u_2, \ldots$ be "independent quantities," obeying the following conditions:

1) $\mathrm{E}u_k = 0$
2) For all natural $m \geq 2$ there exists a constant C_m such that $|\mathrm{E}u_k^m| < C_m$ for all $k \in \mathbb{N}$
3) $\mathrm{E}u_k^2$ "does not get infinitely small, if k grows indefinitely."[59]

Then

$$P\left(\alpha\sqrt{2\sum_{k=1}^{n}\mathrm{E}u_k^2} \leq \sum_{k=1}^{n} u_k \leq \beta\sqrt{2\sum_{k=1}^{n}\mathrm{E}u_k^2}\right) \to \frac{1}{\sqrt{\pi}}\int_{\alpha}^{\beta} \mathrm{e}^{-x^2} dx \quad (n \to \infty)$$

holds for any $\alpha < \beta$.

Indeed, in his 1898 paper Markov did not prove that the moments of each order of the suitably normed sum of random variables converge to those of the normal distribution respectively, as it would have been essential for the application of his main theorem to the case of the CLT. He gave that proof, however, in his already mentioned exchange of letters with Vasilev, which was eventually published in 1899, and also, under somewhat weaker conditions, in his *Textbook on Probability* from its second edition in connection with a "theorem on mathematical expectations."[60]

The content of this theorem was as follows: Let $X_1, X_2, \ldots$ be a sequence of independent random variables, each with expectation $\mathrm{E}X_k$ and variance $\sigma_k^2 > 0$, respectively, where

$$\frac{\sum_{k=1}^{n} \mathrm{E}|(X_k - \mathrm{E}X_k)^r|}{(\sum_{k=1}^{n} \sigma_k^2)^{\frac{r}{2}}} \to 0 \quad (n \to \infty) \qquad (4.52)$$

[59] It does not become clear from this phrasing whether Markov assumed that the sequence $(\mathrm{E}u_k^2)$ was not allowed to converge to 0 or whether he focused on the even stronger assumption that, from a certain k, the quantity $\mathrm{E}u_k^2$ always lies above a certain positive lower bound. As we will see below, Markov ought to have specified the mode of nonconvergence more precisely if he had aimed at the weaker assumption. In his exchange of letters with Vasilev, Markov [1899/2004, 135] instead of 3) used the condition that

$$\frac{\mathrm{E}(u_1 + u_2 + \cdots + u_n)^2}{n}$$

"cannot be arbitrarily small" (see also [Sheynin 1989, 361]).

[60] The proof of this theorem was also separately published in 1907, see [Sheynin 1989, 362]). Essential differences between the original proof in the exchange of letters with Vasilev and the more elaborate versions, which appeared later, do not exist.

and

$$\frac{\sum_{k=1}^{n}(\sigma_k^2)^{r-1}}{(\sum_{k=1}^{n}\sigma_k^2)^{r-1}} \to 0 \quad (n \to \infty) \tag{4.53}$$

for all natural $r \geq 3$. Then, for $n \to \infty$ the expectation of the power

$$\left(\frac{\sum_{k=1}^{n}(X_k - \mathrm{E}X_k)}{\sqrt{2\sum_{k=1}^{n}\sigma_k^2}}\right)^m$$

tends to the limit

$$\frac{1}{\sqrt{\pi}}\int_{-\infty}^{\infty} t^m \mathrm{e}^{-t^2}\,dt$$

for each natural m.

In order to simplify the exposition, it is supposed in the following discussion of Markov's proof (according to [Markov 1912, 77–81]) that $\mathrm{E}X_k = 0$, and Markov's notations $c_k^{(l)} = \mathrm{E}|X_k^l|$ and $B_n = \sum_{k=1}^{n}\mathrm{E}X_k^2$ are used. Quantities depending on multiple indices are expressed in a modern multi-index notation.

Because of the generalized binomial formula we obtain

$$\mathrm{E}\left(\frac{\sum_{k=1}^{n}X_k}{\sqrt{2B_n}}\right)^m = \frac{1}{2^{\frac{m}{2}}B_n^{\frac{m}{2}}}\mathrm{E}\left(\sum_{|\alpha|=m}\frac{m!}{\alpha!}X_1^{\alpha_1}\cdots X_n^{\alpha_n}\right).$$

In this equation $\alpha = (\alpha_1, \ldots, \alpha_n)$ denotes a multi-index from $(\mathbb{N}_0)^n$, where $|\alpha| = \sum_{k=1}^{n}\alpha_k$ and $\alpha! = \alpha_1!\cdots\alpha_n!$. Because the random variables are independent, and because $\mathrm{E}X_k = 0$ for all k, the latter sum equals

$$\frac{1}{2^{\frac{m}{2}}B_n^{\frac{m}{2}}}\sum_{|\alpha|=m;\,\alpha_i\neq 1}\frac{m!}{\alpha!}\mathrm{E}X_1^{\alpha_1}\cdots\mathrm{E}X_n^{\alpha_n}.$$

We therefore reach the following estimate:

$$\begin{aligned}
&\left|\mathrm{E}\left(\frac{\sum_{k=1}^{n}X_k}{\sqrt{2B_n}}\right)^m - \frac{m!}{2^{\frac{m}{2}}B_n^{\frac{m}{2}}}\sum_{\substack{|\alpha|=m\\ \alpha_i=2\vee\alpha_i=0}}\frac{1}{\alpha!}\mathrm{E}X_1^{\alpha_1}\cdots\mathrm{E}X_n^{\alpha_n}\right| \\
&\qquad\leq \frac{m!}{2^{\frac{m}{2}}B_n^{\frac{m}{2}}}\sum_{\substack{|\alpha|=m\\ \alpha_i\geq 3\vee\alpha_i=0}}\frac{c_1^{(\alpha_1)}\cdots c_n^{(\alpha_n)}}{\alpha!} \\
&\qquad= \frac{m!}{2^{\frac{m}{2}}}\sum_{\substack{|\alpha|=m\\ \alpha_i\geq 3\vee\alpha_i=0}}\frac{c_1^{(\alpha_1)}}{B_n^{\frac{\alpha_1}{2}}\alpha_1!}\frac{c_2^{(\alpha_2)}}{B_n^{\frac{\alpha_2}{2}}\alpha_2!}\cdots\frac{c_n^{(\alpha_n)}}{B_n^{\frac{\alpha_n}{2}}\alpha_n!}.
\end{aligned}$$

For $m \geq 3$ we have

$$\sum_{\substack{|\alpha|=m \\ \alpha_i \geq 3 \vee \alpha_i = 0}} \frac{c_1^{(\alpha_1)}}{B_n^{\frac{\alpha_1}{2}} \alpha_1!} \frac{c_2^{(\alpha_2)}}{B_n^{\frac{\alpha_2}{2}} \alpha_2!} \cdots \frac{c_n^{(\alpha_n)}}{B_n^{\frac{\alpha_n}{2}} \alpha_n!}$$

$$\leq \sum_{\substack{\beta_1+\beta_2+\cdots+\beta_r=m \\ \beta_i, r \in \mathbb{N}, \beta_1 \geq \beta_2 \geq \cdots \geq \beta_r \geq 3}} \frac{\sum_{k=1}^n c_k^{(\beta_1)}}{B_n^{\frac{\beta_1}{2}} \beta_1!} \frac{\sum_{k=1}^n c_k^{(\beta_2)}}{B_n^{\frac{\beta_2}{2}} \beta_2!} \cdots \frac{\sum_{k=1}^n c_k^{(\beta_r)}}{B_n^{\frac{\beta_r}{2}} \beta_r!}.$$

Because of (4.52), each of the product terms (whose number is independent of n) tends to 0 as $n \to \infty$. Altogether it follows for $m \geq 3$ and $n \to \infty$:

$$\left| \mathrm{E}\left(\frac{\sum_{k=1}^n X_k}{\sqrt{2B_n}} \right)^m - \frac{m!}{2^{\frac{m}{2}} B_n^{\frac{m}{2}}} \sum_{\substack{|\alpha|=m \\ \alpha_i = 2 \vee \alpha_i = 0}} \frac{1}{\alpha!} \mathrm{E}X_1^{\alpha_1} \cdots \mathrm{E}X_n^{\alpha_n} \right| \to 0. \tag{4.54}$$

In the particular case $m = 2$ the difference on the left side is identical to 0 already for finite n, such that (4.54) is valid even for all $m \geq 2$.

Let m be an odd natural number. Then the joint validity of the conditions $|\alpha| = m$ and $\alpha_i = 2$ or $\alpha_i = 0$ is impossible. Therefore, we have for $m \in \mathbb{N}$ odd:

$$\sum_{\substack{|\alpha|=m \\ \alpha_i = 2 \vee \alpha_i = 0}} \frac{1}{\alpha!} \mathrm{E}X_1^{\alpha_1} \cdots \mathrm{E}X_n^{\alpha_n} = 0.$$

Because of the assumption of vanishing expectations for the single random variables, (4.54) holds for $m = 1$ as well. Altogether, from (4.54) it follows that

$$\mathrm{E}\left(\frac{\sum_{k=1}^n X_k}{\sqrt{2B_n}} \right)^m \to 0 = \frac{1}{\sqrt{\pi}} \int_{-\infty}^{\infty} t^m \mathrm{e}^{-t^2} dt \quad (n \to \infty)$$

for all odd natural numbers m.

Let m be an even number now. Then we have

$$B_n^{\frac{m}{2}} = \left(\sum_{k=1}^n \sigma_k^2 \right)^{\frac{m}{2}} = \sum_{|\alpha|=\frac{m}{2}} \frac{\left(\frac{m}{2}\right)!}{\alpha!} \sigma_1^{2\alpha_1} \sigma_2^{2\alpha_2} \cdots \sigma_n^{2\alpha_n}.$$

This implies

$$\frac{1}{B_n^{\frac{m}{2}}} \left| B_n^{\frac{m}{2}} - \sum_{\substack{|\alpha|=\frac{m}{2} \\ \alpha_i = 1 \vee \alpha_i = 0}} \frac{\left(\frac{m}{2}\right)!}{\alpha!} \sigma_1^{2\alpha_1} \sigma_2^{2\alpha_2} \cdots \sigma_n^{2\alpha_n} \right| = \frac{1}{B_n^{\frac{m}{2}}} \sum_{\substack{|\alpha|=\frac{m}{2} \\ \alpha_i > 1 \vee \alpha_i = 0}} \frac{\left(\frac{m}{2}\right)!}{\alpha!} \sigma_1^{2\alpha_1} \sigma_2^{2\alpha_2} \cdots \sigma_n^{2\alpha_n}.$$

The latter term has the upper bound

$$\left(\frac{m}{2}\right)! \sum_{\substack{\beta_1+\beta_2+\cdots+\beta_r=\frac{m}{2} \\ \beta_i, r\in\mathbb{N}, \beta_1\geq\beta_2\geq\cdots\geq\beta_r\geq 2}} \frac{\sum_{k=1}^{n}\sigma_k^{2\beta_1}}{B_n^{\beta_1}\beta_1!}\frac{\sum_{k=1}^{n}\sigma_k^{2\beta_2}}{B_n^{\beta_2}\beta_2!}\cdots\frac{\sum_{k=1}^{n}\sigma_k^{2\beta_r}}{B_n^{\beta_r}\beta_r!}.$$

On account of the presupposition (4.53) this expression tends to 0 as $n \to \infty$, and therefore, as $n \to \infty$

$$\frac{1}{B_n^{\frac{m}{2}}} \sum_{\substack{|\alpha|=\frac{m}{2} \\ \alpha_i=1\vee\alpha_i=0}} \sigma_1^{2\alpha_1}\sigma_2^{2\alpha_2}\cdots\sigma_n^{2\alpha_n} = \frac{1}{B_n^{\frac{m}{2}}} \sum_{\substack{|\alpha|=m \\ \alpha_i=2\vee\alpha_i=0}} \mathrm{E}X_1^{\alpha_1}\cdots\mathrm{E}X_n^{\alpha_n} \to \frac{1}{\left(\frac{m}{2}\right)!}. \tag{4.55}$$

We have $\alpha! = 2^{\frac{m}{2}}$ if the conditions m even, and $|\alpha| = m$, and $\alpha_i = 2$ or $\alpha_i = 0$ are jointly valid. Thus, for even m, (4.54) and (4.55) imply

$$\mathrm{E}\left(\frac{\sum_{k=1}^{n} X_k}{\sqrt{2B_n}}\right)^m \to \frac{m!}{2^m\left(\frac{m}{2}\right)!} = \frac{1\cdot 3\cdot 5\cdots(m-1)}{2^{\frac{m}{2}}} = \frac{1}{\sqrt{\pi}}\int_{-\infty}^{\infty} t^m e^{-t^2}dt \quad (n\to\infty).$$

Toward the end of his exposition, Markov [1912, 80] noticed that condition (4.53) was superfluous, because it could be deduced from the first condition (4.52) by means of the inequality

$$\left(\sigma_k^2\right)^{r-1} \leq \mathrm{E}(X_k - \mathrm{E}X_k)^{2r-2} \quad (r \geq 3), \tag{4.56}$$

"whose proof does not require great effort." For an arbitrary distribution function F the successive application of the Schwarz inequality actually yields

$$\int_{-\infty}^{\infty} x^2 dF(x) = \int_{-\infty}^{\infty} x^2\cdot 1 dF(x) \leq \left(\int_{-\infty}^{\infty} x^4 dF(x)\right)^{\frac{1}{2}} \leq \cdots \leq \left(\int_{-\infty}^{\infty} x^{2n} dF(x)\right)^{\frac{1}{n}}$$

for $n \geq 2$. After the substitution $n = r - 1$ the inequality (4.56) follows. Markov did not make explicit that condition (4.52) could be deduced from the conditions 2) and 3) in his 1898 version of the CLT. Indeed, for natural r the following inequality holds:

$$\mathrm{E}|u_k^r| \leq K_r,$$

$$K_r = \begin{cases} C_r, & \text{if } r \text{ even} \\ 1 + C_{r+1}, & \text{if } r \text{ odd.} \end{cases}$$

The constants C_r are determined in accordance with Markov's condition 2).[61] It follows

$$\frac{\sum_{k=1}^{n} \mathrm{E}|u_k^r|}{\left(\sum_{k=1}^{n} \sigma_k^2\right)^{\frac{r}{2}}} < \frac{nK_r}{\left(\sum_{k=1}^{n} \sigma_k^2\right)^{\frac{r}{2}}}.$$

If we want for $r \geq 3$ the latter expression tending to 0 as $n \to \infty$, we have to demand

$$\frac{n}{\left(\sum_{k=1}^{n} \sigma_k^2\right)^{\frac{3}{2}}} \to 0 \quad (n \to \infty). \tag{4.57}$$

Therefore, it is possible that a partial sequence of σ_k^2 tends to 0, if only the "additional condition" (4.57) is maintained.

Markov's proof of the CLT was based on two main points. The one was an elegant, but by no means elementary analytical theorem on the interrelation of moment convergence and the convergence of linear partial masses. It was not absolutely necessary for his line of argument that Markov restricted his investigations to a normal limit distribution. The other main point was a theorem on the convergence of moments of normed sums of independent random variables to the corresponding moments of the normal distribution. The proof of this latter theorem was elementary, though rather cumbersome. In his 1898 publication, which reached a broader audience even outside Russia, Markov presented the CLT as a mere corollary of his first theorem, which in its generality went far beyond probability theory. Already the title "Sur les racines de l'équation ..." hints at the fact that this publication primarily dealt with nonstochastic contents. In this stage of Markov's work, the CLT scarcely had the character of an independent research subject. This attitude of Markov would change during the first decade of the 19th century, mainly caused by his competition with Nekrasov and Lyapunov. However, for the time being, the CLT apparently served Markov in a way similar as it did to Chebyshev: mainly as an illustration of methods and results on moments and continued fractions.

4.7 Chebyshev's and Markov's Central Limit Theorem: Starting Point of a New Theory of Probability?

Survey articles on the development and current state of probability theory in Russia written during the Stalinist era like those by Bernshtein [1940/2004a] or

---

[61] If j is even, then it follows from condition 2):

$$\mathrm{E}|u_k^j| = |\mathrm{E}u_k^j| < C_j.$$

Let j be odd, and let F_k be the distribution function of u_k. Then

$$\mathrm{E}|u_k^j| = \int_{-\infty}^{\infty} |x|^j \, dF_k(x) \leq \int_{|x|\leq 1} dF_k(x) + \int_{|x|>1} x^{j+1} dF_k(x) \leq 1 + \mathrm{E}u_k^{j+1} < 1 + C_{j+1}.$$

Kolmogorov [1947/2005], by nature had to praise the achievements of Soviet Science. The arguments for the superiority of Russian contributions also included the reference to a long tradition of probability theory in Russia, beginning already before 1850 with courses on probability theory given at St. Petersburg University by Ankudivich and Bunyakovskii [Bernshtein 1940/2004a, 102 f.; Kolmogorov 1947/2005, 72]. Chebyshev was considered the real founder of the "St. Petersburg school," however. Under Stalin's leadership, mathematics, even pure mathematics, remained relatively unoffended during most times. Still, mathematicians had to take care not to conduct "science for science's sake in an ivory tower" [Lorentz 2002, 195] and not to base their work on "idealistic" philosophy. In view of such problems, Chebyshev, who seemed to have exemplarily reconciled mathematical rigor and orientation toward significant and useful applications, could likewise serve as an authority [Bernshtein 1940/2004a, 104; Kolmogorov 1947/2005, 72 f., 75]. Also regarding Markov as the main successor of Chebyshev, the obligatory affinity to problems of the "real world" was emphasized [Bernshtein 1940/2004a, 108–110; Kolmogorov 1947/2005, 74]. After surveying Chebyshev's, Lyapunov's, and Markov's work—in particular on the CLT—Bernshtein [1940/2004a, 110] even expressed his patriotic feelings (maybe imposed upon him) with the words

> For its transformation from mathematical amusement into a method of natural sciences the theory of probability is mainly obliged to the Petersburg school, which accomplished this fundamental progress leaving west European mathematicians far behind.

Pointing to "mathematical amusement" versus "method of science," Bernshtein possibly wanted to allude to conflicts between mathematicians who contributed to the stochastic theory of genetics, as Kolmogorov and himself, and Stalin's chief geneticist Lysenko, who opposed probabilistic models and stressed the significance of planned exterior influence [Birkner 1996; Lorentz 2002, 216 f.; Roll-Hansen 2008; Sheynin 2009, 113 f.].

Although Bernshtein's and Kolmogorov's expositions on Chebyshev's and Markov's role seem to be (partially) exaggerated, they also repeat assessments already expressed during pre-Soviet times[62] or by non-Russians like von Mises. In his very important paper [1919a] on probabilistic limit theorems (see introductional part of Chapt. 5), von Mises deplored the lack of analytic rigor in most contributions that had appeared up to that time—with the exception of "a small number of works by Russian mathematicians." With this remark, he mainly hinted at Chebyshev's and Markov's contributions to the CLT. Altogether, the need for a (critical) discussion of the role of Chebyshev's and Markov's work, especially on the CLT, seems to be obvious. In this discussion, we shall mainly focus on stochastic concepts, analytic methods, mathematical rigor, and applications.

---

[62] By Vasilev and Lyapunov, for example, see introduction to the present chapter.

4.7.1 Random Variables and Limit Theorems

As we have already seen, Chebyshev did not prove the CLT rigorously. His proof can even be considered a failure, because he did not surmount the difficulties connected with the "usual" device of cutting off series expansions. This fundamental fact notwithstanding, Chebyshev's theorem seems important for two reasons: First, it was—in the same way as his weak law of large numbers of 1866—stated for general "quantities." Second, Chebyshev was the first to express the CLT as a limit theorem proper, and to explicitly state conditions for the validity of the assertion. So, he made the CLT independent from direct relations to specific applications.

Chebyshev clearly distinguished between the "quantities" themselves and their values, as we can see for example in his 1866 paper concerning the Bienyamé–Chebyshev inequality and the weak law of large numbers. His conception of a random variable was considerably more general than Poisson's (see Sect. 2.2.1). In contrast to Poisson, for Chebyshev [1936/2004, 210, 227] it was a matter of course that a linear combination $\sum_{k=1}^{n} \lambda_k \epsilon_k$ of errors or a sum of squared errors $\sum_{k=1}^{n} \epsilon_k^2$ could be conceived as a sum of the random variables $\lambda_k \epsilon_k$ or ϵ_k^2, respectively, and thus could be subjected to the CLT. Markov and Lyapunov took up Chebyshev's notion of random variable in their contributions on probability theory, and thus paved the way for Kolmogorov's development [1933a, 20] of an abstract random variable as a measurable mapping from one sample space into another. We should, however, not forget that Laplace [1781] had already used the designation "quantités variables" as a general term representing different concrete cases, in which real numbers are gained by random experiments.

Until Chebyshev, the CLT mainly had the character of an assertion on an approximation of probabilities of sums by normal distributions, if the number of summands was "large." But it did not give an indication of "how large" this number should be for a sufficiently precise approximation. At several occasions in his work,[63] Markov pointed out the need of estimating the imprecision of approximating formulae, in particular with respect to applications such as those in error theory. Chebyshev's opinion had been very likely the same in this matter. Thus, Chebyshev with his version of a proper limit theorem gave this topic a new, purely analytic quality beyond classical probability.

4.7.2 Analytic Methods and Rigor

The main activities of the St. Petersburg school were in the theory of approximation, of moments, and of probability. As we have seen, in Chebyshev's work the theory of probability was closely connected with the two other disciplines. The same is

[63] [Markov 1899/2004, 139 f.; 1912, I], for example.

true for Markov's work until the turn of the century.[64] The mathematical methods Chebyshev applied were algebraic to a large extent (see Sect. 4.2.3 for a characteristic example) and quite often resorted to continued fractions. Yet, he also cultivated a rich reservoir of analytic methods including those of complex analysis, as we have seen in connection with his proof that a normal distribution is entirely determined by its moments (see Sect. 4.4).

Chebyshev had a "pragmatic" attitude, shared by many of his disciples [Steffens 2006, x], toward the foundations of analysis, and he easily used infinitely small and large quantities in a rather intuitive way. He often passed over from discrete cases to continuous ones without any further explanations: If an assertion was proven in connection with the continued fraction associated with $\sum_{k=1}^{n} \frac{f(x_k)}{x_k - t}$, then it was a matter of course that the analogous assertion for the continued fraction associated with $\int_a^b \frac{f(x)}{x-t} dx$ was true as well. In this sense, at several occasions it seemed more important for Chebyshev to derive and expound methods for problem solving than to deliver complete proofs. Altogether, Chebyshev's methods may be characterized by the headword "algebraic analysis," however with a main focus on continued fractions rather than on power series.

Chebyshev, who had already started analytic research during the 1840s, remained uninfluenced by "Weierstrassian" analytic standards. He refused "philosophizing" on foundational aspects of mathematics, like "what an infinitely small quantity is," because this "does not lead to anything."[65] Thus we should not be too surprised about his frequent interchange of limit processes without giving any justifications in his proof of the CLT.

The mathematical rigor in Chebyshev's work originated from three sources: seeking explicit upper and lower bounds, algebraic arguments at crucial points of proofs, and the reduction of quantities with "infinite" or "continuous" ranges of values to those with finite or discrete ones, where, however, Chebyshev did not care about possible difficulties connected with the mutual transition between such different cases. This program of research and these methods also established the common ground of the St. Petersburg school. Chebyshev's disciples, such as Possé and Markov, went only cautiously beyond the limits of Chebyshev's program and gradually approached the "Weierstrassian" standards of modern analysis. A close relationship to Chebyshev's analytic approach and to his methods and typical problems was maintained in many cases. In contrast to Chebyshev, Markov made the correspondence between the limits of moments and those of distribution functions a central theme, and he advanced the use of such moment methods for dealing with probabilistic limit problems. Lyapunov, Markov's former fellow student, used

[64] For the activities of the St. Petersburg school in approximation theory see [Steffens 2006, Chap. 3].

[65] See Steffens [2006, 74 f.] for the English translation of two quotations from lecture notes of Chebyshev's 1876/77 course on probability theory (different from the course written down by Lyapunov), which was discovered and concisely described by Ermaloeva [1987]. Steffens interprets Chebyshev's words in the sense of a complete disinterest in analytic basics.

"transcendental" methods in his proofs of the CLT, like Fourier transforms, which had been rather neglected by Chebyshev[66] and by most scholars of the St. Petersburg school.

4.7.3 The Role of the Central Limit Theorem in Chebyshev's and Markov's Work

Applications of the CLT within and outside mathematics were rather rare in Chebyshev's work: By the time of issuing his papers in 1846 and 1866, Chebyshev had already started his research on probabilistic limit theorems with the consequence that the CLT was no longer needed for proving (weak) laws of large numbers. He did not discuss at any place in his work stochastic models explaining the occurrence of normal distributions, like the hypothesis of elementary errors. In his 1887 article on the CLT he only touched upon the possibility of establishing the method of least squares by means of the CLT en passant (see Sect. 4.6.2). In his lecture course of 1879/80, Chebyshev [1936/2004, 209–212] first justified the principle of arithmetic mean showing that the "probable error" connected with the estimation of a quantity from (a large number of) direct observations becomes minimal if the arithmetic mean is taken as a compromise value; like Laplace, he tacitly assumed that the CLT even provided an exact normal distribution for linear combinations of numerous errors. He then advanced [1936/2004, 213–217] the Gaussian derivation of the normal law and the therefrom deduced principle of least squares (see Sect. 3.1). Subsequently, he [1936/2004, 217–220] recapitulated Laplace's method, again based on the CLT, for establishing the method of least squares in the case of determining one element from (numerous) linear equations (see Sect. 2.1.5.2). There are no indications that Chebyshev assigned the CLT a particularly important role for stochastic applications beyond this quite restricted field of asymptotic error theory. On the other hand, in his 1887 proof of the CLT, Chebyshev clearly violated the principles of mathematical rigor which had been formulated by himself, and, in this respect, there was scarcely an essential improvement compared with his probability course. Thus, he did not succeed in establishing the CLT as one of the major mathematical theorems independent of its practical applicability. This circumstance has caused considerable irritations among Chebyshev's hagiographers.[67] Yet Chebyshev's lack of rigor can be explained if one supposes in connection with the CLT as his main goal the demonstration of results and methods of moment theory rather

---

[66] Chebyshev [1936/2004, 62–86] in his already mentioned 1879/80 course on probability and related topics discussed Fourier integrals and some implications rather carefully, however.

[67] See, for example, [Maistrov 1974, 205 f.], [Bernshtein 1945/2004a, 82–84], [Kolmogorov 1948], and the summarizing survey in [Sheynin 1989, 361]. Bernshtein (same place) even showed how Chebyshev's arguments could be supplemented under additional assumptions by rigorous considerations on the convergence of the moments of the standardized sum to the corresponding moments of the standard normal distribution. In his deduction, Bernshtein clearly went far beyond Chebyshev's analytic scope, however.

than an entirely rigorous proof. Then, Chebyshev's CLT did not have a completely autonomous status within mathematics, but was rather significant as an illustration of certain aspects of the theory of moments.

Markov started probabilistic research in a stricter sense at a relatively late stage of his career only, despite his quite active teaching of probability theory.[68] Apparently, at first he did not assign limit theorems a larger significance for stochastic applications, which in turn played a significant role in his courses. This is shown by the first and second editions (1900, 1908) of his textbook based on his courses, whose second edition also appeared in a German translation with the title "Wahrscheinlichkeitsrechnung" [Markov 1912].[69] Regarding the CLT, Markov [1912, 69–76] basically recapitulated Poisson's derivation (see Sect. 2.2.5) of an approximate normal distribution. At the end of this section, he pointed out that an estimate for the deviation between the real probability and the Gaussian law was out of reach on the basis of the methods employed, and one could only "suppose" that $\frac{1}{\sqrt{\pi}}\int_\alpha^\beta e^{-x^2}dx$ was the "limit" of the considered probability. For proofs of this "limit theorem" he referred to a list of references, which also included his [1898] and the two articles by Lyapunov [1900; 1901b] (see Sect. 5.1.3).[70] In the next section, Markov [1912, 77–80] expounded his proof that the "mathematical expectation" of the power

$$\left(\frac{\sum X_i - \sum \mathrm{E}X_i}{\sqrt{2\sum \sigma_{X_i}^2}}\right)^m$$

for any natural m tends to

$$\frac{1}{\sqrt{\pi}}\int_{-\infty}^{\infty} t^m e^{-t^2}dt$$

under certain conditions. However, this was only the more elementary part of his complete proof of the CLT. A reason for this insufficient consideration of his own proof of the CLT in the first and second editions of his textbook may have been that Markov originally considered probability theory as a discipline of "applied mathematics." Approximation formulae were needed for applications, which, however, could only be reasonably used if there existed estimates on approximation errors. Such estimates could not be obtained by Markov's methods for proving the CLT. Lyapunov [1900; 1901b], on the other hand, had achieved (quite narrow) bounds for the deviation of the limit term from the exact probability under very weak conditions, but his methods were unsuitable for the exposition in a textbook. Markov could have at least communicated Lyapunov's formula; except for citing the titles of Lyapunov's papers, he did not comment on the latter's results, however. It may

[68] For general surveys of Markov's work in probability and statistics see [Sheynin 1989; 2005b].

[69] To the third edition (1913) considerably new results, especially such regarding limit theorems, were added. The fourth even further enlarged version (1924) was posthumously issued.

[70] At least to the German edition [Markov 1912], a somewhat loose translation of [Markov 1898] was added as a supplement.

be that around the turn of the century Markov increasingly felt a rivalry toward Lyapunov on probabilistic issues. In fact, after 1900 this rivalry inspired Markov to a series of articles on stochastic limit theorems (in particular regarding weakly dependent random variables), in which applications did not play any role (see Sect. 5.1.5). These new results were included in the third and fourth editions of his textbook on probability theory.

As it seems, Markov considered the CLT rather as a corollary of more general moment theoretic results in his first contributions. Like Chebyshev, Markov did not emphasize the CLT as a mathematically autonomous subject within a self-contained theory of probability. With Chebyshev and the younger Markov, the CLT was not yet studied in its own right, neither was application a priority, as it had been in classical probability. The CLT was rather relevant because of the analytic methods which were used in its proof. Chebyshev and Markov very clearly stated conditions and assertions of the CLT, and in a way entirely unusual until their time, they differentiated between "limit theorem" and "approximation." So, they significantly influenced the development toward the modern CLT, for which, however, the reconciliation of "Petersburg" and "transcendental" methods, like characteristic functions, was decisive. Markov himself would continue to contribute to this development even after the turn of the century.

Chapter 5
The Way Toward Modern Probability

Richard von Mises [1919a, 1] wrote in his pioneering paper "Fundamentalsätze der Wahrscheinlichkeitsrechnung":

> The *analytical theorems* of probability theory are lacking—except for a few works by Russian mathematicians—the *precision of formulation and reasoning* which has long been a matter of course in other areas of analysis. And in spite of some valuable approaches, these days there is still an almost complete lack of clarity about the *foundations of probability theory as a mathematical discipline.*

Von Mises dealt with the problem of the axiomatic foundations of probability theory by a frequentistic approach in another article [1919b], which has achieved greater prominence today than the one that preceded it.[1] In his criticism of the inadequate fundamentals of probability theory, von Mises alluded to the lecture David Hilbert had delivered before the 2nd International Congress of Mathematicians, held in Paris in 1900, in which Hilbert had presented "his" 23 problems. The sixth problem related to the "mathematical treatment of the axioms of physics." Following the model of his *Grundlagen der Geometrie* [1899], Hilbert [1900, 47] challenged his colleagues:

> ...*to treat in the same manner, by means of axioms, those physical sciences in which mathematics plays an important part; in the first rank are the theory of probabilities and mechanics.*

Hilbert included probability theory in the study of physics likely because, as he observed in the next passage, he found stochastic considerations in the kinetic theory of gases to be particularly important. In classifying probability theory as a discipline that did not belong to mathematics in a narrower sense, Hilbert was following the prevailing contemporary view. In the tradition of the Laplacian concept, probability theory was still largely regarded as a "universal discipline" at the turn of the century,

[1] Bernhardt [1984] gave a particularly detailed account of contributions surrounding von Mises's work on the foundations of probability theory to the point of demonstrating the consistency of his (somewhat modified) axiom system in the 1960s. For a more philosophical view see [von Plato 1994, 183–189, 233–237]. Further historical material can be found in [Siegmund-Schultze 2006].

H. Fischer, *A History of the Central Limit Theorem*, Sources and Studies in the History of Mathematics and Physical Sciences, DOI 10.1007/978-0-387-87857-7_5,

even if people were no longer willing to share Laplace's optimism regarding its universal applicability.

Toward the end of the 19th century, a "new rigor" and a style that could be traced back to Karl Weierstrass's approach had established themselves in mathematical analysis. Admittedly, most considered this rigor to be unnecessary in the application of analysis in natural sciences and technology, and thus in probability theory as well, since the value of an examination was determined less by a completely consistent mathematical deduction than the closest possible agreement between calculation and experiment. Yet if, like Hilbert [1900, 48], mathematicians "took account not only of those theories coming close to reality, but also of all logically possible theories," they were dependent above all upon absolute analytical rigor in addition to axiomatics.

A willingness to plumb all conceivable possibilities, regardless of their practical usefulness, can be found in stochastics as far back as the mid-19th century in Cauchy's papers on error analysis (see Sect. 2.5). Yet while the disciplines of pure mathematics had made significant progress around the turn of the century in dissociating themselves from external criteria of value and truth, probability theory as an applied discipline maintained resistance into the 1920s and 1930s against "mathematical modernity." This included Émile Borel's camp, whose followers were once again attempting to establish probability theory as a universal scientific discipline appertaining to mathematics only in a wider sense.

The use of the adjective "modern" in this book essentially follows the model of Mehrtens [1990] in characterizing an attitude toward mathematical work which concentrates exclusively on creating structural references and thus pays no regard to external criteria (see Sect. 1.2). However, whereas Mehrtens is primarily interested in the attitudes of mathematicians with respect to fundamental issues, the present examination deals with approaches to mathematical problems where the main focus is not on fundamental principles. It is not the way in which one speaks about mathematics but rather the manner of work within mathematics that is of primary importance here. Thus, there is no contradiction in the fact that many proponents of modern probability theory, such as the aforementioned von Mises, dismissed an exclusively formalistic and, in Mehrtens's words, "modern" notion of mathematical fundamentals, while at the same time working in a very similar way on the modernization of probability theory, with the papers they produced ultimately establishing relationships among largely abstract concepts which no longer required any extra-mathematical reference. The transition to a modern probability theory was not necessarily and exclusively bound to logically satisfying axiomatics and basic concepts founded on measure theory, but instead began with a willingness to sound out stochastic elements in their purely mathematical context to the greatest extent possible and to permit them to have an autonomous inner-mathematical meaning. In the language of sociology of scientific knowledge this means that the essential characteristic of modern probability theory consists in sounding its contingency.

A fragmentary essay by Hausdorff, parts of which have recently been published,[2] shows that autonomy was perceived as substantial for modernity already around 1900. The "principal duty of modern mathematics" Hausdorff saw "in struggling through one's way from heteronomy to autonomy." By "heteronomy" he understood the orientation at external criteria and the obligation to pursue objectives outside mathematics.

If we take the above quote by von Mises seriously, then characteristic features of modern probability theory could only be found in scattered works by Russian mathematicians, if at all. A more precise study of the literature that von Mises referenced in [1919a; 1919b] demonstrates that his main focus was the three papers by Markov, which had been included as an appendix to the German translation of *Probability Theory* [Markov 1912]. The first of these articles [Markov 1898] deals with a part of the proof of the CLT based on considerations relating to moment theory. The other two articles [Markov 1908/12; 1911/12] are concerned with limit theorems for chains of random variables. The first paper [Markov 1898] mainly emphasizes the moment theory aspects and, though it is analytically rigorous, it does not grant the CLT any autonomous mathematical relevance (see Sect. 4.7.3). By contrast, [Markov 1908/12] and [Markov 1911/12] are dedicated to the generalization of a "classical" stochastic structure, namely, independence. Neither of these papers arose as a result of any significant applied approach, but rather simply from the author's interest in generalizing the CLT [Antretter 1989, 9 f.]. There is a break between [Markov 1898] and [Markov 1908/12; 1911/12] which is conspicuous in that the reason why the problem of normal distribution as limit distribution is interesting is no longer primarily because it serves to illustrate analytical theories that are "actually" important, but rather because a particular mathematical autonomy has been achieved.

Alongside other influences yet to be discussed, this break could have been triggered by Lyapunov's proofs [1900; 1901b] of the CLT, of which von Mises was apparently unaware, and in which this fundamental theorem of probability theory was actually handled in a "modern" way for the first time. The articles [Lyapunov 1900; 1901b] mark the advent of a modern probability theory, in which the CLT and the problems associated with it established a link to classical probability theory. One essential element in the development of modern probability theory during the first decades of the century was certainly also the emergence of entirely new stochastic problem complexes, such as stochastic processes and strong laws of large numbers, in addition to the aforementioned axiomatics.[3]

---

[2] See [Purkert 2006b, 570; Hausdorff 2002, 53–55].

[3] Von Plato [1994] elaborated on the importance of these new problems before the backdrop of a world view that was shifting to an indeterministic outlook. For more details on von Plato's characterization of "modern probability theory," see Chap. 8.

5.1 Russian Contributions Between the Turn of the Century and the First World War

Lyapunov's proofs of the CLT around the turn of the century triggered a development toward modern probability theory in Russia which was mainly pursued by Markov (with the first few contributions by Bernshtein, as well) before the First World War. The paper by Markov [1898], which was discussed above, and one by Nekrasov [1898] played an important stimulating role for the articles Lyapunov would pen.

5.1.1 Lyapunov's Way Toward the Central Limit Theorem

Aleksandr Mikhailovich Lyapunov (1857–1918)[4] studied mathematics and physics in St. Petersburg from 1876 to 1880, and then began a university career, which led in 1893 to a professorship in Kharkov. Lyapunov was strongly influenced by Chebyshev but, in contrast to other prominent members of the "St. Petersburg school," did very little work on moment problems, concentrating instead primarily on mathematical physics and particularly on questions of stability. Lyapunov's duties in Kharkov also included lectures on probability theory. It can be assumed that his role model Chebyshev provided orientation in this regard, as well. The notes Lyapunov prepared on a corresponding lecture by Chebyshev from the years 1879–80 were published in 1936 (see also Sect. 4.6.1). Lyapunov himself published only two major works on the subject of probability theory, both containing proofs of the CLT; the already very weak conditions in the first paper were diminished even further by the second.

Lyapunov's motivation for working on the CLT may have been conditioned by his activities as a lecturer [Grigorian 1970–76, 562]. The deduction of the approximative normal distribution of a sum of independent random variables, as given in Chebyshev's lectures, was now far from adequate to deal with the analytical demands of the time being. Markov's proof on the basis of moment theory required a thorough introduction to moment problems and, for this reason, was not considered suitable for lectures that were predominantly oriented toward probability theory itself. Indeed, even Lyapunov's "elementary" considerations in his proofs of the CLT turned out to be so elaborate that it is scarcely possible to recognize a predominant didactic objective in them. However, one significant motivation for Lyapunov might have been the competitive situation within the St. Petersburg school and its wider sphere of influence. As early as the 1890s, Russian mathematicians

[4] Biographical information for Lyapunov can be found in [Grigorian 1970–76], for instance, and above all in the newer biography [Tsykalo 1988], which unfortunately was only published in Russian.

had repeatedly addressed problems of asymptotic error analysis,[5] with the articles by Sleshinskii [1892] (see Sect. 2.5.4) and Nekrasov [1898] being most prominent. Nekrasov played the role of "catalyst" for most of the probabilistic papers Markov produced after the turn of the century; this presumably also applies to Lyapunov's proofs of the CLT.

5.1.2 Nekrasov's Role in the Development of Probability Theory Around 1900

Pavel Alekseevich Nekrasov (1853–1924),[6] who himself was not a member of the St. Petersburg school, had worked as a professor at Moscow University since 1890. From 1883, he was a member of the Moscow Mathematical Society, whose founders had also included Chebyshev. He began to play an important role in this society around 1887, serving at times at vice-president and president of this association, which also published the journal *Matematicheskii Sbornik*. A storm was brewing in 1915 between Nekrasov, who in the meantime had become a chief officer in the Ministry for "Public Enlightenment" (i.e., the Ministry of Education), and several members of the Petersburg Academy, foremost among them Markov and Lyapunov. Their argument had to do with the introduction of subject matter into the high school mathematics curriculum which the members of the academy deemed an abuse of mathematics in its apparent goal "of transforming pure science into a tool bringing religious and political pressure to bear on the rising generation" [Nekrasov 2004, 135]. Soviet historiography portrays Nekrasov's character in a very negative light; he is not mentioned at all in the Great Soviet Encyclopedia. Maistrov [1974, 240] characterizes Nekrasov's contributions to probability theory as "completely unfounded applications of this theory," and he [1974, 242] argues that Nekrasov only "masked his pseudo-scientific deductions with references to probability theory." In fact, Nekrasov's political stance appears to have been formed by bigotry and extreme nationalism.[7] On top of that, he was considered downright cantankerous [Seneta 1984, 70]. The tension between Nekrasov and Markov went back at least as far as 1898 and likely started with a mathematical rather than a political or ideological dispute. The rivalry between the mathematical centers of St. Petersburg and up-and-coming Moscow may have also played a part [Seneta 1984, 70].

To date only one of Nekrasov's mathematical achievements has attracted particular attention and recognition from posterity, namely, his discussion of convergence in the Gauss–Seidel method (1885, 1892) (see [Seneta 1984, 45 f.]). However,

[5] A survey of the contributions on asymptotic error analysis by Russian mathematicians who stood in Markov's and Lyapunov's shadow can be found in [Seneta 1984].

[6] Biographical information on Nekrasov is found in [Seneta 1984, 68–71] and [Sheynin 2003]; see also [Sheynin 2005b, 238–241].

[7] This is especially apparent in Nekrasov's correspondence with P.A. Florenskii (see [Sheynin 1989, 342; 2003, 343]); further material on this subject can be found in [Nekrasov 2004, 109–140].

Nekrasov's mathematical potency is also demonstrated by the fact that he was awarded the "Bunyakovskii Prize" by the Petersburg Academy early in his scientific career. From 1884 on, Nekrasov's main mathematical interest was aimed at complex analysis and the application and generalization of the Laplacian approximation method with regard to complex-valued terms. In his papers on this subject, Nekrasov anticipated the development of the saddle point method, which is usually attributed to Peter Debeye [1909] [Solovev 1997, 9 f.]. Nekrasov's article [1898], in which the author disclosed local and integral limit theorems for sums of independent random variables without providing any proofs, should also be viewed in this methodical context. These were essentially attempts to specify and generalize the Laplacian concept of the discretization of random variables while proving the CLT. Nekrasov paid the most attention to lattice distributed random variables and, in so doing, took into account the possibility that the lattice distance could shrink as a particular function of the number n of random variables, and that the limits considered for their sum could possibly increase as a function of n, in the sense of "large deviations" in today's terminology. Nekrasov anticipated results here that would not be rediscovered until a good half-century later [Seneta 1984, 55–60; Solovev 1997] and that include local and integral CLTs for independent, lattice distributed random variables as a special case. Meanwhile, for his theorems he established highly complicated conditions that were difficult to apply in concrete situations. The subsequent papers from the years 1900–1902, which covered a detailed discussion of the results that had been published in 1898, are over 1000 pages in scope and, owing to their "rambling and unclear presentation" [Seneta 1984, 62], make it practically impossible for the reader to adequately evaluate the work. One of Nekrasov's conditions was the requirement of analyticity of the generating functions $\varphi(z) = \Sigma p_i z^{x_i}$ of each of the discrete random variables with values x_i in a domain of the complex plane $1 - \varepsilon \leq |z| \leq 1 + \varepsilon$ $(\varepsilon > 0)$. This condition was considerably more restrictive than requiring the existence of moments of arbitrary order for these random variables. For the CLT, Lyapunov was able to significantly weaken the latter requirement in particular.

Shortly after the publication of Markov's two articles [1898; 1899] on the CLT, Nekrasov began arguing with him about priority. Nekrasov's remonstrances and Markov's responses occasionally appeared in the mathematical memoranda published by Kazan University, and some were also published in *Sbornik* between 1899 and 1900. They have been recently translated into English [Nekrasov 2004, 22–58]. Moreover, detailed depictions of the battle between Nekrasov on the one hand, and Markov, and, somewhat later, Lyapunov on the other hand are available in [Seneta 1984, 60–65, 75] and [Sheynin 1989, 363 f.; 374], and so a summary of the most salient points is all that is necessary here. Nekrasov attacked Markov because the two papers [Markov 1898; 1899] ostensibly had some very close similarities to his own article [Nekrasov 1898], which had been published first. Markov, he claimed, had gleaned crucial information on the CLT without so much as a word to acknowledge the suggestions and efforts of his colleague. When one considers the completely different objectives and analytical methods employed by Nekrasov and Markov in their work on the CLT, Nekrasov's claim of priority seems rather far-fetched. Interesting, though, is Nekrasov's allusion to having sent Markov and several other mathematicians a very early offprint of his 1898 article. This remark suggests that Markov had not only been stimulated to closer exploring the CLT through Vasilev's biography on Chebyshev, as stated by himself (see Sect. 4.6.4). In fact, in the waning years of the 19th century, the CLT had become a subject which many in Russian mathematics were attempting to approach from different directions.

As Seneta [1984, 61] reports, Nekrasov used another attack on Markov to universally criticize the inexpediency of the "Petersburg methods," i.e., the application of moment theory, to probabilistic limit theorems. Nekrasov identified his actual concern as that of refining the results achieved by Chebyshev and exploring them in greater detail. In expressing this intent, but also in airing his criticism, he astoundingly allied himself with Lyapunov [1900, 359], who had characterized Chebyshev's and Markov's treatment of the CLT as "convoluted and complicated," and who had set himself the task of seeking "more general conditions" for the CLT.

It appears that Nekrasov's work exerted a certain influence on Lyapunov. Even though he wrote that [Nekrasov 1898] did not include any proofs and deviated entirely from his own investigations in terms of the conditions he established, Lyapunov [1900, 361] himself acknowledged that Nekrasov had indicated his analytical method, namely, elementary treatment of generating functions and, in this context, considerations in conjunction with the "Lagrange series" for the series expansions of inverse functions; this in turn pointed to the use of the Laplacian approximation method and the saddle point method. These analytical methods in particular, if in an absolutely different form, played a decisive role in Lyapunov's proof of the CLT.

It was the desire of both Nekrasov and Lyapunov to take the CLT seriously as a distinct mathematical object and not simply regard it as an instance in which specific analytical methods could be applied. Nevertheless, Nekrasov's rendering was not according to the demands of the 20th century. So it remained an easy task for Markov, and later for Lyapunov, to parry the attacks from their rival Nekrasov using suitable counterexamples without ever having to delve into the actual mathematical content of his papers. It is not only due to his political views that Nekrasov was mocked or silenced in post-revolutionary Russia; with the very poor presentation of his own work, he had also made it almost impossible for ensuing generations to conduct any serious study of his achievements in probability theory.

Thus Nekrasov remains relegated to the role of a driving force on the path to the modern state of probabilistic limit theorems, a role he continued to play after the turn of the century. Although Markov and Lyapunov enjoyed a good personal relationship [Grodzenskii 1987, 72 f.], the atmosphere of competition that prevailed after Lyapunov's success with the CLT gave Markov an important motivation to immerse himself more deeply and even more intensively in probability theory. Yet also Nekrasov was responsible for a sizable share of Markov's growing interest in probability theory. As emerges from the correspondence between Markov and Aleksandr Chuprov [Ondar 1981][8] and from Markov's response [1912/2004, 74 f.]

[8] See also [Seneta 1984, 65–68] and, for additional letters and corrections, [Sheynin 1996b, 61–84].

to a new allegation leveled by Nekrasov (see [Nekrasov 2004, 59–72]), Markov was inspired to examine the weak law of large numbers for chains of random variables as a result of comments made by Nekrasov [1902]. Markov [1906/2004] also wanted to use his paper on this subject to disprove Nekrasov's presumption of the necessity of pairwise independence of the random variables under consideration for the weak law of large numbers.

5.1.3 Lyapunov Conditions and Lyapunov Inequality

Lyapunov [1900] proved the following theorem:
Let $x_1, x_2, x_3, \ldots$ be an infinite sequence of independent random variables ("variables indépendantes"), for which the expectations $\mathrm{E}x_i =: \alpha_i$, $\mathrm{E}(x_i - \alpha_i)^2 =: a_i$, and $\mathrm{E}|x_i^3| =: l_i$ exist, respectively. Furthermore, let $A_n := \frac{\sum_{i=1}^n a_i}{n}$ and $L_n^3 := \max_{1\leq i\leq n} l_i$. Under the condition

$$\frac{L_n^2}{A_n} n^{-\frac{1}{3}} \to 0 \quad (n \to \infty), \tag{5.1}$$

for all $z_1 < z_2 \in \mathbb{R}$ the modulus of

$$P\left(z_1\sqrt{2nA_n} < \sum_{i=1}^{n}(x_i - \alpha_i) < z_2\sqrt{2nA_n}\right) - \frac{1}{\sqrt{\pi}}\int_{z_1}^{z_2} \mathrm{e}^{-z^2}\,dz$$

has an upper bound Ω_n independent of z_1, z_2 such that

$$\Omega_n \to 0 \quad (n \to \infty).$$

The condition (5.1) is met, for example, if the absolute moments of third order l_i of each single random variable have a uniform upper bound C^3, and all variances a_i have a uniform lower bound c. Then we have

$$\frac{L_n^2}{A_n} n^{-\frac{1}{3}} \leq \frac{C^2}{c} n^{-\frac{1}{3}} \to 0.$$

Lyapunov did not give any closer specification of the character of the probability distributions under consideration. In the proof he restricted himself to the discussion of random variables which respectively take a finite number of values only. He made plausible, however, that all results were valid even for random variables with an infinite number of possible values, if these random variables could be considered as being generated by means of a limit process from the former ones.

Lyapunov [1901a;b;c][9] was even able to further weaken the conditions for his theorem, by presupposing the existence of the respective expectations α_i, a_i, and

[9] [Lyapunov 1901a;c] only contained the results; these were thoroughly discussed and proved in [Lyapunov 1901b].

$d_i := \mathrm{E}|x_i - \alpha_i|^{2+\delta}$ ($\delta > 0$ arbitrarily small), and by requiring that, instead of (5.1),[10]

$$\frac{(d_1 + d_2 + \cdots + d_n)^2}{(a_1 + a_2 + \cdots + a_n)^{2+\delta}} \to 0. \tag{5.2}$$

The now so-called "Lyapunov inequality" played an important role in its originator's works. The inequality had been applied in the 1900 article (on pp. 372 f.) without proof, in [1901b, 2 f.] Lyapunov gave an explicit formulation, and he also hinted, if in a very concise and somewhat obscure manner, at ideas of proof for the inequality.[11] His "lemma" concerning the inequality was as follows:

Let

$$x', x'', x''', \ldots$$

be a sequence of positive numbers, and let $f(x)$ be any function whose values

$$f(x'), f(x''), f(x'''), \ldots$$

are all positive. If one generally sets

$$f(x') + f(x'') + f(x''') + \cdots = \sum f(x),$$

and by l, m, n understands any numbers which are according to the inequalities

$$l > m > n \geq 0,$$

then one has[12]

$$\left(\sum f(x)x^m\right)^{l-n} < \left(\sum f(x)x^n\right)^{l-m}\left(\sum f(x)x^l\right)^{m-n}. \tag{5.3}$$

For random variables X, which can be considered, in the general case, as "limits" of such variables with a finite number of values, an immediate consequence (not explicitly stated by Lyapunov) of the lemma above is[13]

---

[10] The condition in [Lyapunov 1901a] was, as Lyapunov [1901b, 4] explained, considerably weaker than (5.1), however still somewhat stronger than the "ultimate" version (5.2), which was introduced in Lyapunov [1901b;c].

[11] For a complete proof, which tries to follow Lyapunov hints, see [Uspensky 1937, 265 f.].

[12] In a strict sense, the following inequality has to obey a "$\leq$" rather.

[13] Strictly speaking, the "Lyapunov inequality" (5.3) is a particular case of the "Hölder inequality," which was stated and proved by Otto Hölder in [1889] (in a more specialized form by Leonhard James Rogers in [1888] already), and which is as follows: Let $a_1, \ldots, a_n; b_1, \ldots, b_n$ be nonnegative numbers and let $\frac{1}{r} + \frac{1}{s} = 1$, where r, s are positive; then we have

$$\sum a_i b_i \leq \left(\sum a_i^r\right)^{\frac{1}{r}}\left(\sum b_i^s\right)^{\frac{1}{s}}.$$

Lyapunov's inequality can be derived from Hölder's, by setting

$$a_i = (f(z_i))^{\frac{l-m}{l-n}} z_i^{\frac{n(l-m)}{l-n}}, b_i = (f(z_i))^{\frac{m-n}{l-n}} z_i^{\frac{l(m-n)}{l-n}}, r = \frac{l-n}{l-m}, s = \frac{l-n}{m-n}$$

for positive z_i. In the standard monograph on inequalities [Hardy, Littlewood, & Pólya 1934, 24–27], both Hölder's and Lyapunov's inequality are proved by means of a more general inequality. The most general version of the Hölder and the Lyapunov inequality, respectively, for Stieltjes

$$\left(\mathrm{E}|X|^m\right)^{l-n} \le \left(\mathrm{E}|X|^n\right)^{l-m}\left(\mathrm{E}|X|^l\right)^{m-n}. \tag{5.4}$$

By a repeated and partially tricky application of his inequality, Lyapunov [1901b, 4 f.] showed that condition (5.2) holds for any δ such that $0 \le \delta \le 1$, if absolute third-order moments exist for all random variables and if (5.1) is met. This consideration also implied that (5.2) is a weaker condition than (5.1), because in the former condition the existence of moments $\mathrm{E}|x_i|^3$ is not required.

In principle, Lyapunov's proofs of the CLT used the basics of Poisson's derivation of an approximate normal distribution for sums of independent random variables. His analytic rigor, however, followed the practices of the post-Weierstrassian era to a large extent. On the other hand, Lyapunov was still influenced by conventional St. Petersburg school methods. So he gave an "elementary" exposition on the basis of discrete random variables each of which had finitely many values only, and in all possible cases he used sums rather than integrals. Altogether, these features were connected with a rather long-winded and intricate exposition. Lyapunov [1900, 360 f., 362–364] quite comprehensively discussed preliminary work of other authors and the main difficulties he had to overcome. He particularly emphazised trigonometric methods, especially the application of Dirichlet's discontinuity factor for representing the probability of a sum. Following [Czuber 1891, 254], Lyapunov gave credit to Glaisher [1872a;b] for being the first to use this method when dealing with sums of random variables. At the same time, Lyapunov called the contribution of Cauchy/Sleshinskii to the CLT for linear combinations of errors (see Sect. 2.5.4) "too restrictive" and did not consider this approach to be applicable for more general situations. Nonetheless, after analyzing Lyapunov's proofs, one may conclude that he was strongly influenced by Cauchy's methods. Lyapunov's discussion of possible problems arising from the use of the Dirichlet factor comprised difficulties with the interchange of the order of integration, and—probably clinging to Czuber's [1891, 253–257] description—the by no means evident negligibility of the "tail" when applying the Laplacian method of peaks to the expression obtained by use of the Dirichlet factor (cf. Sect. 2.2.4). In order to overcome this difficulty, Lyapunov introduced, in addition to the considered n random variables, a normally distributed auxiliary variable with zero mean and with a variance vanishing as $n \to \infty$. Possibly, Lyapunov came across this idea when reading Czuber's monograph on error theory, where Crofton's contribution (see Sect. 3.3.2.2) was discussed in detail [Czuber 1891, 91–97]. Lyapunov evaded said problems with interchanging the order of integration by assuming that his random variables could only take finitely many values, except for the auxiliary variable. He [1900, 379] argued that his results would also be valid for random variables that "take infinitely many possible values." Lyapunov justified this by merely stating that one could consider arbitrary

integrals can be derived by suitable limit processes from the original inequalities. It deserves some interest that Feller [1971], in his popular monograph on probability theory, exclusively refers to Hölder's inequality. Loève, in his well-known 1955 book (I refer to the second edition from 1960) on p. 156 proves Hölder's inequality, and on p. 172 poses Lyapunov's as a problem to solve. There exists, as it seems, a "western" tradition of giving priority to Hölder rather than to Lyapunov in context with those inequalities.

random variables as limits of discrete random variables with finitely many values. Thanks to the particular probability density of the auxiliary variable, the representation of the probability for the sum of n random variables plus the auxiliary variable by convolution made certain algebraic-analytic manipulations possible such that explicit application of the Dirichlet factor became superfluous, as Lyapunov [1900, 369] underlined. The general statement that his proof was based on using characteristic functions,[14] apparently goes back to Lyapunov's discussion of the Dirichlet factor and his prevalent consideration of trigonometric terms like $\sum f(z_i)e^{z_i t\sqrt{-1}}$ (f designating a probability function of a random variable with finitely many values z_i). In fact, it is possible to reconstruct Lyapunov's proofs from the point of view of the theory of characteristic functions, as shown in [Uspensky 1937, 289–292]. This argumentation, probably going back to [Cramér 1923] and [Bernshtein 1926, 8–12], has common features with Lyapunov's approach, but in a mere abstract way. Lyapunov never used general concepts such as inversion formula or correspondence between convolution of distributions and products of characteristic functions, not to mention the correspondence between limits of distributions and limits of characteristic functions as they were later elaborated, most notably by Lévy.

A particular characteristic of both proofs of Lyapunov are explicit, though very complicated, expressions for an upper bound Ω_n [Lyapunov 1900, 385; 1901b, 16 f.]. Lyapunov [1900; 1901b] in both papers showed that his bounds are asymptotically of an order of magnitude $\frac{\log n}{\sqrt{n}}$.[15] Doing so he realized Chebyshev's demand for giving explicit error bounds regarding the approximation of the distribution of a (suitably normed) sum by the normal distribution, and he solved an important problem of the "Petersburg" research program, even though he [1900, 386] appeared to be unhappy about the "roughness" of his estimates.

Lyapunov's work on the CLT appears modern insofar as it brought full mathematical autonomy to this important probabilistic problem. To Lyapunov, the CLT was neither of priority for error calculus or distribution statistics nor did it serve to illustrate "really interesting" analytical problems. This is also shown by the fact that it was Lyapunov's [1900, 360] goal to find a "direct" proof with such analytic methods which corresponded to the "true nature" of this theorem rather than moment methods. On the other hand, in his speaking about mathematics, he was still removed from a "modern" point of view ("modern" as defined by Mehrtens). In his obituary for Chebyshev, Lyapunov [1895], maintaining that only such mathematical investigations were valuable which were based on "scientific

[14] See, for example, [Sheynin 1989, 362]. Loève [1978, 295] even wrongly claims that Lyapunov introduced the designation "characteristic function."

[15] Simplifying the original version to some extent, Uspensky [1937, 296] estimated Ω_n, under the condition of finite absolute third-order moments, for n such that $\omega_n := \frac{\sum E|X_i|^3}{(\sum \operatorname{Var} X_i)^{\frac{3}{2}}} < \frac{1}{20}$, in the form

$$\Omega_n = \frac{8}{5}\omega_n\left[\left(\log\frac{1}{3\omega_n}\right)^{\frac{1}{2}} + 1.1\right] + \omega_n^2\log\frac{1}{3\omega_n} + \frac{5}{3}\omega_n^{\frac{3}{2}}e^{-\frac{1}{3}\omega_n^{-\frac{3}{2}}}.$$

An especially simple upper bound Ω_n of the same order of magnitude was found by Cramér [1923] (see Sect. 5.2.8.1).

or practical applications,"[16] presented himself as a sheer "counter-modernist." Furthermore, to the modern reader, Lyapunov's analytic style appears old-fashioned. On the one hand, this is due to his long-winded presentation, because he does not explicitly use characteristic functions. On the other hand, Lyapunov did not always meet the demands of analytical rigor, as shown by his only vague arguments in favor of the validity of his results for generally distributed random variables.

5.1.4 Sketch of Lyapunov's Proof for the Central Limit Theorem

Let (X_k) be a sequence of independent random variables, each taking a finite number of values only. The main goal of both the first [1900] and second [1901b] articles is to prove, under suitable "Lyapunov conditions," that

$$P\left(z_1\sqrt{2\sum_{k=1}^{n}\mathrm{Var}X_k} < \sum_{k=1}^{n}(X_k - \mathrm{E}X_k) < z_2\sqrt{2\sum_{k=1}^{n}\mathrm{Var}X_k}\right) = \frac{1}{\sqrt{\pi}}\int_{z_1}^{z_2}\mathrm{e}^{-z^2}\,dz + \Delta, \quad (5.5)$$

where $\Delta \to 0$, uniformly for all z_1, z_2, if $n \to \infty$. Lyapunov in the second article extensively uses results of the first. The main "trick" of both articles is the introduction of a random variable ξ, independent of $X_1, \ldots, X_n$, which has a zero mean and a variance $\mathrm{Var}\xi = 2\kappa^2$, κ being initially an arbitrary positive constant. The core of both works consists in an estimate of the term $|R(g,h)|$, where

$$R(g,h) := P\Big(-h < \sum_{k=1}^{n}(X_k - \mathrm{E}X_k) + \xi - g < h\Big) - \frac{1}{\sqrt{\pi}}\int_{(g-h)/\sqrt{2\sum_{k=1}^{n}\mathrm{Var}X_k}}^{(g+h)/\sqrt{2\sum_{k=1}^{n}\mathrm{Var}X_k}}\mathrm{e}^{-z^2}\,dz.$$

In this expression, g is defined by

$$g := \frac{z_1 + z_2}{2}\sqrt{2\sum_{k=1}^{n}\mathrm{Var}X_k},$$

and h is any positive number.

If f_k denotes the probability function of the random variable X_k, then we have

$$P\Big(-h < S_n + \xi - g < h\Big) = \frac{1}{2\kappa\sqrt{\pi}}\sum_{v_1 \in W(X_1),\ldots,v_n \in W(X_n)} f_1(v_1)\cdots f_n(v_n)\int_{-h-s_n+g}^{h-s_n+g}\mathrm{e}^{-\frac{x^2}{4\kappa^2}}\,dx \quad (5.6)$$

[16] An English translation of the respective text passage is in [Maistrov 1974, 190].

[Lyapunov 1900, 368], where $S_n := \sum_{k=1}^{n}(X_k - \mathrm{E}X_k)$, and $s_n := \sum_{k=1}^{n}(\upsilon_k - \mathrm{E}X_k)$; $W(X_k)$ denotes the range of all possible values of X_k. By means of

$$\int_0^\infty \frac{\sin 2at}{t} \mathrm{e}^{-t^2} dt = \sqrt{\pi} \int_0^a \mathrm{e}^{-x^2} dx, \tag{5.7}$$

from (5.6) it can be deduced that

$$P(-h < S_n + \xi - g < h)$$
$$= \frac{2}{\pi} \int_0^\infty \frac{\sin ht}{t} \sum_{\upsilon_1 \in W(X_1),\ldots,\upsilon_n \in W(X_n)} f_1(\upsilon_1) \cdots f_n(\upsilon_n) \cos(s_n - g)t\mathrm{e}^{-\kappa^2 t^2} dt \tag{5.8}$$

[Lyapunov 1900, 369]. In the same way as already Poisson in his approach to the CLT (see Sect. 2.2.2), Lyapunov sets

$$\sum_{\upsilon_1 \in W(X_1),\ldots,\upsilon_n \in W(X_n)} f_1(\upsilon_1) \cdots f_n(\upsilon_n) \cos(s_n - g)t$$
$$=: \mathrm{Re} \Lambda_1 \cdots \Lambda_n \mathrm{e}^{-gt\sqrt{-1}}, \tag{5.9}$$

where

$$\Lambda_k := \sum_{\upsilon_k \in W(X_k)} f_k(\upsilon_k) \mathrm{e}^{(\upsilon_k - \mathrm{E}X_k)t\sqrt{-1}}. \tag{5.10}$$

From a modern point of view the quantities Λ_k may be interpreted as the characteristic functions of $X_k - \mathrm{E}X_k$. Lyapunov does not discuss any rules regarding these quantities, however, and he does not explain how to apply them generally in the context of sums of independent random variables. With the abbreviations $\rho_k \mathrm{e}^{\sigma_k \sqrt{-1}} := \Lambda_k$ and $\sigma := \sigma_1 + \cdots + \sigma_n$ from (5.8), under consideration of (5.9) and (5.10), ensues:

$$P(-h < S_n + \xi - g < h) = \frac{2}{\pi} \int_0^\infty \frac{\sin ht}{t} \rho_1 \cdots \rho_n \cos(\sigma - gt) \mathrm{e}^{-\kappa^2 t^2} dt \tag{5.11}$$

[Lyapunov 1900, 368 f.; 1901b, 6]. On account of (5.11) and (5.7) it follows that

$$R(g, h) = \frac{2}{\pi} \int_0^\infty \frac{\sin ht}{t} T dt,$$

where

$$T := \rho_1 \cdots \rho_n \cos(\sigma - gt) \mathrm{e}^{-\kappa^2 t^2} - \cos gt \mathrm{e}^{-t^2 \sum_{k=1}^{n} \mathrm{Var} X_k / 2}.$$

By use of elementary inequalities, Lyapunov [1900, 370–376, 384 f.; 1901b, 9 f.] for $0 < \tau < \tau_1 < \infty$ infers

$$|R(g,h)| < \frac{2}{\pi}\int_\tau^\infty |T|\frac{dt}{t} + \frac{2}{\pi}\int_0^\tau |T|\frac{dt}{t} < \frac{2}{\pi\tau}\int_\tau^{\tau_1} \rho_1\cdots\rho_n dt + \frac{1}{\pi\kappa^2\tau_1^2}e^{-\kappa^2\tau_1^2}+$$
$$+\frac{2}{\pi\tau^2\sum_{k=1}^n \mathrm{Var}X_k}e^{-\tau^2\sum_{k=1}^n \mathrm{Var}X_k/2} + \frac{2\kappa^2}{\pi}\int_0^\tau t\rho_1\cdots\rho_n dt+$$
$$+\frac{2}{\pi}\int_0^\tau \left(|\rho_1\cdots\rho_n - e^{-t^2\sum_{k=1}^n \mathrm{Var}X_k/2}| + |\sigma|e^{-t^2\sum_{k=1}^n \mathrm{Var}X_k/2}\right)\frac{dt}{t}. \tag{5.12}$$

With respect to the hitherto described arguments, the first and second articles are almost identical. With regard to the estimates of the terms $\rho_1\cdots\rho_n$, $|\rho_1\cdots\rho_n - e^{-t^2\sum_{k=1}^n \mathrm{Var}X_k/2}|$, and $|\sigma|$ occur major differences, because of the different conditions in both papers. In the following, only the chief steps of proof in the second article [1901b] (which automatically also covers the results of the first) are analyzed. Furthermore, Lyapunov's abbreviations

$$A := \sum_{k=1}^n \mathrm{Var}X_k \text{ and } D := \sum_{k=1}^n \mathrm{E}|X_k - \mathrm{E}X_k|^{2+\delta}$$

are used.

For τ and τ_1 Lyapunov assumes

$$4\frac{D}{A}\tau_1^\delta < 1,\ \tau < \tau_1,\ D\tau^{2+\delta} < k^{2-\delta}, \tag{5.13}$$

where $0 < \delta \le 1$ is an arbitrary number initially, and k is the positive solution of the equation $k^{2-\delta} = 8(1-k^2)$. Under these assumptions the following estimates hold:

$$\begin{aligned} |\rho_1\cdots\rho_n - e^{-At^2/2}| &< 2Dt^{2+\delta}e^{-(A-4D\tau^\delta)t^2/2} \text{ for } t\in]0;\tau[,\\ \rho_1\cdots\rho_n &< e^{-(A-4D\tau^\delta)t^2/2} \text{ for } t\in]0;\tau[,\\ \rho_1\cdots\rho_n &< e^{-(A-4D\tau_1^\delta)t^2/2} \text{ for } t\in]0;\tau_1[,\\ |\sigma| &< Dt^{2+\delta}e^{2D\tau^\delta t^2} \text{ for } t\in]0;\tau[. \end{aligned} \tag{5.14}$$

In his derivation of these estimates Lyapunov makes essential use of a particular case of "his" inequality (5.4), namely,

$$(\mathrm{Var}X)^{2+\delta} < (\mathrm{E}|X-\mathrm{E}X|^{2+\delta})^2.$$

If one substitutes the inequalities (5.14) in (5.12), and augments the right side of (5.12) by enlarging the upper limits of integration to ∞, then it follows

$$|R(g,h)| < \frac{1}{\pi\kappa^2\tau_1^2}e^{-\kappa^2\tau_1^2} + \frac{4}{\pi q_1 A\tau^2}e^{-q_1A\tau^2/2}+$$
$$+\frac{2\kappa^2}{\pi q A} + \frac{3}{\pi}\left(\frac{2}{q}\right)^{\frac{2+\delta}{2}}\frac{D}{\sqrt{A^{2+\delta}}} =: L, \tag{5.15}$$

where the abbreviations $q := 1 - 4\frac{D}{A}\tau^{\delta}$ and $q_1 := 1 - 4\frac{D}{A}\tau_1^{\delta}$ are used, and the conditions (5.13) are presupposed. Because the right side of (5.15) does neither depend on g nor on h, it can be considered as a uniform upper bound of $|R(g,h)|$ [Lyapunov 1901b, 11–16].

At this place the validity of the second "Lyapunov condition"

$$\exists 0 < \delta \leq 1 : \quad \frac{D^2}{A^{2+\delta}} \to 0 \quad (n \to \infty) \tag{5.16}$$

becomes essential.[17] Taking (5.16) into account, the parameters τ, τ_1, and κ in (5.15) can be chosen, under the constraint (5.13), in such a manner that L tends to 0 as $n \to \infty$ [Lyapunov 1901b, 16–18]. The probability (5.5) can be expressed by a linear combination of several probabilities of the form $P\Big(-h < S_n + \xi - g < h\Big)$ (h being suitably chosen) and a further probability, which depends on the auxiliary variable, and which vanishes for $n \to \infty$. In this way it can be finally justified that the quantity Δ, defined in (5.5), tends to 0, uniformly for all z_1, z_2 [Lyapunov 1900, 364–366, 378–381; 1901b, 7 f.].

5.1.5 Markov's Reaction

Lyapunov's proofs were written in French and published in journals which were also available in Western Europe, at least in greater libraries. Also, abstracts of Lyapunov's most important results had appeared in the *Comptes rendus*, and there was a review of [Lyapunov 1900] in the *Jahrbuch über die Fortschritte der Mathematik* (JFM 31.0228.02). However, there was hardly any attention drawn to his work outside Russia. A significant exception was Georg Bohlmann [1901, 913], who, in his survey of life insurance mathematics in the *Encyklopädie der Mathematischen Wissenschaften*, hinted at the contributions of Chebyshev, Markov, and Lyapunov to the CLT.[18] This bibliographical reference in a primarily application-oriented article might have missed the proper audience, though. Except for marginal exceptions, ambitious analysts outside Russia did not consider probability theory a promising field of activity until the end of World War I. Only then did Lyapunov's contributions receive increasing attention, as can be seen with [Pólya 1920] and [Lindeberg 1922b;c], for example.

Inside Russia, Lyapunovs proofs had especially impact on Markov, as it seems. Only after the turn of the century, in particular between 1906 and 1913, when he was able to do significantly more research after his retirement from teaching—at the age of 49 (!)—was Markov increasingly engaged in probability theory. Nekrasov's influence on this development has already been described above. In a similar way, however, also Lyapunov's—at least indirect—influence on Markov's probabilistic

[17] If (5.2) holds for $\delta > 1$, then this condition is also met for all $\delta \leq 1$.

[18] There is also some evidence that Hausdorff became interested in Lyapunov's work before 1915. Concrete results of this interest, however, cannot be found before 1923 (see Sect. 5.2.5).

work can be observed. One may assume that Markov, despite his friendship with Lyapunov, felt challenged by him [Schneider 1988, 443]. This challenge seems to have consisted in Lyapunov's criticism of Markov's application of moment methods in probability theory, where the former used almost the same arguments as Nekrasov.

Markov [1908/13/2004] actually succeeded in proving the CLT under the Lyapunov condition (5.2) by moment methods. This success was based on the newly introduced device of truncated random variables. If (X_k) is a sequence of random variables, each of which has a zero mean and an infinite range of values, and if (N_n) is a sequence of real numbers tending to ∞, then the random variables X'_{nk}, where

$$X'_{nk} = \begin{cases} X_k & \text{for } |X_k| \le N_n \\ 0 & \text{for } |X_k| > N_n, \end{cases}$$

are called "truncated."[19]

Markov [1908/13/2004, 145–148] showed that, given a sequence of independent random variables (X_k) with zero expectations which obey the Lyapunov condition (5.2), a sequence (N_n) can be found such that

$$\sum_{k=1}^{n} P(X'_{nk} \neq X_k) \to 0 \tag{5.17}$$

and

$$\frac{N_n^2}{B_n} \to 0, \quad \frac{\sum_{k=1}^{n} \mathrm{E}X'^2_{nk}}{B_n} \to 1, \quad \text{where } B_n := \sum_{k=1}^{n} \mathrm{Var}X_k. \tag{5.18}$$

(Here and in the following all limit assertions are to the condition $n \to \infty$.) The truncated random variables have moments of arbitrary order, and, by virtue of (5.18), Markov in a way analogous to his method in [1899] proved that, for all $m \in \mathbb{N}$,

$$\mathrm{E}\left(\frac{\sum_{k=1}^{n} X'_{nk}}{\sqrt{2B_n}}\right)^m \to \frac{1}{\sqrt{\pi}}\int_{-\infty}^{\infty} x^m \mathrm{e}^{-x^2}\,dx. \tag{5.19}$$

On the other hand, Markov [1908/13/2004, 144] from (5.17) inferred that

$$\left| P\left(a < \frac{\sum_{k=1}^{n} X'_{nk}}{\sqrt{2B_n}} < b\right) - P\left(a < \frac{\sum_{k=1}^{n} X_k}{\sqrt{2B_n}} < b\right) \right| \to 0. \tag{5.20}$$

Markov's general theorem on the correspondence between limits of moments and distributions [Markov 1898] (see Sect. 4.6.4) implied

$$P\left(a < \frac{\sum_{k=1}^{n} X'_{nk}}{\sqrt{2B_n}} < b\right) \to \frac{1}{\sqrt{\pi}}\int_a^b \mathrm{e}^{-x^2}\,dx.$$

[19] In a strict sense, for each $n \in \mathbb{N}$, different truncated variables are assigned to the X_k in dependence on the respective value of N_n. Therefore, contrary to Markov's original notation, the truncated variables are specified by a double index.

Taking into account (5.20), finally the CLT

$$P\left(a < \frac{\sum_{k=1}^{n} X_k}{\sqrt{2B_n}} < b\right) \to \frac{1}{\sqrt{\pi}} \int_a^b e^{-x^2} dx$$

followed.[20]

A closer examination of Markov's proof yields—if from the present point of view only—a surprising consequence. The existence of a sequence (N_n) and a sequence $B_n > 0$ (no longer necessarily related to variances) such that (5.17) and (5.18) hold, is the weakest possible condition which implies that the distributions of suitably normed sums $\sum_{k=1}^{n} X_k/\sqrt{B_n}$ of independent random variables which are sufficiently centered around the point of origin (by the condition $EX_k = 0$, for example) tend to the standard normal distribution. In exactly this form, Lévy in 1931 established a sufficient condition for the convergence to the Gaussian law, which, under the restraint of summands that are asymptotically negligible with respect to the total sum, was even necessary, as he proved in 1935. In the case $EX_k = 0$ this convergence actually follows from Markov's considerations, in particular from (5.19) and (5.20). Discussing CLTs for random variables without second-order moments, and considering sums of random variables with a general norming—different from the standard deviation—was still beyond Markov's point of view, however.

Bernshtein [1922; 1926] (see Sect. 5.2.7), without any doubt influenced by Markov's approach, was the first to draw such considerations which would later play an important role in the history of the CLT. More explicitly than Bernshtein, and independently of him, Lévy [1925b] (see, for example, Sect. 5.2.6.7) stressed the possibility of a general, "non-classical" norming. However, when the method of truncation was adopted by Lévy around 1931, he apparently had not concrete knowledge of Markov's and Bernshtein's contributions. In this context, he [1931, 132] rather referred to the work of Kolmogorov and Khinchin, who significantly advanced Markov's methods. Thus, Lévy was at least indirectly affected by the ongoing impact of Markov's idea of truncation.

Presumably, Markov's numerous works on chained random variables were motivated by the author's intention to reduce requirements and extend validity of the weak law of large numbers and the CLT as far as possible. Markov used "his" method of moments also in this field. By modifying the idea of generating functions he showed that the moments of arbitrary order of the normed sum converge to the corresponding moments of the standard normal distribution. Three important contributions to the CLT for homogeneous simple and also complex Markov chains were translated shortly after their appearance from the Russian [Markov 1907/10; 1908/12; 1911/12].[21]

Markov's quest for general conditions is also witnessed by the "Markov condition" for the validity of the weak law of large numbers

---

20 A comprehensive discussion of Markov's proof is given by Uspensky [1937, 383–395].

21 Surveys of pertinent work by Markov can be found in [Maistrov 1974, 215–217; Sheynin 1989, 364–370; Yushkevich 1970–76b, 128 f.; Basharin, Langville, & Naumov 2004; Seneta 2006]. English translations of [Markov 1906/2004; 1908; 1910] are in [Sheynin 2004a].

$$P\left(\left|\frac{\sum_{i=1}^{n}(X_i - \mathrm{E}X_i)}{n}\right| > \varepsilon\right) \to 0 \quad \forall \varepsilon > 0,$$

(X_i) being a sequence of independent random variables. By means of truncated variables Markov proved in the third edition of his book on probability theory (1913) that this law is obeyed if there exists $\delta > 0$ such that $\mathrm{E}|X_i|^{1+\delta}$ has a finite upper bound independent of i.[22] With this result Markov already came rather close to those conditions, found by Kolmogorov in 1926, which were also necessary for the weak law of large numbers under the general assumption of independent random variables with expectations [Maistrov 1974, 261 f.].

5.2 The Central Limit Theorem in the Twenties

Following the interruption caused by the First World War, probability theory began to be discovered as a field for ambitious analysts, even outside Russia. The CLT consequently ceased to be an issue merely for "users," such as astronomers, geodetics specialists, insurance specialists, or economists—who had actually produced quite impressive results in the second half of the 19th century, particularly in the field of error theory, although their work was little noted by mathematicians—and became an object of study within mathematics itself. The impetus gradually developed to move toward a far-reaching generalization of the classic limit theorems. In 1922 the development of the CLT reached one of its first peaks when Bernshtein and Lindeberg established similar sufficient conditions for the theorem, which later also proved to be essentially necessary.

5.2.1 A New Generation

The study of limit theorems for probability distributions was the only aspect of the emerging modern form of probability theory that was linked in any significant way to the results produced in the 19th century. Important figures in promoting this field in the 1920s included Bernshtein, Cramér, Lévy, Lindeberg, von Mises, and Pólya.

Richard von Mises (1883–1953)[23] studied at the "Technische Hochschule" in Vienna from 1901 to 1905, majoring in engineering. Emanuel Czuber, the leading figure in premodern probability theory and statistics in the German-speaking countries, was a professor there. In 1908 von Mises received his doctorate from the "Technische Hochschule" in Vienna, and in the same year he completed his "Habilitation" degree at the "Technische Hochschule" in Brünn, qualifying him to teach as a university lecturer. In 1909 he was appointed associate professor ("außerordentlicher Professor") in applied mathematics at the University

---

[22] For a discussion of Markov's proof see [Uspensky 1937, 191–195].

[23] For von Mises see [Bernhardt 1984; 1985; Siegmund-Schultze 2004]. I thank Reinhard Siegmund-Schultze for informing me about biographical details.

of Strassburg. During the First World War he served in the Austrian air force. Von Mises's academic publications before the First World War were mainly concerned with theoretical and mathematical issues involved in mechanical engineering and hydrodynamics; some degree of interest in statistical questions is also evident in his article on "Kollektivmaßlehre" [1912], however. After the war, following brief periods working at the University of Frankfurt/Main and the "Technische Hochschule" in Dresden, von Mises moved to the University of Berlin in 1920 to become the director of the newly formed institute of applied mathematics. Until the Nazis' seizure of power in 1933 he was engaged in extremely lively and varied activities in the fields of research and teaching.[24] The start of his more detailed studies of probability theory, dating from around 1919, probably also owed something to the fact that this was an area that was relatively poorly researched from the mathematical point of view, so that he saw it as offering opportunities to distinguish himself—as he was afraid of losing his professorship in Strassburg toward the end of the First World War.[25] His study on probabilistic limit theorems [von Mises 1919a] had a substantial influence on developments leading toward modern probability theory in Western Europe, where researchers—including von Mises—were initially largely uninformed about more recent studies by Russian mathematicians (with the exception of Pólya to some extent.)

Georg Pólya (1887–1985)[26] was an exceptionally versatile mathematician, whose main focus was in pure analysis and the theory of numbers. He taught at the "Eidgenössische Technische Hochschule" in Zürich from 1914 to 1940. From 1919 he published, partly in friendly competition with Lévy, some articles on (the later so-called) stable probability laws and the moment theoretic background of the "central limit theorem," which term he coined in 1920. The purely analytical aspect was obviously in the foreground of his work. Pólya had discovered probability calculus as a treasure trove of interesting analytical problems already in his doctoral thesis (1912). He had a remarkably good knowledge of Markov's work which appeared after the turn of the century and by which he was presumably stimulated to a more intensive preoccupation with probability theory, particularly since its analytical background was closely connected to his own interests. In the following period, Pólya dealt with stochastic problems time and again, although he did not really focus on probability theory.

At the beginning of his career Jarl Waldemar Lindeberg (1876–1932)[27] devoted himself to variational calculus and potential theoretical problems as an "adjoint professor" for mathematics at the University of Helsinki, where he had also studied. In 1920 he published his first work on probability calculus, the results of which he fundamentally generalized once again in his famous proof of the CLT under the "Lindeberg condition" (1922). Thereafter, Lindeberg dealt mainly with questions

---

[24] See [Antretter 1989, 32–48, 77 f.] for von Mises's activities in probability and statistics during that period.

[25] The article [von Mises 1919a] was submitted to the *Mathematische Zeitschrift* on 31 August 1918.

[26] For Pólya see [Alexanderson & Lange 1987].

[27] For Lindeberg see [Lindelöf 1934; Elfving 1981].

of correlation theory. He could not use the broad recognition of his probabilistic achievements, which began with a certain temporal delay only, for an improvement of his professional situation, however.[28]

Paul Lévy (1886–1971) had started mathematical work with integral equations, potential theory, and functional analysis.[29] In 1919, as a professor at the "École Polytechnique," Lévy had to give three lectures on error theory, specifically on the role of the Gaussian error law. This was the beginning of his thorough preoccupation with probability theory.[30] Lévy only knew the rather elementary books of Bertrand, Poincaré, and Borel on probability calculus and had no knowledge of Cauchy's, Chebyshev's, Lyapunov's, or Markov's results. Without any wider previous knowledge and without consideration of von Mises' contributions, Lévy developed a theory of characteristic functions and used this for a proof of the CLT under very general conditions. Lévy's article on this proof, however, was published only shortly after Lindeberg's proof, which was done under even weaker conditions. Focusing on the analytical aspects of probabilistic problems, Lévy encountered resistance from Borel, the leading mathematician in France, who wanted to develop probability calculus rather interdisciplinarily, according to the classical point of view. Lévy's self-confidence as a mathematician was so strong that he continued the examination of "his" problems. Perhaps as a consequence of Borel's criticism, he developed the style and methodology of his contributions toward a stronger emphasis on "intuitive" stochastic concepts during the thirties (see Sect. 6.2.2). No mathematician has contributed to limit distributions of sums of independent random variables in such a density as Lévy between 1920 and 1935.

Sergei Natanovich Bernshtein (1880–1968) reconciled the tradition of the St. Petersburg school with Western European mathematical methods.[31] He studied in Paris and Göttingen (1898–1903) and then returned to Russia, where he worked at Kharkov University from 1907. At first he dealt with partial differential equations and approximation theory.[32] Influenced by Markov's contributions, particularly to sums of nonindependent random variables, around 1910 Bernshtein started his work on probabilistic limit theorems. He was particularly interested in sufficient and also necessary conditions for the convergence of suitably normed sums of (not necessarily independent) random variables to the Gaussian distribution. During the twenties he succeeded in publishing fundamental contributions to this field, based on the method of characteristic functions. Although highly esteemed in the Russian mathematical community, with his probabilistic work Bernshtein was later in the shadow of the younger mathematicians Kolmogorov and Khinchin.

---

[28] Cramér [1976, 514] also describes Lindeberg as a master in the art of living, for whom the professional career was not too important.

[29] See the autobiographical notes [Lévy 1970; 1976, 1–6].

[30] Notes on these lectures have recently been published, see [Lévy 1919/2008] and [Barbut & Mazliak 2008].

[31] See [Yushkevich 1970–76a] for biographical details. Seneta [1982] concentrates on Bernshtein's probabilistic work. In these two contributions references can be found to secondary literature, which is more detailed, but published exclusively in Russian.

[32] Especially with regard to this field see [Akhiezer 2000].

Harald Cramér (1893–1985)[33] had first dealt with problems of analytical number theory, from which he acquired the analytical capacity, particularly regarding Fourier methods, for his later work in stochastics. He also worked as an insurance mathematician from 1918 since he did not earn enough as an assistant professor at the University of Stockholm. In connection with these professional activities he developed an interest in probability theory of which particularly the work of Pólya, Lindeberg, and Lévy impressed him. The first material probabilistic contribution of Cramér—an improvement of the upper bound given by Lyapunov for the deviation between the actual distribution of a normed sum of independent random variables and the approximating normal distribution—appeared in 1923. This problem was also important in the estimation of insurance risks in practice. Cramér devoted himself to the problem of improving Lyapunov's upper bound in detail during the following years, apparently not used very much in his insurance activities. For this aim he examined, in the tradition of the hypothesis of elementary errors of Scandinavian statistics, the asymptotic behavior of Charlier and Edgeworth series. Due to the status which Cramér had achieved in the field of stochastics, he was appointed to the newly created chair of actuarial science and mathematical statistics at the University of Stockholm in 1929, where he stayed until his retirement in 1958. The "Stockholm group" organized by Cramér was an important center of stochastic research.

One characteristic that all of these persons had in common was that they were not trained with the main emphasis in probability theory, but brought in their, partly already longstanding, analytical research to the "newfound" area. The analytical methods of differential and integral equations, Fourier analysis, analytical theory of numbers, as well as measure and integration theory proved to be particularly useful for probabilistic problems. On the other hand, this analytical orientation of the early "modern" probabilists also influenced the style and methods of their discussion of stochastic problems.

5.2.2 Von Mises: Laplacian Method of Approximation, Complex and Real Adjunct

Von Mises's "Fundamental Limit Theorems of Probability Theory" ("Fundamentalsätze der Wahrscheinlichkeitsrechnung") on the one hand consisted of limit theorems for distributions of linear combinations of independent random variables, or in von Mises's [1919a, 76 f.] own words "linearen Faltungen von Kollektivs." On the other hand, among these "fundamental theorems" were also limit theorems according to Bayes and Laplace, for the a posteriori probabilities of distribution parameters which were to be determined from the results of a test series.[34] Von Mises's chief

---

[33] See the autobiographic sketches [Cramér 1976] and [Cramér & Wegman 1986].

[34] The simplest case of a limit theorem of this kind had been introduced by [Laplace 1774] (see [Stigler 1986, 131–135; Hald 1998, 167–170]). The problem was to calculate the a posteriori probability that the success probability p in a Bernoulli process consisting of a large number of trials is within a certain interval around the observed relative frequency of success. Based on the

methods were the advancement of the Laplacian approximation principle from the point of view of "modern" analysis and the discussion of distributions by their real or complex "adjuncts," that is, in modern terminology, by their Laplace or Fourier transforms, respectively. Von Mises chiefly made explicit references to the contributions of Laplace, the work of Chebyshev (as described by the Chebyshev biographer Vasilev [1900]), and to Markov's papers (as contained in the appendix of the German translation of the latter's "Probability Theory" [1912]). However, von Mises's knowledge of the work of Chebyshev and Markov was merely superficial, as evidenced by the fact that he cited it only partially and in a thematically completely wrong connection (cf. [von Mises 1919a, 51], for example).

Von Mises's [1919a, 20 f.] newly conceived notion of "distribution" ("Verteilung") as a monotonically increasing function, being right continuous and having limit 0 as $x \to -\infty$ and limit 1 as $x \to \infty$, was important for generality as well as precision of analytic exposition. Apparently one of the first to do so, von Mises represented probabilities, as well as higher moments, by Stieltjes integrals referring to those distributions.[35] The use of Stieltjes integrals, as well as the analytic skill employed in dealing with moments, proves that von Mises was informed about the current development of moment theory, at least in its main features.

Based on Stieltjes integrals, von Mises formulated and proved his local and integral CLTs for real- and vector-valued random variables as statements about convolutions of discrete probability functions, densities, and distribution functions, respectively. So, his exposition was purely analytic and did not resort to probabilistic interpretations and concepts (to which he dedicated the second, substantially shorter section of his work).

Concerning the Laplacian principle of approximation von Mises [1919a, 7–15] established the following theorem:

Let $(f_k)_{k\in\mathbb{N}}$ be a sequence of integrable functions $\mathbb{R}\to\mathbb{C}$ with the following properties (which, as von Mises indicated at several places in his article, could be generalized even more):

(i) for each index k, real numbers a_k and s_k can be found such that $f_k''(a_k)$ exists, $f_k(a_k) = 1$, $f'(a_k) = 0$, and $f_k''(a_k) = -2s_k^2$;

(ii) for all $h \neq 0$ within a neighborhood of zero independent of k, $|\frac{f_k''(a_k+h)-f_k''(a_k)}{h}|$ exists and has an upper bound independent of k;

(iii) each function f_k has the following property: for each $y_0 > 0$ there exists $\delta_k > 0$ such that for all y (with the exception of a Lebesgue null set)

assumption of an a priori equiprobability of all hypothetical success probabilities between 0 and 1, Laplace was able to prove that for $\epsilon(n)$ of an order less than $n^{-\frac{1}{3}}$ and greater than $n^{-\frac{1}{2}}$:

$$P\,(h_n - \epsilon(n) < p < h_n + \epsilon(n)|h_n) \to 1 \quad (n \to \infty).$$

In 1764/65 Thomas Bayes and Richard Price had already published works on this problem; they had only found, however, rather intricate approximations in the case of a very large n [Stigler 1986, 122–131; Dale 1991, 16–51; Hald 1998, 133–154].

[35] Von Mises's "distribution" is identical with "distribution function," which is more common now. However, even today distribution functions are simply called "distributions" when the context is clear.

$$|f_k(a_k + y)| < 1 - \delta_k \quad \text{if } |y| > y_0;$$

(iv) for each function f_k there exist positive numbers α_k and X_k such that $|x^{\alpha_k} f_k(x)| < 1$ for all $|x| > X_k$;
(v) the sequences $(|a_k|)$ and (s_k^2) are bounded, and there exists a real number $s \neq 0$ such that $s_k^2 \geq s^2$ for all k.

Furthermore, let $\psi : \mathbb{R} \to \mathbb{C}$ be a function, integrable on bounded intervals, whose modulus has a finite upper bound. Then, for

$$p_n(u) := \prod_{k=1}^{n} f_k(a_k + \frac{u}{r_n}), \quad r_n = \sqrt{\sum_{k=1}^{n} s_k^2}$$

the following assertion is valid:
$p_n(u)$ converges for almost all $u \in \mathbb{R}$ uniformly to e^{-u^2} and

$$\lim_{n\to\infty} \int_a^b p_n(u)\psi(u)du = \int_a^b \mathrm{e}^{-u^2}\psi(u)du \tag{5.21}$$

for arbitrary a, b with the property $-\infty \leq a < b \leq \infty$.

For the proof of this theorem the Laplacian idea of expanding $\log f_k(a_k + h)$ in powers of h was essential. Including the residual, von Mises considered the terms of this expansion up to the power h^2.

Von Mises [1919a, 17 f.] also sketched the generalization of his approximation principle to functions $f_k:\mathbb{R}^t \to \mathbb{C}$ of several variables $x_1, \ldots, x_t$. Instead of numbers a_k and s_k^2, now t-dimensional vectors $\overline{a}_k$ and matrices $\overline{s}_k \in \mathbb{R}^{t,t}$ were considered, where $f_k(\overline{a}_k) = 1$, $\frac{\partial}{\partial x_i} f_k(\overline{a}_k) = 0$, and $\frac{\partial}{\partial x_i \partial x_j} f_k(\overline{a}_k) = -2s_k^{(i,j)}$ for $i, j = 1, \ldots, t$, and $\sum_{i,j=1}^{t} s_k^{(i,j)} y_i y_j \geq 0$ for all vectors $\overline{y} \in \mathbb{R}^t$. By adjusting conditions (i) to (v) regarding the functions f_k, von Mises found an analog to (5.21), which, however, was only stated for the particular case $\psi = 1$:

$$\lim_{n\to\infty} \left| \int_Z \prod_{k=1}^{n} f_k(\overline{a}_k + \frac{1}{\sqrt{n}}\overline{z})d\overline{z} - \int_Z \mathrm{e}^{-\sum_{i,j=1}^{t} h_n^{(i,j)} z_i z_j} d\overline{z} \right| = 0, \tag{5.22}$$

where Z designated a "finite or infinite part of space" (that is, $\mathbb{R}^t$) and $h_n^{(i,j)} = \frac{\sum_{k=1}^{n} s_k^{(i,j)}}{n}$.[36]

The most prominent probabilistic innovations von Mises delivered with his applications of characteristic functions, which he himself called "complex adjuncts," were to local and integral limit theorems for sums of lattice distributed or

---

[36] The original version [von Mises 1919a, 18] of (5.22) is somewhat confusing (and not entirely correct):

$$\lim_{n=\infty} \int_Z \prod_{k=1}^{n} f_k(\overline{a}_k + \frac{1}{\sqrt{n}}\overline{z})d\overline{z} = \int_Z \mathrm{e}^{-\sum_{i,j=1}^{t} h_n^{(i,j)} z_i z_j} d\overline{z}.$$

continuously distributed random variables. In the case of random variables with densities he succeeded in proving the following theorem:
Let $(v_k)_{k\in\mathbb{N}}$ be a sequence of probability densities (each defined for all real numbers) of uniformly bounded variation.[37] Let $a_k := \int_{-\infty}^{\infty} x v_k(x)dx$ and $\sigma_k^2 := \int_{-\infty}^{\infty} x^2 v_k(x)dx - a_k^2$. Moreover, let $b_n := \sum_{k=1}^{n} a_k$, $r_n := \sqrt{2\sum_{k=1}^{n}\sigma_k^2}$,

$$w_n(u) := \int_{-\infty}^{\infty}\cdots\int_{-\infty}^{\infty} v_n(r_n u + b_n - x_1 - \cdots - x_{n-1})v_{n-1}(x_{n-1})\cdots v_1(x_1)dx_{n-1}\cdots dx_1,$$

and $\overline{w}_n(u) := \frac{1}{2}[w_n(u+\text{o}) + w_n(u-\text{o})]$. If the values of $|a_k|$, σ_k^2, and $|\int_{-\infty}^{\infty}(x-a_k)^3 v_k(x)dx|$ are each always bounded by a constant, independent of k, then:

1) $\overline{w}_n(u)$ converges in each compact set of real numbers u to $\varphi(u) := \frac{1}{\sqrt{\pi}}\mathrm{e}^{-u^2}$;
2) $\int_{-\infty}^{u} w_n(x)dx$ converges in $\mathbb{R}$ uniformly to $\int_{-\infty}^{u}\varphi(x)dx$.

For the proof von Mises [1919a, 31–33] used the basic properties of the complex adjunct

$$f(x) := \int_{-\infty}^{\infty} \mathrm{e}^{(x-a)(z-a)\sqrt{-1}}dV(z)$$

of a distribution V with expectation a, which he [1919a, 26 f.] explained generally. These properties included the correspondence between convolutions of distributions and products of complex adjuncts, as well as connections between derivatives of complex adjuncts and moments. In the special cases of the densities v_k and w_n, described above, he was able to prove by Fourier's integral theorem:

$$\overline{w}_n(z) = \frac{1}{2\pi}\int_{-\infty}^{\infty} \mathrm{Re}\; p_n(u)\mathrm{e}^{-zu\sqrt{-1}}du,$$

where

$$p_n(u) = f_1(a_1 + \frac{u}{r_n})f_2(a_2 + \frac{u}{r_n})\cdots f_n(a_n + \frac{u}{r_n}) \tag{5.23}$$

and

$$f_k(x) = \int_{-\infty}^{\infty} \mathrm{e}^{(x-a_k)(z-a_k)\sqrt{-1}}v_k(z)dz.$$

[37] A real-valued function f has the property of bounded variation if and only if there is a positive constant C such that for all $n \in \mathbb{N}$ and for all arguments $x_1 < x_2 < \cdots < x_n$ of f the expression

$$|f(x_1) - f(x_2)| + |f(x_2) - f(x_3)| + \cdots + |f(x_{n-1}) - f(x_n)|$$

remains less than C. The least upper bound of all these expressions is denoted by "total variation" S of f. Any density function of bounded variation can be represented, according to [von Mises 1919a, 31], as the difference of two monotonically increasing functions whose values are ≥ 0 and $\leq \frac{S}{2}$. A sequence of functions (f_k) is of uniformly bounded variation if and only if all f_k are of bounded variation and their total variations S_k possess an upper bound. The condition of bounded variation played an important role, for example as a prerequisite for the validity of Fourier's integral theorem [Pringsheim 1907].

Now, von Mises applied his theorem on the Laplacian approximation principle in connection with the particular case $\psi(u) := \mathrm{e}^{-zu\sqrt{-1}}$ to the product (5.23). Condition (i) of this theorem was met because $\sigma_k^2 = 2s_k^2$. The validity of (ii) could be justified by the assumption that $|f_k'''(a_k)| = |\int_{-\infty}^{\infty}(x-a_k)^3 v_k(x)dx|$ is uniformly bounded. The evidence for (iii) and (iv) was essentially based on the fact that the functions v_k were supposed to be of uniformly bounded variation. This property also led to a positive lower bound for all variances σ_k^2 from which, because of the boundedness of the sequences $|a_k|$ and σ_k^2, the validity of (v) could be shown. Altogether, taking into account that $\sigma_k^2 = 2s_k^2$, von Mises proved that, for all real z,

$$2\pi \overline{w}_n(z) = \int_{-\infty}^{\infty} \operatorname{Re} p_n(u)\mathrm{e}^{-zu\sqrt{-1}}du \to \int_{-\infty}^{\infty} \operatorname{Re} \mathrm{e}^{\frac{-u^2}{4}}\mathrm{e}^{-zu\sqrt{-1}}du = 2\sqrt{\pi}\mathrm{e}^{-z^2}.$$

An explicit proof of the uniformity of this convergence in all compact sets of z-values was not given by von Mises.[38] However, he used this property in proving the second part of his assertion.

Already prior to his discussion of sums of continuous random variables, von Mises [1919a, 28–31] had treated the case of lattice distributed random variables in an entirely analogous way. The transfer of this proceeding to the case of general distributions, however, was not possible for him. By use of the Laplacian approximation principle he was able to show that the complex adjuncts of the convolutions

$$W_n(z) = \int_{-\infty}^{\infty} \cdots \int_{-\infty}^{\infty} V_n(r_n z + b_n - x_1 - \cdots - x_{n-1})dV_{n-1}(x_{n-1})\cdots dV_1(x_1)$$

(r_n, b_n as above) tend to the complex adjunct of the Gaussian distribution with expectation 0 and variance $\frac{1}{2}$. For this proof he essentially needed the mere existence of the third absolute moments of the single distributions. Concerning the convergence of distributions, however, a complete argumentation, which was analogous to his proofs in cases of continuous or lattice distributed random variables, could not be given in the general case, as von Mises [1919a, 33–35] explained.

In order to treat this general case completely, von Mises [1919a, 35] introduced the "real adjunct" of a distribution V with expectation a and variance σ^2, that means the function

$$g(u) := \mathrm{e}^{-\frac{\sigma^2 u^2}{2}} \int_{-\infty}^{\infty} \mathrm{e}^{-(x-a)u} dV(x),$$

defined in $\mathbb{R}$. If q_n denotes the real adjunct of W_n, then

$$q_n(u) = g_1\left(\frac{u}{r_n}\right) g_2\left(\frac{u}{r_n}\right) \cdots g_n\left(\frac{u}{r_n}\right).$$

Dealing with the general case, von Mises [1919a, 22 f.] presupposed that

(a) for each distribution V_k there exists a positive number c_k^2 such that

[38] A closer examination of von Mises's line of argument in connection with his advancement of Laplace's approximation principle in the particular case $\psi(u) := \mathrm{e}^{-zu\sqrt{-1}}$ shows uniform convergence for all real z.

$$\int_{-\infty}^{\infty} e^{c_k^2 x^2}\, dV_k(x) < \infty$$

(as a consequence, the V_k have finite moments of arbitrary order);
(b) the variance σ_k^2 of each distribution is positive;
(c) for all $C > 0$ there exists a positive number c^2 (independent of k) such that

$$\int_{-\infty}^{\infty} e^{\frac{c^2(x-a_k)^2}{2\sigma_k^2}}\, dV_k(x) < C;$$

(d) $|a_k|$ and σ_k^2 are both bounded sequences, and there exists a positive function $\epsilon(n)$, tending to 0 as $n \to \infty$, such that

$$\frac{n^{\frac{2}{3}}}{\sum_{k=1}^{n} \sigma_k^2} \leq \epsilon(n).$$

As von Mises made explicit, the significance of condition (c) essentially lies in the fact that it implies the moduli of the "moments"

$$M_k^{(m)} := \int_{-\infty}^{\infty} \left(\frac{x - a_k}{\sigma_k \sqrt{2}} \right)^m dV_k(x)$$

having an upper bound independent of k for each order.

From these assumptions it followed that $g_k\left(\frac{u}{\sigma_k\sqrt{2}}\right) = \sum_{m=0}^{\infty} c_k^{(m)} u^m$ for all $u \in \mathbb{R}$, where

$$c_k^{(0)} = 1, \quad c_k^{(1)} = c_k^{(2)} = 0, \tag{5.24}$$

and $|c_k^{(m)}|$ has an upper bound independent of k for each $m \geq 3$ [von Mises 1919a, 36–38]. In the subsequent text, von Mises mainly discussed the problem of whether from the convergence of the coefficients $k_n^{(m)}$ in the power series of the real adjunct q_n of the convolution W_n to the corresponding coefficients of the real adjunct of $\Phi_{0;\frac{1}{2}}$ one could conclude that W_n also tends to this distribution. Because the real adjunct of the normal distribution with expectation 0 and variance σ^2 is equal to

$$\frac{1}{\sigma\sqrt{2\pi}} e^{-\frac{\sigma^2 u^2}{2}} \int_{-\infty}^{\infty} e^{-xu} e^{-\frac{x^2}{2\sigma^2}}\, dx = 1,$$

von Mises had to show first how, in consideration of (5.24), it could be justified that $k_n^{(m)} \to 0$ for $m \geq 3$. In particular, condition (d) was important in this context [von Mises 1919a, 44–46]. Von Mises [1919a, 39] proved that the moments $\int_{-\infty}^{\infty} \left(\frac{x-b_n}{r_n}\right)^m dW_n(x)$ depend linearly on the coefficients $k_n^{(0)}$ to $k_n^{(m)}$. Thus, the following discussion focused mainly on nestimates of the absolute values of the differences between the normal distribution and such distributions whose moments are close to the corresponding moments of the normal distribution. Von Mises [1919a, 40–43, 46–50] treated this problem in a way which was in general analogous to Markov's reasoning (see Sect. 4.6.4). Quite frequently, however, he used step functions which possess, up to a certain order, the same moments as the distributions considered. This method is similar to that of Stieltjes [1918a] in his posthumous variant of the proof of the Chebyshev–Markov inequalities. It cannot be known whether von Mises was influenced by Stieltjes's contribution. As an improvement of Markov's result, he even succeeded in proving that W_n tends uniformly to $\Phi_{0;\frac{1}{2}}$.

Von Mises [1919a, 54–58] also sketched a generalization of his local and integral limit theorems toward multidimensional random variables. He showed that the difference between the complex adjunct of the convolution of multidimensional distributions, normed as in (5.22), and the complex adjunct of the corresponding normal distribution tends to 0 as the number of summands increases. He did not give any proof, however, that from this convergence of characteristic functions the CLT itself could be deduced.

The article [von Mises 1919a] shows a good deal of truly "new" results, particularly regarding local limit theorems for sums of lattice distributed and continuous random variables, presupposing quite weak conditions. Although von Mises's results on the integral CLT for general distributions had already become obsolete in the one-dimensional case through the work of Lyapunov and (concerning moment methods) Markov, his account had an influence—not to be underestimated—on later contributions to the CLT (particularly by Pólya, Lindeberg, and Cramér) due to the variety of analytical methods he employed, and due to his rigorous and general presentation of the analytic aspects of basic probabilistic principles. It was the first work in which the tool of characteristic functions was used in a comprehensive and systematic manner. However, its author did not succeed in proving a theorem on the correspondence of the convergence of distributions and their accompanying characteristic functions in the general case.

For a balanced historical assessment, however, one has to observe that von Mises's [1919a] aim was not solely a purely analytical exposition. He also wanted to connect the most important applications of probability theory with his "theoretical" explanations. To this end he [1919a, 78, 93] formulated two "fundamental theorems" ("Fundamentalsätze"). The first of these two theorems consisted of a renewed enumeration of his local and integral CLTs, but now in the language of "Kollektivs," and focused on the most prominent applications. Similarly, the second summarized von Mises's results on inverse probabilities. With his fundamental theorems von Mises repeated a mode of exposition, which could already be found in the work of Laplace and Poisson, to differentiate between analytical theorems and "real" stochastic contents, whose application-oriented relevance had to be clarified independently of purely mathematical considerations. Siegmund-Schultze [2006] stresses the correspondence of this point of view to von Mises's attitude as an "applied mathematician." However, this "applied" position was not free of inner conflicts. Von Mises, on the one hand, apparently felt obliged, due to the general conditions of research in the the post-Weierstrassian era, to strive for the utmost analytic rigor and generality without considering any aspects outside of mathematics. On the other hand, he wanted the problem of the generality of assumptions to be judged by the necessities of possible applications, as one can see from his controversial discussion with Pólya (see below); thus he stressed also external criteria for the assessment of mathematical work.[39]

5.2.3 *Pólya and Lévy: Laws of Error, Moments and Characteristic Functions*

After the First World War, Pólya published several brief articles on a peculiarity of the Gaussian law (which had already been sporadically mentioned by some authors[40]), namely, the fact that it corresponds—in modern terminology—to a stable distribution. Besides that work, he became well-known through an article on different aspects of the convergence of distributions, in which the CLT also received its name. Lévy started a little later than Pólya with publications of similar content. However, he advanced his examinations much further and more comprehensively. Lévy's book *Calcul des probabilités* (from 1925) presents a collection of the results of its author from 1922 to 1924 and already touches on a large part of those problems concerning sums of independent random variables which did not receive their general setting and solution until about 1940.

Lévy's first probabilistic publication [1922a] already discussed the main concept of the theory of those distributions which he himself called "stable" ("lois stables"). In this article he focused only on symmetric laws. Stable distributions with characteristic function $z \mapsto e^{-a|z|^\alpha}$ $(0 < \alpha < 2, a > 0)$ were possible limit distributions for suitably normed sums of independent identically distributed random variables X_k with $E|X_k|^p < \infty$ for $p \leq \alpha$, and $E|X_k|^p = \infty$ for $p > \alpha$. Thus, stable distributions with $\alpha < 2$ played the same role for sums of independent identically distributed random variables without finite moments of second order as the Gaussian law with the characteristic function $z \mapsto e^{-a|z|^2}$ did for sums of independent identically distributed random variables with finite moments of second order. Stable distributions were also considered as limit distributions for suitably normed sums of independent, but not identically distributed random variables. Thus, already from Lévy's first contributions it became clear that the classical theorem on the Gaussian law as a limit distribution was only one among many "with equal rights."

5.2.3.1 Pólya's First Contributions

Pólya had a remarkable knowledge of Russian sources on probability theory. Markov's proof of the CLT [1908/13/2004], whose French translation Pólya [1914/15] had reviewed, especially stimulated the latter for a closer discussion of probabilistic problems which referred to moment theory. Pólya started with the

[39] For a general discussion of von Mises's attitude toward those "fractures of modernity," see [Siegmund-Schultze 2004].

[40] Especially by Edgeworth [1883; 1905], see Sect. 3.4.2.3, and by Förster [1915], see below.

well-known property of the Gaussian error law that the sum of independent normally distributed errors again has a Gaussian distribution. This property had already motivated Edgeworth to a general, though scarcely noticed, discussion of special error laws, which he called "reproductive" (see Sect. 3.4.2.3). Pólya [1919a] posed the problem of finding a nonnegative function $\varphi \not\equiv 0$ together with positive numbers a, b, c, such that for all real x

$$\frac{1}{c}\varphi\left(\frac{x}{c}\right) = \frac{1}{ab}\int_{-\infty}^{\infty}\varphi\left(\frac{u}{a}\right)\varphi\left(\frac{x-u}{b}\right)du. \tag{5.25}$$

(Later, Lévy called such densities "semistable.") On the function φ Pólya imposed the additional condition that it be bounded on each finite interval, and have moments

$$K_n = \int_{-\infty}^{\infty} x^n \varphi(x)dx$$

of arbitrary order $n \in \mathbb{N}_0$ (all integrals had to be understood in the improper Riemannian sense). As a consequence of (5.25) Pólya showed that $K_0 = 1$ and $K_1 = 0$. For $m \geq 3$ he was able to establish a relation among the moments K_n ($n \leq m$), from which he successively deduced that for solutions φ_1, φ_2 of (5.25) with identical moments of second order:

$$\int_{-\infty}^{\infty} x^l \varphi_1(x)dx = \int_{-\infty}^{\infty} x^l \varphi_2(x)dx \quad \forall l \in \mathbb{N}_0.$$

Pólya now used moment theoretic results achieved by Borel [1901] and Godfrey Harold Hardy [1917] for the proof that the function $\varphi(x) = \frac{h}{\sqrt{\pi}}e^{-h^2x^2}$ was the only solution of (5.25) with $K_2 = \frac{1}{2h^2}$.

In principle, Pólya's problem, as well as his reasoning, had a certain similarity with a contribution of Gustav Förster. Förster [1915] discussed the problem of finding an error function φ with the following property: If the independent errors ϵ_1 and ϵ_2 have the respective density functions $\lambda_1\varphi(\lambda_1 x)$ and $\lambda_2\varphi(\lambda_2 x)$, λ_1, λ_2 being positive constants, then for all positive numbers a, b there exists a (necessarily unique) positive parameter λ_3 such that $\epsilon_3 = a\epsilon_1 + b\epsilon_2$ has the density $\lambda_3\varphi(\lambda_3 x)$. Förster likewise deduced a recursive relation for the moments of higher order and thus justified, if not always in a rigorous manner, that only error laws of the Gaussian type obeyed his condition, which was stronger than Pólya's. Moreover, Förster treated in his work a problem of lesser generality, assuming the analyticity of φ. According to a statement in his [1919b], Pólya initially was not aware of Förster's contribution.

Apparently motivated by [von Mises 1919a], Pólya [1920] devoted himself to analytic aspects, in particular related to moments, of the theorem which he called "zentraler Grenzwertsatz." The theorem itself, however, was only briefly referred to at two places in his article [1920, 171 f., 177]. The problem seemed rather in the foreground under which conditions it was possible to infer the convergence of distribution functions from the convergence of moments of arbitrary order. Concerning the convergence to the normal distribution this problem had already been solved by Markov [1898]. In a general setting, Pólya [1920, 178] maintained that by methods

expounded in Hamburger's work [1919] on the unique determinacy of a distribution function over $\mathbb{R}$ by its moments, the convergence problem could likewise be solved. However, Pólya aimed at a brief and direct proof for the "continuity theorem of the moment problem." This theorem states the following:
If for the moments $t_m = \int_{-\infty}^{\infty} x^m df(x)$ $(m \in \mathbb{N})$ of a continuous distribution function[41] f the condition

$$\limsup_{m\to\infty} \frac{\sqrt[2m]{t_{2m}}}{m} < \infty \tag{5.26}$$

is valid, and if $(f_n)_{n\in\mathbb{N}}$ is a sequence of distribution functions such that

$$\lim_{n\to\infty} \int_{-\infty}^{\infty} x^{\mu} df_n(x) = t_{\mu} \quad \text{for all } \mu \in \mathbb{N},$$

then $f_n(x)$ tends to $f(x)$ uniformly for all $x \in \mathbb{R}$.
The condition (5.26) is equivalent to Hamburger's [1919] uniqueness condition

$$t_m \leq \rho^m (2m)! \quad (m \in \mathbb{N}_0)$$

with an appropriate $\rho>0$. Pólya based his theorem on three auxiliary theorems dealing with i) the relation between pointwise and uniform convergence of distribution functions, ii) the relation between the convergence of the antiderivatives of distribution functions and the distributions themselves, and iii) a method of inferring the convergence of distributions from the convergence of generating functions. The first of these theorems is especially important even today: If a sequence of distribution functions converges to an everywhere continuous distribution function pointwise in a set dense in $\mathbb{R}$, then even the uniform convergence of the sequence in $\mathbb{R}$ follows.[42]

From recent historical research we know, however, that Pólya himself valued the third auxiliary theorem particularly highly. As Siegmund-Schultze [2006] reports, there was a controversy—expressed in an exchange of letters—between Pólya and von Mises in 1919/1920 on von Mises's contributions to the CLT. Pólya's chief criticism was that von Mises's treatment of the integral CLT was inferior to the earlier contributions of Lyapunov and Markov in the general case. Von Mises, on the contrary, emphasized the low application relevance of theorems on random variables with general distributions. On the other hand, however, he stressed the importance of his analytic innovations, especially the application of the complex adjunct to local limit theorems for densities or lattice distributions. As one can see from Siegmund-Schultze's account, Pólya's article of 1920 should be understood as a public response by Pólya to von Mises. Still, it can hardly be inferred from the wording of the article alone that the two mathematicians actually had a scientific quarrel. Among all his arguments which were expounded in letters to von Mises, Pólya repeated only the following in his paper: He criticized, if in a rather implicit and restrained way, that von Mises's reasoning in connection with the real adjunct

[41] Pólya defined distribution functions f as monotonically increasing functions, continuous on the right, with $\lim_{x\to-\infty} f(x) = 0$ and $\lim_{x\to\infty} f(x) = 1$.

[42] Pólya's proof is described in [Uspensky 1937, 386 f.], for example.

was too complicated. Instead of von Mises's moment theoretic considerations, he proposed the application of the third auxiliary theorem of his article:

> Theorem III. I consider the sequence of improper Stieltjes integrals
>
> $$\int_{-\infty}^{\infty} e^{ut}\,df_1(t),\ \int_{-\infty}^{\infty} e^{ut}\,df_2(t),\ldots,\ \int_{-\infty}^{\infty} e^{ut}\,df_n(t),\ldots,$$
>
> where $f_1(t)$, $f_2(t)$, ..., $f_n(t)$, ... denote distribution functions, and where there exists a positive quantity a such that each integral converges for $-a \le u \le a$. It is assumed that for the same values of u
>
> $$\lim_{n=\infty} \int_{-\infty}^{\infty} e^{ut}\,df_n(t) = \int_{-\infty}^{\infty} e^{ut}\,df(t),$$
>
> where $f(t)$ denotes a continuous distribution function. Then
>
> $$\lim_{n=\infty} f_n(x) = f(x)$$
>
> uniformly for all values of x.

Apparently, Pólya attached great expectations to this theorem. Actually, through its application, the proof of the integral CLT under von Mises' conditions can be made easier and shortened considerably, as Pólya [1920, 171] hinted at.[43] Pólya at the same place asserted that Lyapunov's version of the CLT could also be reached by his method. This was only correct in a very benevolent interpretation, however. Pólya's generating functions were defined on the vertical strip $-a \le u \le a$. For this condition, the existence of moments of arbitrarily large order for all given random variables was necessary. Lyapunov's theorem only presupposed the existence of absolute moments of order $2+\delta$. Therefore, only by the use of certain tricks (the introduction of truncated variables, for example) can one prove Lyapunov's theorem by means of Pólya's generating functions. A direct use of these generating functions, however, is not possible. There may be a second reason why Pólya's hopes connected with his Theorem III were not realized. Around the beginning of the twenties the discussion of stable distributions became an increasingly important topic of probability theory. Stable distributions, however, with the exception of the Gaussian distribution, do not have moments of arbitrarily large order and therefore cannot be treated by means of generating functions. In 1922, when Lévy succeeded in proving a theorem on the convergence of characteristic functions even for general distributions, Pólya's generating functions became definitely inferior to characteristic functions (see below).

Pólya's work on error laws and the convergence of distributions remained connected to particular analytical problems rather than being dedicated to a more comprehensive theory. The plan of the publication of a monograph on probability theory, hinted at in [1920, 172], was never realized. Lévy, who around 1922 started publishing on problems similar to Pólya's, discussed the asymptotic behavior of distributions of sums of independent random variables in a far more comprehensive and systematic way.

---

[43] As far as I know, there does not exist any printed version of a proof modified by use of Pólya's lemma. By a line of argument, however, which is very close to the typical textbook proof of the CLT which uses characteristic functions, one can see that Theorem III actually provides a considerable simplification of von Mises's proof of the CLT.

5.2.3.2 The Hypothesis of Elementary Errors as a Motivation for Lévy's First Articles

According to Lévy's own statements [1970, 71; 74], at the beginning of his involvement with probability theory he did not possess any knowledge of classic works, such as the *TAP* of Laplace, or the analytical tools used in these works. He knew just as little about the contributions of Russian authors. Lévy obtained his basic knowledge of probability primarily from the second edition of Poincaré's *Calcul des probabilités* [1912]. The hypothesis of elementary errors, as discussed by Poincaré (see Sect. 4.6.3), formed the original motivation for Lévy's work on sums of independent random variables. In this early period Lévy [1924, 14–17, 37–44; 1925b, 278–294] discussed in detail the error theoretic aspects of his stochastic contributions. In the thirties, however, after the development of probability theory toward a mainly inner mathematical orientation, he no longer considered these problems very important (cf. [Le Cam 1986, 80]), although he never completely gave up employing arguments beyond mathematics (see Sect. 6.2.2).

Unlike Poincaré who had, entirely in accordance with the tradition of the 19th century, treated the hypothesis of elementary errors as an "approximative" assertion on the distribution of a sum consisting of many small components, Lévy [1924, 25] (by way of a hint already [1922b]) specified this hypothesis through a limit theorem. He only considered total errors with variance 1 and expectation 0, although a more general setting could easily be reached. According to Lévy, the elementary errors had the form $\frac{m_i}{M_n} X_i$, where the X_i were to be considered independent random variables, each with variance 1 and expectation 0, and where for the positive numbers m_i, $M_n := \sqrt{\sum_{i=1}^{n} m_i^2}$ the additional condition $\lim_{n\to\infty} \max_{1\le i\le n} \frac{m_i}{M_n} = 0$ had to be valid. In this way, he connected the hypothesis of elementary errors with a CLT of Laplace–Chebyshev type for linear combinations $\sum m_i X_i$ of random variables. This CLT, for which Lévy [1922a;b] found especially weak sufficient conditions (see Sect. 5.2.6.5), played the role of a "théorème fondamental" within his error theory.

Lévy's early discussions of the hypothesis of elementary errors, the Gaussian error law, and the method of least squares deduced from this law, were influenced by the tensions between formalism and intuition, between pure mathematics and practice. In his general explanations Lévy advocated Poincaré's point of view on the foundations of mathematics, which stressed ideas like "experience," "intuition," "harmony," "economy of thinking," and "good sense" (see [Mehrtens 1990, 223–256]). In regard to attempts to discuss the method of least squares axiomatically, particularly by Felix Bernstein and Werner Siegbert Baer [1915], Lévy [1924, 42–44] rejected Hilbert's program of basing all mathematical disciplines on axioms. Rather he aimed to base a theory, in his case error theory, on simple principles in harmony with good sense.

He admitted that error theory as a practical discipline was feasible without great mathematical effort and without complete analytical rigor. For the mathematician, however, this attitude could not be accepted, as Lévy [1924, 17] pointed out. Despite his declared opposition to purely formalistic considerations, Lévy had to see himself, due to his emphasis on mathematical rigor concerning the probabilistic foundation of error theory, in contradiction to Borel, who had become the leading figure of French mathematics since Poincaré's death. Borel held the opinion that considerable mathematical effort would not be worthwhile for establishing the Gaussian error law and for discussing alternative error laws [Le Cam 1986, 82]. He hoped that probability calculus, due to its specific problems, would become a particular discipline within natural and social sciences, but not within mathematics in a narrower sense [Knobloch 1987, 217]. In this respect Lévy had, as one can see from the preface of his book [1925b], to justify his approach to error theory which emphasized purely mathematical aspects. He was convinced of the mathematical relevance of his probabilistic work, due to his self-confidence as an analyst, and therefore took the liberty of not ignoring the opinion of an important authority, but altogether considering it only secondary.

Lévy apparently had to bear certain disadvantages due to the attitude just described. He was not invited to contribute to the *Traité du calcul des probabilités et des ses applications*, a collection of several monographs, which appeared from 1925 and was edited by Borel, according to whom it should prove the "unity and importance" [Borel 1925, vi] of probability as an independent field of science. Only toward the end of the thirties were Lévy's successes in the theory of probability, which had not turned out the way Borel originally intended, rewarded by the latter. For example, Borel included Lévy's *Théorie de l'addition des variables aléatoires* [1937a] in his series *Collection de monographes sur la théorie des probabilités* (supplementing the *Traité*) as the first volume. Lévy's influence on probabilistic activities in the "Institute Henri Poincaré," directed by Borel, which was the leading institution of mathematical research in France from the end of the twenties, advanced significantly during the thirties.[44]

Lévy's first probabilistic investigations concerned the role of Gauss's law and other stable laws in the framework of the hypothesis of elementary errors [Lévy 1922a;b], the generalization of Fourier's integral formula to the case of Fourier transforms expressed by Stieltjes integrals, the "continuity of the correspondence" between distributions and its characteristic functions [Lévy 1922c], and the properties of characteristic functions of certain types of distributions [Lévy 1923a;b;c]. These contributions were merely written in a roughly sketched form, while their error theoretic aspects were explained thoroughly in [Lévy 1924]. A more detailed, if still uncomplete, discussion of these topics is contained in Lévy's book *Calcul des probabilités* [1925b]. In particular, the central chapter 6 of this book on "exceptional laws" presents at several places only rather vague hints, and sometimes even mere assumptions.

The poor reception of his book, later deplored by Lévy [1970, 81], was at least somewhat grounded in its densely written style. Nonetheless, as Cramér [1976, 516] reports, it offered many suggestions for the still small community of mathematicians involved in the development of modern probability theory, and thus played a pioneering role for research in the field of sums of independent random variables

[44] For Lévys own opinion on his relation to Borel, see [Lévy 1970, 82–84]. For an outline of stochastic research in the "Institute Henri Poincaré," see [Siegmund-Schultze 2001, 169–175].

well into the middle of the thirties. This is especially true for Lévy's theory of characteristic functions.

5.2.3.3 Poincaré and the Concept of Characteristic Functions

In the early stage of his probabilistic work, Lévy made consistent use of characteristic functions $\mathbb{R} \ni z \mapsto \mathrm{E}e^{\mathrm{i}zX}$ for the investigation of distributions of random variables X. Lévy [1976, 1] claimed to have been motivated toward this method by a short passage in the second edition of Poincaré's *Calcul des probabilités* [1912, 206–208]. In addition to his moment theoretic treatment of sums of random variables, which was already contained in the first edition (see Sect. 4.6.3), in the second edition Poincaré also discussed "fonctions caractéristiques." With this name he designated the term

$$f(\alpha) = \sum p(x)e^{\alpha x}$$

in the case of discrete "quantities," whose values x occur with probability $p(x)$, and

$$f(\alpha) = \int_{-\infty}^{\infty} \varphi(x)e^{\alpha x}\,dx$$

in the case of "continuous" quantities with density φ. Poincaré did not specify the number set to which the numbers α belonged. Apart from $f(\alpha)$, however, he also considered the function term $f(\mathrm{i}\alpha)$ ($\mathrm{i} = \sqrt{-1}$). Using the latter he deduced on the basis of Fourier's integral theorem (for which he did not discuss any conditions) the inversion formula $2\pi\varphi(x) = \int_{-\infty}^{\infty} f(\mathrm{i}\alpha)e^{-\mathrm{i}\alpha x}\,d\alpha$. Altogether, from today's point of view, Poincaré treated generating functions rather than "modern" characteristic functions.

With his "characteristic functions" Poincaré did not open a new chapter in the analytic methods of probability theory, but in contrast to the authors of the 19th century who had made use of Laplace or Fourier transforms in probability theory, he connected with these terms an autonomous meaning,[45] and sketched out how the convergence to the Gauss density could be reduced to the convergence of the accompanying characteristic functions. However, he did not explicitly treat the problem of whether from the convergence of "characteristic functions" the convergence of probability distributions could be deduced in general.

The interpretation of generating (or characteristic) functions as expectations can already be seen in Hausdorff's 1901 paper (see Sect. 3.4.2.1), although the latter gave the functions $z \mapsto \mathrm{E}e^{zX}$ no special name [Hausdorff 1901, 169]. This interpretation made it particularly easy to justify Cauchy's multiplication theorem, according to which the characteristic function of a sum of independent random variables equals the product of the accompanying characteristic functions.

Poincaré [1912, 211–218] also used characteristic functions for discussing "exceptions" to the Gaussian law. In this context he explained (among other topics)

---

[45] Cauchy got relatively close to this "modern" conception of characteristic functions, see Sect. 2.5.2.

the significance of the error law with the characteristic function $f(i\alpha) = e^{-|\alpha|}$, corresponding to the "Cauchy distribution" which had already been considered by Poisson (see Sect. 2.2.3.1). By his rather comprehensive treatment of "exceptional laws," in particular concerning the problem of how the precision of the arithmetic mean of a larger number of observations depends on the precision of each single observation, Poincaré considerably influenced Lévy's discussion of stable error laws.

5.2.3.4 Lévy's Fundamental Theorems on Characteristic Functions

Particularly important within the theory of characteristic functions were, as Lévy [1970, 75] stated retrospectively, the inversion formula for deducing distributions from characteristic functions, and the "continuity theorem" on the correspondence between the convergence of characteristic functions and distributions. Lévy [1922c] gave sketches of proofs for both fundamental theorems. More elaborate versions can be found in [Lévy 1925b, 163–169, 192–200].

Lévy used the concept of real random variables ("variable, quantité, erreur") only intuitively, without giving a precise definition. "Probability law" ("loi de probabilité") of a variable X Lévy named the mapping $S \mapsto P(X \in S)$ for subsets S of $\mathbb{R}$, which were not further specified. Lévy [1925b, 136] designated the possibility of those mappings "obvious." The probability law of the random variable X was, according to Lévy (e.g., [1925b, 137]), uniquely determined by the distribution function ("fonction des probabilités totales") $F(x)$, defined by

$$F(x) := P(X < x) + \frac{1}{2}P(X = x).$$

Indeed, Lévy was well aware about the difficulties of assigning a probability measure to any arbitrary subset of $\mathbb{R}$, as we can see from his article on probability measures on "abstract sets" [Lévy 1925a], which he apparently considered so important that he included a reprint in the appendix of his 1925 book. Lévy [1925a, 330] explained, though in a not entirely precise way, that a measure defined on all subsets of the interval [0; 1] would be "necessarily very arbitrary." A few lines below, Lévy stated that it would "suffice in practice" to restrict all considerations to Borel sets.[46]

Lévy's inversion formula provided a means for determining distribution functions F from given characteristic functions

$$\varphi(z) = \int_{-\infty}^{\infty} e^{izx}\, dF(x). \tag{5.27}$$

In his deduction of this formula Lévy represented the integral

$$I_c := \int_{-c}^{c} \varphi(z) \int_{0}^{t} e^{-iz\tau}\, d\tau\, dz \tag{5.28}$$

46 Banach & Kuratowski [1929] proved that there does not exist any sigma-additive set function m defined on all subsets of $\mathbb{R}$ such that $m(\{x\}) = 0$ for all $x \in \mathbb{R}$. This implies, for example, the nonexistence of a probability measure defined on all subsets of $\mathbb{R}$ which has a density function.

in two different ways. On the one hand, substituting (5.27) in (5.28) yielded the "Dirichlet integral"

$$I_c = 2\int_{-\infty}^{\infty} [F(x+t) - F(x)] \frac{\sin cx}{x} dx. \tag{5.29}$$

Because of the definition of F, for all real x the equation

$$F(x) = \frac{1}{2}(F(x+\mathrm{o}) + F(x-\mathrm{o}))$$

was valid, and from this Lévy concluded that

$$\lim_{c\to\infty} I_c = 2\pi[F(t) - F(0)]$$

for all real t. On the other hand, integrating (5.28) with respect to τ resulted in

$$I_c = \int_{-c}^{c} \frac{\varphi_0(z)\sin tz + \varphi_1(z)(1-\cos tz)}{z} dz,$$

$$\text{where } \varphi(z) = \varphi_0(z) + \mathrm{i}\varphi_1(z). \tag{5.30}$$

Altogether, Lévy obtained

$$F(t) - F(0) = \frac{1}{2\pi}\int_{-\infty}^{\infty} \frac{\varphi_0(z)\sin tz + \varphi_1(z)(1-\cos tz)}{z} dz,$$

or

$$F(t) - F(0) = \lim_{T\to\infty} \frac{1}{2\pi}\int_{-T}^{T} \frac{1-\mathrm{e}^{-\mathrm{i}tz}}{\mathrm{i}z}\varphi(z) dz.$$

With these considerations Lévy followed the path—very similar to the reasoning of Dirichlet and Cauchy in the 19th century—of representing probabilities through the use of appropriate jump functions. In the context of the inversion formula one advantage of Lévy's definition of distribution function—somewhat strange from today's point of view—becomes clear: possible discontinuities of this function must not be considered separately.

For a closer investigation of the "continuous correspondence" between probability laws and characteristic functions, Lévy in [1925b, 192] (and in a less precise manner already in [1922c]) considered probability laws L_λ depending on a parameter λ which converge to a limit law $\mathscr{L}$ as λ tends to the constant value λ_0. Lévy's definition of convergence of probability laws corresponded to the (weak) convergence of distributions which is now common: Let F_λ be the distribution function of L_λ, and let G be the distribution function of $\mathscr{L}$; by definition L_λ converges to $\mathscr{L}$ as $\lambda \to \lambda_0$ if $\lim_{\lambda\to\lambda_0} F_\lambda(x) = G(x)$ in each point of continuity x of G. In his book [1925b, 200] Lévy even hinted at a second criterion for the convergence of L_λ to $\mathscr{L}$ by introducing a "distance" between probability laws in the following sense: Let A_a and B_a be the intersection points between the graphs (completed, if necessary, in jumps by vertical segments) of F_λ and G, respectively, and the line $x+y=a$. Then, this "distance" is equal to $\sup_{a\in\mathbb{R}} \overline{A_a B_a}$ ($\overline{A_a B_a}$ denoting the Euclidean distance between the two points). Apparently, it was obvious for Lévy that L_λ converges to $\mathscr{L}$ in accordance with the definition above if the "distance" between these laws tends to 0.

Lévy [1925b, 195–200] made the relations between the convergence of distributions L_λ and the accompanying characteristic functions φ_λ explicit with the following theorems:

Theorem 1: If for $\lambda \to \lambda_0$ the laws L_λ tend to the limit law $\mathscr{L}$ with characteristic function ω, then $\varphi_\lambda(z)$ also tends to $\omega(z)$ uniformly in each compact interval of z-values.

Theorem 2: If there exists a characteristic function ω such that $\lim_{\lambda\to\lambda_0}\varphi_\lambda(z) = \omega(z)$ uniformly in each compact interval of z-values, then L_λ tends to the probability law $\mathscr{L}$ which belongs to ω.[47]

Theorem 2 is important especially for applications of characteristic functions to limit theorems.[48] For a proof of this theorem, Lévy used the following idea: The convolution ("composition") of an arbitrary distribution and the special distribution

$$\Phi_a(x) = \int_{-\infty}^{x} e^{-\frac{t^2}{2a^2}} \frac{dt}{a\sqrt{2\pi}}$$

yields an absolutely continuous distribution, whose density has the upper bound $\frac{1}{a\sqrt{2\pi}}$. With a sufficiently small a one can ensure that the "distance" between the convolution and the original distribution becomes arbitrarily small. The same applies for the absolute value of the differences of the accompanying characteristic functions.[49]

By this method the general case could be reduced to the case of a sequence of absolutely continuous distributions whose densities were uniformly bounded above. For such distributions Lévy showed that for a given $\varepsilon > 0$ there exists $C > 0$ such that for all λ and for all $c > C$:

$$\left| 2\pi[F_\lambda(t) - F_\lambda(0)] - 2\int_{-\infty}^{\infty} [F_\lambda(x+t) - F_\lambda(x)] \frac{\sin cx}{x} dx \right| < \frac{2}{3}\pi\varepsilon \tag{5.31}$$

and

---

[47] The formulation and proof of an assertion equivalent to Theorem 2 can already be found in [1922c], whereas Theorem 1 is only contained in the book of 1925.

[48] It has to be taken into account that Lévy demands that the characteristic functions converge locally uniformly to a function of which it is already known that it is a characteristic function. By use of Eduard Helly's [1930] theorem of choice for a sequence of distribution functions, Cramér [1937, 29–31, 121] was able to show (although with a mistake in the first version, which was corrected in the later editions, see [Cramér 1976, 525]), that the pointwise convergence of the characteristic functions to any limit function f is already sufficient for the convergence of the respective distributions to a limit distribution if $f(z)$ is continuous in $z = 0$.

[49] In [1922c] Lévy had only written: "If a is sufficiently small, then the functions which occur in the wording of the assertion, are changed arbitrarily little." In his book of 1925, for a closer specification Lévy introduced the "distance" between distributions described above.

$$\left|2\pi[G(t) - G(0)] - 2\int_{-\infty}^{\infty} [G(x+t) - G(x)]\frac{\sin cx}{x}dx\right| < \frac{2}{3}\pi\varepsilon. \tag{5.32}$$

On the other hand, with the abbreviations $\varphi_{\lambda,0} = \operatorname{Re}\varphi_\lambda$ and $\varphi_{\lambda,1} = \operatorname{Im}\varphi_\lambda$, it resulted from the locally uniform convergence of φ_λ that for λ sufficiently close to λ_0,

$$\left|\int_{-c}^{c} \frac{\varphi_{\lambda,0}(z)\sin tz + \varphi_{\lambda,1}(z)(1-\cos tz)}{z}dz - \right.$$
$$\left. - \int_{-c}^{c} \frac{\omega_0(z)\sin tz + \omega_1(z)(1-\cos tz)}{z}dz\right| < \frac{2}{3}\pi\varepsilon. \tag{5.33}$$

On account of (5.29) and (5.30) it followed from (5.31), (5.32), and (5.33) by use of the triangle inequality:

$$|2\pi[G(t) - G(0)] - 2\pi[F_\lambda(t) - F_\lambda(0)]| < 2\pi\varepsilon,$$

and therefore

$$F_\lambda(t) - F_\lambda(0) \to G(t) - G(0) \quad (\lambda \to \lambda_0) \tag{5.34}$$

for all $t \in \mathbb{R}$.

From this last equation Lévy [1922c, 335] concluded[50] that

$$F_\lambda(t) \to G(t). \tag{5.35}$$

The reader might reach the impression that Lévy's theorems on the "continuous correspondence" between characteristic functions and distributions conform to an historically continuous "story of success" of Fourier methods, which were introduced in the context of probabilistic limit theorems by Laplace, and thereafter remained—with respect to the basic ideas—unchanged. Following Laplace, however, it was his method of approximation to "functions of large numbers," which finally, by gradual refinement and adjustment to contemporary analytical standards, enabled Lyapunov's proofs of the integral CLT. Laplace's method could be exclusively applied in the framework of Gaussian limit distributions, whereas the theorem of Lévy—a mathematician, who did not (!) have, according to his own words [Lévy 1970, 71–75], any knowledge about the contributions of the 19th century up to Lyapunov—made it possible to consider arbitrary limit distributions. Consequently, the classic CLT with the normal distribution as a limit became one theorem among many other "central limit theorems" with in principle equal rights. Therefore, Lévy's fundamental theorems do not represent the completion of a long-lasting development, but rather mark a new start, which was scarcely influenced by previous results.

---

[50] A justification (which in today's courses on probability may be a nice problem for homework) for the conclusion from (5.34) to (5.35) was not given by Lévy.

5.2.3.5 Pólya's Reaction to Lévy's First Articles

Lévy [1922a], in the context of a brief discussion of counterexamples to the CLT, had considered functions

$$\varphi(t) = \mathrm{e}^{-a|t|^{\alpha}} \quad (a > 0,\ 0 < \alpha \leq 2) \tag{5.36}$$

which he referred to as characteristic functions of—as he wrote—"stable laws." At this place, Lévy did not prove that functions of the type (5.36) are actually characteristic functions of probability distributions. In his first characterization of stable laws [1922a, 10], stated only in words, it is required that a sum of mutually independent errors, each obeying the same type of stable law, again obeys this type of stable law. In Lévy's own words, two errors obey the same type of a stable law if one can be reduced to the other by a "change of the unit."

Due to the features in common between Lévy's problems and his own "deduction" of the Gaussian law, Pólya was prompted to a renewed and more general discussion of the integral equation (5.25). There was an exchange of letters on these issues between Pólya and Lévy in 1922 and in 1923. However, it seems that only Lévy's letters, which are kept in the archives of the ETH Zürich,[51] have survived. In these letters Lévy informed Pólya chiefly about his contributions which had appeared in the *Comptes rendus*. In the first two letters (9 April 1922; 23 April 1922) Lévy emphasized the advantages of characteristic functions over generating functions. Especially important for Lévy was the property of characteristic functions to be "always well defined" (first letter), "without any restrictions on the probability law" (second letter), in particular regarding the existence of moments. Already shortly after Lévy had published his article [1922a], Pólya, as Lévy hinted at in the letter from 23 April 1922 and also reported in his autobiography [1970, 78], drew the latter's attention to the work of Cauchy (see Sect. 2.5.2), in which characteristic functions of the type (5.36) had already been discussed. Apparently, Pólya, who in his previous articles had not mentioned Cauchy's account at all, felt motivated to look for earlier work on stable distributions. In fact, one can find, in Czuber's very popular report on the development of probability theory [1899, 183 f.], a hint at characteristic functions of type (5.36) with Cauchy. It seems possible that Pólya learned about Cauchy's contributions from this report. Pólya, however, must also have known Edgeworth's discussion of stable distributions (see Sect. 3.4.2.3), as a letter from Edgeworth to Pólya reveals.[52]

At this point it may be useful to clarify the substantial differences between Cauchy, Lévy, and Pólya in their discussion of stable laws.

Cauchy looked for error laws such that, presupposing a finite number of identically, mutually independent errors of observation, the moduli of the differences between estimated and real values in a linear model meet a special condition of minimality. In this way he came across error densities with characteristic functions

[51] ETH-Bibliothek, Archive, Hs 89: 320–326.

[52] ETH -Bibliothek, Archive, Hs 89: 132. The letter is undated, possibly written in the early 1920s. Edgeworth informs Pólya mainly about his article [Edgeworth 1905].

(5.36). He did not realize, however, that $\alpha \leq 2$ was necessary for φ to be actually the characteristic function of a probability distribution. Instead, he considered the case $\alpha = \infty$ especially important (see Sect. 2.5.2). The property of stability according to Lévy's characterization only played a minor role in Cauchy's contributions.

Lévy searched for counterexamples to the CLT. A (suitably normed) sum of arbitrarily many random variables, each obeying the same distribution according to (5.36) with $\alpha < 2$, can never be normally distributed. In [1923b] Lévy defined, in a slightly more formal manner than described above, stable laws in the following way: A probability law $\mathscr{L}$ is called "stable" if it does not correspond to a degenerate distribution (i.e., a distribution concentrated in one point), and if for independent random variables X_1, X_2, each with probability law $\mathscr{L}$, this condition is valid: For all $a_1, a_2 > 0$ there exists $a > 0$ such that $\frac{1}{a}(a_1X_1 + a_2X_2)$ likewise obeys the law $\mathscr{L}$.[53] As we will see below (footnote 71), the quantity a is necessarily uniquely determined. Lévy [1923b] expressed this fact, without giving any justification, by the words that a was "a function" of a_1 and a_2. When Lévy's attention, according to his own statement [1970, 77], was drawn by a question of one of his students to the problem of distributions with the property just described, he immediately perceived that, in the same way as the Gaussian distribution, any other stable distribution was a candidate for being a limit distribution of sums of independent random variables. This observation had also been the major motivation for Edgeworth's discussion of "reproductive" laws. Lévy, however, was apparently not aware of Edgeworth's ideas. Besides this limit property, stable error laws for Lévy were important because the precision of the arithmetic mean of several observations could be uniquely related to the precision of each single observation if each observation could be characterized by a stable error law of the same type.[54] Even though stable distributions different from the Gaussian distribution could be likewise generated by an additive accumulation of many small elementary errors, Lévy [1924, 41] was convinced that a good arrangement of measurement would only lead to such elementary errors whose accumulation, on the basis of the "théorème fondamental" (that means the classic CLT), would produce a normal distribution.

Pólya searched for distribution functions V with the following property (expressed in Lévy's style for better comparison): There exist (in Lévy's definition we read "for all") positive numbers a_1, a_2 such that, for all independent random variables X_1, X_2 with distribution function V, the normed sum $\frac{1}{a}(a_1X_1 + a_2X_2)$ (a a suitable positive number) is also distributed according to V. In his first articles Pólya additionally required the existence of all moments of the distribution V (which he also assumed to be absolutely continuous). Under these conditions he found the Gaussian distribution as the only solution of his problem.

[53] It must be emphasized here that this definition makes a more restrictive demand on stable distributions than is common today. Probability laws obeying this condition are now called "strictly stable." Only in the 1930s (see Sect. 7.2.2), in the definition of stable laws were possible translations of the origin additionally considered.

[54] For closer details on Lévy's discussion of precision on the basis of stable error laws, see [Sheynin 1996a, 182–188].

In his first papers Pólya employed mainly moment methods for his discussion of the integral equation (5.25), and in a far-reaching accordance with these methods, generating functions (in [1919b]). Apparently stimulated by Lévy's treatment of characteristic functions, Pólya was later able to handle his problem under far more general conditions by using the "new" tool. Now, he aimed at a characterization of all nonnegative functions φ defined in $\mathbb{R}$ with $0 < \int_{-\infty}^{\infty} \varphi(x)dx < \infty$ such that

1) $\int_{-\infty}^{\infty} x^2\varphi(x)dx := \sigma^2 < \infty$,
2) $\varphi(x)$ is bounded in each finite interval,
3)

$$\exists a, b, c > 0 : \quad \frac{1}{c}\varphi\left(\frac{x}{c}\right) = \frac{1}{ab}\int_{-\infty}^{\infty} \varphi\left(\frac{u}{a}\right)\varphi\left(\frac{x-u}{b}\right)du.$$

The consideration of the moments up to second order [1923, 99 f.] yielded the equation $a^2 + b^2 = c^2$, as already explained in [1919a]. Yet, Pólya [1923, 100–103] now used characteristic functions. With the abbreviation $\Phi(x) := \int_{-\infty}^{\infty} e^{ixt}\varphi(t)dt$ condition 3) could be written in the form

$$\Phi(x) = \Phi(\alpha x)\Phi(\beta x), \tag{5.37}$$

where $\alpha = \frac{a}{c}$ and $\beta = \frac{b}{c}$. By an idea hinted at already at the end of his [1919a], Pólya applied (5.37) to $\Phi(\alpha x)$ and $\Phi(\beta x)$ respectively, and in this way obtained:

$$\Phi(x) = \Phi(\alpha^2 x)\Phi(\beta\alpha x)\Phi(\alpha\beta x)\Phi(\beta^2 x).$$

Repeating this procedure n-times ((5.37) corresponds to $n = 1$) finally resulted in

$$\log \Phi(x) = \frac{x^2}{2}\Phi''(0) = -\frac{\sigma^2 x^2}{2}.$$

It could be concluded from this equation that $\varphi(x) = \frac{1}{\sigma\sqrt{2\pi}}e^{-\frac{x^2}{2\sigma^2}}$ meets conditions 1), 2), and 3). Pólya [1923, 103; 105 f.] proved by mainly resorting to condition 2) that $\varphi(x)$ is the only function with these properties.

Pólya [1923, 104 f.] also discussed solutions of his problem which do not meet condition 1), that is, solutions which do not possess finite moments of second order. He succeeded in characterizing solutions of 3) by their Fourier transform $\Phi(x)$ in the following way: Let $\alpha, \beta \in]0; 1[$ and $N > 0$ such that $\alpha^N + \beta^N = 1$. If there exist $\omega > 0$ and natural numbers m, n such that $-\log\alpha = \omega m$, $-\log\beta = \omega n$, and if $\psi(x)$ denotes "any periodic function" with the period ω, then

$$\Phi(x) = e^{-|x|^N\psi(\log|x|)}$$

is a solution of (5.37).[55] That

$$\varphi(x) = \frac{1}{\pi}\int_0^{\infty} \Phi(t)\cos(xt)dt$$

[55] Pólya did not explicitly state whether all solutions of (5.37) were given by this formula.

is a probability density, Pólya proved for the case $0 < N < 1$ and $\psi \in C^2$ in a tricky way using a general theorem on functions, which he had already published in [1918, 378]. From Pólya's consideration it could be concluded, in particular, that the functions $\Phi(x) = e^{-|ax|^N}$, already discussed by Cauchy, Edgeworth, and Lévy, were actually characteristic functions of error densities for $0 < N < 1$.

Pólya apparently felt very challenged by Lévy's probabilistic work. This fact is also highlighted by Pólya's "new proof" for Lévy's theorem on the correspondence between the convergence of characteristic functions and distributions. Once again, Pólya [1923, 106 f.] quoted his "Theorem III" on generating functions (Sect. 5.2.3.1) and stated that his new proof of Lévy's theorem was "closely related" to the proof of "Theorem III." Like Lévy, he [1923, 107 f.] considered a sequence of characteristic functions (Φ_n) accompanying the sequence of distributions (f_n), and a characteristic function Φ accompanying the distribution f. He claimed that from the uniform convergence of $\Phi_n(t)$ to $\Phi(t)$ in all compact intervals of t-values the convergence $f_n(x) \to f(x)$ would follow if f is continuous in x. For a proof Pólya constructed the "rooflike" function D, where $D(y) = D(-y)$ and

$$D(y) = \begin{cases} 1 & \text{for } 0 \le y \le h \\ 1 - \frac{y-h}{2\eta} & \text{for } h \le y \le h + 2\eta \\ 0 & \text{for } y \ge h + 2\eta \end{cases}$$

for arbitrary $h, \eta > 0$. He represented $D(y)$ by a Fourier integral, and he showed that

$$\int_{-\infty}^{\infty} D(t-s)df_n(t) \to \int_{-\infty}^{\infty} D(t-s)df(t) \quad (n \to \infty).$$

Because of the particular form of the function D, Pólya was able to conclude that, for all $s \in \mathbb{R}$ and $h > 0$ with the property that $f(x)$ is continuous in $x = s \pm h$, the relation

$$f_n(s+h) - f_n(s-h) \to f(s+h) - f(s-h),$$

and therefore

$$\lim_{n\to\infty} (f_n(x_1) - f_n(x_2)) = f(x_1) - f(x_2)$$

for "arbitrary points of continuity x_1, x_2 of $f(x)$" was valid.[56] In a letter to Pólya (13 May 1923, ETH-Bibliothek, see above), Lévy praised this new proof as "easier than mine." In his book on probability, however, Lévy [1925b, 197–199] elaborated his own method of proof and did not use Pólya's.

Lévy in his later work readily appreciated Pólya's results on characteristic functions of stable and semistable distributions. To claim that Pólya had been the first to discuss stable distributions in a systematic manner, as did Feller [1945, 821], is misleading, however. On the contrary, Lévy's first articles apparently aroused Pólya's interest in a more general discussion of the integral equation (5.25) and its solutions.

[56] Pólya did not prove that from this latter limit relation the assertion $f_n(x) \to f(x)$ followed. In this context, he only hinted in a very general manner on the fact that the functions considered were distribution functions.

Pólya thereafter tended to other probabilistic problems, particularly in the field of random walks (see [Antretter 1989, 11 f.]). Lévy, however, began a truly comprehensive examination of stable and semistable distributions, which continued until 1937 when his second book on probability theory appeared.

5.2.4 Lindeberg: An Entirely New Method

The complete mathematical work of Lindeberg contains only one truly outstanding, virtually epochal performance: the proof of the CLT under a very weak condition, which under certain "natural" assumptions even proved to be necessary. Lindeberg's arguments were based on an entirely new analytic method, which would later be applied to far more general problems. In [1920] Lindeberg, still without any knowledge of Lyapunov's works, had already proven the CLT for normed sums $\Sigma_{k=1}^{n}\frac{X_k}{r_n}$ of mutually independent random variables X_k, each with distribution U_k, with zero expectation, variance σ_k^2, and finite absolute moment of third order, presupposing that

$$\frac{1}{r_n^3}\sum_{k=1}^{n}\int_{-\infty}^{\infty}|x|^3 dU_k(x)\to 0\quad (n\to\infty),\qquad r_n=\sqrt{\sum_{k=1}^{n}\sigma_k^2}.$$

After certain modifications of his arguments, he was able, in 1922, to publish his famous proof of the CLT under even weaker conditions. He expressed this theorem in several versions. The version which comes closest to Lindeberg's concepts is probably his "Theorem III": Let $U_1, U_2, \ldots, U_n$ be the distribution functions of n mutually independent "probability quantities" $u_1, u_2, \ldots, u_n$, each with expectation 0 and with variance σ_k^2, where $\sum_{k=1}^{n}\sigma_k^2 = 1$. Let

$$U(x):=\int_{-\infty}^{\infty}\int_{-\infty}^{\infty}\cdots\int_{-\infty}^{\infty}U_n(x-t_1-t_2-\cdots-t_{n-1})dU_{n-1}(t_{n-1})\cdots dU_1(t_1). \quad (5.38)$$

Then U is the distribution of the sum of all random variables. Let

$$s(x):=\begin{cases}|x|^3 & \text{if } |x|<1\\ x^2 & \text{else.}\end{cases}$$

Even if the positive number ε is taken arbitrarily small, a positive number η can be chosen such that

$$\left|U(x)-\int_{-\infty}^{x}\frac{e^{-\frac{t^2}{2}}}{\sqrt{2\pi}}dt\right|<\varepsilon \quad (5.39)$$

if

$$\sum_{k=1}^{n}\int_{-\infty}^{\infty}s(x)dU_k(x)<\eta \quad (5.40)$$

[Lindeberg 1922b, 219 f.].[57]

[57] Quotation with slight changes within the mathematical formulae. There are no equation numbers in the original text.

So, Lindeberg proved a theorem which can be applied both to normed partial sums related to simple sequences of random variables and to sums of elements within different rows of a triangular array of random variables.

5.2.4.1 The Proof

Lindeberg considered the convolution of the function U and an auxiliary function f with derivatives up to a certain order. In [1922c, 213] he assumed that $|f'''(x)| \leq k$ for all $x \in \mathbb{R}$ with a suitable positive constant k. In his first paper, Lindeberg [1920] had still taken a normal distribution function for f. Although this trick is strongly reminiscent of Lyapunov's procedure (see Sect. 5.1.4), Lindeberg, who according to his own statement [1922b, 226; 1922c, 211], had initially no knowledge of Lyapunov's works, had developed the idea of an auxiliary distribution independently. However, it is quite possible that Lindeberg was influenced by the account on Crofton's method (see Sect. 3.3.2.2) in the standard monograph on error theory [Czuber 1891, 97–99], in which the use of an auxiliary distribution was described.

Under the general assumptions of "Theorem III," Lindeberg in the first part of his [1922c] discussed distributions with finite third-order moments. He [1922c, 213 f.] started his considerations with a general estimate for arbitrary distributions V with zero expectation and variance σ^2. Let

$$F(x) := \int_{-\infty}^{\infty} f(x-t)dV(t),$$

and, with the abbreviation $\varphi(x,\sigma) := \frac{1}{\sigma\sqrt{2\pi}}\mathrm{e}^{-\frac{x^2}{2\sigma^2}}$, let

$$\Phi(x) := \int_{-\infty}^{\infty} f(x-t)\varphi(t,\sigma)dt.$$

Using the Taylor expansion of f up to the third power (there is a certain similarity with Crofton's procedure again) Lindeberg showed that

$$|F(x) - \Phi(x)| < k\int_{-\infty}^{\infty} |x|^3 dV(x). \tag{5.41}$$

By repeated application of (5.41) to

$$F_1(x) := \int_{-\infty}^{\infty} f(x-t)dU_1(t), \quad F_2(x) := \int_{-\infty}^{\infty} F_1(x-t)dU_2(t), \ldots,$$
$$F_n(x) := \int_{-\infty}^{\infty} F_{n-1}(x-t)dU_n(t)$$

and

$$\Phi_1(x) := \int_{-\infty}^{\infty} f(x-t)\varphi(t,\sigma_1)dt, \quad \Phi_2(x) := \int_{-\infty}^{\infty} \Phi_1(x-t)\varphi(t,\sigma_2)dt, \ldots,$$
$$\Phi_n(x) := \int_{-\infty}^{\infty} \Phi_{n-1}(x-t)\varphi(t,\sigma_n)dt,$$

respectively, Lindeberg [1922c, 214–216] obtained

$$\left|\int_{-\infty}^{\infty} f(x-t)dU(t) - \int_{-\infty}^{\infty} f(x-t)\varphi(x,1)dt\right| < k\sum_{i=1}^{n}\int_{-\infty}^{\infty}|x|^3 dU_i(x). \quad (5.42)$$

With the aid of special, piecewise defined functions f, Lindeberg [1922c, 216 f.] deduced inequalities from (5.42), whose combination yielded the estimate

$$\left|U(x) - \int_{-\infty}^{x}\varphi(t,1)dt\right| < k\sum_{i=1}^{n}|x|^3 dU_i(x) + \frac{4}{\sqrt{2\pi}\sqrt[3]{2k}}.$$

By choosing k such that

$$\frac{1}{\sqrt[3]{2k}} := \left(\sum_{i=1}^{n}\int_{-\infty}^{\infty}|x|^3 dU_i(x)\right)^{\frac{1}{4}},$$

the inequality

$$\left|U(x) - \int_{-\infty}^{x}\varphi(t,1)dt\right| < 3\left(\sum_{i=1}^{n}\int_{-\infty}^{\infty}|x|^3 dU_i(x)\right)^{\frac{1}{4}} \quad (5.43)$$

followed. Lindeberg [1920] had already reached an analogous inequality, but with a far more complicated right-hand side.

In that 1920 paper, as mentioned above, he had used the auxiliary distribution $f(x) = \int_{-\infty}^{x}\varphi(t,\sigma)dt$ with a suitable σ. Now, the consideration of less special auxiliary functions made a more flexible argumentation possible, a substantial simplification of the proof, and an even more general treatment of the CLT in the second part of [1922c], in which the existence of absolute moments of third order no longer had to be presupposed.

Lindeberg [1922c, 220 f.] now assumed that f meets the conditions $|f'''(x)| \le k$ and $|f(x)|$, $|f'(x)|$, $|\frac{f''(x)}{2}| < \frac{k}{24}$ for all $x \in \mathbb{R}$. By modification of the deduction of (5.41) it followed with the abbreviations used there ($s(x)$ as in "Theorem III"):

$$|F(x) - \Phi(x)| < ck\int_{-\infty}^{\infty} s(x)dV(x), \quad \text{where } c < \frac{3}{2}, \quad \text{if } \int_{-\infty}^{\infty} x^2 dV(x) < 1.$$

In the same way as earlier, it was possible to justify an inequality analogous to (5.42) if

$$\int_{-\infty}^{\infty} |x|^3 dU_i(x)$$

was substituted now by the term

$$\frac{3}{2}\int_{-\infty}^{\infty} s(x)dU_i(x),$$

and if it was assumed that

$$\sum_{i=1}^{n}\int_{-\infty}^{\infty} s(x)dU_i(x)$$

was sufficiently small. The inequality corresponding to (5.43) (not explicitly stated by Lindeberg) is

$$\left|U(x) - \int_{-\infty}^{x} \varphi(t,1)dt\right| < 3\sqrt[4]{\frac{3}{2}}\left(\sum_{i=1}^{n}\int_{-\infty}^{\infty} s(x)dU_i(x)\right)^{\frac{1}{4}}. \quad (5.44)$$

From this inequality, Lindeberg's "Theorem III" follows immediately, or, in other words, (5.40) actually implies (5.39).

5.2.4.2 Different Theorems, Different Conditions

On the basis of (5.43), Lindeberg [1922c, 219] stated, as he had already done in [1920, 21], the "classic assertion" that "the sum of a large number of mutually independent small errors obeys the Gaussian law." Lindeberg presupposed the n elementary errors $u_1, \ldots, u_n$, each with zero expectation, to have only values whose moduli remain below a finite upper bound d_n. If U_i denotes the distribution of the ith elementary error, then

$$\sum_{i=1}^{n}\int_{-\infty}^{\infty} x^2 dU_i(r_n x) = 1 \quad \left(r_n = \sqrt{\sum \mathrm{Var} u_i}\right).$$

Because of

$$\sum_{i=1}^{n}\int_{-\infty}^{\infty} |x|^3 dU_i(r_n x) < \sum_{i=1}^{n}\frac{d_n}{r_n}\int_{-\infty}^{\infty} x^2 dU_i(r_n x) = \frac{d_n}{r_n},$$

from (5.43) for all $\varepsilon > 0$ the relation (U as in (5.38))

$$\left|U(r_n x) - \int_{-\infty}^{x}\frac{e^{-\frac{t^2}{2}}}{\sqrt{2\pi}}dt\right| < \varepsilon$$

follows, provided that

$$\frac{d_n}{r_n} \le \left(\frac{\varepsilon}{3}\right)^4.$$

Lindeberg [1922c, 219] now supposed that the "quantities u_μ vary with respect to their number and the form of their distribution functions such that at the same time d_n tends to zero and r_n approaches closer and closer a nonzero value r." If U denotes the distribution of the sum of n quantities of this kind, then the equation

$$\lim_{n\to\infty} U(x) = \int_{-\infty}^{x} \frac{e^{-\frac{t^2}{2r^2}}}{r\sqrt{2\pi}} dt$$

has to be valid uniformly for all x. In this way Lindeberg rather loosely explained a fact which in another paper [1922b] was stated more formally by a limit theorem for the distributions of sums $\sum_{\mu=1}^{n} u_{n\mu}$, related to a triangular array $u_{n\mu}$, $1 \le \mu \le n$, of elementary errors.[58]

Lindeberg [1922b] comprehensively discussed possible specifications of the hypothesis of elementary errors in the framework of different versions of the CLT. One can learn from the paper [1922b] how important the problem of a precise stochastic analysis of the accumulation of small elementary errors was for Lindeberg. At this place, he explicitly favored a triangular array of elementary errors and a limit assertion referring to it. It was quite useless, however, to compare the contributions of the 19th century with limit theorems of the modern fashion. Hagen, Bessel, and all the other mathematicians, astronomers, or geodesists considered a "very large" or an "indefinitely large," but quasi-*fixed* number n of elementary errors; the Gaussian distribution resulted from neglecting the higher terms in a series expansion for large n. In modern reconstruction this enables an interpretation as a limit theorem for row sums of a triangular array as well as an interpretation in the sense of a limit theorem for normed partial sums assigned to a simple sequence of random variables. The latter interpretation, which Lindeberg—if only by rather vague arguments—criticized as not being general enough, was used by Lévy (see Sect. 5.2.3.2).

From (5.44), as Lindeberg [1922c, 225] pointed out at the end of his article, the CLT follows in its usual form. Let (X_i) be a sequence of independent random variables with distributions V_i. For simplicity it is assumed that $EX_i = 0$. Then, for all natural n and for all $i \le n$, the random variables $u_i := \frac{X_i}{r_n}$ $(r_n = \sqrt{\sum \text{Var}X_i})$ with distribution functions $U_i(x) = V_i(r_n x)$ are of the type required for Lindeberg's "Theorem III" (equation numbers (5.39) and (5.40)). From (5.44), under the condition

$$\sum_{i=1}^{n} \int_{-\infty}^{\infty} s\left(\frac{x}{r_n}\right) dV_i(x) \to 0 \quad (n \to \infty), \tag{5.45}$$

it follows that

$$\lim_{n\to\infty} P\left(\frac{\sum_{i=1}^{n} X_i}{r_n} \le x\right) = \int_{-\infty}^{x} \frac{e^{-\frac{t^2}{2}}}{\sqrt{2\pi}} dt.$$

[58] The elementary errors were assumed to have zero expectation, to be independent within each row, and meeting the additional condition $\sum_{\mu=1}^{n} \text{Var}u_{n\mu} \to r^2$.

Lindeberg [1922c, 222–224] also established the following condition, equivalent to (5.45):

$$1 - \int_0^1 \left(\sum_{i=1}^{n} \frac{1}{r_n^2} \int_{|x| \leq \tau r_n} x^2 dV_i(x) \right) d\tau \to 0. \tag{5.46}$$

Lévy [1924, 32], in his discussion of Lindeberg's proof, introduced—without any justification, as happened quite frequently in his papers—the condition

$$\frac{1}{r_n^2} \sum_{i=1}^{n} \int_{|x| > t r_n} x^2 dV_i(x) \to 0 \quad \forall t > 0. \tag{5.47}$$

This condition was adopted by Khinchin [1933, 7] in his survey *Asymptotische Gesetze der Wahrscheinlichkeitsrechnung* and is now commonly called *the* "Lindeberg condition," despite the fact that it cannot be found in any paper written by Lindeberg himself. It may be a nice (nontrivial) problem for the reader to show the equivalence between (5.45) and (5.47). Some suggestions for a proof can be found in Lindeberg's discussion [1922c, 222–224] of the equivalence between (5.45) and (5.46).

5.2.5 Hausdorff's Reception of Lyapunov's, von Mises's, and Lindeberg's Work

Besides his 1901 article already broached in Sections 3.4.2.1 and 5.2.3.3, Felix Hausdorff only wrote a second larger article [1897] with essentially probabilistic content, which would play an important role in the history of risk theory (see Sect. 5.2.8.1). However, even after the turn of the century, he remained very interested in probability theory, especially in its set- and measure theoretic background and in probabilistic limit theorems, as we can see from a passage of his book on set theory [1914], several private notes, and letters [Girlich 1996; Purkert 2006a;b; Siegmund-Schultze 2010]. Hausdorff's activities in moment problems were directly linked with the CLT, as explicitly shown by lecture notes written by himself [1923/2006]. He apparently had good knowledge of current literature which is why he is an important contemporary witness of the reception of probabilistic novelties, in particular during the first quarter of the 20th century.

Concerning Hausdorff's acquaintance with Lyapunov's work on the CLT, a statement by Cramér [1976, 1355] is frequently quoted[59] where he reports that "notes" on Lyapunov's work provided by Hausdorff "had a great influence on my subsequent work."[60] There actually exist, in Hausdorff's "Nachlass,"[61] notes on conditions

[59] See, for example, [Girlich 1996, 50; Chatterji 2006, 740].

[60] In this context, this would be especially true of [Cramér 1923], see Sect. 5.2.8.1.

[61] Universitäts- und Landesbibliothek Bonn, NL Hausdorff, Kapsel 51, Faszikel 1128, Bl. 1.

and assertions of [Lyapunov 1900; 1901a;c] (written before 1915),[62] that are directly taken from the respective reviews JFM 32.0230.02, JFM 31.0228.02, JFM 32.0230.01 in *Jahrbuch über die Fortschritte der Mathematik*, as is made clear by the organization of Hausdorff's notes and wrong bibliographic data, which Hausdorff apparently adopted from the *Jahrbuch* without any modifications.[63] In the *Jahrbuch*, the comprehensive article [Lyapunov 1901b] was only referred to by a bibliographic note (JFM 33.0248.07), but without any review and under specification of a wrong article language (Russian instead of French). Notes by Hausdorff with a deeper examination of Lyapunov's methods are not extant. On the other hand, it seems rather probable that he actually studied Lyapunov's chief papers [1900; 1901b] in their original versions. The respective journals should have been easily available for Hausdorff, both at the University of Bonn (where he taught until 1913 and again from 1921) and the University of Greifswald (1913–1921).[64] On the basis of the at present accessible sources one cannot decide with certainty, however, whether Hausdorff's knowledge about Lyapunov's contributions actually went beyond the reviews in *Jahrbuch*, or whether his knowledge was as comprehensive as suggested by Cramér. As we will see below, Hausdorff, in his above-mentioned 1923 course on probability theory, would give a detailed discussion of Lyapunov's conditions, although without using Lyapunov's methods.

Hausdorff had, in his article of 1901, mainly focused on generating functions. In a certain sense he had also broached characteristic functions, however without using their decisive advantage, existence independent of the one of any moments (see Sect. 3.4.2.1). For this reason, the naive reader might come to the opinion that Hausdorff should have—at least in parts—esteemed von Mises's 1919 account, in which, as we have seen, generating and characteristic functions played a significant role. A letter (6 January 1920)[65] from Hausdorff to Pólya reveals that the contrary was true. Hausdorff wrote about von Mises:

> M. gives very complicated and unnecessarily narrow conditions for the convergence of distributions to the Gaussian exponential law, whereas from works of Chebyshev, Stieltjes, and others by far more general and simpler ones can be obtained; an especially beautiful and little demanding [condition] we owe Lyapunov (1901!).

---

62 "Before 1915" according to the catalog of Hausdorff's "Nachlass" (Universitäts- und Landesbibliothek Bonn). The last entry on the sheet with the notices on Lyapunov also contains a brief description of a 1913 paper by Perron "Math. Ann. 74." Therefore it seems probable that Hausdorff's notices were written around 1913/14/15.

63 In the *Jahrbuch*, the year "1900" (instead of 1901) for the volume 132 of the *Comptes rendus* was incorrectly stated. Hausdorff, in the same erroneous manner, repeated the bibliographic data referring to [Lyapunov 1900; 1901a;c] in a letter to von Mises from 2 November 1919 (see [Hausdorff 2006, 826]) and in his notes on a course in probability theory he gave in summer semester of 1923 [1923/2006, 674] (see below).

64 As both libraries communicated to me, the pertinent volumes of the *Petersburg Memoirs* and the *Bulletin* were available for loan at the time in question. In a letter to Pólya (6 January 1920, ETH-Bibliothek, Archive, Nachlass Pólya, 89: 237), Hausdorff referred to "Liapunoff (1901!)" for a particularly weak condition of the CLT, now with the correct year.

65 Already mentioned in the preceding footnote.

This quotation shows—besides revealing an apparently malicious attitude toward von Mises—that Hausdorff was mainly interested in the integral version of the CLT and its treatment by moment methods. As we have seen, von Mises's conditions for the integral version of the CLT were, if in a somewhat complex way, expressed through generating functions, but they were basically not more restrictive than those needed for a direct inference (without using the truncation trick) from the convergence of moments (or generating functions) to the convergence of distributions. Hausdorff seems to have neglected that Lyapunov owed his success, to a large extent, to the fact that a method alternative to that of moments had been applied, and Hausdorff apparently did not see the full potential of characteristic functions as discussed by von Mises, not to mention their implicit occurrence in Lyapunov's work.

Hausdorff's interest in moment methods is also exemplified by his course on probability theory in summer semester 1923 (Bonn). His own lecture notes, recently edited (see [Hausdorff 1923/2006]), provide a rich collection of very innovative ideas and approaches from the point of view of the early 1920s. In particular, Hausdorff in this course took on, at a very early moment of time already, Lindeberg's methods for proving the CLT. He [1923/2006, 674–678] presupposed "variables" (without explaining this notion) $X_1, \ldots, X_n$ with zero means, second-order moments $a_1^2, \ldots, a_n^2$, and absolute third-order moments $c_1^3, \ldots, c_n^3$. He expounded Lindeberg's line of arguments, with certain modifications,[66] up to this point where the inequality (5.43) was established. From (5.43) Hausdorff [1923/2006, 679] deduced a finitary version of the CLT, which he named "Grenzwerthsatz von Liapunoff" ("Lyapunov's limit theorem"), and which can be summarized as follows: If Φ_n denotes the distribution function[67] of $\sum_{k=1}^{n} \frac{X_k}{\sqrt{2}b_n}$, where $b_n^2 = a_1^2 + \cdots + a_n^2$, and d_n is defined by $d_n = (c_1^3 + \cdots + c_n^3)^{\frac{1}{3}}$, then, with the denotation $\Phi(x) := \frac{1}{\sqrt{\pi}} \int_{-\infty}^{x} e^{-t^2} dt$,[68]

$$|\Phi_n(x) - \Phi(x)| \leq \mu \left(\frac{d_n}{b_n} \right)^{\frac{3}{4}},$$

where μ is a "numerical constant." Hausdorff additionally noticed that the condition

$$\frac{d_n}{b_n} \to 0 \qquad (n \to \infty) \tag{5.48}$$

was sufficient for the (uniform) convergence of $\Phi_n(x)$ to $\Phi(x)$.

Hausdorff [1923/2006, 680 f.] also discussed Lyapunov's more general condition (5.2), which he strangely enough designated as the hitherto most general—leaving out Lindeberg's even weaker condition (5.45). On the one hand, he did not include

[66] For comments on Hausdorff's modifications see [Chatterji 2006, 752 f.]. Chatterji [2007] also elaborated a modified proof of Lindeberg's theorem on the basis of Hausdorff's ideas.

[67] The distribution function $F(x)$ of a random variable X Hausdorff defined by $F(x) := P(X < x)$.

[68] Hausdorff contrary to Lindeberg still used the traditional norming of 19th-century error theory which referred to the normal distribution $\Phi_{0;\frac{1}{2}}$ rather than to the standard normal distribution.

any details on Lyapunov's approach to the CLT. On the other hand, however, he showed by means of the Lyapunov inequality, which he proved very elegantly[69] via the concavity of $\log(p_1 u_1^\alpha + \cdots + p_n u_n^\alpha)$ (p_k and u_k being positive), that Lyapunov's condition (5.2) implied the condition (5.48) that served as a basis for his own proof.

Without further speculations one can only state that Hausdorff, in his 1923 course, was apparently more interested in Lindeberg's method than in the nowadays celebrated Lindeberg condition. We must not forget that the significance of this condition was established in its full dimension only after Feller had proven in 1935 that it was even necessary for the convergence to the Gaussian law. Hausdorff's wrong assessment on the superiority of Lyapunov's condition may have been based on a misunderstanding, perhaps caused by Lindeberg himself, who, in his 1922 paper, particularly emphasized the "classic assertion" (see Sect. 5.2.4.2), which could be derived from (5.43). Finally, even if Hausdorff had been still interested in characteristic functions, Lindeberg's rather elementary and quite easily conceivable method would have been, at least with respect to the primarily didactic goals of a lecture course, more appealing than Lyapunov's rather cumbersome arguments.

Besides his proof of the CLT via Lindeberg's method, Hausdorff [1923/2006, 684–693] in his lecture notes also gave a comprehensive account on the problem of the correspondence between convergence of moments and convergence of distributions. By elementary, however rather intricate methods, Hausdorff proved that for a sequence of monotonically increasing functions (φ_n) and for a further monotonically increasing function φ, where φ_n and φ are defined in $\mathbb{R}$ such that $\varphi_n(-\infty) = \varphi(-\infty) = 0$ and all these general distributions possess moments of arbitrarily high order, the convergence of the moments

$$\int_{-\infty}^{\infty} x^k \, d\varphi_n(x) \to \int_{-\infty}^{\infty} x^k \, d\varphi(x) \quad (n \to \infty)$$

for all $k \in \mathbb{N}_0$ implies the convergence of the distributions

$$\varphi_n(x) \to \varphi(x) \quad (n \to \infty)$$

for all points of continuity x of φ, if φ is uniquely determined by its moments. For this latter property Hausdorff established a necessary and sufficient condition, and he proved that this general theorem could be applied to the particular case of the normal distribution being the limit distribution. This latter assertion was named by Hausdorff both "Grenzwerthsatz von Tschebyscheff" ("Chebyshev's limit theorem") and "zweiter Grenzwerthsatz" ("second limit theorem"), probably because in Chebyshev's 1887 paper (see Sect. 4.6.2) an analogous theorem had been formulated as a second after the weak law of large numbers as the first [Chatterji 2006, 731].[70] "Chebyshev's theorem" Hausdorff [1923/2006, 693] designated "in a certain respect more general" than Lyapunov's, because (if in his own version only) the former could be basically applied to nonindependent variables also.

[69] Hausdorff's proof was essentially based on the same idea which Hölder had used for proving "his" inequality (see footnote 13).

[70] For closer details on this part of Hausdorff's 1923 course, see [Girlich 1996, 48–52; Chatterji 2006, 730–734].

In an attitude similar to that of Pólya, Hausdorff apparently attributed moment methods as compared with characteristic functions the greater significance for probabilistic limit theorems. Lévy's proof of the correspondence of convergence of characteristic functions and of distributions ([1922c], see Sect. 5.2.3.4) remained generally unnoticed for the time being, as can be seen by an inadequate review in the *Jahrbuch* (JFM 48.0600.03). It was a major merit of Hausdorff to realize the significance of Lindeberg's method, but the problem of utmost generality of the conditions for the CLT was apparently not the focus of his probabilistic research. Notwithstanding Lindeberg's achievement, in the early twenties the quest for conditions as weak as possible was the concern of mainly one mathematician: Paul Lévy. This scientist immediately realized the significance of both Lindeberg's methods and Lindeberg's condition.

5.2.6 *Lévy's Discussion of Stable Laws in His* Calcul des probabilités

Regarding his life achievements, Lévy [1970, 81 f.] stated retrospectively:

> Nobody doubts that my works from 1922 to 1925 showed the importance of characteristic function and established the fundamental theorems of its theory. Nobody can contest any more my decisive role for the discussion of stable laws.

In fact, the power of Lévy's characteristic functions becomes especially clear in the context of his discussion of stable distributions, into which he also integrated the CLT. Lévy's early results in this field were comprehensively presented in the 6th chapter "Les lois exceptionelles" of his *Calcul des probabilités*, which appeared in 1925.

Levy's book consists of a mixture—for today's reader a little strange—of traditional stochastic contents and latest results on limit theorems and stable laws. Philosophical questions on the notion of probability are taken into consideration as well as the elements of error theory and the kinetic theory of gases. So, regarding the contents, the conception of the book also highlights the transition of an application-oriented classic probability calculus to a purely mathematical theory. Apparently, Lévy himself was far more interested in the latter aspect, and it was the discussions connected with this aspect which were substantially responsible for the impact of his book.

5.2.6.1 Stable Laws as Limit Laws

A comprehensive discussion of Lévy's notion of stable laws is already contained in Sect. 5.2.3.5. Lévy's definition directly implies the following property of stable laws, as it appears the most important from his own point of view: If there exists a sequence of identically distributed independent random variables $(X_i)_{i\in\mathbb{N}}$, and a sequence of positive numbers $(N_n)_{n\in\mathbb{N}}$, and a distribution function V such that

$$P\left(\frac{\sum_{i=1}^{n} X_i}{N_n} < x\right) + \frac{1}{2}P\left(\frac{\sum_{i=1}^{n} X_i}{N_n} = x\right) \to V(x) \quad (n \to \infty)$$

in all points of continuity x of V, then V is the distribution function of a stable law. Lévy made this property explicit, but still without proof, in [1924, 33], whereas in his book it obtained a prominent role within the chapter on stable laws. The proof given by Lévy [1925b, 252 f.] was based on the same idea which Edgeworth [1905] (see Sect. 3.4.2.3) had already used for the deduction of the characteristic property of those limit distributions to be "reproductive." Lévy considered stable distributions "natural" generalizations of the classic Gaussian law. Therefore, with regard to Lévy, the history of the CLT is closely connected with his more general discussion of stable limit distributions.

5.2.6.2 The Functional Equation of the Characteristic Function of a Stable Law

Lévy's definition of stable law (see Sect. 5.2.3.5) is equivalent to the following property of its characteristic function $\varphi \not\equiv 1$:

$$\forall a_1, a_2 > 0 \exists a > 0 \forall x \in \mathbb{R} : \varphi\left(\frac{a_1}{a}x\right)\varphi\left(\frac{a_2}{a}x\right) = \varphi(x),$$

or with the abbreviation $z := \frac{x}{a}$:[71]

$$\forall a_1, a_2 > 0 \exists a > 0 \forall z \in \mathbb{R} : \varphi(a_1 z)\varphi(a_2 z) = \varphi(az). \tag{5.49}$$

Despite the fact that stable laws were in the foreground of Lévy's work from the very beginning of his probabilistic activities, only in [1925b, 254 f.] did he discuss in a more comprehensive, though still incomplete manner, the set of solutions of (5.49) (presupposing φ to be a characteristic function).

Passing from the characteristic function to its logarithm $\psi(z) = \log\varphi(z)$ (this is possible since $\varphi(z) \neq 0$ is always valid for stable distributions, although Lévy did not make this property explicit), one obtains

$$\forall a_1, a_2 > 0 \exists a > 0 \forall z \in \mathbb{R} : \psi(a_1 z) + \psi(a_2 z) = \psi(az).$$

Through successive application of this relation it can be concluded that there exists a uniquely determined sequence $(N(n))_{n\in\mathbb{N}}$ of positive numbers such that

$$\forall n \in \mathbb{N} \forall t \in \mathbb{R} : n\psi(t) = \psi\left(N(n)t\right). \tag{5.50}$$

[71] By use of the following property it can be shown that a is uniquely determined by a_1 and a_2.

Lévy [1925b, 254] claimed it would be "almost obvious, and by the way easy to prove" (this was one of his quite frequent standard remarks substituting proofs) that $N(n) \to \infty$ and $\frac{N(n)}{N(n+1)} \to 1$ as $n \to \infty$.[72] From this Lévy concluded without any further explanation that for all $\lambda > 0$ one could find sequences of real numbers (n_k) and (n'_k) such that $n_k, n'_k \to \infty$ and $\frac{N(n_k)}{N(n'_k)} \to \lambda$.[73] Because of (5.50), the quotient $\frac{\psi(N(n_k)t)}{\psi(N(n'_k)t)}$ is independent of t. With $t = \frac{t'}{N(n'_k)}$ it follows that $\frac{\psi\left(\frac{N(n_k)}{N(n'_k)}t'\right)}{\psi(t')}$ is independent of t'. ψ is a continuous function, therefore even in the "limit case" the quotient $\frac{\psi(\lambda t')}{\psi(t')}$ is independent of t'. This statement is valid for arbitrary $\lambda > 0$. Without any detailed discussion Lévy [1925b, 255] finally wrote: "This is only possible if $\psi(t)$ is of the form $\psi(t) = -c|t|^\alpha$, where the [complex] coefficient c may depend on the sign of t."[74] From the general properties of characteristic functions, which had to be valid also for $\varphi(x) = \mathrm{e}^{\psi(x)}$ (in particular $|\varphi(x)| \le 1$, $\varphi(0) = 1$, $\varphi(-x) = \overline{\varphi(x)}$, continuity), it followed

$$\psi(t) = -(c_0 + \operatorname{sgn}(t)c_1\mathrm{i})|t|^\alpha, \tag{5.51}$$

where $\alpha > 0$, $c_0 \ge 0$, $c_1 \in \mathbb{R}$. Because degenerate distributions were not under consideration, even the fact $c_0 > 0$ was guaranteed. Lévy [1924, 34; 1925b, 255] justified $\alpha \le 2$ by the property that, for a normed sum $\sum_{i=1}^n \frac{X_i}{N(n)}$ of mutually independent and identically distributed random variables X_i with positive moments

---

[72] The first assertion can be justified by the fact that characteristic functions of stable distributions do not have zeros (see [Lévy 1937a, 94 f.]). For a proof of an assertion analogous to the second, see [Rossberg, Jesiak, & Siegel 1985, 170].

[73] This can be proven in a way similar to the line of argument used for the proof of Riemann's theorem on the interchange of order within a series (1868). Let $\lambda > 0$ be an arbitrary number. Because $N(n) \to \infty$ there exist sequences (p_k) and (q_k) of natural numbers, where $p_k, q_k \to \infty$, such that $\frac{N(p_k)}{N(q_k)} > \lambda$ and $\frac{N(p_k-1)}{N(q_k)} \le \lambda$, as well as $\frac{N(p_k)}{N(q_{k+1})} < \lambda$ and $\frac{N(p_k)}{N(q_{k+1}-1)} \ge \lambda$. Then

$$\left|\frac{N(p_k)}{N(q_k)} - \lambda\right| \le \frac{N(p_k)}{N(q_k)} - \frac{N(p_k-1)}{N(q_k)} =$$

$$= \left(\frac{N(p_k)}{N(p_k-1)} - 1\right)\frac{N(p_k-1)}{N(q_k)} \le \left(\frac{N(p_k)}{N(p_k-1)} - 1\right)\lambda \to 0.$$

The idea of this proof I owe to Günther Wirsching.

[74] A complete proof of this assertion can be reached by use of the theorem (essentially due to Cauchy (1821), see [Aczél 1961, 47]) that the solutions $u \in C^1(\mathbb{R})$ of the functional equation

$$u(\xi + \eta) = \frac{u(\xi)u(\eta)}{k}$$

(k a complex number) are given by $u(\xi) = k\mathrm{e}^{\rho\xi}$, where $\rho \in \mathbb{R}$. In the present case one has to apply this theorem to $u(z) = \psi(\mathrm{e}^z)$ and to observe that (due to a fundamental property of characteristic functions) $\psi(-x) = \overline{\psi(x)}$.

$m_2 < \infty$ of second order, nondegenerate normal distributions are the only possible limit laws different from distributions concentrated in one point.[75]

5.2.6.3 The Laws of Type $L_{\alpha,\beta}$

For a closer specification of the constants c_0 and c_1, Lévy designated certain probability laws as "laws of type $L_{\alpha,\beta}$" if their characteristic function had the form $\mathrm{e}^{\psi(t)}$, where $\psi(t)$ was a function according to (5.51), $c_0 > 0$, and

$$\frac{c_1}{c_0} = \begin{cases} \beta \tan \frac{\pi}{2}\alpha & \text{for } \alpha \in]0;1[\cup]1;2[\\ \beta & \text{for } \alpha \in \{1;2\}. \end{cases} \tag{5.52}$$

Those laws of type $L_{\alpha,\beta}$ for which $c_0\Gamma(\alpha+1) = 1$, Lévy called "reduced laws of type $L_{\alpha,\beta}$." He [1925b, 256] claimed that laws of type $L_{\alpha,\beta}$ different from degenerate distributions exist if and only if $0 < \alpha \leq 2$ and

$$-1 \leq \beta \leq 1 \text{ for } \alpha \in]0;1[\cup]1;2[,$$
$$\beta \in \mathbb{R} \text{ for } \alpha = 1,$$
$$\beta = 0 \text{ for } \alpha = 2.$$

In modern textbooks this assertion—apart from the fact that today the notion of "stable" is used in a slightly more general manner—is proven by aid of the "canonic" representation of the characteristic functions of infinitely divisible distributions. This approach can already be found in Lévy's second book on probability theory [1937a, 198–203].

In 1925 Lévy had to use different arguments because, at this time, there was no idea about infinitely divisible laws.[76] The necessity of $0 < \alpha \leq 2$ had already been shown in his discussion of the functional equation (5.49). Only in the particular cases $\alpha = 1$ and $\alpha = 2$ could the probability laws (Cauchy's and Gauss's law, respectively) be specified by an explicit formula.[77] For $0 < \alpha < 1$ it already followed

---

[75] $m_2 = \infty$ implies, as Lévy [1925b, 174] had shown at a previous place in his book, $\lim_{t\to 0} \frac{1-\mathrm{Re}\varphi(t)}{t^2} = \infty$. This is only possible if $\alpha < 2$.

[76] The concept of infinitely divisible distributions is due to de Finetti (1929, see Sect. 7.2.1).

[77] Lévy [1939, 53–57] later also found the density function (vanishing for nonpositive arguments u)

$$f(u) = \frac{1}{\sqrt{2\pi}} u^{-\frac{3}{2}} \mathrm{e}^{-\frac{1}{2u}} \quad (u > 0),$$

belonging to

$$\psi(t) = -|t|^{\frac{1}{2}}[1 - \mathrm{i}\,\mathrm{sgn}(t)],$$

in connection with his examination of the distribution of the maximum value of the random function $X(s)$ within an s-interval $[0;T]$ if $X(s)$ follows a Brownian motion. As Lévy [1939, 47] noticed, already from a result of Gustav Dötsch [1935, 622], which had been achieved in a non-probabilistic context, it could be inferred that e^{ψ} was the Fourier transform of f. Sometimes also Nikolai Vasilevich Smirnov is credited with the formula for f, who independent of Lévy discovered it in the 1950s; see [Gnedenko & Kolmogorov 1949/68, 171].

from Pólya's discussions that $e^{\psi(x)}$ was the characteristic function of a probability distribution. However, Pólya's method could only be applied for $\alpha < 1$, and, thus, Lévy [1925b, 258–262] had to find new arguments.

He showed that, for all values of β and $\alpha \neq 1, 2$ under consideration, there exists a probability density f with a characteristic function φ such that

$$\left(\varphi\left(\frac{t}{n^{\frac{1}{\alpha}}}\right)\right)^n \to e^{\psi(t)}, \tag{5.53}$$

ψ according to (5.51) and (5.52). Lévy succeeded in proving that the convergence in (5.53) was uniform in each finite interval of t-values, $\psi(t)$ meeting the conditions (5.51) and (5.52). His proof, however, remained incomplete for the time being. Only from Cramér's version of the convergence theorem for characteristic functions (1937, see footnote 48) could it be concluded that the limit e^{ψ} was actually a characteristic function.

Lévy's proof [1925b, 262] of the assertion that $|\beta| \leq 1$ in the case $\alpha \neq 1, 2$ was not correct either. This problem was finally solved only in an article that he wrote together with Khinchin [1936].

5.2.6.4 A Generalization of the Central Limit Theorem

Lévy [1924, 35; 1925b, 257 f.] presented a general theorem on the convergence of the distributions of suitably normed sums of independent, but not necessarily identically distributed, random variables to a stable distribution. By the term "law $\mathscr{L}_{\alpha,\beta}$" he denoted any law with the property that the logarithm $\psi(t)$ of its characteristic function in a neighborhood of $t = 0$ meets the condition

$$\psi(t) = -(c_0 + c_1 \mathrm{sgn}(t)\mathrm{i})|t|^{\alpha}[1 + \omega(t)],$$

where $\lim_{t \to 0} \omega(t) = 0$ and c_0, c_1 are according to (5.52). Any law of this kind was called "reduced" if $c_0\Gamma(\alpha + 1) = 1$. Given α and β, by "famille normale des lois $\mathscr{L}_{\alpha,\beta}$ réduites" Lévy designated the set of reduced laws $\mathscr{L}_{\alpha,\beta}$ for which there exists a nonnegative function $h(t)$ $(-\tau \leq t \leq \tau)$ such that, for all functions ω belonging to elements of this set, the following property is valid:

$$|\omega(t)| \leq h(t) \ \forall t \in [-\tau; \tau] \text{ and } \lim_{t \to 0} h(t) = 0.$$

Without any loss of generality one can assume that h is an even function, growing with increasing $|t|$.

Lévy's major theorem was as follows: Let $\xi_1, \ldots, \xi_n$ be independent random variables, each belonging to the same "famille normale de lois $\mathscr{L}_{\alpha,\beta}$ réduites."[78] Let $a_1, \ldots, a_n$ be positive numbers such that, for $A^{\alpha} := \sum_{k=1}^{n} a_k^{\alpha}$ and a "very small" number η,

$$\frac{\max_{1 \leq k \leq n} a_k}{A} < \eta.$$

[78] In the text [Lévy 1925b, 257 f.], the word "réduite" was omitted erroneously.

Then "the sum

$$\frac{X}{A} = \frac{a_1\xi_1 + a_2\xi_2 + \cdots + a_n\xi_n}{A}$$

obeys a probability law that deviates[79] from the law $L_{\alpha,\beta}$ all the less the smaller the number η is." By "*the* law $L_{\alpha,\beta}$" Lévy at this place apparently meant the reduced law of type $L_{\alpha,\beta}$.

For the proof Lévy considered the logarithm Ψ of the characteristic function ϕ related to the variable $\frac{X}{A}$. If ψ_k designates the logarithm of the characteristic function of ξ_k, then

$$\Psi(t) = \sum_{k=1}^{n} \psi_k\left(\frac{a_k t}{A}\right) = -(c_0 + c_1 \operatorname{sign}(t)\mathrm{i})|t|^\alpha \left[1 + \sum_{k=1}^{n} \frac{a_k^\alpha}{A^\alpha}\omega_k\left(\frac{a_k t}{A}\right)\right].$$

With the abbreviation $C := \sqrt{c_0^2 + c_1^2}$ it follows that

$$\left|\Psi(t) + (c_0 + c_1 \operatorname{sign}(t)\mathrm{i})|t|^\alpha\right| = C|t|^\alpha \sum_{k=1}^{n} \frac{a_k^\alpha}{A^\alpha}\left|\omega_k\left(\frac{a_k t}{A}\right)\right|.$$

Let T be an arbitrary positive number. Then we have $\eta T < \tau$ for a sufficiently small η. Therefore, for all $t \in [-T; T]$,

$$\left|\Psi(t) + (c_0 + c_1 \operatorname{sign}(t)\mathrm{i})|t|^\alpha\right| \le CT^\alpha h(\eta T).$$

From this inequality the uniform convergence of $\phi(t)$ to $\mathrm{e}^{-(c_0 + c_1 \operatorname{sgn}(t)\mathrm{i})|t|^\alpha}$ for $\eta \to 0$ (and thus $n \to \infty$) in each finite interval follows. The assertion ensues immediately from this fact.

5.2.6.5 The "Classic" Central Limit Theorem as a Special Case

The classic CLT, called "theorème fondamental" by Lévy, was especially important in his exposition of error calculus, on the one hand. In the framework of his purely mathematical discussions, however, it was, if tacitly, "only" a particular case of the limit theorem explained in Sect. 5.2.6.4.

In retrospect, Lévy [1970, 108] wrote on his work on the convergence of distributions of sums of independent random variables to the Gaussian distribution:

> I never had luck with the law of Gauss.

In saying this, he characterized the fact that, in regard to the classic CLT, other authors had always beaten him to the publication of similar or even the same results, and could thus take exclusive credit for the achievement.

---

[79] With "deviation" of two probability laws Lévy apparently alluded to his notion of "distance" between distributions, which he had introduced in his discussion of the mutual correspondence between convergence of characteristic functions and convergence of distributions (see Sect. 5.2.3.4).

In 1922 Lévy experienced the first case of such a priority conflict with his previously unequaled weak conditions for convergence to the normal distribution [1922a]. Due to a misspelling, his conditions were misrepresented in a note published in the *Comptes rendus* (27 March 1922). Lévy's statement, corrected and slightly specified compared with the wording of the first publications, was as follows: For a sequence of distribution functions $(F_k)_{k\in\mathbb{N}}$ of independent random variables X_k, each with zero expectation and variance 1, let

$$\forall\varepsilon > 0\exists a > 0\forall k \in \mathbb{N} : \int_{|\xi|\leq a} \xi^2 dF_k(\xi) \geq 1-\varepsilon. \tag{5.54}$$

Let $(m_k)_{k\in\mathbb{N}}$ be a sequence of positive numbers with

$$\frac{\max_{k=1\ldots n} m_k^2}{\sum_{k=1}^n m_k^2} \to 0 \quad (n\to\infty). \tag{5.55}$$

Then

$$\lim_{n\to\infty} P\left(\frac{\sum_{k=1}^n m_k X_k}{\sqrt{\sum_{k=1}^n m_k^2}} \leq x\right) = \frac{1}{\sqrt{2\pi}}\int_{-\infty}^x \mathrm{e}^{-\frac{t^2}{2}}\,dt.$$ [80]

Lévy [1924, 30 f.; 1925b, 207 f.] showed that (5.54) is sufficient for F_k being a "famille normale des lois $\mathscr{L}_{2,0}$ réduites." Thus, the CLT in the version above was "only" a special case of Lévy's theorem on the convergence to distributions of type $L_{\alpha,\beta}$.

On 29 May 1922 Borel, who, at this time, was not too well-intentioned toward Lévy, presented a note written by Lindeberg [1922a] to the Paris Academy. In this note, which also appeared in the *Comptes rendus*, Lindeberg pointed out Lévy's mistake, presented his own condition (5.46) for the CLT, and rightly designated it more general than Lyapunov's (5.2). He also announced a comprehensive article [Lindeberg 1922c] on these issues. This article, which had already been submitted on 20 November 1921 to the *Mathematische Zeitschrift*, appeared shortly after the note. Lévy's revised publication [1922b] on his own conditions was communicated on 26 June 1922; it also contained the proof of the convergence of the characteristic functions related to the normed sums to the characteristic function of the normal distribution under his conditions.[81] This article was superseded by Lindeberg's work, at least concerning the pure results. As Lévy himself admitted [1922b, 13], he had not reached a proof for the continuous correspondence of characteristic functions and distributions at that time. This latter proof ([Lévy 1922c], see Sect. 5.2.3.4) was published in the *Comptes rendus* after having been communicated on 13 November 1922, and finally completed Lévy's discussion of the CLT.

[80] In this sense the statement can be found in [Lévy 1924, 25; 1925b, 234]. Lévy [1924, 27; 1925b, 240] also gave an equivalent "ε-η-formulation" in the style of the theorem described in Sect. 5.2.6.4.

[81] The notion of "famille normale" Lévy introduced only in [1924, 24]. In his autobiography [1970, 78 f.] he, however, reported that the article [1924] had already been submitted in fall 1922.

Lévy [1924, 18] stressed the independence of his and Lindeberg's work. He proved [1924, 32 f.; 1925b, 244–246] the CLT by use of the method of characteristic functions under the modified "Lindeberg condition" (5.47), which was especially appropriate to characteristic functions (see Sect. 5.2.4.2). This proof largely corresponded to the now common standard proof as contained in many textbooks on probability theory.

For a comparison of Lévy's conditions with the modified Lindeberg condition one has to refer to random variables X_k with zero expectations and variances σ_k^2. In this case Lévy's conditions are obtained after the substitution of (5.54) by

$$\forall\varepsilon > 0 \exists a > 0 \forall k \in \mathbb{N} : \frac{1}{\sigma_k^2} \int_{|\xi| \le a\sigma_k} \xi^2 dF_k(\xi) \ge 1 - \varepsilon, \tag{5.56}$$

and after writing σ_k^2 instead of m_k^2 in (5.55). The Lindeberg condition, with the abbreviation $r_n^2 := \sum_{k=1}^{n} \sigma_k^2$, is

$$\forall t > 0 \forall \eta > 0 \exists n_0 \forall n \ge n_0 : \frac{1}{r_n^2} \sum_{k=1}^{n} \int_{|x| \le r_n t} x^2 dF_k(x) \ge 1 - \eta. \tag{5.57}$$

Whereas Lévy's first condition (5.56) aims at a certain uniformity among the single distribution functions, his second condition (5.55) requires each single variance to be small compared with the variance of the entire sum. Lindeberg's condition (5.57) stresses both aspects at the same time, the uniformity required, however, is weaker than in Lévy's condition. In fact, as Lévy [1925b, 244] proved, from both Lévy conditions together the Lindeberg condition can be deduced.

Lévy admired the "ingenious and simple" method of Lindeberg, which—in contrast to the method of characteristic functions—was appropriate for more general problems than just sums of independent random variables. He described this method comprehensively, if in a modified form, in his book [1925b, 246–249].

There were, however, more differences between Lévy and Lindeberg than just those regarding conditions and analytic methods. In his first article [1922a] Lévy had already put the classic CLT in the context of limit theorems on the convergence to stable distributions, and, even more generally, to convolutions of these distributions. This point of view, which was entirely different from Lindeberg's, opened a whole new field of modern probability theory.

5.2.6.6 More Limit Laws

Lévy [1922a] in an example had already hinted at the fact that stable distributions are not the only possible limit distributions of normed sums of independent, but not necessarily identically distributed random variables. In [1925b, 269 f.] he discussed this problem in a more general way. Let n' and n'' be "large numbers" of independent "errors" X_k and Y_k, respectively, where the X_k obey the reduced law $L_{\alpha',0}$, and the Y_k obey the reduced law $L_{\alpha'',0}$ with $\alpha' < \alpha''$. Let $n := n' + n''$. If n'' is of a "higher order of magnitude" than n' such that with the abbreviations $N' = n'^{\frac{1}{\alpha'}}$ and $N'' = n''^{\frac{1}{\alpha''}}$ the fraction $\frac{N''}{N'}$ has a "certain limit" as $n \to \infty$, Lévy claimed that, after an appropriate norming, $\sum_{k=1}^{n'} X_k + \sum_{k=1}^{n''} Y_k$ would have a limit law characterized by

$$\psi(t) = -c_1|t|^{\alpha'} - c_2|t|^{\alpha''},$$

where any $c_1, c_2 > 0$ were possible. Lévy did not justify this somewhat murky assertion. Indeed, the sequences (n'') and (n') can be chosen such that $\frac{N''}{N'} \to \frac{c''}{c'}$, where c' and c'' are arbitrary positive numbers. The logarithm of the characteristic function of

$$\frac{\sum_{k=1}^{n'} X_k + \sum_{k=1}^{n''} Y_k}{\frac{N'}{c'}}$$

tends to

$$\psi(t) = -\frac{c'^{\alpha'}}{\Gamma(\alpha'+1)}|t|^{\alpha'} - \frac{c''^{\alpha''}}{\Gamma(\alpha''+1)}|t|^{\alpha''}$$

as $n \to \infty$. If one sets $c_1 = \frac{c'^{\alpha'}}{\Gamma(\alpha'+1)}$ and $c_2 = \frac{c''^{\alpha''}}{\Gamma(\alpha''+1)}$, then Lévy's assertion follows. Lévy also indicated how, by a generalization of this situation, even distributions with characteristic functions $e^{\sum \psi_k(t)}$ could be limit distributions if ψ_k were the logarithms of characteristic functions of indefinitely many different stable distributions. It seems that, at this stage of his work, Lévy still was convinced of a certain "dominance" of stable laws even regarding limit distributions of sums of non-identically distributed random variables.

5.2.6.7 Domains of Attraction of Stable Distributions

Lévy's "origins" in functional analysis could also be seen by his use of particular expressions in his discussion of the domains of attraction of stable distributions. He assigned all probability laws which could be reduced to each other by a "change of unit" the same point of an "ideal space" [Lévy 1924, 25; 1925b, 238 f.].[82] In this way, to all laws of the type $L_{\alpha,\beta}$ with fixed α and β, but possibly different c_0 (cf. formulae (5.51) and (5.52)), there corresponds exactly one point of space, which Lévy likewise designated by $L_{\alpha,\beta}$. In Lévy's conception the domain of attraction of $L_{\alpha,\beta}$ consists of a certain set of points of the "ideal space," each point corresponding to a whole class of similar laws. For each distribution function V which characterizes a point of the domain of attraction of $L_{\alpha,\beta}$ there exists a sequence of norming constants $(N(n))$ and a distribution function $V_{\alpha,\beta}$ of type $L_{\alpha,\beta}$ such that

$$\lim_{n\to\infty} V^{n\star}(N(n)x) = V_{\alpha,\beta}(x).$$

[82] Two distribution functions V_1 and V_2 can "be reduced to each other" if there exists a positive constant a such that $V_1(x) = V_2(ax)$. According to Pólya [1923, 97] two distribution functions which are related to each other by the latter equation are called "similar."

Later, in the thirties, this definition of domain of attraction was generalized by the somewhat weaker condition that

$$\lim_{n\to\infty} V^{n\star}\left(N(n)x + a_n\right) = V_{\alpha,\beta}(x)$$

with suitable translation constants a_n. From Lévy's generalization of the CLT (see Sect. 5.2.6.4) it follows that "points" to which probability laws of the type $\mathscr{L}_{\alpha,\beta}$ belong are elements of the domain of attraction of $L_{\alpha,\beta}$. Lévy (for example [1925b, 151, 267, 277]) expressed this fact[83] briefly in the following way: "The laws $\mathscr{L}_{\alpha,\beta}$ belong to the domain of attraction of $L_{\alpha,\beta}$."

The portion of text [Lévy 1925b, 266–277] gives information about the author's attempts to determine the domains of attraction of stable laws, a problem which was completely solved only by the end of the thirties. In this context, semistable distributions (originally introduced by Pólya) played an important role.

Lévy [1925b, 266 f.] explained that the "set of points" $\mathscr{L}_{\alpha,\beta}$ is only a proper subset of the domain of attraction of $L_{\alpha,\beta}$ by the following example: Let (X_k) be a sequence of identically distributed independent random variables, each with characteristic function e^{ψ} such that, for $0 < \alpha \leq 2$,

$$\psi(t) = -|t|^{\alpha}\left(\log\frac{1}{|t|}\right)^{\gamma}[1+\omega(t)],$$

where $\lim_{t\to 0}\omega(t) = 0$, $\gamma > 0$. Then, the sum $\sum_{k=1}^{n}\frac{X_k}{N(n)}$ ($N(n) > 0$ arbitrary at first) has the characteristic function e^{ψ_n}, where

$$\psi_n(t) = -n\left|\frac{t}{N(n)}\right|^{\alpha}\left(\log\frac{N(n)}{|t|}\right)^{\gamma}\left[1+\omega\left(\frac{t}{N(n)}\right)\right].$$

If for $n\to\infty$ also $N(n)\to\infty$ with the additional condition $(N(n))^{\alpha}(\log N(n))^{-\gamma}\sim n$, we have $\psi_n(t)\to -|t|^{\alpha}$. In particular, from this example it follows that the domain of attraction of the Gaussian law also consists of distributions with infinite moments of second order.

Lévy was especially interested in a possible characterization of the domain of attraction of $L_{\alpha,\beta}$ through the existence of moments $\mathrm{E}|X|^{\alpha'}$, where $\alpha' \in [0;\alpha[$. To this end he considered random variables X obeying the law $\mathscr{L}$ such that

$$\begin{aligned} \mathrm{E}|X|^{\alpha'} &< \infty \text{ for } \alpha' < \alpha \\ \mathrm{E}|X|^{\alpha'} &= \infty \text{ for } \alpha' > \alpha. \end{aligned} \tag{5.58}$$

He designated it "very probable" [1925b, 267 f.] that $\mathscr{L}$ belongs to the domain of attraction of a "point" $L_{\alpha,\beta}$ if there exists a sequence (X_k) of independent random

[83] In the case $\alpha = 2$, $\beta = 0$ this assertion in modern textbooks (probably since [Cramér 1937/70, 53]) is often called the "Lindeberg–Lévy theorem." Lindeberg himself, however, did not explicitly treat the case of identically distributed random variables.

variables, each obeying $\mathscr{L}$, and a sequence $N(n)$ of positive numbers such that the sequence of distributions of $\frac{\sum_{k=1}^{n} X_k}{N(n)}$ tends to a limit distribution. He was not able, however, to give a proof for this assertion.

It was necessary to assume—in addition to (5.58)—the existence of such a limit distribution, as Lévy showed in his discussion of semistable laws. By "semistable" he denoted probability laws with the characteristic function e^{ψ}, ψ having the property

$$\exists a_1, a_2, A > 0 : \psi(a_1 z) + \psi(a_2 z) = \psi(Az). \tag{5.59}$$

Lévy only made explicit the particular case $a_1 = a_2$. As Pólya (see Sect. 5.2.3.5) had already explained, (5.59) in this particular case has solutions of the form

$$\psi(z) = \psi_{\alpha,\beta}(z) P\left(\log|z|\right), \tag{5.60}$$

where P is a periodic function with the period $\frac{\log 2}{\alpha}$, and $e^{\psi_{\alpha,\beta}}$ is the characteristic function of a law of type $L_{\alpha,\beta}$. For the case $0 < \alpha < 1$ Pólya had already proven the existence of distributions with such ψ. For the whole range $0 < \alpha < 2$ and $-1 \leq \beta \leq 1$, Lévy [1925b, 270–276] proved the existence of semistable distributions with ψ according to (5.60). He constructed—in a way similar to his proof of the existence of stable distributions—random variables X obeying (5.58) for whose characteristic functions f the limit relation

$$\left(f\left(\frac{z}{2^{\frac{h}{\alpha}}}\right)\right)^{2^h} \to e^{\psi(z)} \quad (h \in \mathbb{N},\ h \to \infty)$$

is valid, although in general for $n \in \mathbb{N}$ the limit

$$\lim_{n\to\infty} \left(f\left(\frac{z}{n^{\frac{1}{\alpha}}}\right)\right)^n$$

does not exist. By this argument Lévy showed at the same time that the limit described above generally does not exist. From his considerations it also followed that in the case $\alpha = 2$ only the Gaussian distribution obeys the relation (5.59). Lévy [1925b, 277] erroneously maintained that the domain of attraction of the normal distribution consists of all laws with $E|X|^{\alpha'} < \infty$, for $\alpha' < 2$. Ten years later he [1935b, 369 f.] found a counterexample, and thus he proved his original assumption wrong.

In his book of 1925, Lévy's discussion of the domains of attraction of stable laws was incomplete and partly speculative. However, it was exactly these problems which would continue to play an important role in the development of probability theory during the thirties.

5.2.7 *Bernshtein and His "lemme fondamental"*

In the "annus mirabilis" 1922, in which Lindeberg's and Lévy's fundamental and influential contributions to the CLT appeared, also a little note on this topic was issued by Bernshtein in *Mathematische Annalen.* This note had already been submitted on 2 August 1921, and thus was certainly independent from Lévy's and Lindeberg's articles. Bernshtein [1922, 237] even maintained that his results on the convergence of distributions of sums of not necessarily independent random variables to the Gaussian distribution, published without any proof, had already been achieved by 1917/18. The central role within Bernshtein's theorems is played by a "lemme fondamental," which generalizes the assertion of the CLT toward "almost independent random variables," and can also be applied to sums of random variables which form Markov chains.[84] The wording of this lemma, which in the special case of independent random variables generalized even Lindeberg's assertion, was not entirely clear. Bernshtein's little article went practically unnoticed. A comprehensive account including all proofs appeared only in 1926.

5.2.7.1 The Statement

Compared with the original version of [1922], in [Bernshtein 1926, 21] the "lemme fondamental" received a slightly different wording:

Let $S_n = u_1 + u_2 + \cdots + u_n$, $\mathcal{M}(S_n^2) = B_n$, [by $\mathcal{M}$ Bernshtein always denoted an expectation] $\mathcal{M}(u_1^2) + \mathcal{M}(u_2^2) + \cdots + \mathcal{M}(u_n^2) = B_n'$ (it is always supposed for simplicity of notation that $\mathcal{M}(u_i) = 0$). *If, for each arbitrary set of already known values* $u_1, u_2, \ldots, u_{i-1}$, *the absolute values of the mathematical expectations of* u_i *and* u_i^2 *do not exceed* α_i *and* β_i *respectively, and at the same time the mathematical expectation of* $|u_i^3|$ *remains below* c_i, *then the probability of the inequality*

$$z_0\sqrt{2B_n} < S_n < z_1\sqrt{2B_n}$$

will tend to the limit

$$\frac{1}{\sqrt{\pi}}\int_{z_0}^{z_1} e^{-z^2}\,dz,$$

presupposing that

$$\frac{\sum_1^n \alpha_i}{\sqrt{B_n}}, \quad \frac{\sum_1^n \beta_i}{B_n}, \quad \frac{\sum_1^n c_i}{B_n^{3/2}}$$

tend to 0 together with $\frac{1}{n}$.[85]

Its full generality the "lemme fondamental" received from the following additional remark [Bernshtein 1926, 23], which was already contained analogously in the original wording of [1922, 238].

---

[84] With a little hindsight, Bernshtein's "lemme fondamental" can be included in the history of martingale limit theorems, see [Crépel 1984] and Sect. 7.1.3 of the present book.

[85] The quantity B_n' is of considerable importance for the proof. From the details of the proof one can see that the "expectation of $|u_i|^3$" is a conditional expectation without any doubt.

The conclusion of the lemma equally subsists even if its conditions are not met in those cases in which the quantities u_k attain certain values which have the probabilities ε_k such that $\sum_1^n \varepsilon_k$ tends to 0 together with $\frac{1}{n}$. In this context, all mathematical expectations, which occur in the wording, have to be calculated under the hypothesis that none of these exceptional values are realized.

I shall try now to integrate my interpretation of the additional remark into the formulation of the "lemme fondamental." For simplicity I take the slightly more specialized, but nonetheless sufficiently general point of view that the "exceptional values" lie beyond certain symmetric intervals. Already in [1922, 237] Bernshtein had mentioned that his results were even valid for sums $S_n = u_1^{(n)} + u_2^{(n)} + \cdots + u_n^{(n)}$. For simplicity, however, he always neglected the upper index. In my reconstruction I will also make explicit these double sequences of random variables which Bernshtein actually had in mind.

In modern terminology the "lemme fondamental" might be expressed in the following way:

Let $(U_{ni})_{n,i\in\mathbb{N},1\leq i\leq n}$ be a double sequence of random variables with distributions F_{ni} and ranges of values W_{ni}. For $n, i \in \mathbb{N}$, $1 \leq i \leq n$, and $u_i \in W_{ni}$ let

$$F_{ni}(x|u_1,\ldots,u_{i-1}) := \begin{cases} F_{n1}(x) & i{=}1 \\ P(U_{ni} \leq x|U_{n1} = u_1 \wedge U_{n2}{=}u_2 \wedge \cdots \wedge U_{ni-1}{=}u_{i-1}) & i{\geq}2. \end{cases}$$

Bernshtein did not comment on possible problems concerning existence, uniqueness, or construction of such conditional distributions and expectations.[86] Let (L_{ni}) denote a double sequence with $\infty \geq L_{ni} > 0$ such that

$$\lim_{n\to\infty} \sum_{i=1}^{n} \int_{|x|>L_{ni}} dF_{ni}(x) = 0. \tag{5.61}$$

Let

$$W_n(x_1,\ldots,x_n) := P(U_{n1} \leq x_1 \wedge \cdots \wedge U_{nn} \leq x_n)$$

and

$$B_n := \int_{|x_1|\leq L_{n1}\wedge\cdots\wedge|x_n|\leq L_{nn}} (x_1 + \cdots + x_n)^2 dW_n(x_1,\ldots,x_n).$$

If there exist double sequences (α_{ni}), (β_{ni}), (c_{ni}) of positive numbers such that for all natural n and all natural i, $1 \leq i \leq n$, and for all

86 A theory of conditional distributions and expectations with respect to a finite set of random variables, which would have been necessary for a rigorous treatment of Bernshtein's considerations, was only developed by Kolmogorov [1933/50, 51–56] using the "theorem of Nikodym" [1930]. Bernshtein, however, tacitly as it seems, assumed the unique existence of conditional distributions and expectations of the random variables U_{ni} with respect to *all* relevant values of the random variables $U_{n1}, \ldots, U_{ni-1}$. Therefore, it is not the aim here to discuss Bernshtein's contribution with the full generality of Kolmogorov's concepts.

$$u_1 \in [-L_{n1}; L_{n1}] \cap W_{n1}, u_2 \in [-L_{n2}; L_{n2}] \cap W_{n2}, \ldots, u_{i-1} \in [-L_{ni-1}; L_{ni-1}] \cap W_{ni-1}$$

the following conditions are valid:

$$\left| \int_{|x| \le L_{ni}} x dF_{ni}(x|u_1, \ldots, u_{i-1}) \right| \le \alpha_{ni}, \text{ where } \lim_{n \to \infty} \frac{\sum_{i=1}^{n} \alpha_{ni}}{\sqrt{B_n}} = 0, \quad (5.62)$$

$$\left| \int_{|x| \le L_{ni}} x^2 dF_{ni}(x|u_1, \ldots, u_{i-1}) - \int_{|x| \le L_{ni}} x^2 dF_{ni}(x) \right| \le \beta_{ni},$$
$$\text{where } \lim_{n \to \infty} \frac{\sum_{i=1}^{n} \beta_{ni}}{B_n} = 0, \quad (5.63)$$

and

$$\int_{|x| \le L_{ni}} |x^3| dF_{ni}(x|u_1, \ldots, u_{i-1}) \le c_{ni}, \text{ where } \lim_{n \to \infty} \frac{\sum_{i=1}^{n} c_{ni}}{B_n^{3/2}} = 0; \quad (5.64)$$

then[87]

$$\lim_{n \to \infty} P\left(z_0 \sqrt{2B_n} < \sum_{i=1}^{n} U_{ni} < z_1 \sqrt{2B_n} \right) = \frac{1}{\sqrt{\pi}} \int_{z_0}^{z_1} e^{-z^2} dz. \quad (5.65)$$

Gnedenko and Kolmogorov [1949/68, 130] have hinted at the fact that Bernshtein's lemma together with the additional remark in the particular case of independent variables yields very general sufficient conditions for the convergence of the distributions of normed sums to the normal distribution. These conditions were equivalent to the ones of Feller in 1935 (Lévy is not referred to), as Gnedenko and Kolmogorov maintain. Feller (and Lévy), however, had also proved the necessity of their conditions. I still have to discuss this claim of priority by Gnedenko and Kolmogorov in favor of their older colleague (see Sect. 6.3.2).

As we can see from a footnote in the introduction (p. 12) of his 1926 paper, Bernshtein actually aimed at an application of his lemma (including the additional remark) to independent random variables. In this part of his article Bernshtein explicitly used characteristic functions for a reconstruction of Lyapunov's proof of the CLT. In the footnote he gave an example of identically distributed random variables which do not possess any finite moments of second order, but do, however, meet conditions which are in accord with (5.61) to (5.64). A similar example for the convergence of normed sums to the normal distribution without the existence of moments of second order had already been discussed by Lévy (see Sect. 5.2.6.7), who, however, had not put this example in the context of a general limit theorem.

---

[87] In the exact wording of Bernshtein's conditions the relations "$< c_{ni}$" instead of "$\le c_{ni}$" occur. In Bernshtein's proof, however, only the somewhat weaker version with "$\le$" is used.

5.2.7.2 The Proof

Bernshtein [1926, 21–23] initially proved his lemma without consideration of the additional remark, or, in other words, only for the case $L_{ni} = \infty$. In a first step he deduced that

$$\lim_{n\to\infty} \frac{B_n}{B_n'} = 1. \tag{5.66}$$

The main part of his proof[88] dealt with the discussion of

$$G_{nm}(\xi) = \mathrm{E}e^{i\xi \sum_{k=1}^{m} Y_{nk}}, \text{ where } Y_{nk} = \frac{U_{nk}}{\sqrt{2B_n'}}$$

for $m \leq n$ and $\xi \in \mathbb{R}$. On the basis of (5.62) to (5.64) Bernshtein was able to show that for all values y_j of Y_{nj} $(j = 1, \ldots, k-1)$ and for all $|\xi| < N$ (an arbitrarily large number)

$$\mathrm{E}\left(e^{i\xi Y_{nk}} \middle| Y_{n1} = y_1, \ldots, Y_{nk-1} = y_{k-1}\right) = 1 - \frac{\mathrm{E}U_{nk}^2}{4B_n'}\xi^2 + \delta_{nk},$$

where

$$|\delta_{nk}| < A\left(\frac{\alpha_{nk}}{\sqrt{B_n}} + \frac{\beta_{nk}}{B_n} + \frac{c_{nk}}{B_n^{3/2}}\right) =: \eta_{nk}.$$

A designated a positive constant depending only on N. By use of this estimate Bernshtein concluded:

$$G_{nm}(\xi) = G_{nm-1}(\xi)\left(1 - \frac{\mathrm{E}U_{nm}^2}{4B_n'}\xi^2\right) + \gamma_{nm}, \quad \text{where } |\gamma_{nm}| < \eta_{nm}.$$

It followed:

$$G_{nm}(\xi) = E_{nm}(\xi) + E_{nm}(\xi)\sum_{k=1}^{m} \frac{\gamma_{nk}}{E_{nk}(\xi)},$$

where

$$E_{nk}(\xi) = \left(1 - \frac{\mathrm{E}U_{n1}^2}{4B_n'}\xi^2\right)\cdots\left(1 - \frac{\mathrm{E}U_{nk}^2}{4B_n'}\xi^2\right).$$

For $|\xi| < N$ and sufficiently large n the inequalities $\left|\frac{E_{nm}(\xi)}{E_{nk}(\xi)}\right| \leq 1$ $(k \leq m \leq n)$ were valid, and therefore also

$$|G_{nm}(\xi) - E_{nm}(\xi)| < \sum_{k=1}^{m} |\gamma_{nk}|.$$

Bernshtein referred to conditions (5.62) to (5.64) to show that the right-hand side of this last inequality tends to 0 for all $m \leq n$ if $n \to \infty$. Because, for $n \to \infty$, $E_{nn}(\xi) \to e^{-\frac{\xi^2}{4}}$ uniformly in each finite interval, also $G_{nn}(\xi) \to e^{-\frac{\xi^2}{4}}$ uniformly in each finite interval ensued. Under consideration of condition (5.66) and on account

[88] In the following I refer to the reconstruction of the lemma in modern terminology.

of the theorem of the continuous correspondence between characteristic functions and distributions, the assertion of the lemma without the additional remark was finally proven.

A justification of the additional remark was only indicated by Bernshtein [1926, 23 f.] in a rather vague manner. The complete proof might—according to Markov's treatment of truncated random variables (see Sect. 5.1.5)—run as follows: Let

$$p_n := P\left(|U_{n1}| > L_{n1} \vee |U_{n2}| > L_{n2} \vee \cdots \vee |U_{nn}| > L_{nn}\right).$$

Because of (5.61) we have

$$\lim_{n\to\infty} p_n = 0. \tag{5.67}$$

Now we define

$$U'_{nk} = \begin{cases} U_{nk} \text{ if } |U_{nk}| \le L_{nk} \\ 0 \quad \text{else,} \end{cases}$$

and introduce the abbreviation $S'_n = \sum_{k=1}^n U'_{nk}$. Then

$$\begin{aligned} &P\left(z_0\sqrt{2B_n} < S_n < z_1\sqrt{2B_n}\right) \\ &\quad = P\left(\left[z_0\sqrt{2B_n} < S_n < z_1\sqrt{2B_n}\right] \wedge [\forall 1 \le k \le n : |U_{nk}| \le L_{nk}]\right) + \\ &\quad + P\left(\left[z_0\sqrt{2B_n} < S_n < z_1\sqrt{2B_n}\right] \wedge [\exists 1 \le k \le n : |U_{nk}| > L_{nk}]\right) \\ &\qquad \le P\left(z_0\sqrt{2B_n} < S'_n < z_1\sqrt{2B_n}\right) + p_n. \end{aligned} \tag{5.68}$$

On the other hand, the event "$\alpha < S'_n < \beta$" consists of those cases for which $\alpha < S_n < \beta$ or for which $|U_{nk}| > L_{nk}$ for at least one $1 \le k \le n$. Therefore we have

$$P\left(\alpha < S'_n < \beta\right) \le P\left(\alpha < S_n < \beta\right) + p_n,$$

and thus

$$P\left(z_0\sqrt{2B_n} < S_n < z_1\sqrt{2B_n}\right) \ge P\left(z_0\sqrt{2B_n} < S'_n < z_1\sqrt{2B_n}\right) - p_n. \tag{5.69}$$

From Bernshtein's proof of the lemma without the additional remark

$$P\left(z_0\sqrt{2B_n} < S'_n < z_1\sqrt{2B_n}\right) \to \frac{1}{\sqrt{\pi}}\int_{z_0}^{z_1} \mathrm{e}^{-t^2}\,dt$$

follows. Therefore, because of (5.67) to (5.69), the assertion of the CLT (5.65) can be proven under the more general assumptions of the additional remark as well.

Bernshtein's "lemme fondamental" is a natural extension of the CLT in its most general setting to "almost independent" random variables. Therefore [Bernshtein 1922; 1926] has to be considered as equal to Lindeberg's and Lévy's contributions in 1922, particularly since in Bernshtein's note of 1922 the idea of a CLT for random variables without finite moments is already present.

For the time being, the universal notion of the "lemme fondamental" was not fully appreciated, probably because the major part of the article [Bernshtein 1926] was dedicated to the application of the lemma to random variables which form Markov chains. This application was based on the following principle, explained already in [Bernshtein 1922, 238]. Given a sum

$$S_n := U_{n1} + U_{n2} + \cdots + U_{nn}$$

of any random variables with zero expectations, try to find $2l$ new random variables X_{nk} and Y_{nk}, respectively, such that $l \to \infty$ as $n \to \infty$, and

$$S_n = Y_{n1} + X_{n1} + Y_{n2} + X_{n2} + \cdots + Y_{nl} + X_{nl},$$

where Y_{ni} with $\mathrm{E}Y_{ni} = 0$ are almost independent, and the "order of growth" of $\mathrm{E}\left(\sum_{i=1}^{l} X_{ni}\right)^2$ is for $n \to \infty$ less than that of $\mathrm{E}\left(\sum_{i=1}^{n} U_{ni}\right)^2 =: B_n$. If a representation of S_n is possible in this way, then one can show that $\mathrm{E}\left(\sum_{i=1}^{l} Y_{ni}\right)^2 \sim B_n$, and by aid of the "lemme fondamental" the relation (5.65) can be deduced.

It is remarkable that Bernshtein [1926, 43–59] also extended his statements toward two-dimensional random variables. This included the proof of a CLT for independent random vectors by use of characteristic functions under conditions analogous to Lyapunov's (5.1), presupposing the existence of absolute moments of third order. Bernshtein also succeeded in proving a two-dimensional analog to the "lemme fondamental" and in applying it to two-dimensional random vectors which form Markov chains.

5.2.8 *Cramér: Lyapunov Bounds and Asymptotic Behavior of "Exponential Series"*

5.2.8.1 Risk Theory as a Starting Point

In chapter 9 of his *TAP* on "advantages which depend on the probability of future events," Laplace had considered the situation of n independent games, each having only the possible results "win" and "loss." Win, loss (considered as negative "win") and the respective probabilities were not necessarily the same in each game. By use of the CLT, Laplace for large n calculated the approximate probability that the overall gain (which can also be negative) exceeds certain values (see Sect. 2.1.5.3). This application of the CLT was the point of departure for a renewed risk theory during the second half of the 19th century,[89] which mainly focused on insurance problems, using concepts and notions related to error theory.[90]

[89] The problem of risk regarding games of chance and insurance was part of the history of probability calculus from its earliest beginnings. For the development of risk "theory" until Laplace, see [Daston 1988]. Purkert [2006a] gives a survey of the development of risk theory from about 1850.

[90] This relationship is especially made explicit in [Hausdorff 1897].

Assuming n independent contracts, for which the insurance company has the expenses e_i and takes the premia p_i $(i = 1, \dots, n)$ within a certain time period, $g_i = p_i - e_i$ is the—possibly negative—gain of the insurance company from the contract i. According to the CLT, for example in Poisson's version, under the assumption of a large number of contracts the total gain $\sum g_i$ of the insurance company within the time period considered approximately obeys a normal distribution with expectation $\sum \mathrm{E}g_i$ and variance $\sum \mathrm{Var}g_i$. Estimating the quantities $\sum \mathrm{E}g_i$ and $\sum \mathrm{Var}g_i$ (square of "main risk") as well as the calculation and the assessment of the probability that the overall gain of the insurance company exceeds certain (positive) bounds were fundamental problems of risk theory. The latter problem was linked with the "stability" of an insurance company. In this context, the question about the quality of approximation of exact probabilities by normal distributions played an increasingly important role. At the beginning of the 20th century, Bohlmann [1901, 903] stated that all investigations of risk theory so far did not have any practical relevance. He noticed [1901, 913] that the CLT "in many cases, however always in a purely formal manner without any consideration of the quality of convergence" had been applied in the mathematical insurance theory since Carl Bremiker [1859].

Risk theory paved the way for a more thorough consideration of the CLT and, in general, of stochastic processes during the first decades of the 20th century. Cramér reported being strongly influenced by the work of Filip Lundberg after the turn of the century; it stimulated an embedding of risk theory into the theory of stochastic processes [Cramér & Wegman 1986, 530]. Lundberg is now regarded as the founder of "collective risk theory" which can be characterized by one main feature: Development of the overall gain of an insurance company is—without considering properties or numbers of the underlying individual contracts—modeled by stochastic processes. In his original approach, Lundberg assumed the total gain of an insurance company being subjected to a stochastic process with independent increments, choosing the company's accumulated risk premium as the independent variable rather than time.[91] By about 1950, the collective approach became the predominant approach to risk theory [Purkert 2006a, 520]. This success was essentially due to Cramér, who had been propagating and refining Lundberg's work since about 1930.

During the 1920s, however, Cramér still took the position of traditional "individual risk theory," which, as described above, was based on sums of random gains in single contracts. In his first papers on risk theory, Cramér thoroughly discussed the quality of the approximation of distributions of sums of independent random variables by normal distributions, a problem to which his attention was drawn—at least in part, if not exclusively—due to his professional activities as an insurance mathematician. He criticized the "naive" approach to use the Gaussian error law for the probability distribution of the overall gain within a certain division of an insurance company without precisely examining the deviation from the exact distribution,

---

[91] See [Cramér 1930, 66–84] for a summarizing account.

even in cases where gain comes from an only moderate number of contracts [Cramér 1923, 210]. Up to this time, Lyapunov's upper bounds for the error of approximation were the most exact in the general case. In his first contribution, Cramér [1923] aimed at continuing with Lyapunov's method and improving his results. To this end, he explicitly used characteristic functions.

His major result in [1923] was as follows: Let X_k $(k = 1, \ldots, n)$ be independent random variables with zero expectations, distribution functions V_k (being right continuous), variances ρ_k, moments σ_k of third order, and absolute moments $\overline{\sigma}_k$ of third order. Moreover, let V be the distribution function of $X_1 + \cdots + X_n$, $\rho := \sum_{k=1}^{n} \rho_k$, $\sigma := \sum_{k=1}^{n} \sigma_k$, and $\overline{\sigma} := \sum_{k=1}^{n} \overline{\sigma}_k$. Then, with the abbreviations $S(x) = V(x\sqrt{2\rho})$ and $\Phi(x) = \frac{1}{\sqrt{\pi}} \int_{-\infty}^{x} \mathrm{e}^{-t^2} dt$, for all $x \in \mathbb{R}$:

$$|S(x) - \Phi(x)| < 6 \max \left(1; \log \frac{\rho^{3/2}}{\overline{\sigma}} \right) \frac{\overline{\sigma}}{\rho^{3/2}}. \tag{5.70}$$

With this result, Cramér significantly improved Lyapunov's bound and he further clarified the latter's assertion that this bound was of the order $\frac{\log n}{\sqrt{n}}$. He [1923, 215] also showed that in the general case "the degree of approximation presupposing a sufficiently large number of n cannot be of a 'better' order than $\frac{1}{\sqrt{n}}$."

For a proof of (5.70), Cramér, who in 1923 apparently did not know of Lévy's work,[92] used von Mises's (complex) "adjunct functions," and with the abbreviation $f(x) = \frac{1}{\sqrt{\pi\lambda}} \int_{-\infty}^{x} \mathrm{e}^{-\frac{t^2}{\lambda}} dt$ he discussed the distribution functions

$$\overline{S}(x) = \int_{-\infty}^{\infty} f(x-t) dS(t) \text{ and } \overline{\Phi}(x) = \int_{-\infty}^{\infty} f(x-t) d\Phi(t).$$

$\overline{S}(x) - \overline{\Phi}(x)$ could be represented by an absolutely convergent integral in the following way:

$$\overline{S}(x) - \overline{\Phi}(x) = \frac{1}{\pi} \mathrm{Im} \int_{0}^{\infty} \frac{\mathrm{e}^{itx}}{t} \left(v(t) - \mathrm{e}^{-\frac{t^2}{4}} \right) \mathrm{e}^{-\frac{\lambda t^2}{4}} dt, \tag{5.71}$$

$$\text{where } v(t) = \int_{-\infty}^{\infty} \mathrm{e}^{-itx} dS(x).$$

The assertion (5.70) was proven on the one hand by estimates of $|v(t)|$ and $|\arg v(t)|$, and on the other hand by splitting the domain of integration of the integral in (5.71) and estimating each single integral generated by this procedure (that was the continued influence of Laplace's method of approximation).

[92] Cramér [1923, 212] cites the related articles [Lyapunov 1900; 1901b], [von Mises 1919a], [Pólya 1920], and [Lindeberg 1922b;c].

5.2.8.2 Cramér's Discussion of the Asymptotics of Edgeworth and Charlier A Expansions

In his subsequent papers on Charlier A and Edgeworth series, Cramér used similar methods. He repeatedly expressed (for the first time in [1925, 411]) the opinion, which was not shared by all proponents of these series expansions to the same extent,[93] that from the point of view of the hypothesis of elementary errors the really interesting problem was not the convergence of these series, but the asymptotic properties of the respective expansions cut off after a few terms, if the number of elementary errors tended to infinity. In statistical practice only a few series terms could be taken into account, anyway.

According to Cramér [1925, 412], the asymptotic behavior was especially important for bringing the theory in "connection with the real causal structure of the phenomena to be examined." One would be "inclined to assume" that a statistical quantity was generated by elementary errors if the frequency distribution observed could be properly approximated by the sum of the first terms of an "exponential series." Cramér regarded the hypothesis of elementary errors as an "ideal scheme," which one could assume "rightly or also wrongly."

Cramér considered his discussion of the asymptotic behavior of Charlier and Edgeworth expansions a contribution to the *mathematical* foundation of the hypothesis of elementary errors. It is remarkable, however, that, from the point of view of *natural sciences*, Cramér justified the hypothesis of elementary errors in a rather vague way only. In the second part of his major paper [1928] on the asymptotics of "exponential series" he delivered a comprehensive comparison between empirical frequency distributions from different fields and distributions calculated by means of Edgeworth series. He did not, however, give an assessment that referred to considerations beyond mathematics proper. The reader has the impression that Cramér considered the hypothesis of elementary errors a "comfortable" assumption, which was appropriate for a certain preselection of statistical methods for the adjustment of frequency curves, but nothing more.

Cramér [1928, 158] considered the distribution function of a random variable especially useful as a means for comparing relative frequencies observed and theoretically assumed. Primarily this function, and not the probability density, could be derived from statistical observations. Thus, Cramér granted series expansions for

---

[93] Toyojirô Kameda published large articles [1915; 1925], which were also mentioned by Cramér [1928, 64], on distributions of sums of independent (continuous or lattice distributed) random variables, in which the convergence of series expansions in derivatives of the Gaussian distribution function was discussed. Kameda made extensive use of generating functions $f(\alpha) = \mathrm{E}e^{\alpha X}$ of random variables X, where α was an element of a certain subset of $\mathbb{C}$. In the case $\alpha = it$ ($t \in \mathbb{R}$) this also implied a discussion of characteristic functions. Kameda, who apparently did not know modern contributions on probability theory (the most recent source he cited was [Markov 1912]), considered the problem of the distribution of a sum of independent random variables completely solved if it could be represented by a convergent "Hermitian" series expansion [Kameda 1925, 49f]. Kameda was not interested at all in the asymptotic behavior of these expansions if they were cut off after a few terms.

distribution functions the predominant statistical importance in contrast to series for densities or discrete lattice probabilities.

Cramér's efforts to extend the CLT toward the consideration of series terms in addition to the normal distribution lasted from about 1923 to 1927. This work also comprised, though not as a matter of priority, the problem of the convergence of the respective series. Cramér published some results on the convergence of Charlier A series for densities [1925, 405] and general distributions [1928, 64 f.]. His theorem in the latter setting was as follows: Let F_n be the distribution function of the normed sum of identically distributed independent random variables of a fixed number n, each with distribution V (Cramér used Lévy's definition of distribution function in this work). If

$$\int_{-\infty}^{\infty} \mathrm{e}^{\frac{x^2}{4\alpha_2}} dV(x) < \infty \quad \left(\alpha_2 = \int_{-\infty}^{\infty} t^2 dV(t)\right),$$

then, for all $x \in \mathbb{R}$,

$$F_n(x) = \Phi(x) + \sum_{k=1}^{\infty} \frac{c_\nu}{\nu!} \Phi^{(\nu)}(x) \quad \left(\Phi(x) := \frac{1}{\sqrt{2\pi}} \int_{-\infty}^{x} \mathrm{e}^{-\frac{t^2}{2}} dt\right)$$

with uniform convergence in all finite intervals of continuity of F_n[94] (the coefficients c_k are determined according to the identities (5.72) below).

In [1925] Cramér had already comprehensively discussed the asymptotic behavior of an expansion according to the first terms of a Charlier A series for distribution and density functions of sums of independent random variables. However, he also pointed out the deficiency of this type of expansion: Higher series terms have not always a smaller order of magnitude, a fact which had already been observed by Edgeworth (see Sect. 3.4.2.3). In his later work and also in his survey article [1972], Cramér gave preference to Edgeworth series. These expansions provide, in contrast to A series, the optimal precision depending on the order up to which the absolute moments of the elementary errors exist. Cramér completed his investigation on this topic by the fall of 1927 and first communicated his result to the London Mathematical Society in a short note [Cramér 1927].

Cramér's main paper from 1928 is exemplary for its clarity and comprehensibility, despite the opinion, expressed by Feller [1971, 531], that the field of Charlier and Edgeworth expansions was "proverbial for its messiness." This paper at the same time conveys a tremendous impression of the single steps involved which are necessary for the required estimates in the discussion of the considered series. For simplicity and clarity of presentation I confine myself in the following to a recapitulation of Cramér's treatment of the asymptotic behavior of Edgeworth expansions for distribution functions of normed sums of independent identically distributed random variables.

---

[94] An interval of continuity of a distribution is any open interval in which the distribution function is continuous at the two boundary points.

In accordance with [Cramér 1928, 34–38] we start with the most important notations. Let us consider n independent identically distributed random variables with zero expectation, each with the distribution function V (defined according to Lévy). It is assumed that, for natural $\nu \leq k$ $(k \geq 2)$, there exists

$$\alpha_\nu := \int_{-\infty}^{\infty} t^\nu dV(t), \quad \beta_\nu := \int_{-\infty}^{\infty} |t|^\nu dV(t).$$

We use the abbreviations

$$\rho_\nu := \frac{\beta_\nu^{1/\nu}}{\beta_2^{1/2}},$$

$$W_1(x) := V(x), \quad W_n(x) := \int_{-\infty}^{\infty} W_{n-1}(x-t)dV(t).$$

Then the distribution function F_n of the normed sum is

$$F_n(x) = W_n(\sigma x) \quad (\sigma := \sqrt{n\alpha_2}).$$

In this case there exist for $\nu \leq k$ the moments

$$\mu_\nu := \int_{-\infty}^{\infty} x^\nu dW_n(x)$$

and the coefficients of the A expansion

$$c_\nu := (-1)^\nu \int_{-\infty}^{\infty} H_\nu(x)dF_n(x), \text{ where } (-1)^\nu H_\nu(x)\mathrm{e}^{-\frac{x^2}{2}} = \frac{\mathrm{d}^\nu}{\mathrm{d}x^\nu}\mathrm{e}^{-\frac{x^2}{2}}. \quad (5.72)$$

Let v, w_n, f_n be the "adjuncts" of V, W_n, F_n, respectively, where, for example,

$$v(t) = \int_{-\infty}^{\infty} \mathrm{e}^{-\mathrm{i}tx} dV(x).$$

The semi-invariants γ_ν of V are defined by the following relation, valid for sufficiently small $|z|$,

$$\log\left(1 + \sum_{\nu=2}^{k} \frac{\alpha_\nu}{\nu!} z^\nu\right) = \sum_{\nu=2}^{k} \frac{\gamma_\nu}{\nu!} z^\nu + Lz^{k+1} + \cdots.$$

The semi-invariants of W_n (replace α_ν by μ_ν) are denoted, according to Cramér, by λ_ν. Φ designating the standard normal distribution, the "symbolic polynomial" $P_\nu(\Phi)$ is defined for $\nu = 1, \ldots, k-2$ by

$$\exp\left(\sum_{\nu=1}^{k-2} \frac{\lambda_{\nu+2}(-\Phi)^{\nu+2}}{\sigma^{\nu+2}(\nu+2)!} z^\nu\right) = 1 + \sum_{\nu=1}^{k-2} P_\nu(\Phi) z^\nu + Nz^{k-1} + \cdots,$$

where after the calculation of the polynomial each power Φ^j has to be replaced by the derivative $\Phi^{(j)}$. On the other hand, the polynomial $P_\nu(it)$, where

$$\exp\left(\sum_{\nu=1}^{k-2}\frac{\lambda_{\nu+2}(-it)^{\nu+2}}{\sigma^{\nu+2}(\nu+2)!}z^\nu\right)=1+\sum_{\nu=1}^{k-2}P_\nu(it)z^\nu+Nz^{k-1}+\cdots, \qquad (5.73)$$

is a "true" polynomial in it. Then we have, and this relation is especially important for Cramér's argumentation:

$$P_\nu(\Phi)=\frac{1}{2\pi}\int_{-\infty}^{\infty}\frac{e^{itx}}{it}P_\nu(it)e^{-\frac{t^2}{2}}\,dt.$$

Similar to Landau symbols, Cramér denoted by Θ any number whose modulus is smaller than a constant dependent only on k, and by Λ any number whose modulus is smaller than a constant dependent only on k and V.

With several lemmata Cramér [1928, 42–50] established estimates for the semi-invariants γ_ν and λ_ν, as well as for $f_n(t)$ in a certain neighborhood of $t=0$. These lemmata are as follows:
Lemma 1: $\beta_1\le\beta_2^{1/2}\le\beta_3^{1/3}\le\cdots\le\beta_k^{1/k}$.[95]
Lemma 2: For $2\le\nu\le k$ we have $\gamma_\nu=\Theta\beta_\nu$.
Lemma 3: For $2\le\nu\le k$ we have $\lambda_\nu=n\gamma_\nu$.[96]
Lemma 4: $P_\nu(\Phi)$ is of the form $\sum_{j=1}^{\nu}H_{\nu j}\,\Phi^{(\nu+2j)}$, and for $1\le\nu\le k-2$ we have the estimate $H_{\nu j}=\Theta\rho_k^{\nu+2j}n^{-\frac{\nu}{2}}$.
Lemma 5 is only important for the A series, and therefore passed over here.
On the basis of (5.73), and by use of Lemmata 1–3 Cramér proved
Lemma 6: For $|t|<\frac{1}{\rho_k}\sqrt[6]{n}$ is

$$e^{\frac{t^2}{2}}f_n(t)=1+\sum_{\nu=1}^{k-3}P_\nu(it)+\Theta n^{-\frac{k-2}{2}}\left(|\rho_k t|^k+|\rho_k t|^{3(k-2)}\right).$$

Lemma 7: For $|t|<\frac{1}{4\rho_k^3}\sqrt{n}$ we have

$$|f_n(t)|<e^{-\frac{t^2}{3}}.$$

In [1923] Cramér had already treated the problem of estimating the difference between $F_n(x)$ and the normal distribution $\Phi(x)$ in terms of characteristic functions by application of a "mollifier" (see Sect. 5.2.8.1). For an estimate of

$$R_n(x)=F_n(x)-\Phi(x)-\sum_{\nu=1}^{k-3}P_\nu(\Phi),$$

---

[95] These inequalities immediately follow from the Lyapunov or the Hölder inequality, which fact, however, Cramér apparently did not observe.

[96] According to the "usual" additivity of semi-invariants.

given by the (not absolutely convergent) representation

$$R_n(x) = \lim_{T\to\infty} \frac{1}{2\pi} \int_{-T}^{T} \frac{e^{itx}}{it} r_n(t)dt,$$

where

$$r_n(t) = \int_{-\infty}^{\infty} e^{-itx} dR_n(x) = f_n(t) - e^{-\frac{t^2}{2}} - \sum_{\nu=1}^{k-3} P_\nu(it)e^{-\frac{t^2}{2}},$$

Cramér [1928, 51–56] had to take a similar path. Now, however, he did not mollify by a convolution with the normal distribution, but by use of so-called "generalized Riemann–Liouville integrals," that means, he considered, for $0 < \omega \leq k-1$:

$$I^{(\omega)}R_n(x) := \frac{1}{\Gamma(\omega)} \int_{-\infty}^{x} (x-t)^{\omega-1} R_n(t)dt.$$

In Lemma 8, Cramér proved that

$$I^{(\omega)}R_n(x) = \frac{1}{2\pi} \int_{-\infty}^{\infty} \frac{e^{itx}}{(it)^{\omega+1}} r_n(t)dt, \tag{5.74}$$

where the integral in (5.74) is absolutely convergent.

Even if the proof of Lemma 8 had been already complicated enough, the proof of the next auxiliary theorem turned out to be particularly cumbersome.
Lemma 9: Let

$$Z := \frac{1}{\pi}\mathrm{Re} \int_{\frac{\sqrt{n}}{4\rho_k^3}}^{\infty} \frac{e^{itx}}{(it)^{\omega+1}} f_n(t)dt.$$

Then we have

$$I^{(\omega)}R_n(x) = Z + \Theta\rho_k^{3(k-2)} n^{-\frac{k-2}{2}}.$$

For the proof Cramér split the range of integration of the integral in (5.74) into four parts, and using Lemmata 1–4 he estimated each single integral obtained by this procedure.

Finally, Cramér [1928, 56–58] began the proof of his main theorem. First he proved
Theorem 1: Let β_k be finite for any $k > 3$ (the case $k = 3$ had already been discussed by Cramér [1923] in the context of his improvement of the Lyapunov bound). Then for $n > 1$ and all real x we have

$$I^{(k-3)}\left[F_n(x) - \Phi(x) - \sum_{\nu=1}^{k-3} P_\nu(\Phi)\right] = \Theta\rho_k^{3(k-2)}(\log n)^2 n^{-\frac{k-2}{2}}.$$

For the proof Cramér showed first that for $\omega = k - 3$ the estimate

$$|Z| < \Theta \rho_k^{3\omega} n^{-\frac{\omega}{2}} \int_{-\infty}^{\infty} \min\left(1, \frac{\rho_k^3}{|x-y|\sqrt{n}}\right) dF_n(y)$$

is valid. Then he split the integral on the right-hand side into five single integrals with domains of integration between $-\infty$ and $x-1$, $x-1$ and $x - \frac{\rho_k^3}{\sqrt{n}}$, $x - \frac{\rho_k^3}{\sqrt{n}}$ and $x + \frac{\rho_k^3}{\sqrt{n}}$, $x + \frac{\rho_k^3}{\sqrt{n}}$ and $x+1$, and, finally, between $x+1$ and ∞.[97] For an estimate of the integrals 2, 3, and 4 Cramér used the already established relation (5.70) (now analogously applied to the standard normal distribution); for the sum of the two tail integrals he easily found the upper bound $\frac{\rho_k^3}{\sqrt{n}}$.

For a deduction of the main theorem from his Theorem 1, Cramér [1928, 59–62] had to use substantially recent work by Friedrich Riesz and Godfrey Hardy on Fourier analysis. In this way, he obtained the main theorem, "Theorem 2" in his own numbering:
If there exists β_k for any $k \geq 3$, and if

$$\sup |v(t)| < 1 \tag{5.75}$$

in each closed interval not containing 0, then we have

$$F_n(x) = \Phi(x) + \sum_{\nu=1}^{k-3} P_\nu(\Phi) + \Lambda n^{-\frac{k-2}{2}}.$$

Parallel to the proof of this statement and under the same assumptions, Cramér proved the corresponding asymptotic for the A series:

$$F_n(x) = \Phi(x) + \sum_{\nu=3}^{k-1} \frac{c_\nu}{\nu!} \Phi^{(\nu)}(x) + \Lambda n^{-\frac{\kappa}{2}}, \quad \text{where } \kappa = \left[\frac{k+2}{3}\right].$$

It is remarkable that for $k > 3$ the condition (5.75) is indispensable for the validity of Theorem 2. This condition is not met if V is a lattice distribution. For a sum of bivalent random variables (W_n can be reduced to a binomial distribution in this case) the existence of asymptotic expansions according to Theorem 2 is impossible if $k > 3$, as Cramér [1928, 66 f.] explained. In this case, F_n shows jumps of the asymptotic magnitude $\frac{1}{\sqrt{n}}$.

Cramér [1928, 69–72] also hinted at a generalization of his theorems to non-identically distributed random variables. A closer specification can be found in his book [1937/70, 83–85]. Roughly speaking, a certain uniformity among the moments

97 Cramér did not discuss the case where $\frac{\rho_k^3}{\sqrt{n}} > 1$, which can only happen for small n. In his book [1937/70, 77] he pointed out that for this case only "trivial" considerations were necessary, and it was therefore not "really interesting."

of the single random variables and among the upper bounds of their characteristic functions in closed intervals not containing zero had to be assumed.

For applications within risk theory this case was particularly important where the single distribution functions are not identical, but nonetheless similar. For this special case Cramér [1930] showed the validity of an analog to Theorem 2 under assumptions close to real practice. In particular, Cramér [1930] discussed the use of asymptotic expansions according to Theorem 2, especially Edgeworth expansions with $k = 4$ and $k = 5$, for calculating the approximate probabilities of insurance gains.

For a set of short time policies, however, it was not possible, as Cramér [1930, 64 f.] explained, to approach the corresponding probabilities in the latter way. Win or loss γ_ν with such insurance contracts happen with probabilities p and $1 - p$, respectively, at the end of a short period and do not vary continuously in dependence on time like life insurances. In the following, the discussion is confined to net gains, that is, to gains which result from the rules of a "fair game." Then we have, if s_ν denotes the single amount which has to be paid by the insurance company in the event of loss,

$$\gamma_\nu = \begin{cases} +s_\nu q \text{ with probability } p = 1-q \\ -s_\nu p \text{ with probability } q. \end{cases}$$

The quantities γ_ν therefore are lattice distributed random variables, to which Cramér, in accordance with the current methods available around 1930, could only apply the estimation (5.70). If S_r denotes the arithmetic mean of the sum of all s_ν^r, then on account of (5.70) (applied analogously to the standard normal distribution Φ and to the distribution function F_n of the normed sum),

$$|F_n(x) - \Phi(x)| < 3(1-2pq)\frac{\log n}{\sqrt{npq}}\frac{S_3}{S_2^{3/2}}.$$

Thus, in case of the realistic situation $n = 20000$ and $q = \frac{1}{21}$, Cramér was only able to give the estimate

$$|F_n(x) - \Phi(x)| < 0.898\frac{S_3}{S_2^{3/2}},$$

which was useless for an assessment of the approximation error because of the general relation $\frac{S_3}{S_2^{3/2}} \geq 1$. So, around 1930, no satisfactory statement could be made as to whether the distribution of the complete gain of an insurance with short time policies could always be approximated sufficiently exactly by the normal distribution. In his discussion of contracts with continuously varying "gains" Cramér [1930, 60–63] found in some cases relatively large deviations between the "real" distributions represented by Edgeworth expansions and the corresponding normal distributions.

As Cramér [1935, 5 f.] pointed out, from Theorem 2 the inequality

$$|F_n(x) - \Phi(x)| \leq \frac{A}{\sqrt{n}}$$

(A an appropriate constant) follows if there just exist moments of a sufficiently high order and if the condition (5.75) (or any stronger condition) is valid.[98] That an estimate of this kind could also be reached without the condition (5.75) was only shown by the works of Andrew C. Berry [1941] and Carl Gustav Esseen [1942].[99] If F_n denotes the distribution function of the normed sum of the independent random variables $X_1, \ldots, X_n$, each with zero expectation and finite absolute moments of third order, then according to the results of both mathematicians the following inequality holds:

$$|F_n(x) - \Phi(x)| \leq \frac{C}{\sqrt{n}} \frac{\sum \mathrm{E}|X_k|^3}{(\sum \mathrm{E}X^2)^{3/2}}. \tag{5.76}$$

Thereafter, the constant C could be reduced to values less than 1.[100] According to the very sharp value $C = 0.7915$, found by I.S. Shiganov [1982], in the case of short time policies above we would obtain

$$|F_n(x) - \Phi(x)| < 0.024 \frac{S_3}{S_2^{3/2}}.$$

Esseen and Berry used different procedures of mollification, which, compared with Cramér's methods, were easier to apply. Among his numerous refinements of Cramér's results, Esseen [1945] also improved Cramér's Theorem 2, insofar as, under the assumption of absolute moments up to the order $k \geq 3$ inclusively, one more series term could be taken into account.[101] According to Esseen we have, under Cramér's assumptions,

$$F_n(x) = \Phi(x) + \sum_{\nu=1}^{k-2} P_\nu(\Phi) + \mathrm{o}\left(n^{-\frac{k-2}{2}}\right).$$

Cramér's work represents the first completion of an exceptionally successful development of the CLT within a short period. During the twenties the classic CLT in

[98] In the particular case of the binomial distribution, however, de la Vallée Poussin [1906] had already proven an equivalent inequality.

[99] In most cases we find the reference to [Esseen 1945], which article has the same content as Esseen's doctoral thesis, completed in 1944 at Uppsala University (see [Cramér 1976, 530]), and which also comprises the results from [Esseen 1942] and [Esseen 1943] (the latter work dealt with the error of approximation in the case of lattice distributions). Esseen [1945, 6] declared that he had already completed his proof of (5.76) by the fall of 1940, and that, due to World War II, he had only had the opportunity to learn about Berry's article [1941] by a summary in the *Mathematical Reviews*.

[100] Note that these estimates hold independently of the particular definition of distribution function (continuous on the right; left; Lévy's definition).

[101] Esseen used the distribution with density $H(x) = \frac{3}{8\pi}\left(\sin\frac{x}{4}\right)^4 \left(\frac{x}{4}\right)^{-4}$ as a mollifier for $R_n(x)$. Berry had mollified by the distribution with density $v_T(x) = \frac{1}{\pi}\frac{1-\cos Tx}{Tx^2}$ and then considered the limit $T \to \infty$. Esseen's major results and his methods are discussed in [Gnedenko & Kolmogorov 1949/68, 196–219]. From [Feller 1971, 536–548] we can learn the way (probably due to Hsu [1945]) how Esseen's improvement of Cramér's Theorem 2 can be deduced by Berry's method, which, on the whole, seems to be more advantageous.

its integral version was proven under very weak conditions by a considerable variety of methods (Lévy, Lindeberg), and was generalized toward weakly dependent random variables (Bernshtein). Chebyshev's demand for appropriate estimates of the error of approximation by the normal distribution was fulfilled in a satisfactory, though not optimal, manner (Cramér). In 1919 von Mises had already treated quite general versions of local CLTs for densities and lattice distributions. Lévy, in his book [1925b], presented several ways for a generalized view of the CLT including nonclassic norming as it was also introduced by Bernshtein. Toward the end of the twenties until approximately the midthirties, however, the increasingly numerous group of probability specialists chiefly dedicated themselves to different problems, in particular to stochastic processes.

Chapter 6
Lévy and Feller on Normal Limit Distributions around 1935

Lévy [1924, 17] had already stated that Lindeberg's condition for the CLT accorded particularly well with "la nature des choses." Was he trying to say that, in a certain way, this condition was also necessary for convergence to the normal distribution? In fact, more than 10 years would pass before Lévy and Feller almost simultaneously proved that certain conditions are both sufficient and necessary for the convergence of distributions of suitably normed sums of independent random variables to the normal distribution. These examinations foresaw general normings that no longer assumed the existence of the variance, and it emerged within this framework that, in the classical case, the Lindeberg condition is necessary for the CLT if the influence of the individual random variables on their sum can be asymptotically neglected in a particular sense. The possibility of nonclassical norming, which Bernshtein and Lévy had already addressed in the 1920s, out of a desire to exhaust all analytical possibilities, became all the more interesting the further mathematical probability theory departed from its original areas of application.

6.1 The Prehistory

The considerable gap in time between Lévy's conjecture in 1924 and its proof can be explained by two factors. By 1925, a degree of saturation had been reached in the area of limit distributions of sums of independent random variables. In return, the fields of stochastic processes and strong laws of large numbers (e.g., the law of iterated logarithm, theorems concerning the probability of convergence of series of independent random variables) underwent a feverish period of development in the years after 1925.[1] As Le Cam [1986, 83–85] points out, this development had also always several points in common with the CLT, and it thus prepared the way for a wider discussion of the theorem in the 1930s. Around 1935, mathematicians

[1] Lévy contributed to this development with a certain temporal delay only, see [Bru & Salah 2009, 11–17].

H. Fischer, *A History of the Central Limit Theorem*, Sources and Studies in the History of Mathematics and Physical Sciences, DOI 10.1007/978-0-387-87857-7_6,

could for a first time take stock in the various branches of modern probability theory that had evolved up to this point and—once Kolmogorov [1933a] had clarified the most important principles—could dedicate their attention to a more precise examination of specific problems within the work they had already accomplished. At this point—in contrast to the situation in the early 1920s—they no longer needed to find ways to legitimate any perceived irrelevance regarding the extra-mathematical application of the results they achieved. Evidence that a certain consolidation of the purely mathematical probability theory was taking place around 1935 is also provided in the form of the first monographs, such as [Cramér 1937], [Lévy 1937a], or [Fréchet 1936/38], which appeared in the second half of the decade and retained their significance for many years in further editions printed after the Second World War. Moreover, it is possible, at least where Lévy is concerned, to reconstruct the purely intrinsic problems that accounted for the long delay leading up to his results on necessary conditions for the CLT. The difficulties in which Lévy found himself may have similarly hampered other mathematicians who were also interested in the solution to these problems.

6.1.1 Lévy and the Problem of Un-negligible Summands

In the field of strong laws of large numbers, a significant part was played by necessary and sufficient conditions for the almost sure convergence of a series of independent random variables. Particularly important in this context was Khinchin's concept of "equivalent" sequences of random variables, which in turn was closely linked to Markov's idea of truncated random variables, which had arisen in conjunction with his activities with the CLT.[2] Lévy [1931] took up the relevant work by Khinchin and Kolmogorov [1925] and Kolmogorov [1928][3] again and produced a modified representation and derivation of the convergence criterion. Lévy's own version was as follows:

> Let (X_k) be a sequence of independent random variables. For the existence of a sequence of real numbers a_k such that $\sum_{k=1}^{\infty}(X_k - a_k)$ almost surely converges, it is necessary and sufficient that there exists a sequence (Y_k) being equivalent to the sequence (X_k), for which $\sum_{k=1}^{\infty} \mathrm{Var} Y_k$ converges [Lévy 1931, 133].

Lévy [1931, 139–142] applied the ideas associated with almost sure convergence—which with regard to Markov's truncation trick had emerged from work on the CLT—to the CLT again: The mutually analogous references, on the one hand between the divergence of the sums of all variances and the almost sure divergence of the series of the random variables, on the other hand between the divergence of the sums of all variances and the validity of the assertion of the CLT, apparently led Lévy to consider equivalent random variables in conjunction with the CLT as well.

---

[2] Two sequences (X_k) and (Y_k), each consisting of mutually independent random variables, are designated "equivalent" if the series $\sum_{k=1}^{\infty} P(X_k \neq Y_k)$ converges.

[3] In this article, Kolmogorov had also presented his famous inequality.

In this way, he arrived at the following theorem ((X_k) again represents a sequence of independent random variables):
If there exists a sequence (Y_k) of bounded random variables being equivalent to (X_k) such that $\max_{1\le k\le n}|Y_k| < d_n$ and $d_n^2/\sum_{k=1}^n \mathrm{Var}Y_k \to 0$, then one can find constants A_n and $B_n > 0$ such that the distribution of $\sum_{k=1}^n X_k/B_n - A_n$ tends to the standard normal distribution.[4]
This result was more general than Lindeberg's "Theorem III" (see Sect. 5.2.4), as Lévy [1931, 140] underlined. With these considerations he again took up the idea of general norming that he had previously developed in his book [1925b].[5] Lévy [1931, 141] speculated that it was possible to find even necessary conditions for the convergence to the normal distribution by further refining the basic ideas which had led to the just-stated theorem. In this context he considered normed sums of random variables, each of them additively composed of one part being "very small in relation to the total sum," and one possibly sizable but normally distributed part.

In order to be able to compare the size of an individual random variable to the overall sum, Lévy utilized his newly created term of "dispersion," which along with its "inversion," called "concentration," proved to be especially useful in discussing the convergence of series of random variables. Lévy defined the concentration $f_X(l)$ of the random variable X assigned to the interval length $l > 0$ as follows:[6]

$$f_X(l) := \sup_{-\infty<a<\infty} P(a < X < a + l).$$

"Dispersion of a random variable X" Lévy called a function $\varphi_X : [0; 1[\to \mathbb{R}_0^+$ with

$$\varphi_X(\gamma) := \inf\left\{x \in \mathbb{R}_0^+ \,|\, f_X(x) \ge \gamma\right\}.$$

Roughly speaking, the dispersion is the minimum interval length related to a particular probability, and the concentration is the maximum probability related to a particular interval length.

In his discussion of the above-described problem of random variables composed of one relatively small and one normally distributed part, Lévy [1931, 141] considered sequences of random variables (X_k) and (η_k), where all variables within each sequence were assumed to be independent. He presupposed that

$$X_k = a_k + b_k\xi_k + \eta_k + \eta'_k,$$

where a_k and b_k were constants, ξ_k obeyed a Gaussian law, and η'_k met the condition that $\sum_{k=1}^\infty P(\eta'_k \ne 0)$ was convergent. L_n denoting the dispersion of $\sum_{k=1}^n X_k$

---

[4] Without loss of generality, the random variables Y_k can be assumed to be defined by $Y_k = X_k$ if $|X_k| < d_k$ and $Y_k = 0$ else, under the restraint $\sum_{k=1}^n P(|X_k| > d_k) \to 0$.

[5] Bernshtein's approach (see Sect. 5.2.7), which ultimately also resulted in a nonclassical norming, was apparently unknown to Lévy at this point. Not until [Lévy 1935a, 201] does one find any mention of [Bernshtein 1926], if only with regard to his generalization of the CLT to nonindependent random variables.

[6] Concentration and dispersion first appear in [Lévy 1931, 128 f.]. A precise definition of dispersion can be found in [Lévy 1935b, 351]. Lévy always wrote "max" rather than "sup" in his definitions.

assigned to $\gamma = \frac{1}{2}$, the variables η_k were assumed to be bounded such that $\max_{1 \le k \le n} |\eta_k|/L_n$ tends to 0. Under the additional assumption $L_n \to \infty$, Lévy claimed that the distribution of the suitably normed sum of the X_k would tend to a Gaussian law. He stated that one could prove this "without any difficulties." Lévy (same place) expressed his opinion that a necessary condition for the convergence to the normal distribution might be found by similar ideas.

It emerges from Lévy's text that he had fairly clear ideas about how he could reach such necessary conditions if one conveniently assumed the smallness of all of the random variables X_k (not only the η_k and the η'_k) that contribute to the sum. In principle, his just-described assumptions for the part $\eta + \eta'$ were equivalent to the later stated condition

$$P\left(\max_{1 \le k \le n} |X_k| > \varepsilon L_n\right) \to 0 \quad \forall \varepsilon > 0,$$

which turned out to be even sufficient for convergence to the normal distribution if all random variables X_k could be considered small in the sense

$$\max_{1 \le k \le n} P(|X_k| > \varepsilon L_n) \to 0 \quad \forall \varepsilon > 0.$$

Lévy did not wish to adopt this restrictive position of generic smallness in 1931, however. The reason why Lévy was so interested in random variables with a possible large influence on the total sum, aside from the purely mathematical importance of this matter, could also be linked to the considerable interest he maintained still around 1930 in the hypothesis of elementary errors. This lasting interest is conspicuous in the papers he drafted while embroiled in a minor dispute about error theory with Maurice Fréchet toward the end of the 1920s. For a time (the years 1925 to 1928), Lévy's letters to Fréchet dealt largely with this issue [Barbut, Locker, & Mazliak 2004, 46–50, 122–149]. Not only had Fréchet [1928] faulted the assumption of the additivity of elementary errors, he had also picked up an example by Hausdorff [1901, 152] (see Sect. 3.4.2.1). In Hausdorff's example, the elementary errors have finite variances σ_i^2, such that $\sum_{i=1}^{\infty} \sigma_i^2 < \infty$, and the distribution of the elementary error sum converges to an error law that differs from normal distribution. Fréchet had concluded that if one were to reason that the normal distribution was the error law of the total error, the assumptions about the distributions of elementary errors could certainly not be as general as it initially had appeared. Lévy [1929, 61] pointed out that Fréchet's example was irrelevant to a discussion of the hypothesis of elementary errors insofar as it did not satisfy the usual basic assumption that all elementary errors turned out to be uniformly small.[7]

Is this basic assumption necessary, though? Could it perhaps be possible that (a few) independent elementary errors of significant size could add up to a normally distributed or nearly normally distributed sum though they themselves might deviate considerably from normal distribution? Or do such elementary errors always necessarily have to be at least approximately normally distributed, in which case they

[7] For more details see [Purkert 2006b, 579–583].

can be considered to be sums of a large number of small "sub-elementary errors"? Lévy [1970, 111 f.] reports how acutely these questions affected him: Without the answers, he did not believe it possible to solve the problem of necessary conditions for the convergence of distributions of sums of independent random variables to a normal distribution. This circumstance led him, while preparing the chief paper [1935b] he wrote on this complex of problems, to open the chapter entitled "Étude des sommes de variables aléatoires non enchaînées dépendant de lois variant d'une manière quelconque" with the following "lemme hypothétique":

If the sum $z = x + y$ of two independent random variables is of the Gaussian type, then this likewise applies to each of the terms [Lévy 1935b, 381].

Cramér [1936] eventually succeeded in proving this lemma; a simplified exposition is found in Lévy's book [1937a, 97 f.].

6.1.2 Feller and the Case Which "does not belong to probability theory at all"

Willy Feller (1906–1971)[8] did not come to probability theory until 1934. Born in Zagreb, he had earned his doctorate in 1926 after studying partial differential equations under Richard Courant in Göttingen. He then taught applied mathematics at the University of Kiel from 1928 to 1933. After the National Socialists seized power, Feller left Germany and, in 1934, joined Cramér's Stockholm group. Based on his mathematical training, Feller was extremely well equipped to deal with characteristic functions, and so—apparently motivated by Cramér and after only a fairly cursory reading of Lévy's *Calcul des Probabilités* [1925b]—he was able to wade into the realm of limit distributions of sums of independent random variables in relatively short order. As is the case with many other pioneers of modern probability theory, the influence of analytical methods of mathematical physics is also evident in Feller's work.

In his study of sufficient and necessary conditions for the CLT, Feller [1935, 523], unlike Lévy, excluded all of those cases in which the influence of the "individual components" on the total sum does not asymptotically disappear as the number of random variables increases. He wrote that these cases did "not belong to the problem area of the limit theorem" although he did not provide any further explanation (Feller's concept of the influence or negligibility of the individual summands was equivalent to Lévy's). Feller wrote apodictically [1935, 531] about cases such as the one according to Hausdorff, stating that they did "not belong to probability theory at all." Such a pronouncement may have seemed strange at first, but it does make sense when one considers that the character of his discussion of un-negligible random variables [1935, 531 f.] demonstrates that he, too, had had a conjecture that corresponded to the "lemme hypothétique." According to Feller's comments, the

[8] Biographical information for Feller can be found, e.g., in [Doob 1990].

convergence of the distribution of a suitably normed sum to the normal distribution in the case of un-negligible summands can occur only if the summands have very specific distributions. It is thus not primarily caused by the stochastic behavior of generally distributed random variables. Feller also showed that for a sequence (X_k) of independent random variables which are not all negligible, the convergence of the distribution of $\frac{\sum X_k}{B_n}$ to the standard normal distribution, under the additional condition $B_n \to \infty$, can occur only if the un-negligible summands (they form a subsequence (X_k') of (X_k)) have distributions V_k' such that $V_k'(c_k' x) \to \Phi(x)$ with appropriate c_k'.

It is worth noting that Alan Turing, working totally independently of Lévy and Feller, drafted an unpublished paper about the CLT in 1934 as part of his discussion of the influence of un-negligible summands. In it, he also came to conclusions that are particularly reminiscent of Feller's, albeit only in the case of random variables with existing mean and variance. As Zabell [1995] reported, Turing referred to his conditions for negligibility, which were equivalent to those of Feller and Lévy within the framework of "classical" conditions, as "quasi-necessary conditions." He wrote: "They are not actually necessary, but if they are not fulfilled, U_n [the distribution of $\frac{\sum X_k - \mathrm{E}X_k}{\sqrt{\sum \mathrm{Var} X_k}}$] can only tend to $\Phi(x)$ by a kind of accident" (quoted in Zabell [1995, 487]). Turing's work can be considered evidence of a latent interest, starting no later than the early 1930s, in a "conclusive" formulation of the CLT with regard to sufficient and necessary conditions, an interest that had likely been aroused first and foremost by Lindeberg's papers in 1922 and Lévy's book in 1925. At the same time, also in Turing's case the specific difficulties become apparent that were associated with the general solution to this problem, namely, when un-negligible summands were included.

6.2 Lévy's and Feller's Results and Methods

Lévy's and Feller's 1935 contributions to the CLT differed to a large extent regarding methods and presentation. Whereas Lévy employed his newly devised, rather idiosyncratic, analytical tools of concentration and dispersion, Feller based his considerations on the "well-known" device of characteristic functions.

6.2.1 Lévy's Main Theorems

As we have already seen (Sect. 6.1.1), Lévy quantified the "smallness" of the summands X_k in terms of the dispersion of the total sum $S_n = \sum_{k=1}^n X_k$. A single X_k was called "individually negligible" ("individuellement négligeable") by Lévy [1935b, 351] "if it can be neglected except for cases of arbitrarily small probability in relation to the dispersion of S_n." He gave a little more precise definition in [1937a, 104]: Let $0 < \gamma < 1$ be an arbitrary, however fixed, probability, and L_n the dispersion of S_n assigned to this probability. X_k $(1 \le k \le n)$ is called "individually negligible" if for an "arbitrarily small" positive ε the probability $P(|X_k| > \varepsilon L_n)$ is "small" as well. Lévy's expression "there are individually negligible terms only" or "all terms are individually negligible" means that for all $\varepsilon > 0$ the maximum probability $\max_{1\le k\le n} P(|X_k| > \varepsilon L_n)$ tends to 0 as n tends to ∞.[9]

In order to also cover such cases where the main part of the probability mass of a random variable X_k lies so far away from zero that "negligibility" does not hold even though the spread of this variable is very small, [Lévy 1937a, 104] (in his book only) introduced the general assumption of zero medians (which can be achieved by addition of an appropriate constant in each case) for all random variables under consideration. This means that $P(X_k \ge 0) \ge \frac{1}{2}$ and $P(X_k \le 0) \ge \frac{1}{2}$.[10] Lévy did not give any further comments on this assumption. In fact, in his proofs this additional condition is not needed, as it appears. However, if the summands $X_k + a_k$ with appropriate a_k are (uniformly) negligible with respect to the dispersion of S_n, then the variables X_k are themselves (uniformly) negligible under the condition of zero medians (see Sect. 6.3.1, footnote 25). It may be that Lévy was inspired by Feller's very careful discussions on possible problems with shifting constants (Sect. 6.2.5) to include the requirement of zero medians in his book.

The idiosyncratic term "la loi des grands nombres s'applique" ("the law of large numbers applies") Lévy [1935b, 348] introduced for the fact that

$$\lim_{n\to\infty} P\left(\max_{1\le k\le n} |X_k| > \varepsilon L_n\right) = 0$$

for all $\varepsilon > 0$. Lévy always tacitly assumed $L_n \neq 0$ from a certain number n on. This basic assumption, however, is almost evident, because it is necessary for S_n being asymptotically normally distributed.[11]

Those cases in which random variables X_k are not negligible, and have to be (approximately) normally distributed due to the "lemme hypothétique" if $\frac{S_n - An}{B_n}$ ((A_n) and $(B_n > 0)$ being suitable number sequences) is normally distributed for $n \to \infty$, Lévy [1935b, 381–385] quite comprehensively discussed. For the time being, however, he was only able to give rigorous proofs in cases of negligible random variables. His chief result (Theorem VI, [1935b, 386]) in this context was:

[9] According to a terminology which has been introduced by Loève (e.g., [1950, 328]), and is quite common now, this means that the random variables X_k/L_n are uniformly asymptotically negligible, or, briefly, that they obey the "UAN condition."

[10] One has to observe that in general the median m of a random variable X is not uniquely determined by the condition $P(X \ge m) \ge \frac{1}{2}$ and $P(X \le m) \ge \frac{1}{2}$. In practically all cases this ambiguity does not cause any problems. Nevertheless, if uniqueness is required, one can choose the minimum value of all medians, for example.

[11] As it will be shown below (Sect. 6.2.3.1, Lemma 4), from the convergence of the distributions of $\frac{S_n - A_n}{B_n}$ to the standard normal distribution (A_n and $B_n > 0$ being suitable constants) $\frac{L_n}{B_n} \to \varphi_N(\gamma)$ follows, $\varphi_N(\gamma)$ denoting the dispersion of the standard normal distribution, assigned to the probability level $\gamma \in]0; 1[$. The assumption of $L_{n_k} = 0$ for infinitely many n_k is inconsistent with this assertion.

If there are individually negligible terms only, constant convergence [convergence constant] to Gauss's law can be obtained only in those cases where the law of large numbers applies.

Lévy emphasized "constant convergence," which means convergence of the distribution of $\frac{S_n - A_n}{B_n}$ ((A_n) and ($B_n > 0$) being suitable number sequences) to the standard normal distribution, in contrast to "intermittent convergence" ["convergence intermittente"], where the only demand is that the distributions of a partial sequence of $\left(\frac{S_n - A_n}{B_n}\right)$ tend to Gauss's law.

In his book [1937a, 117–119] Lévy was able to state and justify, on the basis of the "lemme hypothétique," which had been proven in the meantime, the general solution of the problem of approximate normal distribution for sums of random variables for those cases also in which nonnegligible random variables exist:

For that the law on which S_n depends is of a generalized type being different from the Gaussian to a small amount only, a necessary and sufficient condition, after having reduced the median of each term to zero, is:

1° *Each not individually negligible term is of a generalized type which is only little different from that of Gauss;*

2° *The largest of the individually negligible terms is negligible for itself.*

Condition 2° means that for the individually negligible random variables $X_{k_1}, \ldots, X_{k_s}$ (with $1 \leq k_1 < \cdots < k_s \leq n$) the probability $P(\max_{1\leq m\leq s} |X_k| > \varepsilon L_n)$ has to be small for small ε. In the text preceding the formulation of this main theorem, Lévy briefly explained how his intuitive expressions could be translated into a more convenient "ε-δ language."

Particular attention Lévy [1935b, 359–381] paid to sums of independent random variables X_k with an identical distribution function F. In this case he both discussed convergence under conditions of classical norming and the general case of random variables with an infinite variance. For the classical case his "Theorem I" [1935b, 359] was as follows:

The necessary and sufficient condition for that s_n [$= \frac{1}{\sqrt{n}} \sum_{k=1}^{n} X_k$] depends on a law tending for infinite n to the Gaussian is $\mathcal{E}\{x^2\} = 1$.

One has to observe that Lévy at this place assumed $\mathrm{E}X_k = 0$, and with "Gaussian law" meant the standard normal distribution. For the general case of identically distributed independent random variables Lévy put the following theorems:

Theorem II. — *The necessary and sufficient condition for applicability of the law of large numbers is that expression* (10) *tends to 0 for infinite X* [Lévy 1935b, 366].

Expression "(10)" was given by

$$\frac{X^2 \int_{|t|>X} dF(t)}{\int_{|t|\leq X} t^2 dF(t)}.$$

"Theorem V" [Lévy 1935b, 370 f.] complemented Theorem II :

Convergence to the Gaussian type is only possible if the law of large numbers can be applied.

Theorems II and V provide a complete solution of characterizing the range of attraction of the Gaussian law, a problem which had been already discussed in [Lévy 1925b] (see Sect. 5.2.6.7).

6.2.2 *Lévy's "Intuitive" Methods*

Lévy's article [1935b] has to be seen in the context of the intentions of its author to master all instances concerning Gaussian limit distributions of normed sums by a uniform method. Characteristic functions, which had been championed by Lévy during the 1920s, were primarily tailored to sums of independent random variables. Lévy [1935a] therefore used Lindeberg's method, which he admired, for proving central limit theorems for sums of weakly dependent random variables (see Sect. 7.1.3). Yet, even in his contributions to sums of independent random variables during the 1930s Lévy scarcely made use of characteristic functions any more. Growingly important tools became the already introduced stochastic devices "dispersion" and "concentration." The following properties were of particular use for Lévy's discussion of distributions of sums of independent random variables: If X and Y are independent, then we have

$$\varphi_{X+Y}(\gamma) \geq \max(\varphi_X(\gamma), \varphi_Y(\gamma)),$$

and, inversely,

$$f_{X+Y}(l) \leq \min(f_X(l), f_Y(l)).$$

With concentration and dispersion, Lévy created especially "intuitive" devices, which, in contrast to characteristic functions, focused on "probability" as fundamental notion. Stressing the "essence," Lévy cultivated a mode of exposition which was intended to appeal to the reader's "intuition" in the 1930s. In the preface of his book [1937a], he comprehensively demonstrated that "intuition" and "rigor" were by no means in conflict. At the same time, the adjective "intuitive" became an important attribute of quality of probabilistic research to Lévy. In the first period of his stochastic work, he had stressed "common sense" and use within error theory as external quality criterions, and nonetheless had been criticized by Borel, because he considered the analytic effort of Lévy's work inadequate. Now, Lévy referred to the intuition of his notions, and he used a style of writing which, according to his own opinion, suited perfectly for representing the specific peculiarities of probability. This was strongly embraced by Borel. Lévy's book [1937a] was included as the first volume into a book series edited by Borel, and Lévy received several awards on Borel's recommendation [Lévy 1970, 119].

Stressing external criteria for assessing the quality of mathematical work, Lévy remained a representative of "counter-modernity" (in Mehrten's sense), even after having changed style and methods. Already from his early work on, his explanations were arranged in a strongly verbose manner, and gave sketchy ideas rather than complete proofs. During the twenties, he at least used the common language of analysis. Lévy's new "intuitive" style of the thirties, however, was seen by the average mathematician as rather obscure and vague [Doob 1986]. Retrospectively, Lévy [1970, 119] called his book [1937a], where he had given a survey of his contributions to sums of random variables, a considerable success. This self-assessment is, however, contradicted by the apparently little influence of Lévy's new style on later published standard monographs, such as [Gnedenko & Kolmogorov 1949], [Doob 1953], or [Loève 1955], which retained analytical orientation, although considering recent results of measure theory.

6.2.3 Lévy's Proofs

6.2.3.1 Lévy's Unproven Lemmata on Properties of Dispersion

Lévy often used assertions in his discussion of necessary and sufficient conditions for the CLT which remained unproven, as it was characteristic of his "telegraphic" style. As opposed to characteristic functions, there did not exist an elaborated theory of concentration and dispersion, and, naturally, Lévy's main goal was to proceed to the newest results of the CLT (and associated limit theorems) rather than giving a well-organized theory of auxiliary tools. As Lévy's dispersion (in contrast to concentration) is scarcely subject of modern monographs on probability theory, it may be of further help to the reader listing some auxiliary theorems (tacitly used by Lévy) concerning this notion, and referring to them in the discussion of Lévy's work below. In the following, $\varphi_Z(\gamma)$ denotes the dispersion of the random variable Z with respect to the probability level $\gamma \in]0;1[$.

Lemma 1. Let l be a positive real number and let $\gamma \in]0;1[$. Then for any random variable X:

$$\gamma = f_X(l) \Rightarrow l = \varphi_X(\gamma), \quad l = \varphi_X(\gamma) \Rightarrow f_X(l) \leq \gamma,$$

$$f_X(l) > \gamma \Rightarrow \varphi_X(\gamma) < l, \quad \varphi_X(\gamma) > l \Rightarrow f_X(l) < \gamma,$$

$$f_X(l) < \gamma \Rightarrow \varphi_X(\gamma) \geq l, \quad \varphi_X(\gamma) < l \Rightarrow f_X(l) > \gamma.$$

These relations between concentration and dispersion are essential for the proofs of the following properties.

Lemma 2. If Z has a variance, then

$$\varphi_Z(\gamma) \leq \frac{2\sqrt{\mathrm{Var}Z}}{\sqrt{1-\gamma}}. \tag{6.1}$$

This property is a consequence of the Bienaymé–Chebyshev inequality.

Lemma 3. If X and X' are arbitrary random variables (on a common probability space), if $P(X \neq X') \leq \eta$, and if there exists $0 < \gamma < 1$ such that $\gamma + \eta < 1$, then

$$\varphi_X(\gamma) \leq \varphi_{X'}(\gamma + \eta). \tag{6.2}$$

Moreover, if $\mathrm{Var}X'$ exists, then

$$\varphi_X(\gamma) \leq \frac{2\sqrt{\mathrm{Var}X'}}{\sqrt{1-\gamma-\eta}}. \tag{6.3}$$

(6.2) is a consequence of (I designating a real interval)

$$P(X' \in I) = P(X' \in I \wedge X \in I) + P(X' \in I \wedge X \notin I) \leq P(X \in I) + P(X' \neq X)$$

and the definition of dispersion. (6.3) is an immediate corollary of (6.1) and (6.2).

Lemma 4. Let (Z_n) be a sequence of random variables with distributions F_n, and Z be a random variable with a continuous distribution function F, such that $F_n(x) \to F(x)$ for all real x. If $\varphi_Z(\gamma)$ is continuous in a certain $\gamma \in]0;1[$, then

$$\varphi_{Z_n}(\gamma) \to \varphi_Z(\gamma) \quad (n \to \infty). \tag{6.4}$$

This lemma follows from the uniform convergence of F_n to F.[12]

Lemma 5. Let (X_k) be a sequence of independent random variables, and $S_n = \sum_{k=1}^{n} X_k$ such that, with appropriate real numbers $a_n, b_n > 0$ the distribution of $(S_n - a_n)/b_n$ tends to the normal distribution. If L_n denotes the dispersion of S_n, assigned to a certain probability $\gamma \in]0;1[$, then

$$L_n \sim b_n \varphi_N(\gamma) \quad (n \to \infty), \tag{6.5}$$

where N is a random variable obeying the standard normal distribution.

(6.5) is a direct consequence of (6.4).

6.2.3.2 The "Classical Case"

A relatively elementary application of concentration and dispersion, which, however, was characteristic of more complicated cases, Lévy [1935b, 359–361] discussed in his proof of the above-quoted "classical"
Theorem II: Let (X_k) be a sequence of independent, identically distributed random variables. The distribution of $s_n := \frac{\sum_{k=1}^{n} X_k}{\sqrt{n}}$ for $n \to \infty$ tends to the standard normal distribution Φ if and only if $\mathrm{E}X_1^2 = 1$ and $\mathrm{E}X_1 = 0$.

---

[12] In the case of Z_n being sums of independent random variables, more general assertions might be established by virtue of corresponding limit theorems on concentrations [Hengartner & Theodorescu 1973, 84–87]. At this place, however, the goal is to reconstruct Lévy's ideas by considerations as simple as possible.

Taking into account well-known properties of the CLT, Lévy had to show only that, under the given assumptions,

$$P\left(\frac{\sum_{k=1}^{n} X_k}{\sqrt{n}} \leq x\right) \to \Phi(x) \Rightarrow \mathrm{E}X_1^2 < \infty.$$

For arbitrarily large positive X he considered sequences of random variables (X_k') and (X_k''), where $X_k = X_k' + X_k''$ and

$$X_k' := \begin{cases} X_k & \text{if } |X_k| \leq X \\ 0 & \text{else.} \end{cases}$$

Furthermore, Lévy introduced the denotations $\varepsilon := P(|X_1| > X)$, $S_n' := \sum X_k'$, $S_n'' := \sum X_k''$, $m := \sqrt{\mathrm{Var}X_1'}$. He [1935b, 360] made plausible that S_n' for sufficiently large n could be represented with an arbitrarily small error by a sum of $(1-\varepsilon)n$ (or rather the integer part of this number) "nonzero" terms, each distributed in the same way as X_1' (and, accordingly, S_n'' by a sum of εn "nonzero" terms, each distributed in the same way as X_1''). The "possible variations" of the number of "nonzero" terms of S_n' had a standard deviation of $\sqrt{\varepsilon(1-\varepsilon)n}$ and could therefore be neglected in relation to n, according to Lévy. Because each "nonzero" term of S_n' could be determined by a probability law defined in $[-X; X]$, and each "nonzero" term of S_n'' by a law defined in the complement of this interval, for $n \to \infty$ the sums $s_n' := \frac{S_n'}{m\sqrt{(1-\varepsilon)n}}$, and $s_n'' := \frac{S_n''}{m\sqrt{(1-\varepsilon)n}}$ (the factor $\sqrt{1-\varepsilon}$ was needless, in principle) could be interpreted as independent random variables, as Lévy argued.[13]

Lévy's further arguments [1935b, 359–361] on the basis of the just-explained "almost-independence" of s_n' and s_n'' can be described as follows: Since the concentration of a sum of independent random variables is less than or equal to the concentration of each single random variable, it follows for sufficiently large n:

$$P(|s_n| < hm\sqrt{1-\varepsilon}) \leq f_{\frac{s_n}{m\sqrt{1-\varepsilon}}}(2h) = f_{s_n'+s_n''}(2h) \leq f_{s_n'}(2h) + \varepsilon',$$

where $|\varepsilon'|$ can be considered arbitrarily small for sufficiently large X and n. As a consequence of the CLT, for positive h the relation

$$\lim_{n\to\infty} f_{s_n'}(2h) = 2\Phi(h\sqrt{1-\varepsilon}) - 1$$

is true. According to the presupposition, the probability $P(|s_n| < hm\sqrt{1-\varepsilon})$ tends to $2\Phi(hm\sqrt{1-\varepsilon}) - 1$. Therefore

$$2\Phi(hm\sqrt{1-\varepsilon}) - 1 \leq 2\Phi(h\sqrt{1-\varepsilon}) - 1 + \varepsilon'',$$

where $\varepsilon'' > 0$ can be considered arbitrarily small as dependent on X. Because $\Phi(x)$ is strictly monotonic increasing for positive x, we finally obtain

[13] For more details regarding this idea of asymptotic independence, see Sect. 6.2.3.4.

$$\mathrm{Var}X_1 = \lim_{X\to\infty} m^2 \le 1 < \infty,$$

from which the assertion follows.

6.2.3.3 The "loi des grands nombres" as a Sufficient Condition for the Central Limit Theorem

Referring to ideas which he had already expounded in his article [1931], Lévy [1935b, 348] asserted that one knew that the applicability of the "loi des grands nombres" implied (after a suitable norming) the convergence of the distribution of the sum S_n to the Gaussian law. By use of some hints in [Lévy 1931, 139 f.] this assertion could actually be proven. An explicit proof can be found in Lévy's book [1937a, 105 f.] only.

Let X_k be mutually independent random variables, and let L_n be the dispersion of $S_n = \sum_{k=1}^n X_k$ assigned to a fixed probability $\gamma \in]0;1[$. It is assumed that L_n remains positive from a certain n on. Let us further assume that

$$\eta_n(\varepsilon) := P\left(\max_{1\le k\le n} |X_k| > \varepsilon L_n\right) \to 0$$

for all $\varepsilon > 0$ as $n \to \infty$. Let

$$X'_{nk} = \begin{cases} X_k \text{ if } |X_k| \le \varepsilon L_n \\ 0 \quad \text{else,} \end{cases}$$

and let $S'_n := \sum_{k=1}^n X'_{nk}$. Then we have $P(S_n \ne S'_n) \le \eta_n(\varepsilon)$. Thus, if the distribution of $\frac{S'_n - a_n}{b_n}$ with appropriate a_n, b_n is close to the standard normal distribution, then this has to be true also for $\frac{S_n - a_n}{b_n}$ if $\eta_n(\varepsilon)$ is small. Exactly this argument had been used by Bernshtein [1926] in the proof of his "lemme fondamental" (which circumstance Lévy did not hint at).

On account of Lemma 3 (Sect. 6.2.3.1), for $\gamma + \eta_n(\varepsilon) < 1$ the inequality

$$\varepsilon L_n \le \varepsilon \varphi_{S'_n}(\gamma + \eta_n(\varepsilon)) \le \frac{2\varepsilon\sqrt{\mathrm{Var}S'_n}}{\sqrt{1-\gamma-\eta_n(\varepsilon)}} \tag{6.6}$$

holds. Because L_n is positive for sufficiently large n, the same is true for $\mathrm{Var}S'_n$.

Let $Y_{nk} := X'_{nk} - \mathrm{E}X'_{nk}$. Then $\mathrm{Var}\sum_{k=1}^n Y_{nk} = \mathrm{Var}S'_n$, and $|Y_{nk}|$ is bounded above by $2\varepsilon L_n$. On account of (6.6), for sufficiently large n the expression $\frac{2\varepsilon L_n}{\sqrt{\mathrm{Var}\sum Y_{nk}}}$ can be assumed arbitrarily small as dependent on ε. By use of one of Lindeberg's theorems (which Lévy did not hint at explicitly, see Sect. 5.2.4.2 on Lindeberg's discussion of bounded elementary errors), finally Lévy's assertion that for large n the distribution of $\frac{\sum Y_{nk}}{\sqrt{\mathrm{Var}S'_n}}$ is very close to the standard normal distribution can be followed.

6.2.3.4 Lévy's Decomposition Principle

In the "classical case" Lévy had already used an idea which would be also applied in more general situations: The sums S_n' (consisting of variables with values below the bound X only) and S_n'' (consisting of variables with values above X only) can be considered stochastically independent in an asymptotic sense. In his 1935 paper Lévy only gave a few—rather vague—hints; a relatively complete discussion of this idea (even for the general case of not identically distributed random variables) can be found in [Lévy 1937a, 108 f.].

A situation analogous to the following was considered there: Let $X_1, \ldots, X_n$ be independent random variables, and $L > 0$ be a constant, possibly depending on n. Let α_i denote the probability $P(|X_i| > L)$. Because only probability distributions are relevant, one can, without loss of generality, suppose an appropriate probability space on which all random variables that are considered in the following are defined. Then, for $1 \leq i \leq n$ the random variable X_i can be assumed to be generated according to a random experiment of three independent steps:
First, determine Y_i', where $|Y_i'| \leq L$, and

$$P(Y_i' \leq x) = \frac{P(-L \leq X_i \leq x)}{1-\alpha_i} \quad (|x| \leq L).$$

Second, determine Y_i'', where $|Y_i''| > L$, and

$$P(Y_i'' \leq x) = \begin{cases} \frac{P(-\infty < X_i \leq x)}{\alpha_i} & (x < -L) \\ \frac{P(L < X_i \leq x)}{\alpha_i} & (x > L). \end{cases}$$

Third, determine u_i which only takes the values 0 and 1 with the respective probabilities $1 - \alpha_i$ and α_i.
Let $X_i' := (1 - u_i)Y_i'$ and $X_i'' := u_i Y_i''$. Then $X_i = X_i' + X_i''$, and the sum $S_n = \sum X_i$ can be expressed according to

$$S_n = \sum X_i'' + \sum Y_i' - \sum u_i Y_i', \tag{6.7}$$

where $\sum X_i''$ and $\sum Y_i'$ are independent random variables. This representation of S_n is sometimes called "Lévy decomposition" now [Araujo & Giné 1980, 51 f.].

In turn, if we define the truncated random variables X_i' and X_i'' according to

$$X_i' := \begin{cases} X_i \text{ if } |X_i| \leq L \\ 0 \quad \text{else}, \end{cases}$$

$$X_i'' := \begin{cases} X_i \text{ if } |X_i| > L \\ 0 \quad \text{else}, \end{cases}$$

then X_i' and X_i'' can be assumed to be generated by the above-described procedure.

Roughly, on the basis of (6.7) the "classical case" in Sect. 6.2.3.2 might be treated as follows: $L = X$ is considered fixed but arbitrarily large, whereas

$\varepsilon = P(|X_1|>L)$ can be considered arbitrarily small. On account of the Bienaymé–Chebyshev inequality we have for $r>0$:

$$P\left(\left|\sum_{i=1}^{n} u_i - \varepsilon n\right| > r\sqrt{n}\right) \leq \frac{\varepsilon}{r^2}, \quad P\left(\left|\sum_{i=1}^{n} u_i - \varepsilon n\right| \leq r\sqrt{n}\right) \geq 1 - \frac{\varepsilon}{r^2}. \tag{6.8}$$

If $\sum_{i=1}^{n} u_i = \varepsilon n$, then because of the independence of all random variables $Y_1', \ldots, Y_n'; Y_1'', \ldots, Y_n''$, and because Y_i' and Y_i'' respectively have identical distributions, it follows that S_n can be represented by

$$\tilde{S}_n = \sum_{i=1}^{(1-\varepsilon)n} Y_i' + \sum_{i=1}^{\varepsilon n} Y_i'' =: S_n' + S_n''.$$

We recall that it is presupposed that the distribution of $s_n = \frac{1}{\sqrt{n}} S_n$ tends to the standard normal distribution. If $\sum_{i=1}^{n} u_i = \varepsilon n + \rho\sqrt{n}$ with $|\rho| \leq r$, then one can show that, for sufficiently large n and X, the (conditional) distribution of

$$\frac{1}{\sqrt{n}} \sum_{i=1}^{(1-\varepsilon)n-\rho\sqrt{n}} Y_i' + \frac{1}{\sqrt{n}} \sum_{i=1}^{\varepsilon n+\rho\sqrt{n}} Y_i''$$

is, except for an arbitrarily small error (depending on n, ε, and r), identical to the distribution of $\tilde{S}_n/\sqrt{n}$. On account of (6.8), cases with $|\sum_{i=1}^{n} u_i - \varepsilon n| > r\sqrt{n}$ only occur with an arbitrarily small probability depending on ε and r. Altogether, with respect to its limit distribution and with the exception of an arbitrarily small error, S_n can be treated as composed of two independent partial sums S_n' (consisting of mutually independent variables with values only within $[-L; L]$) and S_n'' (consisting of mutually independent variables with values only beyond $[-L; L]$).

The case where L was assumed to indefinitely grow together with n was even more important for Lévy's discussion of necessary conditions for the convergence to the normal distribution. Lévy [1935b, 364] gave some hints also regarding this case, which he treated in a way different from the case of a constant L. A more elaborated discussion, based on (6.7) can be found in his book [1937a, 109]. There, under the general assumption that $\alpha_i < \frac{1}{2}$ for all i, Lévy for arbitrary $\varepsilon' > 0$ with the denotations $\eta := \sum \alpha_i$ and $\alpha' := \max_{1\leq i\leq n} P(|X_i| > \varepsilon' L)$ obtained the estimate

$$P\left(\left|\sum u_i Y_i'\right| > \eta L\sqrt{\varepsilon'}\right) < 2\eta\alpha' + \sqrt{\varepsilon'}. \tag{6.9}$$

Deriving this inequality he first noticed that (for $1 \leq i \leq n$)

$$P(u_i|Y_i'| > \varepsilon' L) = \alpha_i P(|Y_i'| > \varepsilon' L) \leq \alpha_i \frac{P(|X_i| > \varepsilon' L)}{1-\alpha_i} < 2\alpha'\alpha_i,$$

and therefore

$$P\left(\max_{1\leq i\leq n} u_i|Y_i'| > \varepsilon' L\right) < 2\eta\alpha'. \tag{6.10}$$

The event $\sum |u_i Y_i'| > \eta L\sqrt{\varepsilon'}$ is in any case a subset of the unification of the two events

1) $\max_{1\le i\le n} u_i|Y_i'| > \varepsilon' L$,
2) $\sum u_i > \frac{\eta}{\sqrt{\varepsilon'}}$.[14]

The probability of the second event is bounded above by $\sqrt{\varepsilon'}$.[15] Therefore, by observing (6.10) and

$$P\left(\left|\sum u_i Y_i'\right| > \eta L\sqrt{\varepsilon'}\right) \le P\left(\sum |u_i Y_i'| > \eta L\sqrt{\varepsilon'}\right),$$

one obtains (6.9). The significance of this relation lies in the fact that for ε' and α' being small the distribution of $\frac{S_n}{L}$ for sufficiently large n differs only little from the distribution of $\frac{\sum X_i'' + \sum Y_i'}{L}$ if L grows together with n in such a way that η remains bounded.

6.2.3.5 The "loi des grands nombres" as a Necessary Condition in the Case of Identically Distributed Variables

Lévy's line of argument in the case of arbitrarily distributed random variables was based on the same idea as in the case of identically distributed random variables; therefore, and also because the method still might be of some interest, we will discuss this particular case in detail. Let us assume, as Lévy [1935b, 371 f.] did, that for independent random variables X_i with the common distribution function F the distribution of $\frac{\sum_{i=1}^n X_i - A_n}{B_n}$ ($A_n, B_n > 0$ being appropriate norming constants) tends to the standard normal distribution. Generally presupposing that $\int_{-\infty}^{\infty} x^2 dF(x) = \infty$, Lévy did not consider the "classical case" in this context. He had to show now that under the assumption of the convergence to the normal distribution the random variables X_i obey the "loi des grands nombres," which is, as Lévy [1935b, 367] (see Sect. 6.2.1) had proven, in the case under consideration equivalent to the condition[16]

$$\lim_{X\to\infty} \frac{X^2 P(|X_1| > X)}{\int_{|x|\le X} x^2 dF(x)} = 0. \tag{6.11}$$

Lévy's proof starts with the assumption that this condition is not true. Therefore there exist "indefinitely growing" values X and a constant $a \ne 0$ such that:

[14] This can be easily seen by the fact that the intersection of the events $\max_{1\le i\le n} u_i|Y_i'| \le \varepsilon' L$ and $\sum u_i \le \frac{\eta}{\sqrt{\varepsilon'}}$ is a subset of the event $\sum u_i|Y_i'| \le \eta L\sqrt{\varepsilon'}$.

[15] This is an immediate consequence of the general relation $\int_{x>r} x dV(x) \ge r\int_{x>r} dV(x)$ for all probability distributions V and all $r > 0$, which Lévy (of course) did not hint at.

[16] The case of finite moments of second order, which Lévy did not discuss at this place, can be quite easily treated, and also leads to the following condition as a necessary condition for the convergence to the normal distribution, see [Lévy 1937a, 113; Gnedenko & Kolmogorov 1949/68, 172].

$$P(|X_1| > X)X^2 > a^2 Z_X,$$

where $Z_X = \int_{|x|\leq X} x^2 dF(x)$. Instead of this statement, however, Lévy wrote

$$nP(|X_1| > X)X^2 > a^2\sigma'^2_{n;X}, \quad \sigma'^2_{n;X} := n\mathrm{Var}X'_{1;X},$$

where

$$X'_{i;X} = \begin{cases} X_i & \text{if } |X_i| \leq X \\ 0 & \text{else.} \end{cases}$$

The latter statement is equivalent to the former, however, because—as Lévy [1935b, 365 f.] had shown—$\mathrm{Var}X'_{i;X} \sim Z_X$ as $X \to \infty$.

To give the reader an impression of Lévy's idiosyncratic style, the proof [Lévy 1935b, 371 f.] is quoted in Lévy's own words (commentaries are included in square brackets):

> For each of these values $[X]$ we define n such that $\eta = nY$ $[Y = P(|X_1| > X)]$ obtains (accurately to Y) a fixed and very small value. According to the law of small numbers, S''_n $[= \sum_{i=1}^n X''_i$, where $X''_i = X_i$ if $|X_i| > X$ and $X''_i = 0$ else$]$ contains no or exactly one term different from 0 in such cases whose probabilities tend to $\mathrm{e}^{-\eta}$ and $\eta\mathrm{e}^{-\eta}$, both numbers being, for example, above $\frac{3\eta}{4}$. Because in the first case $S''_n = 0$, and in the second $|S''_n| > X$, the probability that S''_n is within a given interval of length X (the concentration assigned to the length X) is below $1 - \frac{3\eta}{4}$.

Lévy now claimed that, as a consequence of his "Lemma II" [1935b, 364], the sum S'_n could be considered as independent of S''_n with an error asymptotically negligible in relation to X. Lemma II was as follows:

> If n and X are indefinitely growing, and if $nY = \eta$ remains finite, S'_n and S''_n in the limit case can be considered as independent random variables.

There were only a few (quite vague) hints regarding the proof of this lemma in Lévy's 1935 paper. As we will see below, all considerations in connection with the lemma can be substantiated on the basis of the estimate (6.9).

After his remark on "Lemma II," Lévy proceeded:

> Except for the fact that X has to be substituted by a smaller number, by $\frac{2X}{3}$, for example, the principle of the augmentation of the dispersion for the transition from S''_n to S_n can be applied. Therefore, for each interval of length
>
> $$\frac{2X}{3} = 2l\sigma'_n$$
>
> $[\sigma'_n = \sigma'_{n;X}]$ the probability that S_n is beyond this interval is at minimum equal to
>
> $$\frac{3\eta}{4} > \frac{3a^2\sigma'^2_n}{4X^2} = \frac{a^2}{12l^2}.$$
>
> Now, if S_n is of a type very little different from that of Gauss, almost the same is true for the sum S'_n, which differs from S_n in cases of probability η only. Therefore, the coefficient of reduction $[= B_n]$ is $\leq \sigma'_n$, accurately to a relative error tending almost to zero with η (values with a very small probability may, if they are large, increase but not diminish the prearranged value [valeur prévue] for $\sigma\{S'_n\}$). The random variable $\frac{S_n}{\sigma'_n}$ therefore has the

form $k\xi$, where ξ obeys a law which is very little different from that of Gauss, and k stays below a function of η which tends to one for $\eta = 0$.

We are able now to choose η as small as we want, therefore $l > \frac{a}{3\sqrt{\eta}}$ gets as large as we wish. The obtained result is thus contradictory to the preceding, according to which the probability that values of $\frac{S_n}{\sigma_n'}$ are beyond any interval of the length $2l$ tends by far less quickly to zero than the law of Gauss implicates it.

It is actually possible to complement Lévy's very concise and at some places not entirely clear arguments to a correct proof with only a few minor (rather technical) modifications, as we will see in the following.

The presupposition of the theorem is that for a sequence of independent random variables X_i, all with the same distribution F, and for suitable sequences of numbers $(B_n > 0)$ and (A_n),

$$P\left(\frac{\sum_{i=1}^{n} X_i - A_n}{B_n} \leq x\right) \to \Phi(x) \quad (n \to \infty). \tag{6.12}$$

The assertion is that

$$\lim_{z\to\infty} \frac{z^2 P(|X_1| > z)}{\int_{|x|\leq z} x^2 dF(x)} = 0. \tag{6.13}$$

The proof starts with the hypothesis that the latter limit relation is wrong. Then there exist a sequence (z_k) of positive real numbers with $z_k \to \infty$ and a number $a \neq 0$ such that

$$P(|X_1| > z_k) z_k^2 > a^2 \int_{|x|\leq z_k} x^2 dF(x). \tag{6.14}$$

For an (arbitrarily chosen) $0 < \varepsilon_1 < \frac{1}{4}$ and for $\eta_k' := P(|X_1| > z_k)$ we define the natural number n_k according to $n_k := \left[\frac{\varepsilon_1}{\eta_k'}\right] + 1$. Then we have $\eta(n_k) := n_k \eta_k' \geq \varepsilon_1$, as well as $\eta(n_k) \to \varepsilon_1$ and $n_k \to \infty$ for $k \to \infty$. For $x = 0, 1, \ldots$ one gets the limits

$$\lim_{k\to\infty} B(n_k; \eta_k'; x) = \frac{\varepsilon_1^x}{x!} \mathrm{e}^{-\varepsilon_1}.$$

Now, in accord with "Lemma II," the sums S_{n_k}' and S_{n_k}'' such that $S_{n_k} = S_{n_k}' + S_{n_k}''$, where $S_{n_k}' = \sum_{i=1}^{n_k} X_{n_k i}'$, $S_{n_k}'' = \sum_{i=1}^{n_k} X_{n_k i}''$, $X_{n_k i}' + X_{n_k i}'' = X_i$, and

$$X_{n_k i}' = \begin{cases} X_i & \text{if } |X_i| \leq z_k \\ 0 & \text{else,} \end{cases}$$

are introduced. Then, the probability that for a certain number n_k the sum S_{n_k}'' consists exclusively of zero terms tends to $\mathrm{e}^{-\varepsilon_1}$, and the probability that this sum consists of exactly one nonzero term tends to $\varepsilon_1 \mathrm{e}^{-\varepsilon_1}$, according to the just-described arguments. These two probabilities are above $\frac{3}{4}\varepsilon_1$ in any case. In order to establish Lévy's following estimate of the concentration of S_{n_k} in a precise manner, it seems useful to introduce $\delta := (\varepsilon_1 \mathrm{e}^{-\varepsilon_1} - \frac{3}{4}\varepsilon_1)/2$. Then for sufficiently large n_k both probabilities lie above $\frac{3}{4}\eta(n_k) + \delta$.

Lévy's estimate for the concentration of S''_{n_k} can be justified (with a slight modification arising from the introduction of δ) as follows: We either have $S''_{n_k}=0$ or $|S''_{n_k}|>z_k$. Two cases have to be considered for $r\in\mathbb{R}$:

1) $0\notin]r;r+z_k[$. Then we have

$$P(r<S''_{n_k}<r+z_k)\leq P(S''_{n_k}\neq 0)=1-P(S''_{n_k}=0).$$

2) $0\in]r;r+z_k[$. Then we have

$$P(r<S''_{n_k}<r+z_k)\leq P(|S''_{n_k}|\leq z_k)=1-P(|S''_{n_k}|>z_k).$$

It follows that

$$f_{S''_{n_k}}(z_k)\leq \max\left(1-P(S''_{n_k}=0);1-P(|S''_{n_k}|>z_k)\right)<1-\frac{3}{4}\eta(n_k)-\delta.$$

In order to use "Lemma II" as specified by (6.9), we must only substitute n by n_k, and L by z_k, with the result

$$P\left(\Big|\sum_{i=1}^{n_k}u_iY_i'\Big|>\sqrt{\varepsilon'}z_k\eta(n_k)\right)<2\eta(n_k)P(|X_1|>\varepsilon' z_k)+\sqrt{\varepsilon'}\quad(\varepsilon'>0)$$

(u_i and Y_i' are now determined with respect to $X'_{n_k i}$ and $X''_{n_k i}$). Lévy's assertion on the independence of S_n' and S_n'' with a "negligible error" probably refers to an estimate of this kind. We are able now to choose ε' such that, for sufficiently large k (or n_k, respectively),

$$P\left(\Big|\sum_{i=1}^{n_k}u_iY_i'\Big|>\sqrt{\varepsilon'}z_k\eta(n_k)\right)<\delta,$$

where δ is defined as above.

Taking into consideration the (elementarily provable) inequality

$$f_{X+Y}(x)\leq f_X(x+2\upsilon)+P(|Y|>\upsilon),$$

which is valid for any (even dependent) random variables X, Y, and for any $\upsilon>0$, and using

$$f_{S''_{n_k}+\sum_{i=1}^{n_k}Y_i'}(x)\leq f_{S''_{n_k}}(x),$$

which results from the independence of S''_{n_k} and $\sum_{i=1}^{n_k}Y_i'$, one obtains

$$f_{S_{n_k}}(x)\leq f_{S''_{n_k}}(x+2\sqrt{\varepsilon'}z_k\eta(n_k))+\delta.$$

Lévy in this context rather refers to the dispersions of S_{n_k} and S''_{n_k}; it seems, however, more convenient to keep considering concentrations.

Because

$$z_k - 2\sqrt{\varepsilon'} z_k \eta(n_k) \geq \frac{2}{3} z_k$$

for sufficiently small ε', we are able to infer, for sufficiently large k,

$$f_{S_{n_k}}\left(\frac{2}{3} z_k\right) \leq f_{S''_{n_k}}(z_k) + \delta < 1 - \frac{3}{4}\eta(n_k),$$

and therefore

$$P\left(\left|\frac{S_{n_k} - A_{n_k}}{\sigma'_{n_k}}\right| \geq \frac{1}{3}\frac{z_k}{\sigma'_{n_k}}\right) > \frac{3}{4}\eta(n_k), \tag{6.15}$$

where $\sigma'^2_{n_k} = n_k \operatorname{Var} X'_{n_k 1}$.

Because of $\eta(n_k) > \frac{a^2 \sigma'^2_{n_k}}{z_k^2}$ (see (6.14)), for $l_k := \frac{z_k}{3\sigma'_{n_k}}$ the inequality

$$\frac{3\eta(n_k)}{4} > \frac{a^2}{12 l_k^2} \tag{6.16}$$

is valid. From (6.15) it follows that, for sufficiently large k,

$$P\left(\left|\frac{S_{n_k} - A_{n_k}}{\sigma'_{n_k}}\right| \geq l_k\right) > \frac{a^2}{12 l_k^2}. \tag{6.17}$$

A rigorous justification of Lévy's assertion that "the coefficient of reduction is $\leq \sigma'_n$, accurately to a relative error tending almost to zero with η" seems hardly possible without some additional effort. By use of characteristic functions (which Lévy avoided in his 1935 paper, however) one can show [Fischer 2000, 235], for example, that

$$\frac{\sigma'^2_{nk}}{B^2_{nk}} \geq 1 + \varepsilon(n_k) - 4\eta(n_k),$$

where $\varepsilon(n_k) \to 0$ for $n_k \to \infty$. Yet, for proceeding the proof in accord with Lévy's basic ideas, a weaker statement suffices: On account of Lemmata 3 and 5 (Sect. 6.2.3.1), for arbitrarily small $\varepsilon > 0$ and sufficiently large n_k we obtain

$$\frac{\sigma'_{n_k}}{B_{n_k}} \geq \varphi_N(\gamma)(1 - \varepsilon)\sqrt{1 - \gamma - \eta(n_k)}/2,$$

where N denotes a random variable with a standard normal distribution, and $\gamma \in]0; 1[$ is a fixed probability such that $\gamma + \eta(n_k) < 1$. Therefore, there exists a constant $\beta > 0$ such that

$$\frac{\sigma'_{n_k}}{B_{n_k}} > \beta \tag{6.18}$$

for sufficiently large n_k. Because of

$$1 - \Phi(x) \sim \frac{1}{x\sqrt{2\pi}} e^{-\frac{x^2}{2}} \quad (x \to \infty)$$

there exists $l_0 > 0$ such that

$$2(1 - \Phi(\beta l_0)) < \frac{a^2}{12 l_0^2}. \tag{6.19}$$

(6.16) yields $l_k > \frac{a}{3\sqrt{\eta(n_k)}}$. If ε_1 is chosen sufficiently small, then on account of $\eta(n_k) \to \varepsilon_1$ the inequality

$$l_k \geq l_0 \tag{6.20}$$

is valid for sufficiently large k. From (6.12) follows that

$$P\left(\left|\frac{S_{n_k} - A_{n_k}}{B_{n_k}}\right| \geq \beta l_0\right) \to 2(1 - \Phi(\beta l_0)).$$

Therefore, because of (6.19), the inequality

$$P\left(\left|\frac{S_{n_k} - A_{n_k}}{B_{n_k}}\right| \geq \beta l_0\right) < \frac{a^2}{12 l_0^2}$$

holds for sufficiently large k. Altogether, for sufficiently large k, and under consideration of (6.18) and (6.20), we obtain

$$P\left(\left|\frac{S_{n_k} - A_{n_k}}{\sigma'_{n_k}}\right| \geq l_k\right) = P\left(\left|\frac{S_{n_k} - A_{n_k}}{B_{n_k}}\right| \geq \frac{\sigma'_{n_k}}{B_{n_k}} l_k\right) \leq P\left(\left|\frac{S_{n_k} - A_{n_k}}{B_{n_k}}\right| \geq \beta l_0\right) < \frac{a^2}{12 l_0^2},$$

which contradicts (6.17). The assertion (6.13) is therefore true.

6.2.3.6 The "loi des grands nombres" as a Necessary Condition in the General Case of Negligible Variables

In the general case of not identically distributed random variables, which were assumed to be negligible with respect to the total sum, Lévy tried to maintain the basic ideas of his proof for identically distributed variables. A far-reaching analogy to the latter case was not possible any more, however, due to the fact that a condition equivalent to the validity of the "loi des grands nombres," which was usable as easily as (6.11), was not possible, and it was not certain in the general case that for indefinitely growing X always natural numbers n existed such that $\sum_{k=1}^{n} P(|X_k| > X)$ could be considered arbitrarily small. The equivalent for the "loi des grands nombres" which Lévy [1935b, 385 f.] used in the general case, was as follows:

For each pair of positive numbers $\varepsilon, \varepsilon'$ and for all sufficiently large natural numbers n one can find a positive $X(n)$ such that, with the denotations

$$\sigma_n' := \sqrt{\operatorname{Var}\sum_{k=1}^{n} X_{nk}'}, \quad X_{nk}' := \begin{cases} X_k & \text{if } |X_k| \le X(n) \\ 0 & \text{else,} \end{cases}$$

the inequalities

$$1)\ \frac{X(n)}{\sigma_n'} \le \varepsilon' \quad \text{and} \quad 2)\ \eta(n) := \sum_{k=1}^{n} P(|X_k| > X(n)) \le \varepsilon \tag{6.21}$$

are valid.

Lévy did not prove the equivalence of (6.21) and his original condition

$$\eta_n(\varepsilon) = P\left(\max_{1\le k\le n} |X_k| > \varepsilon L_n\right) \to 0 \quad \forall \varepsilon > 0 \tag{6.22}$$

(L_n being the dispersion of S_n assigned to the fixed probability $\gamma \in]0;1[$). By use of Lemmata 3 and 5 (Sect. 6.2.3.1), and by observing that (6.21) is sufficient for $\frac{S_n - \sum_{k=1}^{n} \mathrm{E}X_{nk}'}{\sigma_n'}$ having a normal limit distribution, it can be shown rather easily [Fischer 2000, 237] that (6.21) is equivalent to

$$\lim_{n\to\infty} \sum_{k=1}^{n} P(|X_k| > \delta L_n) = 0 \quad \forall \delta > 0. \tag{6.23}$$

This condition is in turn (for the very elementary proof see [Lévy 1937a, 104 f.]) equivalent to (6.22).

Lévy based his considerations on the assumption that for a sequence of independent random variables (X_i), $S_n := \sum_{i=1}^{n} X_i$, and $L_n := \varphi_{S_n}(\gamma)$ with an arbitrary however fixed $\gamma \in]0;1[$:

$$P\left(\frac{S_n - A_n}{B_n} \le x\right) \to \Phi(x) \text{ and } \max_{1\le i\le n} P(|X_i| > \varepsilon L_n) \to 0 \quad \forall \varepsilon > 0 \tag{6.24}$$

for $n \to \infty$ with suitable A_n and $B_n > 0$. He [1935b, 386–388] gave a sketch of proof that these conditions imply the "loi des grands nombres" in the version (6.21), again in a very "intuitive" style. Again it is possible to establish a rigorous proof on the basis of Lévy's arguments—with slight modifications concerning technical details at some places [Fischer 2000, 238–244]. Lévy [1935b, 388] supplemented his exposition by the following footnote:

> One observes that we have by no means applied Lemma III [= lemme hypothétique, see Sect. 6.1.1]; but, if it is true, it allows to simplify the reasoning (...)

Two years later, Lévy [1937a, 107–109] was actually able, by use of the "lemme hypothétique," which had been proven in the meantime by Cramér, to give a considerably shortened and simplified proof. Therefore, we will only describe the major steps of the 1935 proof (including modifications which seem indispensable), and, for the convenience of the reader, use a more formal mathematical language than Lévy did.

Presupposing (6.24), Lévy started with the hypothesis that (6.21) is not true. Then it can be shown that there even exist sequences $(m_k) \subset \mathbb{N}$ (strictly monotonic) and $(z_k) \subset \mathbb{R}^+$, and positive constants $\varepsilon < \frac{1}{2}, \varepsilon', \delta_0$ such that simultaneously

$$z_k > \varepsilon'\sigma'_{m_k},\ \eta(m_k) := \sum_{i=1}^{m_k} P(|X_i| > z_k) \geq \varepsilon,$$

$$\eta(m_k) \leq \frac{1}{2},\ z_k > \delta_0 L_{m_k} > 0, \quad (6.25)$$

where

$$\sigma'^2_{m_k} := \sum_{i=1}^{m_k} \mathrm{Var} X'_i, \quad X'_i := \begin{cases} X_i & \text{if } |X_i| \leq z_k \\ 0 & \text{else.} \end{cases}$$

Let

$$S''_{m_k} := \sum_{i=1}^{m_k} X''_i, \quad X''_i := \begin{cases} X_i & \text{if } |X_i| > z_k \\ 0 & \text{else,} \end{cases}$$

and let $p_0(m_k)$ be the probability that S''_{m_k} consists exclusively of zero terms, and $p_1(m_k)$ the corresponding probability of exactly one nonzero term. Then, by an elementary consideration, one can show that both probabilities are above $\frac{\varepsilon}{2}$. In analogy to the case of identically distributed random variables it can be proven that, for sufficiently large m_k,

$$f_{S''_{m_k}}(z_k) < 1 - \frac{\varepsilon}{2}, \quad (6.26)$$

and by use of the decomposition principle,

$$f_{S_{m_k}}\left(\frac{2z_k}{3}\right) < 1 - \frac{\varepsilon_1}{2}, \quad 0 < \varepsilon_1 < \varepsilon. \quad (6.27)$$

Lévy in this context used dispersions rather than concentrations, and, in contrast to the case of identically distributed random variables, he disregarded the problem of a necessary reduction of the interval length with respect to the concentration of S_{m_k}.

As a consequence of (6.27), one is able to show that both $\frac{z_k}{B_{m_k}}$ and $\frac{z_k}{L_{m_k}}$ have an upper bound independent of k (Lemma 5!). The same has to be true, by virtue of Lemma 3, for $\frac{z_k}{\sigma'_{m_k}}$. Lévy did not make explicit this step, which, however, is the basis for the subsequent.

Because $\frac{z_k}{\sigma'_{m_k}}$ is bounded above, one obtains

$$\max_{1 \leq \nu \leq m_k} \frac{\mathrm{Var} X'_{m_k \nu}}{\sigma'^2_{m_k}} \to 0 \quad (m_k \to \infty). \quad (6.28)$$

Lévy in this context only wrote:"σ'_ν varies in an almost continuous mode with ν."

(6.28) implies the possibility of partitioning the random variables $X_1, \ldots, X_{m_k}$, without any change of order, into p groups such that the variance of the sum of those random variables within each group is approximately equal to $\frac{\sigma_{m_k}^{\prime 2}}{p}$, where the error of approximation vanishes relatively to $\sigma_{m_k}^{\prime 2}$ as m_k grows. There exists at least one group (from X_{r_k+1} to X_{s_k}) for which

$$\sum_{i=r_k+1}^{s_k} P(|X_i| > z_k) \geq \frac{\eta(m_k)}{p} \geq \frac{\varepsilon}{p}.$$

For this group in analogy to (6.27), with $\frac{\varepsilon_1}{p}$ instead of ε_1, it can be shown (if m_k is sufficiently large)

$$f_{\sum_{i=r_k+1}^{s_k} X_i}\left(\frac{2z_k}{3}\right) < 1 - \frac{\varepsilon_1}{2p}, \tag{6.29}$$

and, under consideration of the first part of (6.25),

$$f_{\frac{\sqrt{p}}{\sigma'_{m_k}} \sum_{i=r_k+1}^{s_k} X_i}\left(\frac{2}{3}\varepsilon'\sqrt{p}\right) < 1 - \frac{\varepsilon_1}{2p}. \tag{6.30}$$

Let $((r_k, s_k))$ be the sequence of pairs of natural numbers which correspond to the group $X_{r_k+1}, \ldots, X_{s_k}$ as above, and let $((r_{k'}, s_{k'}))$ be any subsequence. Presupposing a sufficiently large however fixed p one can show, on the basis of (6.30) and by use of arguments similar to those employed in the case of identically distributed random variables, that the standard normal distribution as a limit distribution of $\frac{1}{b_{k'}}\left(\sum_{i=r_{k'}+1}^{s_{k'}} X_i - a_{k'}\right)$ ($a_{k'}, b_{k'}$ any norming constants) is impossible.

From (6.25) it follows that these r_k and s_k which are different from zero grow indefinitely with k. If there are infinitely many r_k with $r_k = 0$, then there exists a subsequence $(S_{s_{k'}}) \subset (S_{s_k})$ which is not normally distributed in the limit, and this contradicts the assumption (6.24). If $r_k > 0$ from a certain k, then (6.29) and (6.25) imply that $S_{s_k} - S_{r_k}$ "cannot be neglected," as Lévy writes, "in the investigation" of S_{s_k}. This means, in more precise terms, that the sequence of the distributions of $\frac{S_{s_k}-S_{r_k}-(A_{s_k}-A_{r_k})}{B_{s_k}}$ or any subsequence of it cannot tend to a degenerate distribution. Because at least for a subsequence $(S_{r_{k'}})$ of (S_{r_k}) the limit distribution of $\frac{S_{r_{k'}}-A_{r_{k'}}}{B_{s_{k'}}}$ is a (possibly degenerate) normal distribution (with a variance between zero and one), and because

$$\frac{S_{s_{k'}} - A_{s_{k'}}}{B_{s_{k'}}} = \frac{S_{r_{k'}} - A_{r_{k'}}}{B_{s_{k'}}} + \frac{S_{s_{k'}} - S_{r_{k'}} - (A_{s_{k'}} - A_{r_{k'}})}{B_{s_{k'}}},$$

the two members of the right side being stochastically independent, we finally reach a contradiction to (6.24): The right side, in contrast to the left, cannot obey a standard normal distribution in the limit, because the distribution of its second member does not tend to a normal or to a degenerate distribution.

In the last part of his 1935 proof Lévy tacitly used a weakened version of the "lemme hypothétique," which can be easily proven by use of characteristic functions: If the random variable Z is the sum of the independent random variables

X and Y, and Z is Gaussian, and X is a (possibly degenerate) Gaussian variable, then Y is a (possibly degenerate) Gaussian variable. As already mentioned, Lévy at the end of his 1935 proof argued that the full version of the "lemme hypothétique" would have allowed a simplified argumentation.

Strictly speaking, an extension of the "lemme hypothétique" would have been needed, as stated and proven by Lévy [1935b, 382 f.; 1937a, 100 f.]: If $X_n + Y_n = Z_n$, X_n and Y_n being independent, and if the distribution of Z_n tends for $n \to \infty$ to the standard normal distribution, then the distribution of $X_n + a_n$, where $a_n \in \mathbb{R}$ is chosen such that a median of $X_n + a_n$ is zero, tends to a (possibly degenerate) normal distribution.

Now, starting with the hypothesis that (6.21) is not true, we could proceed in the same way as Lévy originally did in 1935, up to the discussion of $p_0(m_k)$, $p_1(m_k)$, and $f_{S''_{m_k}}(z_k)$. Since $0 < \varepsilon \leq \eta(m_k) \leq \frac{1}{2}$, it follows by Lévy's decomposition principle that S'_{m_k} and S''_{m_k} are virtually independent for sufficiently large m_k. By the generalized version of the "lemme hypothétique," $s''_{m_k} := \frac{S''_{m_k} - A_{m_k}}{B_{m_k}}$ has to obey a law close to a (possibly degenerate) Gaussian if the distribution of $\frac{S_{m_k} - A_{m_k}}{B_{m_k}}$ tends to the standard normal distribution. From

$$P(S''_{m_k} = 0) = p_0(m_k) > \frac{\varepsilon}{2}$$

it follows that the distribution of s''_{m_k} is close[17] to a degenerate distribution for large m_k. Then, for sufficiently large m_k and for any positive x, the concentration of $s''_{m_k}(x)$ is arbitrarily close to 1. This contradicts, however, the estimate of $f_{S''_{m_k}}(z_k)$ given by (6.26) (note that, due to the fourth part of (6.25), $\frac{z_k}{B_{m_k}}$ is always greater than a positive constant).

In his 1937 book, Lévy [1937a, 107–109] based his considerations on the variant (6.23) of the "loi des grands nombres" (he did not use this designation any more), and on the hypothesis that this condition was not true despite the validity of (6.24). Then there existed $\delta > 0$ and $\varepsilon > 0$ such that for an indefinitely growing sequence n_k of natural numbers

$$\eta(n_k) := \sum_{i=1}^{n_k} P(|X_i| > \varepsilon L_{n_k}) \geq \delta. \tag{6.31}$$

Lévy used his decomposition principle, and in accord with (6.7) (L being substituted by εL_{n_k}) he represented the sum S_{n_k} by

$$S_{n_k} = \sum_{i=1}^{n_k} X''_i + \sum_{i=1}^{n_k} Y'_i - \sum_{i=1}^{n_k} u_i Y'_i.$$

[17] "Close" with respect to the Lévy distance, for example.

By an estimate analogous to (6.9) he showed that the third term on the right side could be neglected for $n_k \to \infty$ if η_{n_k} was bounded above. For the case that η_{n_k} was growing indefinitely with n_k, Lévy [1937a, 109] recommended to choose n' such that $\eta(n')$ was "between 1 and 2, for example," and represent S_{n_k} by

$$S_{n_k} = \sum_{i=1}^{n'} X_i'' + \left(\sum_{i=1}^{n'} Y_i' + \sum_{i=n'+1}^{n_k} X_i \right) - \sum_{i=1}^{n'} u_i Y_i'.$$

This was again a representation of S_{n_k} by three summands, where the first and the second were independent, and the third could be neglected for $n_k \to \infty$.

In the case of $\eta(n_k)$ being bounded above, Lévy on the basis of (6.31) showed that the probabilities $p_0(n_k)$ and $p_1(n_k)$ (see above) had a positive lower bound, δ_1, say. Due to the extension of the "lemme hypothétique" and because of $p_0(n_k) \geq \delta_1$, for large n_k the distribution of $\sum_{i=1}^{n_k} X_i''/B_{n_k}$ had to be close to a Gaussian law with a "small parameter," as Lévy wrote. This behavior, however, contradicted the property

$$P\left(|\sum_{i=1}^{n_k} X_i''| > \varepsilon L_{n_k} \right) > p_1(n_k) \geq \delta_1.$$

The case of an indefinitely growing $\eta(n_k)$ was analogously treated by considering $\sum_{i=1}^{n'} X_i''$ instead of $\sum_{i=1}^{n_k} X_i''$.

If one considers the total effort for proving the necessity of the "loi des grands nombres," on the one hand in the original 1935 version (whose exposition was only sketchy), on the other hand by use of the (generalized) "lemme hypothétique" in Lévy's 1937 book, then both ways of reasoning seem to be equally laborious. The "lemme hypothétique" for itself, now named "Cramér's theorem," as expounded in [Lévy 1937a, 97–101], required some intricate arguments regarding complex functions, and its extension to a limit process, which relied on compactness criteria for distributions, was by no means trivial. However, also Feller's solution of the central limit problem was, despite its use of more common probabilistic notions and tools, and despite its more elaborate exposition, far from being easy.

6.2.4 Feller's Theorems

In contrast to Lévy, Feller was rather reserved in his use of specialized probabilistic notions, and his ideas could be basically understood even by readers who were not familiar with probability theory. Unlike Lévy he also explicitly presented methods for determining the required norming constants. The fact that Feller chose characteristic functions as his main tool had the advantage that his arguments were familiar to a broader audience, but several estimates could be only reached by rather cumbersome considerations.

Feller generally based his considerations on a sequence of distribution functions (V_n), all continuous on the right, and the respective convolution functions $W_n = V_1 \star V_2 \star \cdots \star V_n$. He [1935, 522] characterized the chief subject of his investigation by the following question:

> Do there exist, for a given sequence of distribution functions $\{V_n(x)\}$, two number sequences $\{a_n\}$ and $\{c_n\}$[18] such that $W_n(a_n x + c_n) \to \Phi(x)$ [the standard normal distribution], and, if this is the case, how can such number sequences be determined?

Feller, too, presupposed the negligibility of the single "components" V_k with respect to the total convolution W_n: Basically, what he demanded was that there exist suitable b_k such that, for each $x \neq 0$,

$$\max_{1\le k\le n} |V_k(a_n x + b_k) - E(x)| \to 0 \quad (n \to \infty), \tag{6.32}$$

where

$$E(x) = \begin{cases} 0 \text{ for } x < 0 \\ 1 \text{ else.} \end{cases}$$

This demand, which was, however, expressed by Feller [1935, 523] in a less explicit way only, is equivalent to the condition that for the random variables X_k obeying the distributions V_k:

$$\max_{1\le k\le n} P(|X_k - b_k| > \varepsilon a_n) \to 0 \quad \forall \varepsilon > 0.$$

Feller used the words "the sequence $\{V_k(x + b_k)\}$ together with the norming factors $\{a_n\}$ belongs to $\Phi(x)$" if the limit relation $W_n(a_n x + c_n) \to \Phi(x)$ $(c_n = \frac{1}{a_n}\sum_{k=1}^n b_k)$ and the condition (6.32) are simultaneously met.

The solution of the problem above—designated as "criterion" by Feller [1935, 526 f.]—was as follows: Let (V_k) be a sequence of distributions, all with a zero median. For each $\delta > 0$ let

$$p_n(\delta) := \min\left\{ r \in \mathbb{R}_0^+ \,\middle|\, \sum_{\nu=1}^n \int_{|x|>r} dV_\nu(x) \le \delta \right\}.$$

Then the presupposition[19]

$$\forall \delta > 0 \quad \lim_{n\to\infty} \frac{1}{p_n^2(\delta)} \sum_{\nu=1}^n \int_{|x|\le p_n(\delta)} x^2 dV_\nu(x) = \infty \tag{6.33}$$

is necessary and sufficient for the existence of sequences $(a_n > 0)$, (b_k) such that the sequence $(V_k(x + b_k))$ together with the norming factors a_n belongs to $\Phi(x)$. The constants a_n and b_k can be obtained in this way: Because of (6.33) a sequence (δ_n) tending to 0 exists for which

---

[18] Note that Feller's sequences $\{a_n\}$ and $\{c_n\}$ correspond to Lévy's sequences B_n and A_n, respectively.

[19] Feller wrote "$|x| < p_n(\delta)$" in (6.33) instead of "$|x| \le p_n(\delta)$." His proof [1935, 552 f.] for the "criterion" is only valid, however, in case of "$\le$".

$$\lim_{n\to\infty} \frac{1}{p_n^2(\delta_n)} \sum_{\nu=1}^{n} \int_{|x|\le p_n(\delta_n)} x^2 dV_\nu(x) = \infty.$$

One sets[20]

$$a_n^2 = \sum_{\nu=1}^{n} \left\{ \int_{|x|\le p_n(\delta_n)} x^2 dV_\nu(x) - \left(\int_{|x|\le p_n(\delta_n)} x dV_\nu(x) \right)^2 \right\}$$

and determines b_k according to

$$\int_{|x|<a_k} (x - b_k) dV_k(x) = 0. \tag{6.34}$$

At first, however, Feller [1935, 533–551] proved a theorem—designated as "main theorem" in the following—which is based on the special assumption that the a_n are already given, and all b_k are equal to 0.

Main Theorem: Let (V_k) be a sequence of distribution functions (not necessarily with zero medians). $(V_k(x))$ together with the positive norming factors a_n belongs to $\Phi(x)$ if and only if for each $\eta > 0$ the following three conditions are simultaneously met:

(I) $$\lim_{n\to\infty} \sum_{\nu=1}^{n} \int_{|x|>\eta a_n} dV_\nu(x) = 0$$

(II) $$\lim_{n\to\infty} \frac{1}{a_n^2} \sum_{\nu=1}^{n} \left\{ \int_{|x|<\eta a_n} x^2 dV_\nu(x) - \left(\int_{|x|<\eta a_n} x dV_\nu(x) \right)^2 \right\} = 1$$

(III) $$\lim_{n\to\infty} \frac{1}{a_n} \sum_{\nu=1}^{n} \int_{|x|<\eta a_n} x dV_\nu(x) = 0.$$

Feller's article became especially prominent by its discussion of the Lindeberg condition. Feller [1935, 541–543] for independent random variables X_k, each with zero expectation, distribution V_k, and variance σ_k^2, explicitly showed that

$$\lim_{n\to\infty} \frac{1}{s_n^2} \sum_{k=1}^{n} \int_{|x|<\eta s_n} x^2 dV_k(x) = 1 \quad \forall \eta > 0 \tag{6.35}$$

is necessary for the assertion that $(V_k(x))$ together with the norming factors $s_n > 0$ ($s_n^2 = \sum_{k=1}^n \sigma_k^2$) belongs to $\Phi(x)$. One can actually deduce this assertion from the "main theorem" [Fischer 2000, 246] as well, as Feller [1935, 542] only briefly noticed.

Like Lévy, also Feller [1935, 554 f.] was concerned with the particular case of independent identically distributed random variables with a nondegenerate distribution V. From the "criterion" he deduced a theorem which was as follows:
Let for $\zeta > 0$

[20] Also in the following formula Feller wrote "<" instead of a correct "≤".

$$Z(\zeta) := \min\left\{z \in \mathbb{R}_0^+ \,\middle|\, \int_{|x|>z} dV(x) \leq \zeta\right\}.$$

Necessary and sufficient for the existence of sequences $(a_n > 0)$ and (c_n) such that $V^{\star n}(a_n x + c_n) \to \Phi(x)$, is the validity of the condition

$$\lim_{\zeta\to 0} \frac{1}{\zeta Z^2(\zeta)} \int_{|x|\leq Z(\zeta)} x^2 dV(x) = \infty. \tag{6.36}$$

6.2.5 *Feller's Proofs*

6.2.5.1 Auxiliary Theorems

Feller based his proofs on some auxiliary theorems, which he proved in the first part of his article. The following numbering of these theorems does not refer, however, to a corresponding numbering in Feller's paper.

Auxiliary Theorem 1 [Feller 1935, 532 f.]: Both the sequences of distribution functions $(V_k(x))$ and $(V_k(x + b_k))$ belong, together with the same positive norming factors a_n, to $\Phi(x)$ if

$$\lim_{n\to\infty} \frac{1}{a_n} \sum_{k=1}^{n} b_k = 0. \tag{6.37}$$

Auxiliary Theorem 2 [Feller 1935, 533–536]: Let $(a_n > 0)$, (b_n) be sequences of real numbers such that

$$\lim_{n\to\infty} \frac{b_n}{a_n} = 0. \tag{6.38}$$

The conditions (I) and (II) of the "main theorem" are valid for the sequence of distribution functions $(V_k(x))$ if and only if (I) and (II) are valid for the sequence of distribution functions $(V_k(x + b_k))$.

Auxiliary Theorem 3 [Feller 1935, 537]: Let b_k be in accord with (6.34), and let $V_k(x)$ be distribution functions meeting the conditions (I) and (II). Let $V_k^*(x) := V_k(x + b_k)$. Then for any $\eta > 0$:

$$\text{(I')} \qquad \lim_{n\to\infty} \sum_{\nu=1}^{n} \int_{|x|>\eta a_n} dV_\nu^*(x) = 0$$

$$\text{(II')} \quad \lim_{n\to\infty} \frac{1}{a_n^2} \sum_{\nu=1}^{n} \int_{|x|<\eta a_n} x^2 dV_\nu^*(x) = 1$$

$$\text{(III')} \quad \lim_{n\to\infty} \frac{1}{a_n} \sum_{\nu=1}^{n} \left|\int_{|x|<\eta a_n} x dV_\nu^*(x)\right| = 0.$$

As Feller showed in the main part of his paper, conditions (I'), (II'), (III') with V_k^* replaced by V_k are sufficient for

$$V_1 \star V_2 \star \cdots \star V_n(a_n x) \to \Phi(x) \text{ and } \max_{1\le k\le n} \int_{|x|>\varepsilon a_n} dV_k(x) \to 0 \;\forall \varepsilon > 0. \quad (6.39)$$

If these conditions are true for $V_k(x)$ (instead of V_k^*), they must also be true for the "symmetric" distributions $1 - V_k(-x)$, and the convergence to the normal distribution still holds if a subsequence $(V_{k'}(x))$ of $(V_k(x))$ is substituted by $1 - V_{k'}(-x)$. Apparently, this circumstance motivated Feller [1935, 525] to introduce the designation "a sequence of distributions $(V_k(x))$ belongs to $\Phi(x)$ in a narrower sense" if (6.39) remains valid for the case that any subsequence $(V_{k'}(x))$ of $(V_k(x))$ is substituted by $(1 - V_{k'}(-x))$ (the sequence (a_n) remaining unchanged). On the other hand, if the convergence to the normal distribution is "in a narrower sense," then (III) can be replaced by (III') (with V_k^* replaced by V_k), because in case of negative summands it is possible to switch from $V_k(x)$ to $1 - V_k(-x)$.[21] Then, as a consequence, also (II) can be substituted by (II').

Auxiliary Theorem 4 [Feller 1935, 537 f.]: Let (V_k) be a sequence of distributions, and let b_k be defined according to (6.34). Then

$$\lim_{n\to\infty} \frac{1}{a_n} \sum_{k=1}^{n} b_k = 0$$

(this is (6.37)!) if and only if for all positive η the relation

$$\lim_{n\to\infty} \frac{1}{a_n} \sum_{k=1}^{n} \int_{|x|<\eta a_n} x\,dV_k(x) = 0$$

(this is condition (III)!) holds.

6.2.5.2 Main Theorem

By virtue of Lévy's theorem on the continuous correspondence between characteristic functions and distributions, Feller expressed the "main theorem" in the form that the joint validity of (I), (II), and (III) is equivalent to

$$\prod_{\nu=1}^{n} v_\nu\left(\frac{t}{a_n}\right) \to \mathrm{e}^{-\frac{t^2}{2}} \quad \left(v_\nu(t) = \int_{-\infty}^{\infty} \mathrm{e}^{\mathrm{i}xt}\,dV_\nu(x)\right) \qquad (6.40)$$

together with

$$\max_{1\le\nu\le n} \left| v_\nu\left(\frac{t}{a_n}\right) - 1 \right| \to 0 \quad (n \to \infty), \qquad (6.41)$$

uniformly in each bounded t-interval, respectively.

---

[21] Note that (6.39) also implies $a_n \to \infty$. For fixed η and fixed ν the summands $\int_{|x|>\eta a_n} x\,dV_\nu(x)$ therefore have the same sign from a certain number n on.

In the first part of the proof for the "main theorem," Feller [1935, 538–540] presupposed (I), (II), and (III), and he showed (6.40) and (6.41). To this aim he essentially used Lévy's method for proving the CLT under the Lindeberg condition by means of characteristic functions, however in a modified form, because in the series expansion of $\log \upsilon_\nu$ derivatives of υ_ν could not be used since the existence of moments was not assumed any longer.

The problem of the first-order moments of the truncated random variables which occurred in (II) and complicated the situation, Feller solved in the following way: Instead of the sequence (V_k) he considered the sequence (V_k^*) defined by $V_k^*(x) := V_k(x + b_k)$, where b_k was determined according to (6.34). By virtue of auxiliary theorem 3 the sequence $(V_k^*(x))$ meets the conditions (I'), (II'), and (III'). Auxiliary theorems 4 and 1 imply that $(V_k(x))$ together with the norming factors a_n belongs to $\Phi(x)$ if and only if $(V_k^*(x))$ together with the norming factors a_n belongs to $\Phi(x)$. In proving that (I), (II), (III) are sufficient for the assertion of the "main theorem," Feller without loss of generality could therefore assume that (V_k) even possessed the properties (I'), (II'), and (III').

The essential steps in Feller's further argumentation were as follows: First, (6.41) was an immediate consequence of (I') [Feller 1935, 538 f.]. Aiming at the subsequent discussion of the series expansion of $\log \upsilon_\nu \left(\frac{t}{a_n}\right)$, Feller showed (basically on account of (6.41)) that $\sum_{\nu=1}^{n} \left|1 - \upsilon_\nu \left(\frac{t}{a_n}\right)\right|$ has in each bounded t-interval an upper bound independent of n. For $n \to \infty$ this property yielded, because of

$$\log \upsilon_\nu \left(\frac{t}{a_n}\right) = -\left(1 - \upsilon_\nu \left(\frac{t}{a_n}\right)\right) + \mathrm{o}\left(1 - \upsilon_\nu \left(\frac{t}{a_n}\right)\right),$$

the limit relation

$$\left|\sum_{\nu=1}^{n} \log \upsilon_\nu \left(\frac{t}{a_n}\right) + \sum_{\nu=1}^{n} \left\{1 - \upsilon_\nu \left(\frac{t}{a_n}\right)\right\}\right| \to 0,$$

uniformly in each bounded t-interval. Therefore, presupposing (I'), (II'), and (III'), one had to prove only that

$$\left|\sum_{\nu=1}^{n} \int_{-\infty}^{\infty} (1 - \mathrm{e}^{\frac{\mathrm{i}xt}{a_n}}) dV_\nu(x) - \frac{t^2}{2}\right| \to 0 \tag{6.42}$$

uniformly in each bounded t-interval. Feller [1935, 540] showed this by an estimate of the left side of (6.42), which was reached by means of expanding $\mathrm{e}^{\frac{\mathrm{i}xt}{a_n}}$ up to the third power in xt.

For his proof of the necessity of (I), (II), and (III), Feller [1935, 543–551] presupposed (6.40) and (6.41), and he based his considerations on the expansion

$$\log \upsilon_\nu\left(\frac{t}{a_n}\right) = -\int_{-\infty}^{\infty}(1-\mathrm{e}^{\frac{\mathrm{i}xt}{a_n}})dV_\nu(x) - \frac{1}{2}\left(\int_{-\infty}^{\infty}(1-\mathrm{e}^{\frac{\mathrm{i}xt}{a_n}})dV_\nu(x)\right)^2 + \\ + \mathrm{o}\left(\left|\int_{-\infty}^{\infty}(1-\mathrm{e}^{\frac{\mathrm{i}xt}{a_n}})dV_\nu(x)\right|^2\right). \quad (6.43)$$

For a reasonable application of this representation it was necessary to reduce the general case to the case of zero medians, because only then an upper bound of

$$\sum_{\nu=1}^{n}\left|\int_{-\infty}^{\infty}(1-\mathrm{e}^{\frac{\mathrm{i}xt}{a_n}})dV_\nu(x)\right|^2, \quad (6.44)$$

independent of n, could be established. To this aim, Feller first defined the sequence (b_k) according to the demand of a zero median of $V_k^*(x) := V_k(x + b_k)$. Then, on the basis of the presupposed condition (6.41), he [1935, 543] also proved that, for each $\eta > 0$,

$$\lim_{n\to\infty}\int_{|x|>\eta a_n} dV_n(x) = 0. \quad (6.45)$$

Without any further explanations, Feller from this equation followed that (6.38) was true.[22] As a consequence of auxiliary theorem 2, $(V_k(x))$ obeyed the conditions (I) and (II) if and only if $(V_k^*(x))$ obeyed these conditions. Feller [1935, 544] proved that (6.38) implies the validity of (6.41) even for the characteristic functions $\upsilon_\nu^*(t) = \upsilon_\nu(t)\mathrm{e}^{-\mathrm{i}b_\nu t}$ of the distributions V_ν^*, whereas (6.40) yields $\prod_{\nu=1}^{n}\left|\upsilon_\nu^*(\frac{t}{a_n})\right| \to \mathrm{e}^{-\frac{t^2}{2}}$. In order to conclude that (I) and (II) are necessary for the assertion that $(V_\nu(x))$ belongs to $\Phi(x)$, it therefore sufficed to prove that (I) and (II) are true for each sequence of distribution functions V_ν with zero medians, if for the characteristic functions of these distributions

$$\prod_{\nu=1}^{n}\left|\upsilon_\nu\left(\frac{t}{a_n}\right)\right| \to \mathrm{e}^{-\frac{t^2}{2}} \quad (6.46)$$

and

$$\max_{1\le\nu\le n}\left|\upsilon_\nu\left(\frac{t}{a_n}\right) - 1\right| \to 0 \quad (6.47)$$

are true, uniformly in each bounded t-interval, respectively.

Generally presupposing in the sequel zero medians for the single distributions, Feller from (6.46) and (6.47) inferred that

---

[22] This conclusion is not a trivial one, however. Let us assume that there exists $\varepsilon > 0$ and a subsequence $\left(\frac{b_{n_k}}{a_{n_k}}\right)$ such that for all n_k above a certain number n_0: $|b_{n_k}| > \varepsilon a_{n_k}$. Then, since $(X_n - b_n)$ has a zero median, we have for all $n_k > n_0$,

$$P(|X_{n_k}|>\varepsilon a_{n_k}) \ge P(|X_{n_k}|\ge|b_{n_k}|) \ge \min\left(P(X_{n_k}-b_{n_k}\ge 0), P(X_{n_k}-b_{n_k}\le 0)\right) \ge \frac{1}{2}.$$

The latter statement contradicts (6.45).

$$\sum_{\nu=1}^{n} \operatorname{Re} \log \upsilon_\nu \left(\frac{t}{a_n}\right) \to -\frac{t^2}{2} \tag{6.48}$$

uniformly in each bounded t-interval. By elementary estimates for the summands in (6.44) (using in particular the zero median property of all distributions), and taking into account (6.43), (6.46), and (6.47), Feller concluded that this sum in each bounded t-interval has an upper bound independent of n, and therefore by virtue of (6.43),

$$\left| \sum_{\nu=1}^{n} \log \upsilon_\nu \left(\frac{t}{a_n}\right) + \right.$$
$$\left. + \sum_{\nu=1}^{n} \left\{ \int_{-\infty}^{\infty} (1 - e^{\frac{ixt}{a_n}}) dV_\nu(x) + \frac{1}{2} \left(\int_{-\infty}^{\infty} (1 - e^{\frac{ixt}{a_n}}) dV_\nu(x) \right)^2 \right\} \right| \to 0 \tag{6.49}$$

uniformly in each bounded t-interval. Using his estimates, Feller further inferred from (6.47) that

$$\sum_{\nu=1}^{n} \left(\int_{-\infty}^{\infty} (1 - \cos \frac{xt}{a_n}) dV_\nu(x) \right)^2 \to 0.$$

Altogether, by combining (6.48), (6.49), and the latter limit relation, Feller showed that

$$\sum_{\nu=1}^{n} \left\{ \int_{-\infty}^{\infty} (1 - \cos \frac{xt}{a_n}) dV_\nu(x) - \frac{1}{2} \left(\int_{-\infty}^{\infty} \sin \frac{xt}{a_n} dV_\nu(x) \right)^2 \right\} \to \frac{t^2}{2} \tag{6.50}$$

uniformly in each bounded t-interval.

Feller [1935, 547] designated (6.50) "the fundamental relation from which the necessity of (I) and (II) can be derived." In fact, several intricate, if elementary, estimates were still needed. Feller in particular derived the inequalities

$$\sum_{\nu=1}^{n} \int_{|x|>\eta a_n} dV_\nu(x) < M'$$

and

$$\frac{1}{a_n^2} \sum_{\nu=1}^{n} \int_{|x|<\eta a_n} x^2 dV_\nu(x) < K,$$

where the upper bounds M' and K are dependent on η but independent of n. By use of these estimates and the relation (6.45), from (6.50) the less complex relation

$$\sum_{\nu=1}^{n} \left\{ \int_{-\infty}^{\infty} (1 - \cos \frac{xt}{a_n}) dV_\nu(x) - \frac{t^2}{2a_n^2} \left(\int_{|x|<\eta a_n} x dV_\nu(x) \right)^2 \right\} \to \frac{t^2}{2} \tag{6.51}$$

could be followed. For $\varepsilon, \eta > 0$ Feller assumed the number $\tau = \tau(\varepsilon, \eta)$ so small that for $|t| < \tau$ and $|x| < \eta a_n$ the inequality

$$1 - \cos \frac{xt}{a_n} \geq (1 - \varepsilon) \frac{x^2 t^2}{2a_n^2}$$

holds. Using this inequality he showed that for sufficiently small $|t|$ the left side of (6.51) is at least equal to

$$\frac{t^2}{2a_n^2} \sum_{\nu=1}^{n} \left\{ \int_{|x|<\eta a_n} x^2 dV_\nu(x) - \left(\int_{|x|<\eta a_n} x dV_\nu(x) \right)^2 \right\} - \frac{t^2}{2} K\varepsilon.$$

From this latter estimate, since ε could be considered as arbitrarily small, the inequality

$$\limsup_{n\to\infty} \frac{1}{a_n^2} \sum_{\nu=1}^{n} \left\{ \int_{|x|<\eta a_n} x^2 dV_\nu(x) - \left(\int_{|x|<\eta a_n} x dV_\nu(x) \right)^2 \right\} \leq 1 \qquad (6.52)$$

ensued. By estimating the left side of (6.51) from above, Feller in a similar way obtained

$$\left(1 - \liminf_{n\to\infty} \frac{1}{a_n^2} \sum_{\nu=1}^{n} \left\{ \int_{|x|<\eta a_n} x^2 dV_\nu(x) - \right.\right.$$
$$\left.\left. - \left(\int_{|x|<\eta a_n} x dV_\nu(x) \right)^2 \right\} \right) \frac{t^2}{2} \leq 2M'. \qquad (6.53)$$

Because, due to (6.52), the left side of (6.53) cannot get negative, and because (6.53) holds even for arbitrarily large t,

$$\liminf_{n\to\infty} \frac{1}{a_n^2} \sum_{\nu=1}^{n} \left\{ \int_{|x|<\eta a_n} x^2 d V_\nu(x) - \left(\int_{|x|<\eta a_n} x d V_\nu(x) \right)^2 \right\} = 1$$

ensued. Finally, condition (II) was an immediate consequence of this equation together with (6.52).

Feller's discussion of the necessity of (I) and (III) was based on the already shown necessity of (II) (the assumption of zero medians was not needed any longer). (I) was an almost direct consequence of (II) [Feller 1935, 550]. By auxiliary theorem 3, from (I) and (II) the validity of (I'), (II'), and (III') for $V_k^*(x) = V_k(x + b_k)$ could be followed, if b_k was according to (6.34). As a consequence of (I'), (II'), and (III') $(V_k^*(x))$ belonged to $\Phi(x)$, and therefore Feller was able to follow, through auxiliary theorem 1, the relation (6.37), and from this, by auxiliary theorem 4, the validity of (III).

6.2.5.3 Criterion

For a proof of the "criterion," Feller generally presupposed zero medians, and he first showed that (6.33) is a necessary consequence of

$$V_1(a_n x + b_1) \star \cdots \star V_n(a_n x + b_n) \to \Phi(x) \tag{6.54}$$

together with

$$\forall x \neq 0: \quad \max_{1 \le k \le n} |V_k(a_k x + b_k) - E(x)| \to 0, \tag{6.55}$$

a_n and b_n being suitable constants. On the basis of (6.54) and (6.55), by virtue of the "main theorem," conditions (I) and (II) for $V_k(x + b_k)$ could be verified. Since all distributions had zero medians, (6.55) yielded relation (6.38). By means of auxiliary theorem 2, from the validity of (I) and (II) for $V_k(x + b_k)$ the validity of both conditions for $V_k(x)$ ensued. Almost directly from the latter fact, Feller [1935, 552] eventually followed the assertion (6.33).

In order to show that (6.33) is sufficient for (6.54) together with (6.55), Feller proved the following, even more general theorem:
Let (q_n) be a sequence of positive numbers and (V_n) a sequence of distributions (not necessarily with a zero medians) such that

$$\lim_{n\to\infty} \frac{1}{q_n^2} \sum_{\nu=1}^{n} \int_{|x|<q_n} x^2 dV_\nu(x) = \infty \quad \text{and} \quad \lim_{n\to\infty} \sum_{\nu=1}^{n} \int_{|x|>q_n} dV_\nu(x) = 0. \tag{6.56}$$

Then (V_n) together with (a_n), where

$$a_n^2 = \sum_{\nu=1}^{n} \left(\int_{|x|\le q_n} x^2 dV_\nu(x) - \left(\int_{|x|\le q_n} x dV_\nu(x) \right)^2 \right),$$

is in accord with the conditions (I) and (II).
For the proof of this theorem, Feller [1935, 553] established, as a consequence of (6.56), the relation $\frac{q_n}{a_n} \to 0$, which immediately implied (I). Furthermore, from this limit relation, for each $\eta > 0$ and for sufficiently large n, the inequality $q_n < \eta a_n$ ensued. By elementary estimates Feller showed that

$$\left| \frac{1}{a_n^2} \sum_{\nu=1}^{n} \left(\int_{|x|<\eta a_n} x^2 dV_\nu(x) - \left(\int_{|x|<\eta a_n} x dV_\nu(x) \right)^2 \right) - 1 \right| \le 3\eta^2 \sum_{\nu=1}^{n} \int_{|x|>q_n} dV_\nu(x),$$

wherefrom, due to the second part of (6.56), property (II) followed. Feller finally referred to auxiliary theorem 3 concerning the property that $V_k(x + b_k)$ even meets the conditions (I'), (II'), and (III') if b_k is defined according to (6.34). The condition (6.56) therefore actually implied the assertions (6.54) and (6.55), even in the "narrower sense" of convergence.

6.2.5.4 Necessity of Lindeberg Condition

Feller's 1935 work became especially prominent by its separate discussion of the Lindeberg condition. Feller showed that this condition (6.35) in the particular case $a_n^2 = s_n^2 = \sum \text{Var}X_k$ and $\text{E}X_k = 0$ was a consequence of (6.40) together with (6.41): At first one concludes from (6.40) and (6.41) that, uniformly in each bounded t-interval,

$$\sum_{\nu=1}^{n} \int_{-\infty}^{\infty} (1 - \mathrm{e}^{\frac{\mathrm{i}xt}{s_n}}) dV_\nu(x) \to \frac{t^2}{2},$$

wherefrom

$$\lim_{n\to\infty} \left| \frac{t^2}{2} - \sum_{\nu=1}^{n} \int_{-\infty}^{\infty} \left(1 - \cos\frac{xt}{s_n}\right) dV_\nu(x) \right| = 0 \tag{6.57}$$

ensues. By virtue of the Bienaymé–Chebyshev inequality we have for each positive η

$$\sum_{\nu=1}^{n} \int_{|x|\geq \eta s_n} dV_\nu(x) \leq \frac{1}{\eta^2}.$$

Taking into account this inequality one infers from (6.57):

$$\limsup_{n\to\infty} \left| \frac{t^2}{2} - \sum_{\nu=1}^{n} \int_{|x|<\eta s_n} \left(1 - \cos\frac{xt}{s_n}\right) dV_\nu(x) \right| \leq \frac{2}{\eta^2}. \tag{6.58}$$

If one combines (6.58) with the estimate

$$\sum_{\nu=1}^{n} \int_{|x|<\eta s_n} \left(1 - \cos\frac{xt}{s_n}\right) dV_\nu(x) \leq \frac{t^2}{2s_n^2} \sum_{\nu=1}^{n} \int_{|x|<\eta s_n} x^2 dV_\nu(x) \leq \frac{t^2}{2},$$

then one obtains

$$0 \leq \frac{t^2}{2} \limsup_{n\to\infty} \left\{ 1 - \frac{1}{s_n^2} \sum_{\nu=1}^{n} \int_{|x|<\eta s_n} x^2 dV_\nu(x) \right\} \leq \frac{2}{\eta^2}.$$

Because the expression within the braces is ≥ 0, and because the latter inequality holds for all real t, the assertion (6.35) follows.

With this proof Feller definitely brought the classical central limit problem, as it could be essentially led back to Laplace, to a completion. His discussion of the Lindeberg condition could be comprehended quite easily, and this was probably one of the main reasons why Feller's results for the CLT were always given more attention than Lévy's.

6.3 A Question of Priority?

"I never had luck with the law of Gauss." This statement by Lévy, already quoted elsewhere in this text (see Sect. 5.2.6.5), seems particularly applicable to his work [1935b] in view of its competitive relationship with that of Feller [1935]. Lévy claimed that the reason why Feller was generally championed for the definitive solution of the CLT, at least where it concerned sequences of independent and uniformly negligible random variables with respect to the total sum, was because the latter's work had been published a bit earlier than his own. As we have already seen (Sect. 6.2.3), there were no significant errors in Lévy's very tersely outlined and not entirely precise proofs that could be held responsible for this disregard of his work.

Le Cam [1986, 85] carefully traced the chronology surrounding the publication of the two articles by Lévy and Feller. According to these findings, Lévy submitted his work significantly earlier than Feller. Moreover, Lévy had already distributed private drafts to a few colleagues, including Hadamard, Borel, and Khinchin (see [Lévy 1970, 108]) in June of 1935. Therefore, strictly speaking, Lévy is entitled to the priority because he had made his paper available to a "public," however small, in the form of a "preprint" considerably earlier than Feller. The issue of the *Journal des Mathématiques* containing Lévy's essay was delivered in December 1935. The corresponding issue of *Mathematische Zeitschrift* with Feller's article should have appeared at about the same time, but the exact delivery date can no longer be determined. For this reason, it is not at all certain that Feller's work actually was published before Lévy's. Thus the relatively modest interest in Lévy's article compared to Feller's can hardly be explained by a later publication date.

In his book on sums of random variables [1937a, 107 (footnote)], Lévy expressly referred to the fact that he had already submitted his paper on necessary conditions for the CLT in the fall of 1934, and suggested that Feller had merely "rediscovered" the theorem. However, in the only monograph on limit theorems besides Lévy's book to be published between 1935 and the outbreak of World War II, and to be observed by a broader audience,[23] Cramér [1937/70, 63] mentions only Feller's name when discussing this subject. Even though Feller's article developed during his residence with Cramér's "Stockholm group" and though Cramér also had a close personal relationship with his colleague Feller (see [Cramér 1976]), Cramér nevertheless usually had high praise for Lévy's work, as is evident from his frequent positive citations of Lévy's achievements. Finally, in their work on limit distributions of sums of independent random variables, which has remained a definitive standard reference practically to this day, Gnedenko and Kolmogorov [1949/68, 130] mention only Feller in connection with necessary conditions. So Lévy was evidently unable to substantiate his claims of priority even from the—usually more generous—retrospective point of view.

Yet is it even possible to speak meaningfully of "priority" when dealing with two works that differ so acutely in style and methods, but also in several particular

[23] Khinchin's survey of sums of independent random variables [1938] remained unnoticed outside Russia.

results? Even Lévy [1970, 108 (footnote)] acknowledges this, along with the easier "applicability" of Feller's results. He stresses, though: ... but basically, we are talking about almost exactly the same theorem." For this latter statement—already expressed in his 1937 book—Lévy never provided any further arguments.

6.3.1 Lévy's and Feller's Results: A Comparison

When the "criterion" of Feller (6.33) and Leǘy's version (6.21) are brought into the same form, then a striking similarity occurs between both assertions.

Let (X_k) be a sequence of independent random variables whose distributions V_k all have a median 0.[24] The main assertion of Feller corresponds to:
There exist sequences $(a_n > 0)$ and (b_k) of real numbers such that

$$P\left(\frac{1}{a_n}\sum_{k=1}^{n}(X_k - b_k) \le x\right) \to \Phi(x)$$
$$\text{and } \max_{1\le k\le n} P(|X_k - b_k| > \varepsilon a_n) \to 0 \quad \forall \varepsilon > 0 \quad (6.59)$$

as $n \to \infty$ if and only if

$$\forall\delta > 0\forall\eta > 0\exists n(\delta,\eta)\forall n \ge n(\delta,\eta) : \quad \frac{p_n^2(\delta)}{\sum_{k=1}^{n}\int_{|x|\le p_n(\delta)} x^2 dV_k(x)} < \eta, \quad (6.60)$$

$$\text{where} \quad p_n(\delta) = \min\left\{r \in \mathbb{R}_0^+ \,|\, P(|X_k| > r) \le \delta\right\}.$$

Lévy's main assertion can be expressed as follows:
Let L_n be the dispersion of $\sum_{k=1}^{n} X_k$ assigned to an arbitrary, however fixed, probability $\gamma \in]0;1[$. There exist sequences $(a_n > 0)$ and (b_k) of real numbers such that

$$P\left(\frac{1}{a_n}\sum_{k=1}^{n}(X_k - b_k) \le x\right) \to \Phi(x)$$
$$\text{and } \max_{1\le k\le n} P(|X_k| > \varepsilon L_n) \to 0 \quad \forall \varepsilon > 0 \quad (6.61)$$

as $n \to \infty$ if and only if

$$\forall\delta > 0\forall\eta > 0\exists n(\delta,\eta)\forall n \ge n(\delta,\eta)\exists X(n) > 0 :$$
$$\frac{X^2(n)}{\sum_{k=1}^{n}\left(\int_{|x|\le X(n)} x^2 dV_k(x) - \left(\int_{|x|\le X(n)} x\, dV_k(x)\right)^2\right)} < \eta \quad (6.62)$$

[24] With respect to the zero median property in connection with Lévy's account, see Sect. 6.2.1.

and

$$\sum_{k=1}^{n} P\left(|X_k| > X(n)\right) < \delta.$$

Despite the far-reaching formal conformity of (6.60) and (6.62), a direct proof for the equivalence of these two conditions seems to be rather difficult. Still, the equivalence of the respective assertions of Feller and Lévy concerning the convergence to the normal distribution can be quite readily seen as follows:
From Feller's constraint

$$\max_{1\leq k\leq n} P(|X_k - b_k| > \varepsilon a_n) \to 0,$$

due to the zero median property of all distributions under consideration, the relation

$$\max_{1\leq k\leq n} P(|X_k| > \varepsilon a_n) \to 0$$

follows.[25] Because the orders of magnitude of a_n and L_n are asymptotically equal, Levy's constraint follows.

In turn, Lévy's (6.61) implies the validity of the "loi des grands nombres"; we use it in the version

$$\sum_{k=1}^{n} P(|X_k| > \varepsilon L_n) \to 0 \quad \forall\varepsilon > 0.$$

$L_n = \mathrm{O}(a_n)$ yields

$$\sum_{k=1}^{n} P(|X_k| > \varepsilon a_n) \to 0 \quad \forall\varepsilon > 0. \tag{6.63}$$

The latter relation is Feller's condition (I), exactly. Since in Lévy's proof of the CLT under the condition $P(\max_{1\leq k\leq n} |X_k| > \varepsilon L_n) \to 0$ (which is equivalent to (6.63)), the norming constants which correspond to a_n are determined according to

$$a_n^2 \sim \sum_{k=1}^{n}\left(\int_{|x|\leq\varepsilon L_n} x^2 dV_k(x) - \left(\int_{|x|\leq\varepsilon L_n} x dV_k(x)\right)^2\right)$$

($\varepsilon > 0$ being arbitrarily small), and because a_n and L_n are of the same order of magnitude for $n \to \infty$, Feller's condition (II) ensues. As Feller [1935, 533–537] has shown, from (I) and (II) the validity of conditions (I'), (II'), and (III') for $V_k^*(x) =$

[25] By an argument very similar to that in footnote 22 one shows that

$$\lim_{n\to\infty} \max_{1\leq k\leq n} \frac{b_k}{a_n} = 0,$$

wherefrom the assertion follows.

$V_k(x+b_k)$ follows, if, as usual for him, b_k is defined in accord with (6.34). Finally, (I'), (II'), and (III') together imply the assertion (6.59), as we have seen in Sect. 6.2.5.2.

Thus, Lévy and Feller actually discussed equivalent problems concerning the convergence of the distribution of suitably normed sums to the Gaussian law. Basically, the contemporary reader was able to find out this fact, but only if he studied the proof details very carefully and comprehensively, a task which might have been, due to Lévy's idiosyncratic style and his sketchy exposition, rather tedious.

6.3.2 Another Question of Priority

Gnedenko [1997, 468] claimed that Bernshtein [1926] was the first posing the problem "when can such constants $B_n > 0$ and A_n be found that the distribution function of sums $(S_n - A_n)/B_n$ converges to the normal distribution?", and that he also gave sufficient conditions for the convergence to the normal distribution in this setting. Gnedenko continued: "Some eight years later Feller showed that the same conditions are not only sufficient but also necessary under the condition that all terms in the sum are uniformly small." With these words, Gnedenko renewed and even expanded a claim that Bernshtein [1945/2004a, 89, 94] himself had laid.

As we have seen, Bernshtein had indicated the possibility of a general norming for sums of random variables in his papers of 1922 and 1926. By using Feller's methods, it can be shown that, under the additional assumption of smallness of the single summands (second part of (6.59)), Bernshtein's general conditions (those referring to suitably truncated random variables) for his *lemme fondamental* are also necessary for the convergence to the normal distribution if convergence is conceived in the "narrower sense" of Feller. Bernshtein's "lemme fondamental" can be formulated for a sequence (X_k) of independent random variables in the following way: Let (K_n) be a sequence of positive numbers, and let

$$r_n := \sqrt{\sum_{k=1}^{n} \int_{|x|\leq K_n} x^2 dF_k(x)}.$$

Under the condition that for $n \to \infty$

$$\sum_{k=1}^{n} \int_{|x|>K_n} dF_k(x) \to 0, \tag{6.64}$$

$$\frac{1}{r_n} \sum_{k=1}^{n} \left| \int_{|x|\leq K_n} x dF_k(x) \right| \to 0, \tag{6.65}$$

and

$$\frac{1}{r_n^3}\sum_{k=1}^{n}\int_{|x|\le K_n}|x|^3 dF_k(x)\to 0, \tag{6.66}$$

$F_1 \star F_2 \star \cdots \star F_n(r_n x)$ converges to $\Phi(x)$ as $n\to\infty$.

If we presuppose in turn that there exists a positive sequence (a_n) such that

$$F_1 \star F_2 \star \cdots \star F_n(a_n x) \to \Phi(x) \quad \text{and} \quad \max_{1\le k\le n} P(|X_k|>\varepsilon a_n)\to 0 \quad \forall \varepsilon>0, \tag{6.67}$$

and if we suppose the convergence to the normal distribution in (6.67) being "in a narrower sense" in Feller's terminology, then (see Sect. 6.2.5.1) the conditions

$$\frac{1}{a_n}\sum_{k=1}^{n}\left|\int_{|x|\le\varepsilon a_n} x dF_k(x)\right|\to 0 \quad \forall\varepsilon>0,$$

$$\sum_{k=1}^{n}\int_{|x|>\varepsilon a_n} dF_k(x)\to 0 \quad \forall\varepsilon>0,$$

and

$$\left|\frac{1}{a_n^2}\sum_{k=1}^{n}\int_{|x|\le\varepsilon a_n} x^2 dF_k(x)-1\right|\to 0 \quad \forall\varepsilon>0$$

follow. One can easily show now that there even exists a null sequence (ε_n) such that

$$\sum_{k=1}^{n}\int_{|x|>\varepsilon_n a_n} dF_k(x)\to 0,$$

$$\frac{1}{a_n}\sum_{k=1}^{n}\left|\int_{|x|\le\varepsilon_n a_n} x dF_k(x)\right|\to 0,$$

and

$$\frac{1}{a_n^2}\sum_{k=1}^{n}\int_{|x|\le\varepsilon_n a_n} x^2 dF_k(x)\to 1.$$

If one sets

$$K_n := \varepsilon_n a_n,$$

then immediately $r_n \sim a_n$ and Bernshtein's conditions (6.64) and (6.65) follow. We also have

$$\frac{1}{r_n^3}\int_{|x|\le K_n}|x|^3 dF_k(x)\le\frac{a_n^3}{r_n^3}\varepsilon_n\frac{1}{a_n^2}\int_{|x|\le\varepsilon_n a_n} x^2 dF_k(x)\to 1\cdot 0\cdot 1=0.$$

Therefore, Bernshtein's condition (6.66) holds as well.

Although Bernshtein's conditions are necessary in a quite general sense for the convergence to the normal distribution, a comparison between Feller's and Lévy's 1935 work on the one hand and the brief remarks in his 1922 and 1926 documents on the other do not seem to be appropriate. A general problem on sums of independent random variables, linked with a corresponding research program as described by Gnedenko, cannot be found in Bernshtein's contributions. On the contrary, only Feller expressed the central limit problem so clearly as indicated by Gnedenko, and there is no evidence that he was in any way influenced by Bernshtein. We should rather see Bernshtein's retrospective claim and its affirmation by Gnedenko as an example of typical statements which were to confirm the superiority of Soviet science.

6.3.3 A Question of Methods and Style

The fact, deplored by Lévy, that Feller's 1935 article had gained much greater acclaim than his own [1935b] can surely be traced back in part to Feller's extensive discussion of the Lindeberg condition and his resultant success in bringing to a certain conclusion the long-lasting mathematical development of the classical CLT in its integral version, i.e., for distribution functions. Feller's work made the reader aware that an important question had been answered once and for all, while at the same time a fresh start was being made by considering sums of independent random variables which reach a particular limit distribution by means of suitable norming regardless of whether any moments exist. Thereafter, Feller himself eagerly cultivated the myth that the idea of a general "nonclassical" norming was essentially his alone (e.g., [Feller 1945, 818]) and he acknowledged Lévy's contributions to the general CLT only in the case of identically distributed random variables [Feller 1937a, 304; 1945, 820]. In a paper that supplemented his 1935 article with further necessary and sufficient conditions, Feller [1937a, 306–309] provided a "criterion" that was practically identical to the one proposed by Lévy (6.62) without ever mentioning this fact.

However, one other circumstance may have played an even greater role in the fact that almost all involved granted Feller priority for the "definitive" solution to the problem of convergence to the normal distribution: Using arguments that occasionally were cumbersome but fairly complete and not just vaguely indicated, the strictly analytical orientation of his article made it possible, even for someone who knew nothing about probability theory, to understand problems and solution strategies that were far from elementary, even by today's standards. Though Feller may have largely avoided stochastic concepts, his style corresponded extensively to the manner of representation favored by other major proponents of probability theory in the 1930s, such as Cramér, Khinchin, and Kolmogorov. It was this common, analytically shaped style that would also shape the momentous Gnedenko and Kolmogorov monograph on sums of independent random variables, which was published in Russian in 1949 and was followed by numerous translations.

By contrast, Lévy the former analyst—atypical in his avoidance of epsilontics—had forsworn the traditional analytical methodology and now preferred "intuitive" methods of representation and conclusion, ones that were consistent with the "true essence" of probability theory. Though Lévy's and Feller's articles may exhibit similarities in terms of basic ideas, they were obeying very different concepts regarding their methods.

Whereas Feller was principally interested in the suitable manipulation and estimation of integrals for the study of characteristic functions, Lévy's main goal was calculating the probabilities in question himself and doing so by methods that were also important in other branches of probability theory which relied much more heavily on concepts of measure theory, such as in connection with strong laws of large numbers. This concerns Lévy's concept of concentration and dispersion, for instance. Here Lévy used elementary but—unlike the characteristic functions—less elaborated concepts with which mathematicians were usually not familiar. His emphasis on "intuitive" phrasing and notions corresponded to a representation in which many explanations were merely sketched out; this feature of Lévy's work did not change significantly in his 1937 book.

How much Lévy's main result regarding sufficient and necessary conditions for convergence of distributions of normed sums to the normal distribution was disregarded by the mathematical community, is also evidenced by the history of the reception of a criterion by Khinchin [1938], very similar to Lévy's "loi des grands nombres," which had been derived in context with Khinchin's research on infinitely divisible limit distributions for sums of negligible random variables (see [Gnedenko & Kolmogorov 1949/68, 126]). Carried over to the case of normed sums $s_n = \sum_{k=1}^{n} \frac{\xi_k - a_k}{B_n}$ of independent random variables ξ_k with

$$\sup_{1 \leq k \leq n} P(|\xi_k - a_k| > \varepsilon B_n) \to 0 \quad \forall \varepsilon > 0,$$

Khinchin's criterion, as reproduced by Gnedenko [1959/2004, 171], is as follows:

> If a limiting distribution for the normed sums s_n exists, then for it to be normal, it is necessary and sufficient that the terms satisfy one single condition, viz., that, as $n \to \infty$
>
> $$P(\sup |\xi_k - a_k| \geq \varepsilon B_n) \to 0, \quad 1 \leq k \leq n \tag{6.68}$$
>
> [for all $\varepsilon > 0$].

Gnedenko (same place) at least conceded that "a similar, and even a somewhat more general formulation" could be found in Lévy's book [1937a]; he did not specify, however, this reference by giving page numbers. As we can see from the portion of text in [Gnedenko & Kolmogorov 1949/68, 127 f.] discussing the same topic, Gnedenko with his hint at "a more general formulation" referred to a result from the theory of stochastic processes [Lévy 1937a, 166–172], which is actually related to Khinchin's criterion. Anyway though, Gnedenko did not give any comment on the striking analogy between Lévy's "loi des grands nombres," which under the above-described assumptions could be formulated according to

$$P(\sup_{1\le k\le n}|\xi_k - a_k| > \varepsilon L_n) \to 0 \quad (n \to \infty),$$

and Khinchin's (6.68). Gnedenko's neglect of Lévy's achievement (which is also apparent in the monograph [Gnedenko & Kolmogorov 1949]) may be taken as an indication that—with regard to the CLT and related problems—Lévy's approach via concentration and dispersion was not very attractive to him. The essential part of Gnedenko and Kolmogorov's 1949 book was constituted by a thorough discussion of infinitely divisible limit distributions for sums of asymptotically negligible random variables, which arose from Gnedenko's contributions around the last years of the 1930s. It is only natural that Gnedenko, who had made almost exclusive use of characteristic functions in this work, should not have been too interested in expounding different methods. On the other hand, Wolfgang Doeblin [1939], who closely collaborated with Lévy, actually succeeded in obtaining results very similar to those of Gnedenko by employing methods related to concentration and dispersion. Doeblin died in 1940 already, and his approach was not followed up for the time being (see Sect. 7.2.1).

Altogether, the relatively poor reception of Lévy's contributions surrounding the problem of necessary and sufficient criteria for the CLT was certainly due to the fact that almost all other contributors to this field preferred the "analytical" method of characteristic functions, on the one hand. On the other hand, however, one may also presume that Lévy's idiosyncratic presentation and style, which also shaped his 1937 book and its second—almost unmodified—edition of 1954, impeded a greater impact of his innovative methods.

Chapter 7
Generalizations

Lévy's and Feller's theorems of 1935 served as a paradigm for further work on sums of independent one- or multidimensional random variables, on the one hand. This strand of development largely preserved the "traditional" analytic orientation. On the other hand, generalizations toward martingales and random elements in metric spaces triggered a growing influence of measure theory even on the "classical" limit problems of probability.

7.1 Lévy on Sums of Nonindependent Random Variables

Around 1935, Lévy also showed a growing interest in sums of nonindependent random variables.[1] In connection with the CLT his main goals were, on the one hand, to further weaken the conditions of Bernshtein's "lemme fondamental," and, on the other, to adapt the treatment of problems on chained variables as far as possible to the treatment of corresponding problems on independent random variables.

7.1.1 Measure-Theoretic Background

Whereas problems dealing with the distributions of sums of independent random variables could be tackled on the basis of rather elementary concepts of real analysis, such as monotonically increasing functions (distribution functions) and convolutions of these functions, nonindependent variables required the use of more sophisticated measure-theoretic concepts. In his booklet *Grundbegriffe der Wahrscheinlichkeitsrechnung*, Andrei Nikolaevich Kolmogorov (1903–1987) introduced the standard which we are used to now, of basing probability theory—at least

[1] For lots of details on Lévy's pertinent work and his relations to other mathematicians in this respect, like Ville and Jessen, see *Electronic Journal for History of Probability and Statistics* **5**, June 2009.

H. Fischer, *A History of the Central Limit Theorem*, Sources and Studies in the History of Mathematics and Physical Sciences, DOI 10.1007/978-0-387-87857-7_7,

with respect to its "purely mathematical development" [Kolmogorov 1933/50, 3, fn 4]—on a field of abstract sets and a probability measure defined on it.[2] As we can see from some passing comments (e.g., [Lévy 1937a, 17]), Lévy credited himself for essential contributions (in particular [Lévy 1925a], see Sect. 5.2.3.4) to measure-theoretic aspects of probability theory.[3] On the other hand, he did not, at least during the 1930s, use the theory of probability spaces in its full generality, even if one can find a concise survey of measure theory on abstract sets in the second chapter of his 1937 book. Instead, Lévy tried to base his discussions, at least in the case of sums of random variables, on Borel subsets of $[0;1]^n$, where n was a (possibly infinite) natural number. For the construction of probability measures on $[0;1]^{\mathbb{N}}$ he [Lévy 1935a, 203; 1937a, 21 f.] referred to Daniell [1919], Jessen [1929], and Steinhaus [1930], but not to Kolmogorov. Altogether, Lévy's exposition of measure-theoretic issues was rather sketchy and not very well-organized. He did not define the notion of random variable, instead he used this designation in a purely intuitive way in context with (well-defined) one- or multidimensional distribution functions.

Lévy's standpoint regarding fundamental concepts of probability was formalistic insofar as

> The calculus of probability, the number of variables being finite or infinite, is in principle, from an abstract point of view, nothing else than a comfortable language for presenting certain results of measure theory in the sense of M. Lebesgue [Lévy 1935a, 203].

Even if the phrasing was apparently exaggerated, it shows Lévy's intention to put his probabilistic work within the framework of concepts and problems of Lebesgue's integration theory, and thus within the scope of real analysis. This provides a plausible explanation as to why Lévy did not take on Kolmogorov's theory in its full generality. Moreover, as one of the leading probabilists of the time, Lévy certainly wanted to give his readers an exposition in his own, idiosyncratic style.

If we consider other influential expositions (besides Kolmogorov's 1933 booklet) of "modern" probability for an assessment of Lévy's attitude toward measure theory in the framework of contemporary probability theory, namely, Cramér's *Random Variables and Probability Distributions* [1937] and Fréchet's *Géneralités sur les Probabilités* [1936/38, T. 1], Lévy's position with respect to axiomatics and "abstract" measure theory is situated, as regards content, rather close to Cramér, and somewhere between Kolmogorov and Fréchet. Despite seminal contributions to a theory of integration over abstract sets [1915] and to several types of convergence of sequences of random variables [1930], Fréchet, in his explanations of the fundamentals of probability theory, used the conventional language of random experiments ("épreuves"), even if, with a little hindsight, his principles, notions, and definitions can be translated into a set- (and measure-) theoretic language à la Kolmogorov [Hochkirchen 1999, 260–262]. Although Cramér (mainly in the first part of his book) directly referred to Kolmogorov and his axioms, he did not develop

---

[2] For comprehensive discussions of Kolmogorov's account and its measure-theoretic context, see [von Plato 1994, Chapt. 7; 2005; Hochkirchen 1999; Shafer & Vovk 2005; 2006].

[3] For more details on Lévy's self-assessment concerning his own achievements compared with Kolmogorov's, see [Shafer & Vovk 2005, 55].

the concept of random variable in its full generality, but restricted his considerations to the sample space $\mathbb{R}^k$ and the sigma-algebra of its Borel subsets.[4] Due to his exclusive focus on independent variables, he did not discuss conditional distributions or expectations. Lévy likewise saw probability as a sigma-additive nonnegative measure, and similar to Cramér, he mainly focused on subsets of $\mathbb{R}^n$. Regarding conditional distributions and expectations, however, he went substantially beyond Cramér's (and also Fréchet's) exposition.

7.1.2 Conditional Distribution and Expectation

Lévy's perception of basic notions in the case of nonindependent random variables, in particular regarding the conditional expectation of a random variable dependent on others, can primarily be seen from his 1937 book, especially § 12 and § 23.[5] Instead of dealing with a random variable X with a distribution function $F(x) = P(X < x)$,[6] he frequently used the trick to consider a random variable ξ, uniformly distributed in $]0; 1[$, and to represent X by ξ through $X = F^{-1}(\xi)$, where $F^{-1}(y) := \inf\{x|F(x) \geq y\}$.[7] In this way, probabilities and expectations related to X could be expressed by probabilities and expectations related to the uniformly distributed variable ξ. In other words, integrals $\int_{-\infty}^{\infty} \varphi(x)dF(x)$ could be reduced to Lebesgue integrals $\int_0^1 \varphi(F^{-1}(t))dt$. This procedure followed a basic idea of measure and integration theory, which had been applied in several ways during the first decades of the 20th century ([Riesz 1910] is an early example, see [Hawkins 1975, 191]).

Aiming at a general definition of conditional distribution, Lévy in § 23 started with the remark that, for any event B and any random variable X, the probability $P(B \wedge X < x)$ (x a real number) could be expressed as $P(B \wedge X < x) = F_1(x)$, where F_1 is a distribution function. Because of the fundamental properties of conditional probability, Lévy was able to state that, for arbitrary real numbers $a < b$,

$$F_1(b) - F_1(a) = P\left(B|a \leq X < b\right)\left(F(b) - F(a)\right), \qquad (7.1)$$

where F is the distribution of X. From this equation Lévy [1937a, 68] inferred that $F_1(x)$ could be "considered as a function of $\xi = F(x)$," with a "well-defined derivative except for a set of measure zero." Conceived as a function of $x \in \mathbb{R}$, Lévy named this derivative $g(x)$, and he stated that $g(x)$ was well-defined for all real

[4] This approach remained quite common for a relatively long time, see Sect. 7.3.2, in particular footnote 46.

[5] In [Lévy 1935a, 206; 1935b, 389 f.] one can only find a few rather vague hints.

[6] In his 1937 book, Lévy tried to deal with distribution functions independent of their special definition regarding the behavior at jumps (right continuous, left continuous, intermediate value, see [Lévy 1937a, 28 f.]). In his explanations surrounding conditional probabilities he showed a certain preference for the definition according to $P(X < x)$, however.

[7] This formula is an interpretation of what Lévy [1937a, 30] expressed in words only.

x "except for a set of probability zero."[8] He designated $g(x)$ as the "conditional probability of B under the hypothesis $X = x$." And finally he argued that

$$P(B \wedge X < x) = \int_{t \in]-\infty;x[} g(t)dF(t) \tag{7.2}$$

"immediately results from this definition."

Lévy did not explicitly refer to the theorem of Radon–Nikodym, as Kolmogorov [1933/50, 48] had done in this context. Already on the basis of Radon's version [1913] of this theorem, which concerned (generalized) Stieltjes integrals (the now so-called Lebesgue–Stieltjes integrals) over subsets of $\mathbb{R}^n$,[9] the existence of $g(x)$ with the property (7.2) was a direct consequence of (7.1), because this equation immediately implies absolute continuity of F_1 with respect to F. Most probably, however, only an expert of integration theory should have been able to understand Lévy's specific arguments. In the second edition of his *Leçons sur l'intégration* [1928], on page 297, Lebesgue had introduced a generalized notion of the derivative f of a monotonic left continuous function F with respect to a monotonic left continuous function α by defining

$$f(x) := \lim_{h \downarrow 0} \frac{F(x+h) - F(x)}{\alpha(x+h) - \alpha(x)}.$$

Earlier in his book Lebesgue [1928, 286–288] had already proven the one-dimensional version of Radon's theorem, which with the just introduced notation asserts that

$$F(b) - F(a) = \int_a^b f(x)d\alpha(x)$$

if F is absolutely continuous with respect to α. Apparently, Lévy alluded to these portions of Lebesgue's book, but he did not make this known to the reader. Previously in his own book [1937a, 34], Lévy had only given a concise account of the "classic" Riemann–Stieltjes integral.

In a note, Levy [1936b] (see also [1937a, 69 f.]) even showed that, in turn, for any given $g : \mathbb{R} \to [0; 1]$ which is measurable with respect to a distribution F (in the sense that for any $\beta \in [0; 1]$ the set $\{x | g(x) < \beta\}$ has a well-defined probability with regard to F), the integral

$$\int_{-\infty}^{\infty} g(x)dF(x) \tag{7.3}$$

represents the probability $P(B)$ of an event B, such that $g(x)$ can be interpreted as $g(x) = P(B|X = x)$, where X is distributed according to F. Again, the integral (7.3) cannot be understood, for a general g, on the basis of the elementary Riemann–Stieltjes theory, but only on the basis of more general concepts. Lévy's

[8] Probability zero with respect to F.

[9] For Radon's work and his "unification" of Lebesgue and Stieltjes integrals, see [Hawkins 1975, 186–194].

note shows his concern regarding the construction of multidimensional distributions via transition probabilities, and in this respect, a clear orientation toward probabilistic applications beyond a purely formalistic point of view.

Lévy [1937a, 71–73] introduced conditional expectations in the following way: Let X, Y be real-valued random variables, $\tilde{F}(x, y) := P(X < x \wedge Y < y)$ the joint distribution function, and $\varphi(x, y)$ a function defined on $\mathbb{R}^2$ such that

$$\mathscr{M}\varphi(X,Y) := \int_{\mathbb{R}^2} \varphi(x, y)d\tilde{F}(x, y)$$

exists. Let $G(x, y) := P(Y < y|X = x)$ and $F(x) := P(X < x)$, then (by (7.2)) one obtains

$$\tilde{F}(x, y) = \int\limits_{t\in]-\infty;x[} G(t, y)dF(t).$$

Instead of the pair of random variables (X, Y), Lévy now considered the pair of random variables (ξ, η), uniformly distributed within the square $0 < \xi < 1$, $0 < \eta < 1$, and he represented (X, Y) through $X = F^{-1}(\xi)$ (for the definition of F^{-1} see above) and $Y = G^{-1}(F^{-1}(\xi), \eta)$ (defined analogously to F^{-1}, however with respect to η for fixed ξ). By this transformation of random variables, and by setting $\varphi(X, Y) = \Phi(\xi, \eta)$, Lévy was able to infer that

$$\mathscr{M}\varphi(X,Y) = \int_0^1 \int_0^1 \Phi(s,t)dsdt. \tag{7.4}$$

On account of Fubini's theorem, the right-hand side of the latter formula could be expressed by

$$\int_0^1 \int_0^1 \Phi(s,t)dsdt = \int_0^1 \left[\int_0^1 \Phi(s,t)dt\right] ds. \tag{7.5}$$

Finally, Lévy connected the left-hand side of (7.4) and the right-hand side of (7.5) in the form

$$\mathscr{M}\varphi(X,Y) = \mathscr{M}\left\{\mathscr{M}_X\{\varphi(X,Y)\}\right\},$$

thus giving an (indirect) definition of the conditional expectation $\mathscr{M}_X\{\varphi(X, Y)\}$ of $\varphi(X, Y)$ if "X is known."

7.1.3 Lévy's Central Limit Theorem for Martingales

As we have seen in Sect. 5.2.7, Bernshtein's conditions in his work on the CLT from 1922/26 already had a far-reaching generality, as in the special case of independent and uniformly small summands these conditions were also necessary. On the other hand, the need for a relative closeness of conditional and total mean squares in Bernshtein's conditions challenged the search for weaker assumptions, at least in particular cases. As Lévy showed, Bernshtein's presupposition on the mean squares could actually be generalized at a far-reaching level, but only by an "extension" of the assertion of the CLT.

Lévy considered a sequence (X_ν) of random variables with the fundamental property that

$$EX_1 = 0, \quad E(X_\nu | X_1, \ldots, X_{\nu-1}) = 0 \quad (\nu = 2, 3, \ldots). \tag{7.6}$$

Compared with Bernshtein's condition of a relative closeness of the conditional expectation of each single random variable to zero, Lévy's assumption was not substantially more restrictive. At several places ([1937a, 242], for example), Lévy illustrated property (7.6) by means of a fair game, consisting of a succession of trials, where the rules are amended after each trial depending on the results of the preceding trials. This game is sometimes called a "martingale" in French; Lévy did not, however, explicitly use this designation in connection with sequences of random variables obeying (7.6).

His essential idea was not to consider "classical" sums $\sum_{\nu=1}^{n} X_\nu$ $(n \in \mathbb{N})$ of these random variables, but sums $\sum_{\nu=1}^{N} X_\nu$, where the upper index limit N itself was a random variable, depending in a certain sense on the conditional mean squares of the sequence (X_ν). For any positive t (which Lévy interpreted as the time needed for a certain number of trials), the random number $N(t)$ was defined by the condition

$$N(t) = \min\{n \in \mathbb{N} | \sigma_1^2 + \sigma_2^2 + \cdots + \sigma_n^2 \geq t\}, \tag{7.7}$$

where

$$\sigma_\nu^2 := E(X_\nu^2 | X_1, \ldots, X_{\nu-1}).$$

To make sure that for any positive t the random variable (7.7) was well-defined, Lévy demanded that

$$P\left(\sum_{\nu=1}^{\infty} \sigma_\nu^2 = \infty\right) = 1. \tag{7.8}$$

Strictly put, Lévy based his considerations on random sums built up in the form

$$S(t) := \sum_{\nu=1}^{N(t)-1} X_\nu + c(t) X_{N(t)}, \tag{7.9}$$

where the random variable $c(t)$ $(0 < c(t) \leq 1)$ was defined by the condition

$$\sigma_1^2 + \cdots + \sigma_{N(t)-1}^2 + c(t)^2 \sigma_{N(t)}^2 = t. \tag{7.10}$$

Presupposing (7.6) and (7.8) Lévy succeeded in establishing that

$$\lim_{t \to \infty} P(S(t) < x\sqrt{t}) = \Phi_{0,1}(x), \tag{7.11}$$

on the basis of additional conditions which were substantially weaker than Bernshtein's. From a modern point of view, $S(t)$ can be considered as a martingale, and thus Lévy's accounts are often—and rightly—subsumed under the early history of martingales. A self-contained theory of martingales is due to Doob, who in his first contribution [1940, 458] to this field referred to Jean Ville [1939]. This mathematician, apparently inspired by Lévy's work on nonindependent random variables (see [Ville 1939, 2, 83]), had introduced the notion and designation "martingale" in his doctoral thesis, which was dedicated to the discussion of von Mises's theory of collectives (see [von Plato 1994, 195–197; Mazliak 2009]).

Lévy [1934b;c] had already announced two sets of conditions (each including (7.6) and (7.8), of course) for (7.11); proofs were given in [Lévy 1935a]. Further generalizations were discussed in [Lévy 1935b]. A summarizing exposition, which does not cover Lévy's results in their full generality, can be found in [Lévy 1937a]. As in his treatment of sums of independent random variables, Lévy also for non-independent random variables tried to reduce more general situations by truncation arguments to the case of bounded random variables, in which the existence of a positive constant U was assumed, such that

$$|X_\nu| < U \quad \forall \nu \in \mathbb{N}. \tag{7.12}$$

In this latter case, the idea of proof was inspired by Lindeberg's method. Besides the given sequence of random variables X_ν, Lévy also considered auxiliary random variables ξ_ν and an additional random variable Z, each of them being mutually independent, and also independent of X_ν. All random variables ξ_ν were assumed to obey a standard normal distribution, and the distribution function F of Z should be smooth, with bounded derivatives up to the third order [Lévy 1935a, 219; 1937a, 239]. For simplifying the rather complicated situation resulting from (7.10), Lévy [1937a, 243] suggested renaming $c(t)X_{N(t)}$ by $X_{N(t)}$. Therefore, without loss of generality, it could be assumed in the proof that $c(t) = 1$ and

$$\sigma_1^2 + \cdots + \sigma_{N(t)}^2 = t. \tag{7.13}$$

Compared with the classical situation, it was an entirely new aspect that the number of terms within the respective partial sums of the X_ν depended in a certain way on the random variables themselves. To master this problem, Lévy considered, if in a rather informal way, random variables X'_ν defined by

$$X'_\nu := \begin{cases} \frac{X_\nu}{\sqrt{t}} & \text{if } \nu \leq N(t) \\ 0 & \text{else.} \end{cases}$$

As a consequence of this definition, X'_ν only depends on $X_1, \ldots, X_\nu$, because the condition $\nu \leq N(t)$ is—assuming the simplification (7.13)—equivalent to $\sigma_1^2 + \cdots + \sigma_\nu^2 \leq t$. Moreover, the conditional mean squares $\sigma_\nu'^2 := \mathrm{E}(X_\nu'^2 | X_1, \ldots, X_{\nu-1})$ are given by

$$\sigma_\nu'^2 = \begin{cases} \frac{\sigma_\nu^2}{t} & \text{if } \nu \leq N(t) \\ 0 & \text{else.} \end{cases}$$

In his proof, Lévy's main goal was to obtain estimates for the distributions of the sums

$$\sum_{k=1}^{n} X'_k + \sum_{j=n+1}^{\infty} \sigma'_j \xi_j + Z =: S'_n + R'_n + Z \quad (n \in \mathbb{N}).$$

Taking into account the assumptions just described on the independence between the X_ν and the auxiliary variables, he represented the relevant (conditional) distribution functions by convolutions, in which the part $P(R'_n + Z < x)$ was expanded up to the third order; using (7.6), he finally obtained the estimate

$$\left|P(X'_n + R'_n + Z < x|X_1, \ldots, X_{n-1}) - P(R'_{n-1} + Z < x)\right| \le \frac{KU}{\sqrt{t}} \sigma'^2_n, \quad (7.14)$$

K being a constant depending on the upper bound of $|F'''|$, and U being the constant corresponding to (7.12). In (7.14) he substituted x by $x - S'_{n-1}$, and, using the equation

$$P(S'_n + R'_n + Z < x) - P(S_{n-1} + R'_{n-1} + Z < x)$$
$$= \mathrm{E}\left(P(S'_n + R'_n + Z < x|X_1, \ldots, X_{n-1}) - P(S_{n-1} + R'_{n-1} + Z < x|X_1, \ldots, X_{n-1})\right),$$

he arrived at

$$|P(S'_n + R'_n + Z < x) - P(S'_{n-1} + R'_{n-1} + Z < x)| \le \frac{KU}{\sqrt{t}} \mathrm{E}\sigma'^2_n.$$

A telescope procedure yielded

$$\left|P(\sum_{\nu=1}^{n} X'_\nu + Z < x) - P(\sum_{\nu=1}^{n} \sigma'_\nu \xi_\nu + Z < x)\right| \le \frac{KU}{\sqrt{t}} \mathrm{E}\left(\sum_{\nu=1}^{n} \sigma'^2_\nu\right).$$

Because of the limit relations

$$\lim_{n\to\infty} P(\sum_{\nu=1}^{n} X'_\nu + Z < x) = P(\frac{S(t)}{\sqrt{t}} + Z < x),$$

$$\lim_{n\to\infty} P(\sum_{\nu=1}^{n} \sigma'_\nu \xi_\nu + Z < x) = P(\sum_{\nu=1}^{N(t)} \frac{\sigma_\nu \xi_\nu}{\sqrt{t}} + Z < x),$$

$$\mathrm{E}\left(\sum_{\nu=1}^{\infty} \sigma'^2_\nu\right) = \frac{1}{t} \mathrm{E}\left(\sum_{\nu=1}^{N(t)} \sigma^2_\nu\right),$$

the estimate

$$\left|P(\frac{S(t)}{\sqrt{t}} + Z < x) - P(\sum_{\nu=1}^{N(t)} \frac{\sigma_\nu \xi_\nu}{\sqrt{t}} + Z < x)\right| \le \frac{KU}{t\sqrt{t}} \mathrm{E}\left(\sum_{\nu=1}^{N(t)} \sigma^2_\nu\right) = \frac{KU}{\sqrt{t}}$$

had to be valid. From this, it followed that

$$\lim_{t\to\infty} P\left(\frac{S(t)}{\sqrt{t}} + Z < x\right) = P(\xi + Z < x),$$

ξ designating any random variable with a standard normal distribution. Because Z could be chosen in such a way that its influence on the probabilities under consideration was arbitrarily small,[10] the CLT (7.11) under the basic assumption (7.12) could finally be concluded. For similar reasons, the CLT was not only valid for the "rounded" sums $S(t)$ according to (7.9), but also for the "complete" sums $\sum_{\nu=1}^{N(t)} X_\nu$, which fact, however, Lévy only hinted at in his [1935b, 391].

Lévy discussed several extensions and modifications of the version of the martingale CLT for bounded random variables just described. In [Lévy 1934c; 1935a, 221] he established a condition which reminds one to a certain extent of Lindeberg's:

$$\forall \varepsilon > 0 : P\left(\lim_{n\to\infty}\left[\sum_{\nu=1}^{n} P(|X_\nu| > \varepsilon\sigma | X_1, \ldots, X_{\nu-1})\right]_{\sigma=b_n} = 0\right) = 1$$
$$(b_n^2 := \sigma_1^2 + \cdots + \sigma_n^2). \quad (7.15)$$

Further conditions were needed, however, for proving (7.11) by a truncation procedure: Let

$$X_\nu = X'_\nu + X''_\nu, \text{ where } X'_\nu := \begin{cases} X_\nu & \text{if } |X_\nu| \le \varepsilon b_n \\ 0 & \text{else.} \end{cases}$$

Lévy demanded that, for all $\varepsilon > 0$,

$$P\left(\lim_{n\to\infty}\frac{1}{b_n}\sum_{\nu=1}^{n} \mathrm{E}(X''_\nu | X_1, \ldots, X_{\nu-1}) = 0\right) = 1,$$
$$P\left(\lim_{n\to\infty}\frac{1}{b_n^2}\sum_{\nu=1}^{n} \mathrm{E}(X''^2_\nu | X_1, \ldots, X_{\nu-1}) = 0\right) = 1,$$
$$P\left(\lim_{n\to\infty}\frac{1}{b_n^2}\sum_{\nu=1}^{n} (\mathrm{E}(X''_\nu | X_1, \ldots, X_{\nu-1}))^2 = 0\right) = 1. \quad (7.16)$$

He [1935a, 222–224] proved that the martingale CLT remained true under the conditions (7.6), (7.8), (7.15), and (7.16).

The idea of also providing a general norming for sums of nonindependent random variables without (conditional) mean squares (as already discussed by Bernshtein) was treated by Lévy in the last section of his [1935b]. For $X > 0$ "sufficiently large" he [1935b, 391] introduced the truncated random variables X'_ν (equal to X_ν or 0, depending on whether or not $|X_\nu| \le X$), and

[10] For a detailed discussion see [Lévy 1937a, 241].

$$b_n' := \sum_{\nu=1}^{n} \mathrm{E}\left(X_\nu' - \mathrm{E}(X_\nu'|X_1,\ldots,X_{\nu-1})|X_1,\ldots,X_{\nu-1}\right)^2,$$

$$\eta_n := P(|X_\nu| > X|X_1,\ldots,X_{\nu-1}).$$

The random number

$$N'(t) = \min\{n \in \mathbb{N}|b_n' \geq t\}$$

was defined analogous to (7.7), and for any positive $\varepsilon > 0$, the existence of an arbitrarily small $\gamma > 0$ such that

$$P(\eta_{N'(t)} > \varepsilon) < \gamma \tag{7.17}$$

was essential to ensure that "the examination of $S(t) = S_n$ $[= S_{N'(t)}]$ is reduced to the one of S_n' $[= S_{N'(t)}']$" [Lévy 1935b, 392]. If t was growing, a continual readjustment of X as dependent on t was necessary for (7.17). In order to apply his martingale CLT for bounded variables to the present situation, the condition

$$\mathrm{E}(X_\nu'|X_1,\ldots,X_{\nu-1}) = 0 \tag{7.18}$$

(or a slightly weaker condition) was important. Lévy [1935b, 393 f.] discussed several assumptions by which (7.18) could be substituted in such a way that the basic ideas of the proof were maintained.

In his "Theorem VII" he assumed, for "simplifying the exposition," that the conditional law $\mathscr{L}_{\nu-1}^{(\nu)}$ of each random variable X_ν, depending on the preceding $X_1,\ldots,X_{\nu-1}$, was symmetric. The condition that $N'(t)$ tended to infinity as $t \to \infty$, "except for cases of probability zero," was, as Lévy [1935a, 211–217] had shown, guaranteed by the condition that the probability of the convergence of $\sum_{\nu=1}^{\infty} X_\nu$ was zero. Presupposing the symmetry of $\mathscr{L}_{\nu-1}^{(\nu)}$ and the impossibility of the convergence of $\sum_{\nu=1}^{\infty} X_\nu$, Lévy [1935b, 393] finally stated:

> *... if, for all sufficiently large t, one can determine X such that, at the same time, ε, $\eta'[=\frac{X}{\sqrt{t}}]$, and γ* [see (7.17)] *become arbitrarily small, then $\frac{S(t)}{\sqrt{t}}$ depends on a law which tends, for infinite t, to that of Gauss.*

Lévy had obtained his far-reaching generalizations of Bernshtein's version of the CLT for chained random variables only by considering the number of variables within the partial sums as random. For a direct comparison between his results and Bernshtein's, however, the discussion of "classical" sums was necessary. Lévy [1935a, 230–232; 1935b, 396–401] actually provided a rather comprehensive account on this issue. Roughly speaking, despite his effort to show that his conditions "are simpler than those obtained by M. S. Bernstein" and even "surpass" them "from certain points of view" [Lévy 1935a, 230], it finally became plausible from his explanations that a further substantial weakening of Bernshtein's conditions was hardly possible in the case of the convergence of distributions of ordinary sums

$S_n := \sum_{\nu=1}^{n} X_\nu$ to the normal distribution. In particular, Lévy [1935a, 232; 1935b, 397] showed that, for a sequence of random variables with finite mean squares obeying (7.6) and the "loi des grands nombres," the condition of a relative closeness of conditional and total mean squares was sufficient and necessary for the distribution of S_n being in the limit of a "Gaussian type," if the single conditional laws $\mathscr{L}_{\nu-1}^{(\nu)}$ were symmetrical and depended only on $|X_1|, \ldots, |X_{\nu-1}|$.[11]

Lévy [1935b, 394–396] also stated the converse of the main theorem described above and sketched a proof. The assertion was that, assuming the "condition of symmetry" of $\mathscr{L}_{\nu-1}^{(\nu)}$ (and tacitly (7.6) as well as (7.8)), the assumption of ε, η', and γ becoming arbitrarily small for sufficiently large X was also necessary for (7.11), if all random variables were individually negligible. As Le Cam [1986, 89 f.] has analyzed, Lévy's arguments were partially erroneous. This circumstance might be one of the reasons why the—in principle very important—article [Lévy 1935b] was not included in the third volume of Lévy's works (containing contributions to sums of random variables), which was planned while Lévy was still alive.

7.2 Further Limit Problems

Feller [1945, 821] characterized "his" achievement in 1935 in the matter of the CLT as a "starting point for many examinations." This is true, for example, for results regarding the weak law of large numbers

$$P\left(\left|\frac{S_n - c_n}{a_n}\right| > \varepsilon\right) \to 0$$

with appropriate $a_n > 0$ and c_n, as achieved by Khinchin [1936] and more generally by Feller [1937b]. Yet Feller's remark, quoted above, suggested to the reader that his (and possibly Lévy's) work on the convergence of distributions of normed sums to the Gaussian law had not only resolved a longstanding question once and for all, but also that a paradigm shift toward "nonclassical" norming had been implemented, with the result being that all other limit problems that were discussed in the second half of the 1930s had been "natural" generalizations of the CLT for normed sums.

This is not entirely accurate, however. First of all, the idea of limit theorems for random variables without variance or even expectation is already recognizable in the

[11] In the 1940s, Michel Loève generalized Bernshtein's theorem insofar as he substituted the assumptions (5.62), (5.63), (5.64) concerning the uniform bounds α_{ni}, β_{ni}, c_{ni} by assumptions on expectations and only considered the normed sums S_n/B_n'. (5.62) was, for example, replaced by

$$\lim_{n\to\infty} \frac{1}{\sqrt{B_n'}} \sum_{i=1}^{n} \mathrm{E}|\mathrm{E}(X_{ni}|U_1', \ldots, U_{i-1}')| \to 0,$$

where $U_1', \ldots, U_{i-1}'$ are the truncated variables (the designations are the same as in Sect. 5.2.7). This work was the starting point for Loève's far-reaching investigations on limit distributions of sums of dependent random variables, see [Loève 1950, 331 f.] for an abstract.

1920s in the work of Bernshtein and Lévy, and so the solution of the central limit problem in 1935 constituted less a new beginning than a first conclusion, although it certainly served as stimulus for even more generalized views. Secondly, and this point is likely more important: A decisive impetus to study innovative and more general limit theorems came from a direction that had only arisen during the modern period of probability theory, namely, the theory of stochastic processes. Above all, the study of stochastic processes with independent increments, which had grown in stature since the early 1930s and had been undertaken independently of classical subjects of probability theory, led to results which could incorporate all that had thitherto been achieved with the CLT. This theorem could thus be examined from a more general point of view.

7.2.1 Stochastic Processes with Independent Increments

Sums of an increasing number of independent random variables can be conceived as stochastic processes with discrete time parameter. Beginning in the late 1920s, their "natural" generalization—stochastic processes with continuous time parameter—became a major area of research in probability theory.[12] Bruno de Finetti (1906–1985) [1929a;b;c] considered a family $(Z(\tau))_{\tau\in[0;T[}$ of random variables with the property that, for every $\tau_1, \tau_2 \in [0; T[$ with $\tau_1 < \tau_2$, there exists a random variable $U_{\tau_1\tau_2}$ that is independent of all $Z(\tau)$ $(\tau \leq \tau_1)$, such that $Z(\tau_2) = Z(\tau_1) + U_{\tau_1\tau_2}$. Today stochastic processes of this type are called "processes with independent increments" (or "i.i. processes"). De Finetti was particularly exhaustive in discussing the special case of an increment $U_{\tau_1\tau_2}$, the distribution function of which depends only on $\tau_2-\tau_1$ but not on the specific position of this time interval on the time scale. The associated processes are known today as "processes with stationary independent increments" (or "s.i.i. processes"). The "modern" notation including the abbreviations given here had already appeared in Cramér's book *Random Variables* in 1937. In the 1930s, though, most people used the adjective "homogeneous" rather than the adjective "stationary." Lévy [1937a, 158] referred to the i.i. processes as "Intégrales a élements aléatoires indépendantes" because the form

$$Z(\tau) = Z(0) + \int_0^\tau dZ(t)$$

recommends itself for $Z(\tau)$.

In the course of his search for general solutions for these processes, de Finetti had begun to examine "infinitely divisible distributions" [von Plato 1994, 261–264]. In

---

[12] Special cases had already been discussed by Louis Bachelier [1900], Albert Einstein [1905], Norbert Wiener [1921], and others. De Finetti [1929a;b;c] started with a general theory; Kolmogorov [1931a; 1933a] discussed measure-theoretic principles. A detailed discussion of the historical development of stochastic processes with continuous time parameter is found in [von Plato 1994].

modern terminology, a distribution F is infinitely divisible if for every natural number n there exists a distribution F_n such that $F = F_n^{\star n}$. According to the original usage of the 1930s, however, distributions V were designated infinitely divisible if for all $n \in \mathbb{N}$ there exist distributions $V_{n1}, \ldots, V_{nn}$ such that

$$V = V_{n1} \star \cdots \star V_{nn},$$

with the additional condition that, in a particular sense, the influence of the individual components V_{nk} on the overall distribution V disappears asymptotically. Nevertheless, as we shall see, the original and the modern parlance both express the same substance. When the phrase "infinitely divisible" appears hereafter, it is used in the sense that has become conventional today.

If f_τ denotes the family of characteristic functions of $Z(\tau)$, then, for every $n \in \mathbb{N}$, in an s.i.i. process in which $Z(0) = 0$ the following must be true:

$$f_\tau(t) = \left[f_{\tau/n}(t)\right]^n. \tag{7.19}$$

Accordingly, the distribution function F_τ is infinitely divisible. De Finetti could infer from (7.19) that, in s.i.i. processes, the distribution F_τ of $Z(\tau)$ is continuously dependent on τ.[13]

Kolmogorov [1932] provided a general formula for the characteristic functions of the distributions F_τ in an s.i.i. process under the condition that all distributions have finite variance. Because each infinitely divisible distribution can be considered as a distribution of one of the random variables $Z(\tau)$ in an s.i.i. process, Kolmogorov had therefore also found a general representation for infinitely divisible distributions with finite variance. Lévy [1934a] even managed to derive a general representation of the characteristic functions of the distributions F_τ in an i.i. process in which these distributions are continuously dependent on τ, without making any assumptions about the existence of moments, and he was able to establish a formula for the family of distributions in a general i.i. process by means of this continuous special case. Lévy's approach was to break the random variables $Z(\tau)$ down into individual independent components which—when regarded as random functions of the argument τ—each have a particular continuity or discontinuity behavior.

Based on Lévy's formula for the characteristic functions of the distributions F_τ that were continuously dependent on the time parameter in i.i. processes, it followed that these distributions also had to be infinitely divisible (in the modern sense) or, as Lévy [1937a, 186] put it, that they "can always be obtained by a homogeneous process." For i.i. processes $Z(\tau)$ with $Z(0) = 0$ we have $Z(\tau) = \sum_{k=1}^{n} U_{nk}$ for all $n \in \mathbb{N}$, where $U_{nk} = Z(\frac{k}{n}\tau) - Z(\frac{k-1}{n}\tau)$. If the distributions F_τ are continuously dependent on the time parameter, then the elements U_{nk} become asymptotically infinitesimal in the sense of

$$\lim_{n\to\infty} \max_{1\le k\le n} P(|U_{nk}| > \varepsilon) = 0 \quad \forall \varepsilon > 0. \tag{7.20}$$

[13] The "continuity" can be established, for instance, by the condition $\lim_{\tau\to t} F_\tau(x) = F_t(x)$ in every point of continuity x of F_t.

The latter relation played an important role in Lévy's derivation of the characteristic functions of those distributions F_τ in i.i. processes that are continuous in τ.

Unlike Lévy, Khinchin [1937] based his examination of the τ-continuous distributions F_τ of i.i. processes directly upon the study of arbitrary triangular arrays of row-wise independent random variables U_{nk}, $n \in \mathbb{N}$, $1 \leq k \leq n$. He showed that under the condition (7.20) the class of limit distributions of the sums $\sum_{k=1}^{n} U_{nk}$ is identical to the class of infinitely divisible distributions (in the modern sense). At Kolmogorov's suggestion, G.M. Bavli [1936] had previously derived a corresponding statement for random variables with finite variance. Thanks to this property of infinite divisibility, Khinchin could offer an alternative to Lévy's results in his representation of the characteristic functions of the distributions F_τ of i.i. processes that are continuous in τ. From today's point of view, it is also clear from Khinchin's results that distributions are "infinitely divisible" in the original sense of the 1930s if and only if they possess this property in the modern conventional sense.

The study of i.i. processes could thus largely be reduced to the examination of triangular arrays with row-wise independent elements. Then again, normed partial sums of sequences of independent random variables could be regarded as row sums of triangular arrays. Accordingly, triangular arrays constituted the interface between the "Laplacian" summation problem on the one hand and "modern" stochastic processes on the other. Although double sequence of random variables had already been examined on a case-by-case basis— by Lindeberg and Bernshtein, for instance—for the purpose of generalizing or specifying classical problems regarding limit distributions of sums of independent random variables, it was the stochastic processes with continuous time parameter that yielded the decisive incentive to take a closer look at triangular arrays. Once Boris Vladimirovich Gnedenko (1912–1995) [1939a] had discovered necessary and sufficient conditions for the convergence of distributions of row sums of triangular arrays with linearly independent and asymptotically infinitesimal elements to a given infinitely divisible distribution, the Lévy–Feller limit problem of 1935 and its logical expansion toward nonnormal limit distributions could in large part be mastered as an application of a more general theory concerning triangular arrays. So beginning with [Gnedenko & Kolmogorov 1949], a treatment of the Lévy–Feller limit problem for normed sums that considerably differs from the arguments of the original papers has established itself.

As we have already seen (Sect. 6.3.3), Gnedenko worked within the framework of characteristic functions. On the other hand, Wolfgang Doeblin (1915–1940) [1939] was able to adapt Lévy's ideas concerning concentration and dispersion to establish necessary and sufficient conditions for the convergence of distributions of sums with independent and asymptotically negligible summands to infinitely divisible distributions. An important role played an inequality that Doeblin and Lévy [1936] had jointly proven: Let $X_1, \ldots, X_n$ be independent random variables, and let the dispersion assigned to the probability $0 < \alpha < 1$ of each of these random variables be above $2l$, l a positive number. Let β be another positive probability. Then for $n > N$, where N depends on α and β only, the dispersion $\varphi_{S_n}(\beta)$ of the sum S_n of these random variables assigned to the probability β obeys the inequality

$$\varphi_{S_n}(\beta) > kl\sqrt{n},$$

where k is a constant only depending on α and β. Already in 1940, Doeblin was killed as a soldier in World War II. Only with a considerable time interval in between, work on a similar basis was resumed, in particular by Kolmogorov [1956; 1958; 1963], who directly referred to Doeblin, and Lucien Le Cam [1965; 1970]. From today's point of view, this approach is by no means inferior to the one by characteristic functions, and this all the more as it allows to derive bounds for the deviations between distributions of sums of independent random variables and approximating infinitely divisible laws, and it can be analogously carried over to problems concerning sums of independent random elements in Banach spaces.[14]

7.2.2 Limit Laws of Normed Sums

The essential questions surrounding limit distributions of sums of independent random variables were solved during the period between 1935 and the start of the Second World War. This phase is characterized both by the coalescence of problems and methods gained through generalization of classical problems and by new problems and solution ideas in the field of stochastic processes.

Beyond the standard problem of limit distributions in triangular arrays, the subject of "limit distributions of normed sums of independent random variables" includes an additional matter: One must find not only the possible limit distributions and the sufficient and necessary conditions for the convergence to them, but also suitable sequences of norming constants so that said convergence can actually occur.

Between about 1935 and 1937—despite the many merits of the other authors mentioned here, such as Feller or Kolmogorov—Lévy and Khinchin were the most important actors in this field.

Following his academic studies in Moscow, Aleksandr Yakovletivich Khinchin (1894–1959),[15] like so many probability theorists early in the modern period, had started his mathematical career with analysis as the main focus of his research. Since the late 1920s, Khinchin had been increasingly interested in limit theorems of probability theory. His outstanding result regarding infinitely divisible distributions as limit distributions of row sums of triangular arrays [Khinchin 1937] was described in the preceding section. Especially notable within the framework of convergence to the normal distribution is Khinchin's characterization of the domain of attraction of the Gaussian law [Khinchin 1935], which appeared at the same time as, but independently of, the articles by Feller and Lévy, and established exactly the same criterion as Lévy for the associated distributions F, namely,

$$\lim_{X\to\infty} \frac{X^2\int_{|t|>X} dF(t)}{\int_{|t|\leq X} t^2 dF(t)} = 0$$

[14] For a comprehensive survey of this approach to limit distributions of sums of independent real- and Banach-valued random variables, see [Araujo & Giné 1980].

[15] Biographical information for Khinchin and a complete list of his works (which is unfortunately erroneous) is included in [Cramér 1962].

(Feller's characterization (6.36) differs slightly from this one). Despite the competition between Lévy and Khinchin, the two men shared an intense and almost friendly exchange of scientific findings, although this was hampered by Khinchin's apparent difficulties with the Stalinist regime (see [Lévy 1970, 109; Sheynin 2009, 111 f.]).

The few remaining open questions regarding the representation of stable laws were answered by Lévy and Khinchin, who collaborated in this field, and now were able to carry out their examinations in the light of the new theory of infinitely divisible distributions, of which the stable laws formed a special case. It was in this context that they also considered the issue of the influence of shifting constants, which Lévy had rated as insignificant in his book in 1925. They called "quasi-stable" those distributions V—being different from degenerate distributions—for which the following condition is true:

$$\forall c_1, c_2 > 0 \exists c > 0 \exists c' : \quad V\left(\frac{x}{c_1}\right) \star V\left(\frac{x}{c_2}\right) = V\left(\frac{x-c'}{c}\right).$$

As Lévy and Khinchin discovered simultaneously,[16] the characteristic functions φ of quasi-stable distribution functions V can be represented by $\varphi(z) = \mathrm{e}^{\psi(z)}$, wherein

$$\psi(z) = -c|z|^\alpha \left(1 + i\beta \operatorname{sign}(z) \tan(\frac{\pi}{2}\alpha)\right) + miz$$
$$(c > 0, |\beta| \leq 1, m \in \mathbb{R}, \alpha \in]0;1[\cup]1;2])$$

or

$$\psi(z) = -c\left(\frac{\pi}{2}|z| + i\beta z \log|z|\right) + miz \quad (c > 0, |\beta| \leq 1, m \in \mathbb{R}).$$

For $\alpha \neq 1$, $V(x+m)$ is simply the distribution function of a law of the type $L_{\alpha,\beta}$ (see Sect. 5.2.6.3). Modern usage generalizes the terminology of the 1920s and 1930s, and likewise refers to the formerly "quasi-stable" distributions as "stable" distributions.

The cooperation between Khinchin and Lévy is particularly well-documented in the discovery of the class of "lois limites" (known as "class L" today) for normed sums of independent and, in a particular sense, asymptotically negligible random variables. Specifically, Lévy and Khinchin conceived of "lois limites" as all of the limit laws of sums in the form $\sum_{k=1}^n X_k/c_n$, where the sequence of positive norming constants increases infinitely, such that $\lim_{n\to\infty} \frac{c_{n+1}}{c_n} = 1$.[17] As Levy [1936a, 265] acknowledged, Khinchin had written him a letter in June 1936 in which he

[16] This result was apparently first published and proven in [Lévy 1937a, 208–211]. In a footnote, however, Lévy expressly recognized Khinchin's efforts.

[17] As Feller [1935, 530–532] proved, the latter condition is equivalent to the condition $\lim_{n\to\infty} \max_{1\leq k\leq n} P(|X_k| > \varepsilon c_n) = 0 \ \forall \varepsilon > 0$, assuming a limit law exists. According to Levy [1936a, 265], Khinchin preferred the likewise equivalent condition (for the proof of equivalence see [Lévy 1937b, 270–275]) $\lim_{n\to\infty} \max_{1\leq k\leq n} P\left(|X_k| > \varepsilon \left|\sum_{h=1}^n X_h\right|\right) = 0 \ \forall \varepsilon > 0$. The condition of uniform asymptotic negligibility of all single random variables with respect to the total sum clearly emerges from these two conditions.

presented the problem of "lois limites" and reported that he had proven that these laws were all "infinitely divisible" (in the original 1930s-era meaning).

Levy [1936a] proposed the following theorem; he published his proof in [1937a, 192–197], admittedly with the explicit use of several more of Khinchin's epistolary suggestions:

> *The necessary and sufficient condition for the fact that $\mathscr{L}$ represents a "loi limite" is:* 1^0 $\mathscr{L}$ *is an infinitely divisible law (discovered by Mr. Khinchin to be a necessary condition);* 2^0 *the variable S, which obeys the law $\mathscr{L}$, can for any λ between 0 and 1 be brought to the form $\lambda(X+Y)$, where X and Y are independent, X is dependent on the law $\mathscr{L}$, and Y is dependent upon an infinitely divisible law* [Levy 1936a, 265].

For the characteristic function f of the law $\mathscr{L}$, this condition means that for all $\lambda \in]0;1[$ there exists a characteristic function g_λ of a distribution that is "infinitely divisible" in the 1930s sense, with

$$f\left(\frac{x}{\lambda}\right) = g_\lambda(x) f(x).$$

With the aid of his general representation for distributions of i.i. processes, Levy [1936a] was successful in providing a "canonical representation" for the characteristic functions of class L.

Remaining unanswered for the time being was the question of which necessary and sufficient conditions the random variables (negligible with respect to the total sum) and norming constants must satisfy for convergence to a class L law. This set of problems included the special case of characterizing the domains of attraction of stable laws, which so far, since the pioneering work that appeared in Lévy's book in 1925, had met with complete success only for the Gaussian distribution. Gnedenko and Doeblin, on the basis of their methods and their results regarding the convergence of distributions of row-sums in triangular arrays to infinitely divisible distributions, solved these latter problems in a very short time just before the Second World War (particularly in the papers [Gnedenko 1939b] and [Doeblin 1940]).

The history of the CLT and its generalizations certainly did not end with the outbreak of World War II, not even in the classical case of independent summands and the normal distribution as a limit law. The demand for error bounds that were as precise as possible in order to estimate the quality of the approximation through the normal distribution was met very quickly, as has been described (see Sect. 5.2.8.2). However, the extension of results obtained for distributions to densities and discrete probabilities in the sense of local limit theorems had not been intensively pursued for the time being. The same held true in the case of limit theorems for "large deviations,"[18] despite individual achievements like the ones by Khinchin [1929] and Cramér [1938]. What was lacking even more were solutions to all of these problems for nonnormal limit laws.

---

[18] This includes statements about the asymptotic behavior of the distribution functions of (normed) sums of random variables for arguments that tend to infinity themselves along with the number of random variables. For example, if (X_k) is a sequence of independent random variables with means 0 and variances σ_k^2, and if (x_n) is a sequence of numbers with $x_n \to \infty$, then we are interested in the behavior of

The requirement that summands be independent had been already eased, a movement that had been driven—after the pioneering work of Markov—primarily by Bernshtein and Lévy in the 1920s and 1930s. Alongside this, new motivations emerged after the war for a generalized view of the CLT, based on the one hand on using the findings that had already been obtained and generalizing them to random variables with values in metric spaces, and, on the other hand, on application-oriented matters such as those connected with so-called "invariance principles."

7.3 Extensions of the Central Limit Theorem to Stochastic Processes and Random Elements in Metric Spaces

After the Second World War, probabilistic work was continued on a broad scale, and in this context the CLT was further extended and generalized. Two important instances are explained in the following: the application of the CLT to limit theorems for Brownian movement, and the generalization of the CLT toward random variables with values in Hilbert spaces. Both examples clearly show fields of interest already occasionally discussed in the 1930s, which became especially important in the 1950s: the approach to stochastic processes in the sense of random elements in function spaces, further generalizations of basic notions, such as expectation or convergence, and in this context the growing impact of functional analysis on probability theory.

7.3.1 *Invariance Principles and Donsker's Theorem*

As we have already seen, the CLT was ever more connected with the theory of stochastic processes from the beginning of the 1930s. Initially, most of the results were achieved in the framework of the traditional theory of finite-dimensional random variables. This situation definitely changed toward the end of the 1940s, especially with Donsker's proof of a theorem on the convergence of the distributions of certain random functions depending on a sequence of independent random variables to the Wiener measure. In the setting which is considered in the following, the Wiener measure is a probability measure on the function space

$$\mathscr{C} := \{f \in C^0([0;1]) \,|\, f(0) = 0\}$$

$$\frac{1 - P\left(\frac{\sum_{k=1}^{n} X_k}{\sqrt{\sum_{k=1}^{n}\sigma_k^2}} \leq x_n\right)}{1 - \Phi(x_n)}$$

as $n \to \infty$. Under certain conditions, these fractions converge to 1.

together with the sigma algebra generated by the L^∞ topology, where the marginal distributions of this measure are given in accord with Brownian motion[19] by

$$P(a_j \leq f(t_j) \leq b_j, j = 1, \ldots, n) = \pi^{-n/2} \left(t_1(t_2 - t_1) \ldots (t_n - t_{n-1})\right)^{-1/2} \times$$
$$\times \int_{a_n}^{b_n} \ldots \int_{a_1}^{b_1} \exp\left(-\xi_1^2/t_1 - \sum_{k=2}^{n} (\xi_k - \xi_{k-1})^2/(t_k - t_{k-1}) \right) d\xi_1 \ldots d\xi_n, \quad (7.21)$$

if $0 < t_1 < t_2 < \cdots < t_n \leq 1$. If $\mathscr{C}$ is endowed with the Wiener measure as a "standard" measure, then (at least since the early 1940s) the designation "Wiener Space" is used.

7.3.1.1 Wiener Measure and Wiener Integral

Donsker's theorem dealt with distributions in a space of continuous functions, and therefore a space of infinite dimension. Existence and construction of distributions on a space of type $\mathbb{R}^M$, where M is an arbitrary index set,[20] was guaranteed—at least in principle—by Kolmogorov's "Fundamental Theorem" [1933/50, 29], which stated the unique existence of a probability distribution on the sigma algebra generated by "Borel cylinder sets"[21] if finite-dimensional distributions are given in such a way that they correspond to marginal distributions. For spaces $\mathbb{R}^M$, where M is a nondenumerable index set, such as the space of all real functions defined on [0; 1], the drawback of Kolmogorov's approach was, as he himself [1933/50, 28] admitted, that typical (sub-) sets of continuous or bounded functions did not belong to the sigma algebra generated by the just-described Borel cylinder sets. Therefore, in the important case of Brownian motion, for example, a probabilistic discussion of boundedness, continuity, or other properties of random functions could not be achieved by the direct application of Kolmogorov's theorem. Not even the possibility of a unique representation of a continuous function $f \in C^0([0; 1])$ by a denumerable sequence of real "coordinates" (the most simple way is to use the sequence $\{f(0), f(1), f(\frac{1}{2}), f(\frac{1}{4}), f(\frac{3}{4}), f(\frac{1}{8}), f(\frac{3}{8}), \ldots\}$) provides an immediate approach to the Wiener measure on $\mathscr{C}$ by Kolmogorov's device.[22]

Wiener had given several more or less methodically different accounts of Brownian motion and random functions—his chief papers are [Wiener 1923; 1924; 1930,

---

[19] For the early history of the probabilistic treatment of Brownian motion, which was mainly connected with the names of Bachelier, Einstein, and von Smoluchowski, see [Courtault 2002] and [Brush 1968]. For a survey of the mathematical development until ca. 1950, see [Kahane 1998].

[20] $\mathbb{R}^M$ designates the set of all mappings $f : M \to \mathbb{R}$.

[21] A Borel cylinder set is the preimage (in the space under consideration $\mathbb{R}^M$) of an orthogonal projection whose image is a Borel set in $\mathbb{R}^s$, s an arbitrary natural number.

[22] In the case of the just-mentioned "dyadic" representation of functions the difficulties are based on the fact that sets $\{(x_i) \in \mathbb{R}^{\mathbb{N}} \| x_i| \leq \alpha\}$ for positive α always contain sequences of coordinates which do not correspond to uniformly bounded continuous functions.

esp. 217–226; Paley & Wiener 1934, Chapt. IX][23]—and in this context had, if in a rather implicit and incomplete way, discussed the existence of a probability measure on $\mathscr{C}$, which was in accordance with (7.21). Apparently, the third (1930) of these accounts became especially influential upon integration in function spaces during the 1940s.

In his first works on Brownian motion, Wiener for integration in the function space $\mathscr{C}$ had used Daniell's notion of integral, the theory of which, however, was still incomplete. Wiener [1930, 218] wrote that it had "seemed" to him "more desirable" now to employ a method for establishing integration in function space which resorted to the well-known theory of the Lebesgue integral, for which "the literature contains a much greater wealth of proved theorems ... than of theorems concerning the Daniell integral." To this aim he established a correspondence between functions in $\mathscr{C}$ and points in the interval [0; 1], and thus reduced the problem of measure in function space to Lebesgue measure on [0; 1], a quite common idea, whose principle can be traced back to Radon's work [1913, 48–57].[24] For arbitrary (not only for continuous) real-valued functions $f(t)$ on the unit interval $0 \leq t \leq 1$ and vanishing at $t = 0$, Wiener [1930, 219] for each $n \in \mathbb{N}$ defined $(2^n)^{2^n}$ "quasi-intervals" as follows: For

$$(m_1, m_2, \ldots, m_{2^n}) \in \{-2^{n-1}, -2^{n-1}+1, \ldots, 2^{n-1}-1\}^{2^n}$$

let the quasi-interval $Q(n; m_1, \ldots, m_{2^n})$ be the set of all the functions f just-defined, for which the following inequalities are valid:

$$\left.\begin{array}{l} \tan\left(\frac{m_j\pi}{2^n}\right) < f\left(\frac{j}{2^n}\right) \leq \tan\left(\frac{(m_j+1)\pi}{2^n}\right) \text{ if } m_j > -2^{n-1} \\ \tan\left(\frac{m_j\pi}{2^n}\right) \leq f\left(\frac{j}{2^n}\right) \leq \tan\left(\frac{(m_j+1)\pi}{2^n}\right) \text{ if } m_j = -2^{n-1} \end{array}\right\} \quad (j = 1, \ldots, 2^n).^{25}$$

The quasi-intervals of order n form a partition of the set of all those functions f. To each quasi-interval Wiener assigned a measure $\mu(Q(n; m_1, \ldots, m_{2^n}))$, which was given by the right-hand side of (7.21) with n substituted by 2^n, $t_j = \frac{j}{2^n}$, $a_j = \tan\left(\frac{m_j\pi}{2^n}\right)$, and $b_j = \tan\left(\frac{(m_j+1)\pi}{2^n}\right)$. Each of those functions f can be closed up by an infinite sequence (Q_ν) of quasi-intervals successively contained within one another and with a respective measure μ tending to zero. The quasi-intervals are mapped to (half-open) subintervals of [0; 1] in such a way that all subintervals assigned to the quasi-intervals of order n form a partition of [0; 1], and subsequent quasi-intervals $Q_{\nu+1} \subset Q_\nu$ correspond to subsequent subintervals $S_{\nu+1} \subset S_\nu$ with lengths $\mu(Q_{\nu+1})$ and $\mu(Q_\nu)$, respectively. On the basis of these "nested intervals,"

---

[23] For a discussion of [Wiener 1923] see [Chatterji 1993, 157–163; Bourbaki 1994, 240–242]. A somewhat more elaborate exposition of Wiener's 1930 and 1934 contributions can be found in [Wiener, Siegel, Ranking, & Martin 1966].

[24] The following description is close to the exposition in [Wiener, Siegel, Ranking, & Martin 1966, 17–20, 37–45], which differs from the original one only regarding (a more convenient) notation and some additional explanations.

[25] $\tan(-\frac{\pi}{2})$ has to be interpreted as $-\infty$, and $\tan(\frac{\pi}{2})$ as $+\infty$.

Wiener succeeded in showing that, except for a set of points of Lebesgue measure zero, to any given point in the interval [0; 1] belongs one and only one continuous function $f \in \mathscr{C}$ with

$$|f(t') - f(t'')| \leq h|t' - t''|^{1/4} \tag{7.22}$$

for some $h > 0$ and all $t', t'' \in [0; 1]$. He also proved that, except for a set of functions which can be enclosed in a denumerable union of quasi-intervals of arbitrarily small total probability, all functions defined in [0; 1] (and a fortiori all functions in $\mathscr{C}$) satisfy the condition (7.22) and belong to one of the just-described points in [0; 1]. Because of this correspondence between almost all points in [0; 1] and almost all functions in $\mathscr{C}$ (which are even Hölder continuous), any "functional"[26] $g : \mathscr{C} \to \mathbb{R}$ determines a function $\tilde{g} : [0; 1] \to \mathbb{R}$ (uniquely except for a set of arguments of Lebesgue measure zero), which may be Lebesgue summable. If the latter condition is satisfied, then the "Wiener integral" $\int_{\mathscr{C}} g(x)dw(x)$[27] is defined by $\int_0^1 \tilde{g}(x)dx$. The extension of Wiener measure and Wiener integral to the space of continuous functions defined on $\mathbb{R}_0^+$ or $\mathbb{R}$ is possible, as Wiener [1930, 226] indicated, by a further, essentially bijective, mapping from $\mathscr{C}$ to the space under consideration.

Wiener restricted his own discussion of "his" integral to cases where g was continuous (with respect to the L^∞ topology, in [1923] and [1924]), or $g(f)$ depended only on a finite number of function values $f(t_1), \ldots, f(t_n)$. In his contributions Wiener neither—explicitly—treated the problem of determining the probability measure of nontrivial subsets of $\mathscr{C}$, such as $\{f \in \mathscr{C} : \|f\|_\infty \leq \alpha\} =: M_\alpha$, nor did he establish the Wiener measure as a measure on a sigma algebra of $\mathscr{C}$. At least he hinted, in his [1923, 167], at the possibility of determining the measure of a set by integrating the characteristic function of this set over $\mathscr{C}$, but without discussing any specific applications of this principle. Taking into account the construction of the Wiener integral as sketched above, the Wiener measure of M_α can actually be established as follows: First, show that

$$M(n; \alpha) := \left\{ f \in \mathscr{C} : -\alpha \leq f\left(\frac{j}{2^n}\right) \leq \alpha,\ j = 1, \ldots, 2^n \right\} \quad (n \in \mathbb{N})$$

corresponds to a Lebesgue measurable subset $T(n; \alpha)$ of [0; 1] with a Lebesgue measure given by (7.21) with n substituted by 2^n, $a_j = -\alpha$, $b_j = \alpha$, $t_j = \frac{j}{2^n}$. Then approximate the characteristic function of the image $\widetilde{M}_\alpha \subset [0; 1]$ of M_α by the characteristic functions of $T(n; \alpha)$. The theorem of dominated convergence for Lebesgue integrals implies the existence of a Lebesgue measure of $\widetilde{M}_\alpha$ and therefore a (positive) "Wiener measure" of M_α, such that

$$\lim_{n \to \infty} P(M(n; \alpha)) = P(M_\alpha).$$

---

[26] Originally, the designation "functional" was not restricted to linear mappings.

[27] During the 1940s, the period which will be focused at in the following, the commonly used notation for Wiener integral was "$\int_C^W f(x)d_w x$."

On the basis of this limit relation one could apply well-known techniques using parabolic differential equations (see [Khinchin 1933, Chapt. III]), for obtaining the formula

$$P\,(M_\alpha) = \frac{4}{\pi}\sum_{m=0}^{\infty}\frac{(-1)^m}{2m+1}\exp\left(\frac{-(2m+1)^2\pi^2}{16\alpha^2}\right), \qquad (7.23)$$

which in principle had already been achieved, based on an intuitive physical idea, by Reinhold Fürth in his work on Brownian motion [1917, 182 f.]. Another way to gain formula (7.23) was found by Erdös and Kac in their work on the so-called "invariance principle" [1946] (see Sect. 7.3.1.3).

7.3.1.2 Cameron and Martin

Wiener's approach from 1930 was taken up by Robert Cameron (1908–1989, Fig. 7.1)[28] and William "Ted" Martin (1911–2004, Fig. 7.2)[29] in a series of papers published during the 1940s. Cameron, after having received his Ph.D. from Cornell University in 1932, obtained a National Research Council Fellowship, which enabled him to do further research at Brown University and at the Princeton Institute for Advanced Studies until 1935. Martin, who received his Ph.D. in 1934 at the University of Illinois, also came to Princeton in 1935 as a National Research Council Fellow. As it seems, however, both mathematicians became acquainted with each other only at MIT, where, Cameron in 1935, and one year later Martin, obtained positions, first as instructors, then as professors. Naturally, both Cameron and Martin were influenced by Norbert Wiener at MIT, especially as far as complex analysis, Fourier analysis, and integration in function spaces were concerned. A first jointly

Fig. 7.1 Robert Cameron (left) together with John Olmsted (1957)

[28] For biographic details see [Aspray 1985; Loud 2005].

[29] For biographic details see [MIT 2004].

Fig. 7.2 William "Ted" Martin

written article (on complex analysis) dates from 1938 [Cameron & Martin 1938]. Whereas Martin remained at the Institute (except for a stint at Syracuse University 1943–1946) until his retirement in 1976—for a long time he had been head of the math department—Cameron moved to the University of Minnesota (U of M) in 1945. At U of M, Cameron unfolded a very active research, and he was advisor to a large number of doctoral candidates, among whom a considerable part worked on problems concerning Wiener space and Wiener measure.[30]

In their work on Wiener measure, Cameron and Martin did not aim at a general theory (which did not exist in published form at that time). Rather, they focused on particular properties, such as the existence and characterization of a positive Wiener measure of $\{f \in \mathscr{C} | \int_0^1 (f(x))^2 dx < a\}$ $(a > 0)$ [Cameron & Martin 1944a; 1945a], or, for each function $f(\lambda)$ defined for all positive λ and satisfying $0 \leq f(\lambda) \leq 1$, the existence of a set $E \subset \mathscr{C}$ such that the Wiener measure of $\lambda E := \{\lambda \cdot g | g \in E\}$ is equal to $f(\lambda)$ for all $\lambda > 0$ [1947]. Cameron's and Martin's work on the behavior of the Wiener measure under translations would later experience far-reaching generalizations: In its original setting the now so-called Cameron–Martin formula [1944b] referred to the following situation: For any $x_0 \in \mathscr{C}$ having a derivative x_0' of bounded variation the translation $\mathrm{T}\Gamma$ of a Wiener measurable set $\Gamma \subset \mathscr{C}$ was defined by

$$\mathrm{T} : \Gamma \ni y \mapsto y - x_0 =: x \in \mathrm{T}\Gamma.$$

The main theorem was that for each function $F : \mathscr{C} \to \mathbb{R}$ for which either term

$$\int_\Gamma F(y) dw(y), \quad \int_{\mathrm{T}\Gamma} F(x + x_0) \exp\left(-2 \int_0^1 x_0'(t) dx(t)\right) dw(x)$$

[30] With respect to Cameron, the Mathematics Genealogy Project currently lists 53 Ph.D. students and 166 descendants, the latter mainly being disciples of Donsker (`http://genealogy.math.ndsu.nodak.edu`).

exists,[31] the other term exists as well, and the following formula is valid:

$$\int_{\Gamma} F(y)dw(y)$$
$$= \exp\left(-\int_0^1 [x_0'(t)]^2 dt\right) \int_{T\Gamma} F(x+x_0) \exp\left(-2\int_0^1 x_0'(t)dx(t)\right) dw(x).$$

For the basic properties of the Wiener measure and the Wiener integral the two authors in each of their papers referred to [Wiener 1930, 214–234], which source, as already mentioned, was far from representing a reasonably complete theory. As a side note: the correct page numbers would have been 216–236. It is an interesting detail that Donsker in his works on the functional CLT [1949, 9; 1951, 1] still referred to the wrong page numbers. One may therefore conclude that knowledge of the Wiener measure and the Wiener integral was mainly transferred through communication in seminars and lecture courses.[32] At several places Cameron and Martin used theorems on Wiener integrals which apparently were part of a mathematical folklore within a little group of experts, such as an analog to the monotone convergence theorem [Cameron & Martin 1944b, 391].

7.3.1.3 The Invariance Principle

Mark Kac (1914–1984) is one of the most prominent mathematicians of the 20th century, therefore biographic details seem unnecessary here. Around 1945—Kac worked at Cornell University—he published a few papers, partly in collaboration with Paul Erdös (1913–1996), on the now so-called "invariance principle." To cite Donsker [1949, iii; 1951, 1], whom that designation seems to be due to, this principle holds if, in context with a probabilistic limit problem, a limiting distribution exists and is independent of the particular distributions of the random variables involved. Kac [1946] discovered the invariance principle while dealing with a probabilistic limit theorem in close connection with Cameron and Martin's work [1944a] on the determination of the Wiener measure of $\{f \in \mathscr{C} \mid \int_0^1 (f(x))^2 dx < a\}$. As he relates in his autobiography [1985, 113], Kac had been stimulated to pursue this investigation by Martin, who was the chairman of the mathematics department at

---

31 As Cameron & Martin [1944b, 390] hinted at, the integral $\int_0^1 f(t)dx(t)$ exists for any function $f : [0;1] \to \mathbb{R}$ of bounded variation and any function $x \in \mathscr{C}$ in a Riemann–Stieltjes sense, chiefly due to the fact that for any system $0 = t_0 < t_1 < t_2 < \cdots < t_n = 1$:

$$\sum_{j=1}^{n} f(t_j)\left(x(t_j) - x(t_{j-1})\right) = f(1)x(1) - \sum_{k=1}^{n-1} x(t_k)\left(f(t_{k+1}) - f(t_k)\right).$$

32 This conjecture is endorsed by a remark of Donsker [1949, v], who, for "a detailed examination of the properties of Wiener measure and the Wiener integral" refers to "unpublished class notes of Professor R. H. Cameron."

Syracuse University at this time, and whom Kac saw "regularly." Together with Erdös (who also occasionally worked at Syracuse University) he proved by use of the invariance principle some more theorems on the asymptotic behavior of functions of partial sums of independent random variables, by first showing that the invariance principle was valid, and thereafter by choosing convenient distributions for the single random variables such that the determination of the specific limiting distribution was possible. In all cases the one- or multidimensional version of the CLT was the basis for showing the validity of the invariance principle.

For example, Erdös and Kac in their first jointly published paper [1946] presented simple and elementary proofs of four limit theorems concerning the asymptotic behavior of the partial sums $s_k = X_1 + \cdots + X_k$ of independent random variables $X_1, X_2, \ldots$, each with zero expectation and variance 1 such that the CLT was applicable. The two first of them dealt with the limits (as $n \to \infty$) of

$$P\left(\max(s_1, s_2, \ldots, s_n) < \alpha n^{1/2}\right) \quad \text{and} \quad P\left(\max(|s_1|, |s_2|, \ldots, |s_n|) < \alpha n^{1/2}\right).$$ [33]

In this context, Erdös and Kac [1946, 293] referred to previous work of Louis Bachelier (covering particular cases), which they, however, designated as "un-rigorous."[34] At the same place they also hinted at the possibility of proving those limit theorems by use of parabolic equations, as expounded by Khinchin [1933],[35] but they characterized their own method as more elementary. Erdös and Kac's third limit theorem dealt with the limit of

$$P\left(s_1^2 + s_2^2 + \cdots + s_n^2 < \alpha n^2\right).$$

The fourth theorem had already been treated in Kac's first paper on the matter (see above), and concerned the limit of

$$P\left(|s_1| + |s_2| + \cdots + |s_n| < \alpha n^{3/2}\right).$$

In a further paper Erdös and Kac [1947] proved that the average number of positive s_k $(1 \le k \le n)$ follows the arcsine law in the limit.

Retrospectively, Kac [1985, 115] pointed out that he and Erdös only proved "a number of special cases." He placed greater value on a 1949 paper, in which he characterized the Laplace transform of the distribution function

$$\sigma(\alpha; t) := P\left(\int_0^t V(x(\tau))d\tau < \alpha\right),$$

---

33 Assuming for the X_i a normal distribution, from the second limit theorem the formula (7.23) for the Wiener measure of uniformly bounded subsets of $\mathscr{C}$ could be derived.

34 As Bachelier [1937, 6 f., 17 f.] recapitulated in a summarizing account, which was also cited by Erdös and Kac, he had derived the limiting values for both probabilities in the particular case of two-valued random variables already between 1901 and 1908.

35 Kolmogorov [1931b] had applied Green functions related to the heat equation in order to determine asymptotic formulae for the probability that all partial sums s_k remain within certain intervals possibly varying with k; this approach was modified and generalized by Khinchin [1933, 54–59].

$x(t)$ designating the elements of the Wiener space of all continuous functions with $0 \leq t < \infty$ and $x(0) = 0$, and V a piecewise continuous, nonnegative function on $\mathbb{R}$, by a differential equation quite similar to Schrödinger's equation in quantum mechanics. As Kac [1985, 116] reports, he was decisively stimulated to do this work by Richard Feynman's idea of using path integrals for the construction of propagators in quantum mechanics. Kac [1949b, 3] clearly indicated the connection of his present work to his investigations on the invariance principle: One could suspect that, for partial sums s_k as above,

$$\lim_{n\to\infty} P\left(\frac{1}{n}\sum_{k\leq nt} V\left(\frac{s_k}{\sqrt{n}}\right) < \alpha\right) = \sigma(\alpha;t). \tag{7.24}$$

In their papers on the invariance principle Kac and Erdös had actually proven[36] the validity of (7.24) in particular cases, such as $V(x) = x^2$ or $V(x) = |x|$. Still, a sufficiently general discussion of the asymptotic relation between partial sums of random variables and random paths in Wiener spaces had not been carried out. This was essentially Donsker's merit.

7.3.1.4 Donsker's General Invariance Principle

Monroe David Donsker (1925–1991, Fig. 7.3) graduated from U of M in 1944, and at the same university received his master's degree with a thesis on "A Set of Equivalent Axiomatic Systems for Topology" in 1946. In 1949 he completed his doctoral dissertation, supervised by Cameron, with the title "The Invariance Principle for

Fig. 7.3 Monroe Donsker (left) together with Henry McKean (1971)

[36] This follows from the fact that, if the invariance principle holds, one can take as a special distribution for each random variable the standard normal distribution.

Wiener Functionals."[37] Donsker's work on the invariance principle had a considerable impact on the career of its originator (as well as on far-reaching generalizations, naturally). Among all of the disciples of Cameron, Donsker became the most prominent. From 1962 until his death he had a professorship at the Courant Institute of New York University, with a main focus on probability theory.[38]

In its original version of the doctoral dissertation of 1949, "Donsker's theorem" [1949, 93] was as follows:[39]
Let $X_1, X_2, X_3, \ldots$ be a sequence of independent identically distributed random variables, each of them having the distribution ϕ with mean 0 and variance 1. For natural numbers $1 \leq j \leq n$ and vectors $(u_1, \ldots, u_n) \in \mathbb{R}^n$ the normed sums $S_{jn}^* := \frac{1}{\sqrt{2n}} \sum_{i=1}^{j} X_i$ and $s_{jn}^* := \frac{1}{\sqrt{2n}} \sum_{i=1}^{j} u_i$ are defined.[40] For $0 \leq t \leq 1$ and for $w_i \in \mathbb{R}$ $(i = 1, \ldots, n)$ the polygon x_n which connects the points $(0|0)$, $(\frac{1}{n}|w_1), \ldots$, $(1|w_n)$ is defined by

$$x_n(t; w_1, w_2, \ldots, w_n) := \begin{cases} w_{j-1} + (nt - j + 1)(w_j - w_{j-1}) & \text{for } \frac{j-1}{n} < t \leq \frac{j}{n} \\ 0 & \text{for } t = 0. \end{cases}$$

Let $F : \mathscr{C} \to \mathbb{R}$ be a bounded function uniformly continuous with respect to the L^∞ topology. Then

$$\lim_{n\to\infty} \int_{-\infty}^{\infty} \cdots \int_{-\infty}^{\infty} F\left(x_n(\cdot\,; s_{1n}^*, s_{2n}^*, \ldots, s_{nn}^*)\right) d\phi(u_1) \cdots d\phi(u_n) = \int_{\mathscr{C}} F(x) dw(x). \quad (7.25)$$

When writing his thesis, Donsker could not be familiar with the newly developed concept of "weak convergence" of distributions, which later during the 1950s turned out to be especially well suited for investigating convergence of distributions in "abstract spaces." According to Aleksandr Danilovich Aleksandrov [1943, 169, 172] (whom this notion seems to be due to), a sequence of "additive set functions" μ_n defined on a sigma algebra generated by the topology of a topological space X weakly converges to a set function μ defined on the same algebra if and only if

$$\int_X f(x) d\mu_n(x) \to \int_X f(x) d\mu(x) \quad (n \to \infty) \quad (7.26)$$

for all continuous and bounded functions $f : X \to \mathbb{R}$. This notion of convergence was, as it seems, adapted to probability theory and further propagated by Russian mathematicians since the end of the 1940s.[41] The modern theory of weak

[37] See the library catalogue of U of M for Donsker's master's and Ph.D. theses.

[38] For biographic details see [NYT 1991].

[39] In the following, the somewhat more convenient notation of Donsker's 1951 article is used.

[40] Donsker did not specify the normed sums by double indices; this seems, however, more convenient for a better understanding.

[41] An early description of this concept can be found in [Gnedenko & Kolmogorov 1949, Chapt. 2].

convergence actually shows that (7.25) is equivalent to this version of Donsker's theorem which states the weak convergence of random polygons arising from partial sums to a Wiener process,[42] and which is usually ascribed to Prokhorov [1953] (who also weakened the presupposition in the sense of a Lindeberg condition, which proceeding would, however, have been in Donsker's reach as well). After stating (7.25), Donsker abruptly ended his dissertation without giving any further comments or applications. In the introduction to the thesis he [1949, iii, v] had hinted at the "invariance principle" as explained above, and set the goal to reach an "invariance property" for a "large class of Wiener functionals." Indeed, by a rather obvious argument involving characteristic functions it would have been possible to show how closely his result and the limit theorems of Kac and Erdös were related to each other. This connection was only established in Donsker's 1951 paper, where the chief ideas of his dissertation were expounded in a somewhat modified form.

Donsker's article of 1951 was written at Cornell, where he worked in a postdoc position with Kac (see [Kac 1985, 115]). During this time, Donsker not only succeeded in giving more streamlined arguments at several places, but also achieved more generalized and extended results. "Donsker's Theorem," as it was later referred to and occurred explicitly in the 1951 paper only, was as follows: Let the random variables X_j and the partial sums S^*_{jn} and s^*_{jn} be as above. For $0 \leq t \leq 1$ and for $w_i \in \mathbb{R}$ $(i = 1, \ldots, n)$ the step functions

$$x_{(n)}(t; w_1, w_2, \ldots, w_n) := \begin{cases} w_i & \text{for } \frac{i-1}{n} < t \leq \frac{i}{n} \\ w_1 & \text{for } t = 0 \end{cases}$$

are introduced. Let R be the space of all functions defined on $[0; 1]$ which are continuous except possibly for a finite number of finite jumps, and let $F : R \to \mathbb{R}$ be a function which is continuous with respect to the L^∞ topology at almost all (Wiener measure) points of $\mathscr{C}$. Then for every $\alpha \in \mathbb{R}$ at which

$$\sigma(\alpha) := P\big(\{x \in \mathscr{C} | F(x) < \alpha\}\big)$$

is continuous we have

$$\lim_{n\to\infty} P\left(F\big(x_{(n)}(\cdot\,; S^*_{1n}, S^*_{2n}, \ldots, S^*_{nn})\big) < \alpha\right) = \sigma(\alpha).$$

As we see, Donsker did not consider polygons in 1951 any more, but restricted his account to step functions, which had already played the decisive role in the proofs of his dissertation. In this way, a considerably shortened exposition was possible.

The common core of both Donsker's thesis and his 1951 paper was as follows: Donsker [1949, 1–57; 1951, 2–5] first proved a "simple" version of his theorem. Let k be a fixed natural number, and let $\alpha_j \leq \beta_j$ $(j = 1, \ldots, k)$ be real numbers. By R_n denote the event that, for all $j = 1, \ldots, k$,

---

[42] If X is a metric space, then it is sufficient for weak convergence that (7.26) is true for all bounded and uniformly (!) continuous f.

$$\alpha_j \leq x_{(n)}(t; S^*_{1n}, \ldots, S^*_{nn}) \leq \beta_j \quad \text{if} \quad \frac{j-1}{k} < t \leq \frac{j}{k},$$

and by E denote the subset of all $x \in \mathscr{C}$ such that

$$\alpha_j \leq x(t) \leq \beta_j \quad \text{if} \quad \frac{j-1}{k} < t \leq \frac{j}{k}.$$

Then we have

$$\lim_{n\to\infty} P(R_n) = P(E). \tag{7.27}$$

The proof was based on the multidimensional CLT, but several elementary, if quite intricate, considerations were additionally needed. The large difference between 57 pages in the thesis and only 5 in the 1951 paper was partly due to the fact that Donsker had modified some arguments in the meantime. Of course, too, a doctoral dissertation is usually more detailed than an article. Furthermore, in his 1951 account Donsker skipped some "technical" details, for example regarding properties of Wiener measure, which he had discussed in his thesis at considerable length. For example, on pages 39–46 of the thesis Donsker proved that the set Q of functions $x(t)$ in $\mathscr{C}$ with

$$\lambda_j \leq x(t) < \gamma_j \quad \text{for} \quad \frac{j-1}{k} < t \leq \frac{j}{k} \quad (j = 1, \ldots, k)$$

has the same Wiener measure as the set $\overline{Q}$ of functions $x(t)$ in $\mathscr{C}$ with

$$\lambda_j \leq x(t) \leq \gamma_j \quad \text{for} \quad \frac{j-1}{k} < t \leq \frac{j}{k} \quad (j = 1, \ldots, k).$$

If a thoroughly elaborated theory of Wiener measure had existed at that time, Donsker's laborious considerations would have been needless.

In his thesis, Donsker [1949, 57–79] also proved an analog to (7.27) for random polygons $x_n(t; S^*_{1n}, \ldots, S^*_{nn})$. Corresponding considerations are missing in the 1951 paper.

The final parts of both the doctoral dissertation and the 1951 article were based on very similar ideas. The most significant difference between the two works was that the polygons x_n in the thesis were substituted by step functions $x_{(n)}$ in the 1951 paper. Therefore, the following description, which refers to the latter, yields a description of the former if polygons are inserted instead of step functions. Donsker [1951, 5 f.] for $j = 1, \ldots, k$ and $m = 1, \ldots, n$ introduced the abbreviations (x designates a bounded function defined on [0; 1]):

$$p_j(x) := \sup_{\frac{j-1}{k} < t \leq \frac{j}{k}} x(t), \qquad q_j(x) := \inf_{\frac{j-1}{k} < t \leq \frac{j}{k}} x(t),$$

$$s^*_{mn} := \frac{1}{\sqrt{2n}} \sum_{i=1}^{m} u_i \quad ((u_1, \ldots, u_n) \in \mathbb{R}^n)),$$

$$p_j^{(n)} := \sup_{\frac{j-1}{k} < t \leq \frac{j}{k}} x_{(n)}(t; s^*_{1n}, \ldots, s^*_{nn}), \quad q_j^{(n)} := \inf_{\frac{j-1}{k} < t \leq \frac{j}{k}} x_{(n)}(t; s^*_{1n}, \ldots, s^*_{nn}),$$

and he stated that with B designating the $2k$-dimensional "interval"

$$B := \{(\tau_1, \ldots, \tau_{2k}) | -\infty < \tau_i \leq \beta_i, \alpha_i \leq \tau_{k+i} < \infty \quad (i = 1, \ldots, k)\},$$

the limit relation (7.27) could be rewritten in the form

$$\lim_{n\to\infty} \int_{-\infty}^{\infty} \cdots \int_{-\infty}^{\infty} \chi_B \left(p_1^{(n)}, \ldots, p_k^{(n)}, q_1^{(n)}, \ldots, q_k^{(n)}\right) d\phi(u_1) \cdots d\phi(u_n)$$
$$= \int_{\mathscr{C}} \chi_B(p_1(x), \ldots, p_k(x), q_1(x), \ldots, q_k(x)) dw(x), \quad (7.28)$$

where χ_B denoted the characteristic function of B and ϕ was the distribution function common to the random variables X_j.

On the basis of (7.28) a more general situation could be treated (substantial remarks concerning the proof can only be found in Donsker's dissertation, in the first chapter [Donsker 1949, 3–8] already): If $f : \mathbb{R}^{2k} \to \mathbb{R}$ is bounded, Borel measurable, and Riemann integrable on every finite $2k$-dimensional interval, then

$$\lim_{n\to\infty} \int_{-\infty}^{\infty} \cdots \int_{-\infty}^{\infty} f\left(p_1^{(n)}, \ldots, p_k^{(n)}, q_1^{(n)}, \ldots, q_k^{(n)}\right) d\phi(u_1) \cdots d\phi(u_n)$$
$$= \int_{\mathscr{C}} f(p_1(x), \ldots, p_k(x), q_1(x), \ldots, q_k(x)) dw(x). \quad (7.29)$$

The next step of proof was to show that sufficiently general "functionals" F defined on R (or $\mathscr{C}$ in the thesis, for the definition of R see the formulation of Donsker's 1951 theorem above) could be approximated (in a certain sense with respect to Wiener measure) by particular "functionals"

$$f(p_1(\cdot), \ldots, p_k(\cdot), q_1(\cdot), \ldots, q_k(\cdot))$$

defined on R (or on $\mathscr{C}$). The set of such "functionals" with f meeting the conditions for (7.29) he named Q. His central assertion [Donsker 1951, 6 f.] was as follows (an analogous theorem can be found in the dissertation [Donsker 1949, 84]):[43]

> Let $F(g)$ be bounded and uniformly continuous in the uniform topology on R. Then, there exists a pair of sequences of functionals $\{F_k^*(g)\}$ and $\{F_k^{**}(g)\}$ all belonging to Q such that for each k and all g in R
>
> $$F_k^{**}(g) \leq F(g) \leq F_k^*(g) \quad (7.30)$$
>
> and such that
>
> $$\lim_{n\to\infty} \int_{\mathscr{C}} (F_k^*(x) - F_k^{**}(x)) dw(x) = 0. \quad (7.31)$$

For the proof Donsker [1951, 7–9] (analogously [Donsker 1949, 84–92]) set for $g \in R$:

$$g_k^*(t) := \sup_{\frac{j-1}{k} < u \leq \frac{j}{k}} g(u) \quad \text{for } \frac{j-1}{k} < t \leq \frac{j}{k}$$
$$g_k^{**}(t) := \inf_{\frac{j-1}{k} < u \leq \frac{j}{k}} g(u) \quad \text{for } \frac{j-1}{k} < t \leq \frac{j}{k} \qquad (j = 1, \ldots, k),$$

[43] Quotation with different equation numbers and with a different notation of Wiener integral compared with Donsker's text.

and with M_g denoting the set of functions $h \in R$ such that $g_k^{**}(t) \leq h(t) \leq g_k^*(t)$ he defined

$$F_k^*(g) := \sup_{h \in M_g} F(h) \qquad F_k^{**}(g) := \inf_{h \in M_g} F(h).$$

The validity of (7.30) followed directly from the definition of $F_k^*(g)$ and $F_k^{**}(g)$. Because of the assumptions on the continuity of F and $x \in \mathscr{C}$ it could easily be shown that the integrand in (7.31) tends to 0 for each fixed $x \in \mathscr{C}$. The boundedness of F_k^* and F_k^{**} as a consequence of the boundedness of F finally yielded (7.31) (an analog of Lebesgue's theorem on dominated convergence for Wiener integrals was tacitly assumed). To prove that F_k^* and F_k^{**} had the required properties with respect to measurability and integrability, and therefore actually belonged to Q, was an elementary, but quite cumbersome task.

As a direct consequence of (7.29), (7.30), and (7.31), Donsker [1951, 10 f.] was able to follow that, for any "functional" $F : R \to \mathbb{R}$ that is bounded and uniformly continuous in the L^∞ topology on R,

$$\lim_{n\to\infty} \int_{-\infty}^{\infty} \cdots \int_{-\infty}^{\infty} F\left(x_{(n)}(\cdot\,; s_1^*, \ldots, s_n^*)\right) d\phi(u_1) \cdots d\phi(u_n) = \int_{\mathscr{C}} F(x) dw(x). \quad (7.32)$$

In the doctoral dissertation, the final assertion (7.25) followed in a similar way.

As already stated, further considerations can only be found in the 1951 article. There [1951, 10 f.], by an approximation argument, the validity of (7.32) for "functionals" F bounded on R and continuous in the L^∞ topology at almost all (with respect to Wiener measure) points of $\mathscr{C}$ was justified. Now, because it was obvious that the validity of (7.32) could be extended to complex-valued functions, the limit relation

$$\lim_{n\to\infty} \int_{-\infty}^{\infty} \cdots \int_{-\infty}^{\infty} \exp\left(izF\left(x_n(\cdot\,; s_1^*, \ldots, s_n^*)\right)\right) d\phi(u_1) \cdots d\phi(u_n) = \int_{\mathscr{C}} \exp\left(izF(x)\right) dw(x)$$

for all real z and all (even unbounded) F defined on R and continuous in the L^∞ topology at almost all points of $\mathscr{C}$ followed. Because F could be treated as a real-valued random variable, the continuity theorem for characteristic functions immediately yielded "Donsker's theorem."

With this theorem of 1951, Donsker had all at once given a universal approach to several functional limit theorems, not only those of Kac and Erdös, but also those of other mathematicians, such as Robert Fortet [1949] ($F(x) = x^{2m}$ and Abraham Mark [1949] (a disciple of Kac, $F(x) = \min_{0\leq t\leq 1}(x(t) - x(\alpha t))$ $(0 < \alpha < 1)$). More important, however, was the fact that Donsker clearly highlighted relations between the asymptotic behavior of discrete processes and Brownian motion, which had been suspected for a rather long time, especially in connection with random walks. Already Khinchin's 1933 monograph on asymptotic laws at many places showed ideas referring to those relations.

Taking up an idea of Joseph Leo Doob [1949], Donsker [1952] published an extension of his theorem to empirical processes. If $X_1, X_2, \ldots$ are mutually independent, identically distributed random variables with distribution function $F(\lambda)$, and $\nu_n(\lambda)$ is the number of the variables $X_1, \ldots, X_n$ with outcomes $\leq \lambda$, then, by a result of Kolmogorov [1933b], the distribution of

$$\sup_{\lambda\in\mathbb{R}} \left(\frac{\nu_n(\lambda)}{n} - F(\lambda)\right) \tag{7.33}$$

had to be independent of $F(\lambda)$ if F was assumed to be continuous on $\mathbb{R}$. Therefore, for the sake of convenience, Donsker could assume for $F(\lambda)$ the special distribution $F(\lambda) = \lambda$ $(0 \leq \lambda \leq 1)$.

Donsker now considered the space $\hat{\mathscr{C}} := \{x \in \mathscr{C} | x(1) = 0\}$ endowed with a probability distribution (with respect to the sigma algebra generated by the L^∞ topology) whose finite-dimensional marginal distributions are given by

$$\begin{aligned}
&P(a_j \leq x(t_j) \leq b_j, j = 1, \ldots, n)\\
&= (2\pi)^{-n/2}\Big(t_1(1-t_1)(t_2-t_1)(1-(t_2-t_1))\ldots(t_n-t_{n-1})(1-(t_n-t_{n-1}))\Big)^{-1/2} \times\\
&\times \int_{a_1}^{b_1}\cdots\int_{a_n}^{b_n} \exp\left(-\frac{\xi_1^2}{2t_1(1-t_1)} - \sum_{k=2}^{n}\frac{(\xi_k-\xi_{k-1})^2}{2(t_k-t_{k-1})(1-(t_k-t_{k-1}))}\right) d\xi_1\ldots d\xi_n,
\end{aligned} \tag{7.34}$$

if $0 < t_1 < t_2 < \cdots < t_n < 1$. The measure on $\hat{\mathscr{C}}$ characterized by (7.34) is identical to the conditional Wiener measure on $\mathscr{C}$ under the condition $x(1) = 0$ ((7.21) being rescaled by substituting the limits of integration a_j and b_j by $a_j/\sqrt{2}$ and $b_j/\sqrt{2}$, respectively).[44]

At the beginning of his considerations, Donsker for

$$x_n(t) := \sqrt{n}\left(\frac{\nu_n(t)}{n} - t\right) \quad (0 \leq t \leq 1)$$

was able to state that, in correspondence to the multidimensional CLT, for fixed j and $0 < t_1 < \cdots < t_j < 1$:

$$\lim_{n\to\infty} P(x_n(t_i) \leq \alpha_i;\, i = 1, \ldots, j) = P\left(\{x \in \hat{\mathscr{C}} | x(t_i) \leq \alpha_i;\, i = 1, \ldots, j\}\right).$$

Using results by Kac [1949a] and Chung [1949] he [1952, 279 f.] succeeded in showing that even

$$\lim_{n\to\infty} P\left(\sup_{0\leq t\leq 1} x_n(t) \leq \alpha\right) = P\left(\max_{0\leq t\leq 1} x(t) \leq \alpha\right), \tag{7.35}$$

[44] A random element with values in $\hat{\mathscr{C}}$ is also called a "Brownian bridge" today.

where the probability on the right was according to the conditional Wiener measure on $\hat{\mathscr{C}}$. This result yielded, after applying Kolmogorov's theorem that the distribution of (7.33) was independent of F, a general limit theorem for the deviations between theoretical and empirical distribution functions. More important, however, was Donsker's extension of (7.35) to a limit theorem for general "functionals" of x_n and $x \in \hat{\mathscr{C}}$, respectively, which was in perfect analogy to his main theorem above, and which could be proven by the same methods: Let the space R and the "functional" F be as above in the context of Donsker's main theorem (with the exception that $\mathscr{C}$ is to be substituted by $\hat{\mathscr{C}}$). Then

$$\lim_{n\to\infty} P\big(F(x_n) \leq \alpha\big) = P\big(\{x \in \hat{\mathscr{C}} | F(x) \leq \alpha\}\big)$$

at all points of continuity of the distribution function on the right.

Did Donsker with his theorems open a new chapter in the history of the CLT? On the one hand, his considerations were based on the classical (multidimensional) CLT; on the other hand, however, his results represented a shift from the classical paradigm of finite-dimensional random variables and limit distributions toward random variables with values in function spaces and distributions defined on these spaces. In this latter respect Donsker's work actually showed a first "unification" of the "modern" theory of stochastic processes and the classical theory of sums of (independent) random variables.

7.3.2 The Central Limit Theorem for Sums of Random Elements in Hilbert Spaces

At almost the same time as Donsker, though independent of him, research on limit theorems in abstract Banach and Hilbert spaces was started by Robert Fortet and his disciple Edith Mourier (born 1920, Fig. 7.4). Just like Donsker, also Fortet [1949] in one of his first pertinent papers referred to Erdös and Kac's contributions to invariance principles. Fortet was especially interested in applications of stochastic processes to physics, for example to signal processing, statistical optics, noise and turbulence, as also highlighted in his book *Théorie des fonctions aléatoires* [1953], written in collaboration with theoretical physicist André Blanc-Lapierre.[45] Also Wiener's work on harmonic analysis referred to those applications [Masani 1990, Chapts. 7, 9], as well as some contributions by Kac and his research group. Among Fortet and Mourier's most important contributions during the first half of the 1950s were results concerning laws of large numbers for random elements with values in Banach spaces [Fortet & Mourier 1952; 1953; 1954], and particularly, a version of the CLT for random elements in separable Hilbert spaces, which Mourier [1953a;b] derived in her doctoral thesis.

[45] Material regarding the French stochastic community at that time, especially with respect to Fortet and Fréchet, and persons they closely worked together with, is found in [Bru 2002].

Fig. 7.4 Edith Mourier (1971)

Concerning the fundamentals of probability in metric spaces, Mourier, who began to publish on those topics from about 1949, based herself on Fréchet's approach [1948]. A comprehensive account, which summarized Mourier's results up to this time, and which was essentially identical to her doctoral thesis, appeared in 1953 [Mourier 1953a]. Mourier generalized Fréchet's notion of expectation of random elements in (real or complex) Banach spaces $\mathscr{X}$ to the case when there was a (probability-) measure defined on a certain sigma algebra $\mathscr{F}$ of $\mathscr{X}$ (not necessarily generated by the open sets with respect to the norm of the Banach space) such that each element of the dual space $\mathscr{X}^*$ (the space of all linear and continuous—real or complex—functionals of X) was measurable with respect to $\mathscr{F}$. Fréchet had considered less general sigma algebras only. If several random elements $X_1, \ldots, X_n$ were given, Mourier—in a way methodically characteristic of many contributions of that time—considered the product space $\mathscr{X}^n$, endowed with the sigma algebra $\mathscr{F}^n$ and with an appropriate probability measure on it, as the common domain of definition of these random elements.[46] In turn, each of the random elements was conceived as a projection from $\mathscr{X}^n$ onto $\mathscr{X}$ (see [Mourier 1953a, 166], for example). Her notion of expectation was as follows: For a given random element X in $\mathscr{X}$, the element $\mathrm{E}(X) \in \mathscr{X}$ is called the "expectation" of X, if and only if, for all $x^* \in \mathscr{X}^*$, the "regular" expectation $\mathrm{E}x^*(X)$ of the random variable $x^*(X)$ exists, and

$$x^*(\mathrm{E}(X)) = \mathrm{E}x^*(X).$$

Mourier [1953a, 164–169] showed that this generalized notion of expectation has the usual linearity properties, and she proved that for Banach spaces $\mathscr{X}$ which are reflexive and separable, the norm $\|X\|$ of a random element X was a random variable, and the existence of $\mathrm{E}\|X\|$ implied the existence of $\mathrm{E}(X)$. Mourier

[46] It is an interesting detail that [Gnedenko & Kolmogorov 1949, § 1] designated such an approach "too deficient," even in cases of finite-dimensional random variables, and recommended to base all investigations, as explained in [Kolmogorov 1933a], on an abstract probability space which was to serve as a common domain of definition for all random variables considered.

[1953a, 169–172] also showed that her definition of expectation (which was, as she explicitly explained, equivalent to Pettis's notion of integral [Pettis 1938]) was consistent with Frechet's (which was based on Bochner integrals [Bochner 1933]).

A considerable part of Mourier's 1953 article was on strong laws of large numbers. These contributions became rather popular and are still connected with her name (see [Gänssler & Stute 1977, 337], for example). More important for our subject, however, is her discussion of characteristic functions of random elements. Basing herself on work by Le Cam [1947] (the conceptual idea of which, however, had been anticipated by Kolmogorov [1935][47]) she introduced the characteristic function φ_X of a random element X in a real Banach space $\mathscr{X}$ as a function $\mathscr{X}^* \to \mathbb{C}$ by

$$\varphi_X(x^*) := \mathrm{E}\exp(\mathrm{i}x^*(X)).$$

Mourier's main result [1953a, 226] in this context was a convergence theorem for characteristic functions with the following content: Let $\mathscr{X}$ be a separable and reflexive real Banach space, and X_n be a sequence of random elements in this space. If the sequence of characteristic functions $\varphi_{X_n} : \mathscr{X}^* \to \mathbb{C}$ tends to a function $\varphi : \mathscr{X}^* \to \mathbb{C}$ uniformly for all $x^* \in \mathscr{X}^*$ with $\|x^*\| \leq A$ (for some positive A), and if there exists an $\alpha > 0$ such that $\mathrm{E}\|X_n\|^\alpha$ is uniformly bounded, then the function φ is the characteristic function of a certain random element in $\mathscr{X}$.

On the basis of this theorem, Mourier [1953a, 242] considered the characteristic functions φ_{Z_n} of the normed sums

$$Z_n := \frac{1}{\sqrt{n}}(Y_1 + \cdots + Y_n),$$

where Y_k are identically distributed and independent random elements with values in a separable real Hilbert space (endowed with the sigma algebra generated by its metric) such that $\mathrm{E}Y_k = 0$ and $\mathrm{E}\|Y_k\|^2 = s^2 < \infty$. She showed that φ_{Z_n} tends to the characteristic function of a "Laplacian" random element in this Hilbert space as $n \to \infty$. Mourier used the adjective "Laplacian" to designate (with reference to Fréchet [1951]) those random elements Z in a real Banach space with $x^*(Z)$ being normally distributed for all x^* of the dual space. Mourier first proved that $\mathrm{Var}\|Z_n\|$ exists and is equal to s^2. Then she expanded φ_{Z_n} in the form

$$\varphi_{Z_n}(x^*) = \left[\varphi_{Y_1}\left(\frac{x^*}{\sqrt{n}}\right)\right]^n = \left(1 - \frac{1}{2n}\mathrm{E}(x^*(Y_1))^2 + \frac{1}{n}\|x^*\|^2\omega\left(\frac{x^*}{\sqrt{n}}\right)\right)^n,$$

where $\omega(x^*) \to 0$ if $\|x^*\|$ tends to 0. By use of the just-described convergence theorem the assertion could be followed.

[47] Kolmogorov generalized the idea of "Laplace transform" (he rather meant Fourier transform) toward "real and completely additive" functions defined in Banach spaces exactly in the same way as later Mourier. He hinted at applications in probability theory, especially at a "generalization" of the "theorem of Laplace," but he did not give any detailed explanations. Le Cam apparently did not know Kolmogorov's contribution, and his implications were less general than Kolmogorov's.

Now Mourier [1953a, 243] claimed, without giving any further explanations, that she had equally ("également") proven the following theorem (her "Théorème 5"):

If $Y_1, Y_2, \ldots, Y_n$ are independent random elements of the same law with values in a separable Hilbert space, and if $\mathrm{E}\|Y_i\|^2 = s^2$, then

$$\frac{1}{\sqrt{n}}(Y_1 + \cdots + Y_n)$$

for $n \to \infty$ tends to a Laplacian random element in the sense of Bernoulli.

Specifying the mode of convergence, she used a somewhat strange terminology, which was, however, possibly influenced by a word choice of Lévy [1937a, 52], who (for real-valued random variables) had designated convergence in law as "convergence au point de vue de Bernoulli." As we can see from the subsequent publication [Fortet & Mourier 1954, 28 f.], Mourier actually meant convergence of the distribution function of $\|Z_n'\|^2$ ($Z_n' = Z_n - \mathrm{E}Z_n$) to the distribution function of $\|Y\|^2$, where Y was the Laplacian random element assigned to the limit of the characteristic functions $\varphi_{Z_n'}$. This convergence can be shown by the following considerations: Let (x_k) be an orthonormal Schauder basis of the real Hilbert space, and

$$Z_n = \sum_{k=1}^{\infty} A_{nk} x_k, \quad Y = \sum_{k=1}^{\infty} A_k x_k, \quad Y_1 = \sum_{k=1}^{\infty} B_k x_k.$$

Then also

$$\|Z_n\|^2 = \sum_{k=1}^{\infty} A_{nk}^2, \quad \|Y\|^2 = \sum_{k=1}^{\infty} A_k^2, \quad \|Y_1\|^2 = \sum_{k=1}^{\infty} B_k^2.$$

Without loss of generality one can assume that $\mathrm{E}Y_1 = 0$. The convergence of the characteristic functions $\varphi_{Z_n}(x^*) \to \varphi_Y(x^*)$ with $x^* = < t_1 x_1 + \cdots + t_s x_s, \cdot >$ ($< \cdot, \cdot >$ designating the scalar product of the Hilbert space and $t_1, \ldots, t_s \in \mathbb{R}$) and the theorem on the "continuous" correspondence between characteristic functions and distributions in the case of finite-dimensional random variables imply that

$$\lim_{n\to\infty} P(\sum_{k=1}^{s} A_{nk}^2 \leq a) = P(\sum_{k=1}^{s} A_k^2 \leq a)$$

for all fixed $s \in \mathbb{N}$ and all $a > 0$. Because $E\|Y_1\|^2 < \infty$ and by virtue of the independence of $Y_1, Y_2, \ldots$ the equation

$$\mathrm{E}\sum_{k>s} A_{nk}^2 = \sum_{k>s} \mathrm{E}B_k^2$$

follows,[48] the right-hand side tending to 0 as $s \to \infty$. As an immediate consequence we have

[48] As Mourier [1953a, 242] showed, for independent random elements X_1, X_2 in a Hilbert space with zero expectations $\mathrm{E}\|X_1 + X_2\|^2 = \mathrm{E}\|X_1\|^2 + \mathrm{E}\|X_2\|^2$ holds.

$$P(\sum_{k>s} A_{nk}^2 > \varepsilon) \to 0 \quad (s \to \infty)$$

uniformly in n for any positive ε, which completes the argumentation.

Fortet and Mourier [1954, 29 f.] even showed that $\varphi_{Z_n}(x^*) \to \varphi_Y(x^*)$ implies that the distribution of $f(Z_n)$ tends to $f(Y)$ for all real functions f which are defined on the Hilbert space under consideration and are uniformly continuous in each bounded ball of that Hilbert space. These results were further generalized in [Fortet & Mourier 1955] toward random elements in certain separable and reflexive Banach spaces such as $L^\alpha(\mathbb{R})$ ($\alpha \geq 2$). By use of this kind of CLT in Banach spaces, Fortet and Mourier [1955] succeeded in deriving limit relations like (7.24), thus giving an alternative approach to invariance principles.

Activities starting at the end of the 1940s generalizing classical limit theorems were very lively, and therefore it is not very astonishing that different mathematicians came to similar results. In the preceding sections, a survey of only two decisive trends could be given, which led to the theory of limit theorems for distributions on metric or even more general topological spaces. As we have seen, motivation for research could result from particular problems, partly related to practical applications, as well as from "structural" questions regarding "abstract" measure spaces. In Donsker's work as well as in Mourier and Fortet's contributions, several components of the theory of convergence of distributions of random elements in metric spaces already occurred. The diffusion of Alexandrov's comprehensive treatise on convergence of measures from 1943, already referred to in Sect. 7.3.1.4, was initially impeded by World War II. Therefore, the various aspects of convergence of distributions were only perceived by a broader audience outside Russia from the beginning of the 1950s. In this context, the "Conference on Probability Theory and Mathematical Statistics" in Berlin from 19 to 22 October 1954 played a significant role, see [Kolmogorov & Prokhorov 1956]. Yuri Vasilevich Prokhorov's celebrated article [1956] on "Convergence of Random Processes and Limit Theorems in Probability Theory" inspired a great number of authors to further investigations in this field. Until the mid-1960s, a rather complete theory of probabilistic limit theorems in metric spaces was elaborated, as shown by the monographs of Kalyanapuram Rangachari Parthasarathy [1967] and Patrick Billingsley [1968], for example.

Many notions, concepts, and results of the "classical" theory for sums of independent real-valued random variables can be transferred to random elements in "abstract spaces." This applies, for example, to infinitely divisible distributions, to characteristic functions, or to versions of the Lindeberg condition. However, there was a big shift between the contributions just described and the work of the early contributors to the CLT in a twofold sense: Firstly, the classical applications of the CLT, such as error theory, were hardly relevant any more for a science in which nondeterministic considerations regarding the stochastic point of view prevailed. Secondly, the analytic character and the typical devices of analytic probability theory had to be, to a large extent, replaced by sophisticated measure-theoretic considerations. This circumstance is exemplified by the fate of characteristic functions: Despite far-reaching analogies between properties of characteristic functions defined on dual spaces and "classical" characteristic functions, the most important property in the case of finite-dimensional random variables, the continuous correspondence between characteristic functions and distributions, could not be extended in sufficient generality to those cases where random variables have values in infinite-dimensional spaces [Araujo & Giné 1980, 30].

Chapter 8
Conclusion: The Central Limit Theorem as a Link Between Classical and Modern Probability Theory

During the period between the two world wars, modern probability theory emerged as a mathematical subdiscipline—with the typical formation of concepts, main theorems and methods—by integrating the subfields of axiomatics (including aspects of measure theory), strong laws of large numbers, stochastic processes, and limit theorems for distributions of sums of random variables, which at first were related by little more than the shared generic term "probability." Only this last field of limit theorems could point to any significant contributions in the 18th and particularly the 19th centuries, and it played an especially important role during the transition from classical to modern probability theory. In the following summary of the most important aspects of this book, that role will be examined more closely. In so doing, it is particularly important to pursue the question of how the CLT can be understood as "classical" content of probability theory and what types of changes it underwent while crossing over to modern probability.

Extensive interest in probabilities of sums of independent random variables already existed in the 18th century. It began with problems relating to games of chance, such as the throwing of several dice, and ranged to problems of the theory of errors, which increasingly gained in visibility from 1750 on. Using methods of generating functions, it was possible to establish exact formulae for the probabilities of sums, at least for the algebraically simplest forms of probability functions of the individual summands that were tacitly assumed to be independent. However, even a moderate number of summed random variables made it impossible to perform a numerical analysis of these very complicated result terms. It proved very difficult, however, to move significantly past de Moivre's approximation to the binomial distribution in the 18th century.

This is why it was such a major accomplishment when, around 1810, Laplace was able to show that every sum of a considerable number of independent, identically distributed random variables, under conditions that, in practice, were always fulfilled, had to be normally distributed in apparently very good approximation. This result, which Poisson would later also generalize to include sums of non-identically distributed random variables using modified Laplacian methods, abruptly expanded the application spectrum for probability theory in an entirely remarkable way and

H. Fischer, *A History of the Central Limit Theorem*, Sources and Studies in the History of Mathematics and Physical Sciences, DOI 10.1007/978-0-387-87857-7_8,

shaped Laplace's main work in stochastics, *Théorie analytique des probabilités*, the first edition of which appeared in 1812. The universality of Laplace's approximation, which in the 19th century generally had the status of an unchallenged natural law, serves as justification for speaking not about *a* but rather *the* CLT, even if, strictly speaking, the word "limit" did not quite apply to Laplace and Poisson.

In modern notation, Laplace established the following facts; they were not presented generally at any point in his work, but they were derived using the same method in the specific application: Let $X_1, \ldots, X_n$ be independent, identically distributed and bounded random variables, which are continuous or discrete. If n is a large number, then, for $\mu := \mathrm{E}X_1$ and $\sigma^2 := \mathrm{E}(X_1 - \mu)^2$,

$$P(n\mu + r_1\sqrt{n} \leq X_1 + \cdots + X_n \leq n\mu + r_2\sqrt{n}) \approx \frac{1}{\sqrt{2\pi}} \int_{r_1}^{r_2} \frac{1}{\sigma} \mathrm{e}^{-\frac{t^2}{2\sigma^2}} dt. \quad (8.1)$$

In deriving this approximation, Laplace provided the "raw form" for the treatment of CLTs based on the method of "characteristic functions."

As emerges from Laplace's *TAP*, three areas were of particular importance when applying the approximation (8.1) or associated relations for linear combinations of identically distributed observation errors: the probabilistic justification of the method of least squares for fitting parameters to error-distorted observations, the examination of apparent or hidden regularities in nature, and the discussion of "advantages that depend upon the probability of future events"—an early approach to risk theory.

Laplace was the most prominent probabilist during the era that stretched from the systematic evaluation of games of chance—which had its advent in the letters between Pierre Fermat and Blaise Pascal in 1654—until the close of the 19th century. Lorraine Daston [1988] has described the epoch encompassing Laplace and his immediate successors as "classical probability theory." The main goal of this "theory" was to assist people in making "reasonable decisions," and it existed first and foremost on the basis of its applications; in other words, it was part of "mathesis mixta" and thus represented a discipline of mathematics only in the broadest sense.

Daston's use of the adjective "classical" is in partial correspondence to an understanding of probability theory as it was still widely practiced in the early 20th century, and it also refers to the longstanding dominance of modeling and application problems in this discipline. Still, her concept has to be contrasted with an understanding of the word "classical" in the sense of "traditional" or "established" but also in the sense of "exemplary" or "standard," as it is often employed by the modern "working mathematician." These "pragmatic" meanings correspond to Kolmogorov's characterization [1933/50, 8 f.] of independence as a "classical" subject of study. K. L. Chung, the translator of the second English edition of Gnedenko and Kolmogorov's *Limit Distributions* [1949/68], also expresses a similar view when he writes of the "classical beauty of this definitive work" in the preface (p. iii).

In fact, Laplace's accomplishments in probability theory can specifically be called "classical" in Daston's sense insofar as he was working entirely within the framework of classical probability theory to develop stochastics into a universal method to which all scientific fields could be made accessible. In this respect, the CLT played a leading part as a "tool of common sense." Yet Daston's characterization of classical probability theory as a theory which hardly possessed mathematical structures that were specific or generally relevant, and which was thus little more than a "sum of its applications," proved to be no longer true for Laplace's stochastic work. His "analytical" probability theory already transcended the range of its applications due to the relevance of its mathematical methods. This trend along with the polarization between analytical methods and applications was further bolstered by his successor Poisson.

Following Laplace's death, the research program surrounding classical probability theory came under intense criticism. It was directed primarily at applications to the field of human decisions, such as in legal proceedings. In reaction, though, people in fields where probabilistic approaches had not been fundamentally rejected, such as the theory of errors, began to discuss whether specific arguments had been based on hypotheses that were arbitrary and thus subject to attack. The transition phase after Laplace's lifetime lasted until the first decade of the twentieth century.

Error theory was this subdiscipline of probability theory that underwent the most extensive mathematization in the post-Laplace period. This field in particular provoked repeated critical examinations during the 19th and early 20th centuries, at least some of which involved demanding analytical considerations. Several papers took substantial steps toward a rather abstract—some might even say nitpicky—view of the theory of errors. This applies in large measure to the articles Cauchy published during his dispute with Bienaymé about the fundamentals of the method of least squares in 1853. The CLT was particularly important for the justification according to Laplace of this method. It was here that Cauchy provided the outline for a rigorous proof of a CLT for linear combinations of errors of observation, though under rather restrictive conditions. For applications of least squares to a "considerable" number of observations, Laplace had tacitly assumed that the distribution of the deviation between true value and estimator was by all means very close to a normal distribution. Cauchy wished to conduct a critical examination of this assumption. His efforts essentially failed, however.

Laplace's reasons for claiming the superiority of least squares among all methods for estimating unknown parameters from observed values were limited to a large number of observations. However, the two justifications (1809, 1823) of the method of least squares by Gauss were already valid for a small number of observations. The first of these two different approaches was based on the assumption of a "Gaussian" error law, meaning a normal distribution for errors of observation, and enjoyed considerable popularity throughout the 19th century likely because it concentrated on a specific probability distribution, without which several in-depth studies of error theory would have been impossible to conduct. In a departure from Gauss's original line of argument, the first justification was modified by Hagen and Bessel (1837/38) in such a way that the normal distribution of errors was derived from the "hypothesis of elementary errors." This hypothesis is understood to be the assumption that every observational error consists of a sum of a very large number of small independent "elementary errors." Even if an elementary error hypothesis does not explicitly appear anywhere in Laplace's work, his central "limit" theorem had laid the intellectual groundwork for this idea, which others, especially Quetelet, soon carried over to all mass phenomena in biology and social sciences. The widest effect Laplace's universal approximation of distributions of sums of independent random variables had, manifested itself in so-called "Queteletism," the school of thought which presumed a normal distribution behind every statistical ensemble. The mathematical examination of the hypothesis of elementary errors was carried on into the 1930s and was still playing a not insignificant role in motivating Lévy's preoccupation with the central limit problem. Not only here can the reverberations of Laplace's stamp on classical probability theory be traced far into the 20th century; the approach to risk theory that Laplace established was also vital in Cramér's work on extended versions of the CLT with asymptotic series expansions in the 1920s.

Laplace had consistently stressed the relevance of analytical methods of probability theory, primarily with regard to "approximations of formula functions of large numbers." This is evident in compressed form in the preface to the first edition of his *TAP*, where he called examinations of this type the "most delicate, difficult and useful" part of probability theory and expressed the hope that, owing in part to their specific analytical aspects, his achievements might arouse "the attention of the geometers." Laplace's most important tool in deriving asymptotic probabilities was his method for approximating integrals which depend on a very large number (presented for the first time in [Laplace 1774]). In this way, viewed from an analytical perspective, statements that are today interpreted as limit theorems became an appendix to the theory of definite integrals and could serve for illustrating it. It was in this sense that Dirichlet presented his lectures on probability theory in the 1830s and 1840s. Dirichlet's lectures thus represent an early example of a development in which the CLT also acquired inner-mathematical significance.

Such a development is also apparent in articles by Chebyshev and Markov. Chebyshev is the starting point for the history of CLTs for sums of independent random variables in their narrower meaning as limit theorems. Even if the occasional author in the first half of the 19th century, in discussing an assertion related to (8.1), happened to mention that the difference between the right and left side equaled "exactly" zero for an "infinitely large" number of random variables, this did not constitute a statement about the convergence of a clearly defined sequence of distributions that are assigned to sums of random variables. To the practice-oriented classical probability theory, a limit theorem such as this would not have been very interesting, anyway. Chebyshev, for whom probabilistic applications were often of secondary importance, considered partial sums $\sum_{k=1}^{n} X_k$ with respect to a given sequence (X_k) of independent random variables, and he established sufficient conditions for

$$P\left(r_1\sqrt{2\sum_{k=1}^{n}\mathrm{Var}X_k} \leq \sum_{k=1}^{n}(X_k - \mathrm{E}X_k) \leq r_2\sqrt{2\sum_{k=1}^{n}\mathrm{Var}X_k}\right) \to \frac{1}{\sqrt{\pi}}\int_{r_1}^{r_2} \mathrm{e}^{-x^2}\,dx \quad (n \to \infty). \qquad (8.2)$$

Scarcely any characteristics of classical probability theory as described by Daston can be detected in Chebyshev's work. Rather, the CLT seems to have served as an example within his theory of moments. At the same time, Chebyshev formulated the CLT in its "classical" form, and here the adjective "classical" is being employed in its "pragmatic" usage, which was already described.

For Chebychev's disciple Markov, the CLT (8.2) also initially served as a tool for illustrating interesting relationships within the theory of moments. By contrast, the two papers (1900/01) by Lyapunov on a CLT in the form (8.2) included a decisive step toward abstraction and the incipient autonomy of this theorem. Lyapunov's proofs consisted of a particularly elementary but also rather intricate reconstruction of the Poisson approach before the backdrop of criteria for analytical rigor, which had become customary after Weierstrass. On the other hand, Lyapunov managed to render the proposition more exact in a form already required by Chebyshev, by explicitly indicating a uniform bound for the difference of the two sides of (8.2) with a finite number n of random variables.

Lyapunov's papers also set themselves apart from the work of his predecessors by virtue of a new mathematical objective. Lyapunov [1900, 359–361] identified his two main goals as those of providing a "direct" proof of the CLT that Chebyshev and Markov had derived as one application of specific theories about moments, and of weakening the conditions as much as possible. Lypunov's interest in this theorem thus sprang neither from inner-mathematical nor outer-mathematical applications. Instead, he wanted to work on this theorem using "direct" and "elementary" methods in order to better explain the internal relationships.

To some extent, then, Lyapunov's aims correspond to the characterization of "modern mathematics" given by Mehrtens [1990], and also to the more general description of "modernity" as provided by the sociology of scientific knowledge, e.g., [Luhmann 1992]. The most significant features of "modern mathematics," as explained in Sect. 1.2, are autonomy, self-reference, and contingency.

Despite the extensive internal orientation of his articles on the CLT, Lyapunov continued to stress practical applicability as an important criterion of quality for mathematical accomplishment. In this discrepancy between working within mathematics and talking about mathematics he was not different from most of the authors in the modern period of probability theory. Practically all proponents of modern probability theory, such as von Mises, Lévy, Cramér, Khinchin, and even Kolmogorov who in 1933 had axiomatically placed probability theory on a measure theory basis, rejected an exclusively formalistic course for mathematics, and in so doing indicated a mindset that Mehrtens has termed "counter-modern." Despite their different attitudes regarding fundamental questions, they were working in very similar ways on a modern probability theory in the sense that they were establishing relationships between largely abstract concepts that did not need to have an outer-mathematical meaning. Even if one did not wish to depart entirely from external criteria of meaning or truth when talking about one's own mathematical activity, the actual mathematical work could evolve essentially free of such restrictions.

It was primarily after the First World War that the aforementioned von Mises, Lévy, and Cramér, and likewise Lindeberg and Bernshtein discovered the "classical" limit theorems of probability theory as a rewarding and promising field of activity. Others taking part in the continued development of CLTs and related questions in the 1930s included Feller, Khinchin, and Kolmogorov. All of these mathematicians had originally shared a common focus on the domain of analysis, mostly applied analysis, and had achieved some initial and occasionally remarkable results in this field. How successful their activities were in the "new field" of modern probability theory, and the rapid upsurge this discipline experienced as a result, is shown by the fact that, with the exception of Bernshtein and Kolmogorov, none of the authors discussed here returned to his original area of work to any larger extent. This circle even produced its first students in the years before World War II, such as Doeblin and Gnedenko, who from the very beginning of their mathematical careers worked almost exclusively in a probabilistic milieu. The central role occupied by questions of convergence to the normal distribution in the emergence of probability theory as a subdiscipline of modern mathematics between 1920 and 1940 is also demonstrated by the immediate and common adoption of the "central limit theorem" nomenclature, after Pólya had introduced this designation in 1920.

Independent of Mehrtens' ideas for a general description of modern mathematics, von Plato [1994] characterized the development of "modern probability," as it took place between the two world wars, from two points of view: Beginning with Borel's strong law of large numbers for relative frequencies [1909], new problems about infinite sets appeared which could no longer be mastered by the simple continuation of finite considerations, e.g., the transition from discrete probabilities to densities, as it had been the case with the "classical" theorems of probability theory. The measure theory fundament of probability theory began to develop as a consequence. Secondly, the basis of interpretation of probabilistic results in the natural sciences migrated to an indeterministic worldview. Von Plato [1994] calls this jump from classical to modern probability theory a "probabilistic revolution," which he sees as part of the overall metamorphosis from "classical" to "modern" science. The further development of "classical" content, which was vital to the formation of the discipline of modern probability theory before the Second World War, is mentioned only in passing by von Plato. In fact, the classical problems were not simply included in the theory; they exerted a significant influence on the development of the modern components of the theory.

The effects of this complex of classical problems on the growth of modern probability theory can be seen in the history of the CLT as early as ca. 1925. The obligation to absolutely adhere to analytical rigor—which admittedly was not supported by all "probabilists"—and to strive for the weakest possible conditions fostered the eventual dissociation of probability theory from its applications and its emancipation to become a mathematically autonomous discipline. In his emphasis on mathematical rigor in the area of the probabilistic foundation of error calculus, Lévy found himself contradicting Borel, who had become France's leading mathematician following the death of Poincaré, and was of the opinion that expending significant mathematical effort in order to establish the Gaussian law of errors and to examine exceptions to this law was not worthwhile. Yet due to his self-confidence as an analyst, Lévy was convinced of the mathematical relevance of his work in probability theory. For this reason and despite a number of hardships he had to bear, he granted himself the freedom to ignore the opinions of an important authority.

With the study of limit properties of distributions of sums of independent random variables, a component of probability theory began to establish in the 1920s that was shaped by "classical" analysis and that remains valuable to this day, as is also evident from newer monographs ([Rossberg, Jesiak, & Siegel 1985] or [Petrov 1995]). Following mainly from the work by von Mises [1919a], sums of random variables could be characterized by convolutions of distribution functions (i.e., of monotonically increasing functions with values between 0 and 1). With the aid of Lévy's theory of characteristic functions, it was possible to precisely examine and characterize the convergence behavior of such convolutions. In this sense, the CLT was a continuation of the analytical theories of monotonic functions and of Fourier transform. On the other hand, due to the examination of associated Stieltjes integrals, the study of distribution functions provided important incentives to exploit probability theory from a measure theory perspective. In contrast to other analytical resources, such as moment methods, characteristic functions became a tool that was specifically adapted to the requirements of probability theory while at the same time also conforming to traditional analytical practices. Lévy's theorem on the "continuous correspondence" of distributions and of characteristic functions replaced the "classical" approach of discussing normal limit distributions by the Laplacian approximation method, and allowed nonnormal limit distributions as well as "nonclassical" normings to be considered for sums of random variables without variance. Both aspects are extensively discussed in Lévy's book from 1925.

In the late 1920s and early 1930s, studies of strong laws of large numbers intensified. A significant part was played here by necessary and sufficient conditions for the almost sure convergence of series of independent random variables. Particularly important in this context was Khinchin's concept of "equivalent" sequences of random variables, that in turn was closely linked to Markov's idea of truncated random variables, which had arisen in conjunction with his activities with the CLT. Lévy [1931] took up the relevant work by Khinchin and Kolmogorov (1925, 1928) again and, by availing himself of his newly devised auxiliary variables of concentration and dispersion, produced a modified representation and derivation of the convergence criterion that differed from Kolmogorov's 1928 article. The mutually analogous references, on the one hand between the divergence of the sum of all variances and the almost sure divergence of the series of the random variables, on the other hand between the divergence of the sum of all variances and the validity of the CLT, apparently led Lévy to consider equivalent random variables in conjunction with the CLT as well. In this way, he arrived at a statement in 1931 which ultimately resulted in his solution of the central limit problem in 1935.

Both Lévy's and Feller's solutions of the central limit problem brought necessary and sufficient conditions for convergence to the normal distribution with uniformly "small" summands and with generally "nonclassical" norming. Even if their kinship with Laplace's original versions of the CLT is still immediately recognizable in a formal sense, there are far-ranging discrepancies between Laplace's "classical" and Lévy's and Feller's "modern" accounts in all other aspects: Whereas Laplace interpreted the CLT as an entire complex consisting of mathematical statement, analytical methods, and concretization in natural and social sciences, the limit problem according to Lévy and Feller gains its importance solely from references within "modern" mathematics. From a modern standpoint, the limit problem of 1935 can certainly also be considered "classical" in two senses: On the one hand, it serves as a "classical" model for further limit problems, such as those relating to infinitely divisible limit distributions or regarding the generalization of sums of independent random variables to martingales. On the other hand, it retains the predominantly "classical" analytical bias, at least in its "classical" version for independent summands and when characteristic functions are used as a main tool.

By the start of World War II, extending the limit problem to nonnormal limit distributions had created a close relationship between "modern" stochastic processes with independent increments and "classical" limit problems. As long as stochastic processes were considered to be "only" families of random variables $(X_t)_{t\in[0;1]}$, and accordingly the main focus was on the distribution functions F_t of the individual random variables X_t, the familiar analytical scope would largely be maintained.

However, this methodical orientation ultimately changed when considerations of stochastic processes were included in the form of random functions. Norbert Wiener had already begun to advance this concept in the context of Brownian motion in the early 1920s. In this case it was at least "intuitively clear" from the study of random walks, as reflected, e.g., in Khinchin's 1933 monograph, that these discrete processes must converge to a Brownian motion not only in each single point of time but even in a certain "functional" sense, when the time between each two successive steps is reduced and the number of steps is simultaneously increased. However, an exact proof of such a relationship was first provided by Donsker in 1949. Donsker built upon "modern" work that had emerged in the 1940s in relation to the Wiener measure and the Wiener integral as well as upon limit theorems of a rather "classical" application-related type for special "functionals" of partial sums which were associated with sequences of independent random variables. Thus in a manner of speaking, Donsker's theorem resulted in the "unification" of the "classical" problem field surrounding sums of independent random variables and the measure theory approach to stochastic processes.

Working at nearly the same time as Donsker, Mourier was proceeding from structural considerations about random elements in metric spaces in order to come up with the first version of a CLT for sums of independent random elements in a Hilbert space. Mourier's conclusion amounted to disengagement with several "classical" paradigms at once: With her CLT, Mourier fulfilled entirely different goals than Laplace had done within the framework of the specific problems addressed by his "classical" probability theory; the "classical" perspective of sums of independent random variables as part of the doctrine of the monotononic increasing functions could not be carried forward to random elements in more general spaces and had to be replaced by considerations substantially related to measure theory and functional analysis; in accordance with this development, the significance of characteristic functions diminished: Although the concept of characteristic functions can in principle be transferred to random elements in more general vector spaces, the "continuity" in the correspondence between distributions and characteristic functions no longer exists in such a universality in which it does in the finite-dimensional case.

Alongside structure-oriented and generalizing approaches, practical relevance once again became a driving factor in the fields related to sums of random variables around the year 1950, as we have already seen in the case of Donsker's. Whereas, for instance, the mathematically rigorous analysis of the properties of Brownian motion was regarded as largely meaningless for physics in the 1940s [Kac 1985, 112], just a few years later the corresponding mathematical studies were considered to be so important for military applications in signal processing or optics that they—insofar as they originated in the U.S.—were very often promoted by the Office of Naval Research, which was founded in 1946. The bolstered interest in collaboration in these areas between mathematics and physics is demonstrated by the monumental monograph *Théorie des fonctions aléatoires* by Blanc-Lapierre and Fortet [1953].

Gnedenko and Kolmogorov published their original Russian edition of the classic *Limit Distributions of Sums of Independent Random Variables* in 1949, and it later became widely available in several editions and many languages. This monograph contains all of the results that had been achieved with regard to the CLT and its generalizations for real-valued independent random variables, mostly before the Second World War. The exposition is based almost exclusively on the theory of characteristic functions. It appears to be self-contained and straightforward, and it thus illuminates those "stochastic structures" that are relevant in connection with sums of independent random variables, such as "independence," "limit distributions," "modern limit problem," "convergence to the normal distribution," "infinitely divisible distributions."[1] Even if the work by Gnedenko and Kolmogorov represented a summary of the measure theory foundation of probability theory and a highly modern discussion of the various types of convergence, it clearly indicates how important the "classical" analytical process was to the development of "modern" probability theory. Nevertheless, Lévy's contributions to the central limit problem are disregarded, at least in part, by the specific type of representation; a discussion of the tools concentration and dispersion, which are likewise of analytical character but are more directly associated with the notion of probability, cannot be found. The absence of current references to stochastic processes, in particular Brownian motion, is likewise conspicuous.

If one so desires, the book by Gnedenko and Kolmogorov can be considered the "ultimate" representation of a "classical-modern" subfield of probability theory whose basic structures were already in place around 1940, even if many individual problems like those involving local limit theorems, large deviations, special properties of infinitely divisible and in particular stable distributions remained unsolved for the time being. On the other hand, when one attempts to describe the history of the CLT between ca. 1945 and 1955, one could spontaneously tend to speak of a "postmodern" development. Of course, probability theory remained "modern," in

[1] The phrase "stochastic structures" was introduced by Michel Loève [1978, 291] in order to characterize "properties of probability distributions." In so doing he apparently wished to allude to similarities between structures within modern probability theory and Bourbakist concepts.

all the ways already described, and in particular according to Luhmann's idea of plumbing out contingency. However, if *Limit Distributions* from 1949 is interpreted as a "grand narrative," in the words of Jean-François Lyotard [1979/84], which utilizes a consistent language and form of argumentation (distribution functions and characteristic functions) and follows a consistent ideology (exploring "classical" stochastic structures in a mathematical discourse), then Lévy's shift to a method involving concentration and dispersion, and particularly the postwar developments characterized by a mix of methods and new interactions between theory and application, can be seen as trends with postmodern features.

If we can gain anything from this interpretation, then the CLT, in its history from Laplace to the middle of the 20th century, will have been a link between various cultural and intellectual characteristics of probability theory: In classical probability theory as a "mathesis mixta," the CLT was a kind of "natural law" which established the order of the normal distribution from the chaotic interplay of the individual random variables. From the mid-19th century, the CLT took on increased inner-mathematical significance, though at first it served more as an illustration of specific analytical theories and techniques. With Lyupanov, the CLT became an autonomous mathematical object that was studied for its own sake and for its bearing upon other inner-mathematical subjects. While fundamentally retaining "classical" structures, such as the independence of summed random variables, the CLT—generalized occasionally by nonnormal limit laws and weakening of independence—became a "classical" subject of the modern period of mathematics. Even if—or perhaps because?—developments after the Second World War resulted in a multitude of other "postmodern" branches regarding problems and methods, the CLT still represents one of the central rubrics in the *Zentralblatt* and the *Mathematical Reviews.*

For this book, however, the postmodern interpretation has a further consequence: Because postmodern developments do not have a "leitmotif" and cannot be presented by means of a uniform narration, it is reasonable to suggest that the present tale now also come to an end.

References

Aczél, Janos 1961. *Vorlesungen über Funktionalgleichungen und ihre Anwendungen.* Basel–Stuttgart: Birkhäuser.

Adams, William J. 2009. *The Life and Times of the Central Limit Theorem*, 2nd edn. Providence, Rhode Island: American Mathematical Society. The 1st edn. appeared in 1974.

Airy, George Biddell 1861. *On the Algebraical and Numerical Theory of Errors of Observations and the Combination of Observations.* London: Macmillan.

Airy, George Biddell 1879. 3rd edition of [Airy 1861]. London: Macmillan.

Akhiezer, Naum Ilich 1998. Function Theory According to Chebyshev. In *Mathematics of the 19th Century*, Vol. 3, A. N. Kolmogorov & A. P. Yushkevich (eds.), pp. 1–82, Basel: Birkhäuser.

Akhiezer, Naum Ilich 2000. *Das Akademiemitglied S. N. Bernstein und seine Arbeiten zur konstruktiven Funktionentheorie.* Originally published in Russian (1955). Giessen: Mathematisches Seminar.

Aleksandrov, Aleksandr Danilovich 1943. Additive Set Functions (Part 3). *Matematicheskii Sbornik* **55**, 169–238.

Alexanderson, Gerald L. & Lange, L. H. 1987. George Pólya. *Bulletin of the London Mathematical Society* **19**, 559–608.

Antretter, Georg 1989. *Von der Ergodenhypothese zu stochastischen Prozessen.* Schriftliche Hausarbeit für die Zulassung zur Ersten Staatsprüfung für das Lehramt an Gymnasien in Bayern. München: Ludwig-Maximilians-Universität.

Araujo, Aloisio & Giné, Evarist 1980. *The Central Limit Theorem for Real and Banach Valued Random Variables.* New York: Wiley.

Aspray, W. 1985. Interview with Robert Cameron. In: The Princeton Mathematics Community in the 1930s; An Oral-History Project, Charles C. Gillispie (ed.), Princeton.
`http://www.princeton.edu/~mudd/finding_aids/mathoral/pmc04.htm`

Bachelier, Louis 1900. *Théorie de la spéculation.* Paris: Gauthier–Villars.

Bachelier, Louis 1937. *Les lois des grands nombres du calcul des probabilités.* Paris: Gauthier-Villars.

Banach, Stefan & Kuratowski, Kazimierz 1929. Sur une généralisation du problème de la mesure. *Fundamenta Mathematicae* **14**, 127–131.

Barbut, Marc, Locker, Bernard, & Mazliak, Laurent 2004. *50 ans de correspondance en 107 lettres, Paul Lévy, Maurice Fréchet.* Paris: Hermann.

Barbut, Marc & Mazliak, Laurent 2008. Commentary on the Notes for Paul Lévy's Lectures on the Probability Calculus at the Ecole Polytechnique. *Electronic Journal for History of Probability and Statistics* **4**, June 2008.

Barth, Friedrich & Haller, Rudolf 1994. *Stochastik, Leistungskurs*, 4th edn. München: Ehrenwirth.

Basharin, Gely P., Langville, Amy N., & Naumov, Valeriy A. 2004. The Life and Work of A. A. Markov. *Linear Algebra and its Applications* **386**, 3–26.

Bavli, G. M. 1936. Über einige Verallgemeinerungen der Grenzwertsätze der Wahrscheinlichkeitsrechnung. *Matematicheskii Sbornik (Moskva) (2)* **43**, 917–930.

Belhoste, Bruno 1991. *Augustin-Louis Cauchy*. New York: Springer.

Bernhardt, Hannelore 1984. *Richard von Mises und sein Beitrag zur Grundlegung der Wahrscheinlichkeitsrechnung im 20. Jahrhundert*, Habilitationsschrift. Berlin: Humboldt-Universität.

Bernhardt, Hannelore 1985. Von Mises, Richard Martin Edler. In [Johnson & Kotz 1982–1989, Vol. 5, pp. 502–504].

Bernkopf, Michael 1970–76. Stieltjes, Thomas Jan. In [Gillispie 1970–1976, Vol. 13, pp. 55–58].

Bernoulli, Daniel 1778. Diiudicatio maxime probabilis plurium observationum discrepantium atque verisimillima inductio inde formanda. *Acta academiae scientiarum imperialis Petropolitanae* (1777, 1), 3–23. Reprinted in [D. Bernoulli 1982, pp. 361–375].

Bernoulli, Daniel 1780. Specimen philosophicum de compensationibus horologicis et veriori mensura temporis. *Acta academiae scientiarum imperialis Petropolitanae* (1777, 2), 109–128. Reprinted in [D. Bernoulli 1982, pp. 376–390].

Bernoulli, Daniel 1982. *Die Werke des Daniel Bernoulli*, D. Speiser, Hsg. Bearbeitet und kommentiert von L. P. Bouckaert und B. L. van der Waerden. Basel: Birkhäuser.

Bernoulli, Jakob 1713. *Ars conjectandi* (posthumously published). Basel: Thurnisius. Reprinted in *Die Werke von Jakob Bernoulli*, Vol. 3, Basel: Birkhäuser, 1975.

Bernshtein, Sergei Natanovich 1922. Sur le théorème limite du calcul des probabilités. *Mathematische Annalen* **85**, 237–241.

Bernshtein, Sergei Natanovich 1926. Sur l'extension du théorème limite du calcul des probabilités aux sommes des quantités dépendantes. *Mathematische Annalen* **97**, 1–59.

Bernshtein, Sergei Natanovich 1940/2004a. The Petersburg School of the Theory of Probability. Published originally in Russian in Uchenye Zapiski LGU No. 55 (ser. math. sci. No. 10), 3–11 (1940). All references are to the English translation in [Sheynin 2004a, pp. 101–110].

Bernshtein, Sergei Natanovich 1945/2004a. Chebyshev's Work in the Theory of Probability. Published originally in Russian in 1945. All references are to the English translation in [Sheynin 2004a, pp. 64–97].

Bernshtein, Sergei Natanovich 1947/2001. Chebyshev's Influence on the Development of Mathematics. Published originally in Russian in Uchenye Zapiski MGU **91**, 35–45 (1947). All references are to the English translation by O. B. Sheynin in *Mathematical Scientist* **26**, 63–73, 2001.

Bernstein, Felix & Baer, Werner Siegbert 1915. Ein Axiomensystem der Methode der kleinsten Quadrate. *Mathematische Annalen* **76**, 284–294.

Berry, Andrew C. 1941. The Accuracy of the Gaussian Approximation to the Sum of Independent Variates. *Transactions of the American Mathematical Society* **49**, 122–136.

Bertrand, Joseph Louis François 1889. *Calcul des probabilités*. Paris: Gauthier-Villars.

Bessel, Friedrich Wilhelm 1818. *Fundamenta astronomiae pro anno 1755 deducta ex observationibus viri incomparabilis James Bradley in specula astronomica Grenovicensi per annos 1750–1762 institutis*. Königsberg: Nicolovius.

Bessel, Friedrich Wilhelm 1830. Letter to C. G. Jacobi, 7 January 1830. Archiv der Berlin-Brandenburgischen Akademie der Wissenschaften, F. W. Bessel, Briefband 15.

Bessel, Friedrich Wilhelm 1836. Letter to G. Hagen, 6 March 1836. Archiv der Berlin-Brandenburgischen Akademie der Wissenschaften, Nachlass Bessel 411.

Bessel, Friedrich Wilhelm 1838a. Letter to C. G. Jacobi, 14 August 1838. Archiv der Berlin-Brandenburgischen Akademie der Wissenschaften, F. W. Bessel, Briefband 15.

Bessel, Friedrich Wilhelm 1838b. Letter to C. G. Jacobi, 24 August 1838. Archiv der Berlin-Brandenburgischen Akademie der Wissenschaften, F. W. Bessel, Briefband 15.

Bessel, Friedrich Wilhelm 1838c. Untersuchungen über die Wahrscheinlichkeit der Beobachtungsfehler. *Astronomische Nachrichten* **15**, Col. 369–404. All references are to the reprint in *Abhandlungen*, Vol. 2, R. Engelmann (ed.), pp. 372–391, Leipzig: Engelmann, 1876.

Bienaymé, Irenée Jules 1852. Sur la probabilité des erreurs d'après la méthode des moindres carrés. *Journal de mathématiques pures et appliquées (1)* **17**, 33–78.

Bienaymé, Irenée Jules 1853a. Remarques sur les différences qui distinguent l'interpolation de M. Cauchy de la méthode des moindres carrés, et qui assurent la supériorité de cette méthode. *Comptes rendus hebdomadaires des séances de l' Académie des Sciences de Paris* **37**, 5–13.

Bienaymé, Irenée Jules 1853b. Remarks on [Cauchy 1853b;c]. *Comptes rendus hebdomadaires des séances de l' Académie des Sciences de Paris* **37**, 68 f.

Bienaymé, Irenée Jules 1853c. Remarques à l'occasion des Notes insérées par M. Cauchy dans les Comptes rendus de deux des séances précédentes. *Comptes rendus hebdomadaires des séances de l' Académie des Sciences de Paris* **37**, 197 f.

Bienaymé, Irenée Jules 1853d. Remarks on [Cauchy 1853e]. *Comptes rendus hebdomadaires des séances de l' Académie des Sciences de Paris* **37**, 206.

Bienaymé, Irenée Jules 1853e. Considérations à l'appui de la découverte de Laplace sur la loi de probabilité dans la méthode des moindres carrés. *Comptes rendus hebdomadaires des séances de l'Académie des Sciences de Paris* **37**, 309–324.

Billingsley, Patrick 1968. *Convergence of Probability Measures*. New York: Wiley.

Birkner, Dunja 1996. Kolmogorov im Stalinistischen Rußland. In R. Seising & T. Fischer (eds.), *Wissenschaft und Öffentlichkeit*, pp. 41–54, Frankfurt am Main: Peter Lang.

Blanc-Lapierre, André & Fortet, Robert 1953. *Théorie des fonctions aléatoires*. Avec un chapitre sur la mécanique des fluides par J. Kampé de Fériet. Paris: Masson.

Blumenberg, Hans 1987. *Die Sorge geht über den Fluß*. Frankfurt: Suhrkamp.

Bochner, Salomon 1933. Integration von Funktionen, deren Werte die Elemente eines Vektorraumes sind. *Fundamenta Mathematicae* **20**, 262–276.

Bohlmann, Georg 1901. Lebensversicherungs-Mathematik. In *Encyklopädie der Mathematischen Wissenschaften*, I, 2, W. F. Meyer (ed.), Artikel ID4b. Leipzig: Teubner, 1900–1904.

Borel, Émile 1901. *Séries divergeantes*. Paris: Gauthier-Villars.

Borel, Émile 1909. Les probabilités denombrables et leurs applications arithmétiques. *Rendiconti del Circolo Matematico di Palermo* **27**, 247–271.

Borel, Émile 1925. *Les principes de la théorie des probabilités* (= *Traité du calcul des probabilités et de ses applications*, Tome I, Fasc. I). Paris: Gauthier-Villars.

Bourbaki, Nicolas 1994. *Elements of the History of Mathematics*. New York: Springer.

Bouvard, Alexis 1821. *Tables astronomiques publiées par le Bureau des Longitudes de France, contenant les tables de Jupiter, de Saturne et d'Uranus, construites d'après la théorie de la Mécanique céleste*. Paris: Bachelier.

Bowley, Arthur Lyon 1928. *F. Y. Edgeworth's Contributions to Mathematical Statistics*. London: Royal Statistical Society.

Brasseur, Jean Baptiste 1856. Review on "Note contenant une démonstration nouvelle du théorème de Bernouilli par M. A. Meyer". *Bulletins de l'Académie Royale des Sciences, des Lettres et des beaux Arts de Belgique* **23**, 349 f.

Bremiker, Carl 1859. *Das Risiko bei Lebensversicherungen*. Berlin: Nicolai'sche Verlagshandlung.

Brezinski, Claude 1991. *History of Continued Fractions and Padé Approximations*. Berlin: Springer.

Bru, Bernard 1981. Poisson, le calcul des probabilités et l'instruction publique. In [Costabel, Dugac, & Métivier 1981, pp. 51–94].

Bru, Bernard 2002. L'œuvre scientifique de Robert Fortet. In *Ecrits sur les processus aléatoires*, M. Brissaud (ed.), pp. 19–50. Paris: Lavoisier.

Bru, Bernard & Eid, Salah 2009. Jessen's Theorem and Lévy's Lemma: A Correspondence. *Electronic Journal for History of Probability and Statistics* **5**, June 2009.

Bruhns, Christian 1869. *Johann Franz Encke, sein Leben und Wirken*. Leipzig: Günther.

Bruns, Heinrich 1897. Ueber die Darstellung von Fehlergesetzen. *Astronomische Nachrichten* **143**, Col. 229–334.

Brush, Stephen G. 1968. A History of Random Processes. I: Brownian Movement from Brown to Perrin. *Archive for History of Exact Sciences* **5**, 1–36.

Burckhardt, Heinrich 1914. Trigonometrische Reihen und Integrale (bis etwa 1850). In *Encyklopädie der Mathematischen Wissenschaften* II, 1, 2, Artikel IIA12, H. Burckhardt, R. Fricke, & W. Wirtinger (eds.). Leipzig: Teubner, 1904–1916.

Butzer, P. L., Schmidt, M., & Stark, E. L. 1988. Observations on the history of central B-splines. *Archive for History of Exact Sciences* **39**, 137–156.

Callens, Stéphane 1997. *Les Maîtres de l'erreur. Mesure et probabilité au XIXe siècle.* Paris: Presses Universitaires de France.

Cameron, Robert & Martin, William 1938. Analytic Continuation of Diagonals and Hadamard Compositions of Multiple Power Series. *Transactions of the American Mathematical Society* **44**, 1–7.

Cameron, Robert & Martin, William 1944a. The Wiener Measure of Hilbert Neighborhoods in the Space of Real Continuous Functions. *Journal of Mathematics and Physics* **23**, 195–209.

Cameron, Robert & Martin, William 1944b. Transformations of Wiener Integrals under Translations. *Annals of Mathematics* **45**, 386–396.

Cameron, Robert & Martin, William 1945a. Evaluation of Various Wiener Integrals by Use of Certain Sturm–Liouville Differential Equations. *Bulletin of the American Mathematical Society* **51**, 73–90.

Cameron, Robert & Martin, William 1945b. Transformations of Wiener Integrals under a General Class of Linear Transformations. *Transactions of the American Mathematical Society* **58**, 184–219.

Cameron, Robert & Martin, William 1947. The Behavior of Measure and Measurability under Change of Scale in Wiener Space. *Bulletin of the American Mathematical Society* **53**, 130–137.

Cauchy, Augustin Louis 1818. Seconde note sur les fonctions réciproques. *Bulletin de la Societé Philomatique*, ohne Bandangabe, 121–124. Reprinted in *Œuvres complètes (2)* **2**, pp. 228–232. Paris: Gauthier-Villars, 1958.

Cauchy, Augustin Louis 1825. *Sur les intégrales définies prises entre des limites imaginaires.* Paris: De Bure. All references are to the reprint in *Œuvres complètes* **15**(2), pp. 41–89. Paris: Gauthier-Villars, 1974.

Cauchy, Augustin Louis 1826. Sur diverses relations qui existent entre les résidus des fonctions et les intégrales définies. *Exercises de mathématiques* **1**, 95–113. All references are to the reprint in *Œuvres complètes (2)* **6**, pp. 124–145. Paris: Gauthier-Villars, 1887.

Cauchy, Augustin Louis 1827. Théorie de la propagation des ondes à la surface d'un fluide pesant d'une profondeur indéfinie. *Mémoires presentées à l'Académie Royale par divers savants (2)* **1**, 4–318. Reprinted in *Œuvres complètes (1)* **1**, pp. 4–313. Paris: Gauthier-Villars, 1882. Submitted in 1815 already.

Cauchy, Augustin Louis 1835/37. Mémoire sur l'interpolation. *Journal de mathématiques pures et appliquées* **2**, 193–205. All references are to the reprint in *Œuvres complètes (2)* **2**, pp. 5–17. Paris: Gauthier-Villars, 1958. Lithographic version already published in 1835.

Cauchy, Augustin Louis 1844. Note sur les intégrales euleriennes. *Comptes rendus hebdomadaires des séances de l'Académie des Sciences de Paris* **19**, 67–72. All references are to the reprint in *Œuvres complètes (1)* **8**, pp. 258–264. Paris: Gauthier-Villars, 1893.

Cauchy, Augustin Louis 1845. Mémoire sur les approximations des fonctions de très grands nombres. *Comptes rendus hebdomadaires des séances de l'Académie des Sciences* **20**, 691–735. All references are to the reprint in *Œuvres complètes (1)* **9**, pp. 84–121. Paris: Gauthier-Villars, 1896.

Cauchy, Augustin Louis 1847a. Mémoire sur la détermination des orbites des planètes et des comètes. *Comptes rendus hebdomadaires des séances de l'Académie des Sciences* **25**, 401–413. Reprinted in [Cauchy 1897, pp. 374–389].

Cauchy, Augustin Louis 1847b. Second Mémoire sur la détermination des orbites des planètes et des comètes. *Comptes rendus hebdomadaires des séances de l'Académie des Sciences* **25**, 475–478. Reprinted in [Cauchy 1897, pp. 389–393].

Cauchy, Augustin Louis 1849. Recherches nouvelles sur les séries et sur les approximations des fonctions de très grands nombres. *Comptes rendus hebdomadaires des séances de l'Académie des Sciences de Paris* **29**, 42–47. All references are to the reprint in *Œuvres complètes (1)* **11**, pp. 134–140. Paris: Gauthier-Villars, 1899.

Cauchy, Augustin Louis 1853a. Mémoire sur l'evaluation d'inconnues déterminées par un grand nombre d'equations approximatives du premier degré. *Comptes rendus hebdomadaires des séances de l'Académie des Sciences* **36**, 1114–1122. All references are to the reprint in [Cauchy 1900, pp. 36–46].

Cauchy, Augustin Louis 1853b. Mémoire sur l'interpolation, ou remarques sur les remarques de M. Jules Bienaymé. *Comptes rendus hebdomadaires des séances de l'Académie des Sciences* **37**, 64–68. All references are to the reprint in [Cauchy 1900, pp. 63–68].

Cauchy, Augustin Louis 1853c. Sur la nouvelle méthode d'interpolation comparée à la méthode des moindres carrés. *Comptes rendus hebdomadaires des séances de l'Académie des Sciences* **37**, 100–109. All references are to the reprint in [Cauchy 1900, pp. 68–79].

Cauchy, Augustin Louis 1853d. Mémoire sur les coefficients limitateurs ou restricteurs. *Comptes rendus hebdomadaires des séances de l'Académie des Sciences* **37**, 150–162. All references are to the reprint in [Cauchy 1900, pp. 79–94].

Cauchy, Augustin Louis 1853e. Sur les résultats moyens d'observations de même nature, et sur les résultats les plus probables. *Comptes rendus hebdomadaires des séances de l'Académie des Sciences* **37**, 198–206. All references are to the reprint in [Cauchy 1900, pp. 94–104].

Cauchy, Augustin Louis 1853f. Sur la probabilité des erreurs qui affectent des résultats moyens d'observations de même nature. *Comptes rendus hebdomadaires des séances de l'Académie des Sciences* **37**, 264–272. All references are to the reprint in [Cauchy 1900, pp. 104–114].

Cauchy, Augustin Louis 1853g'. Remarks on [Bienaymé 1853e]. *Comptes rendus hebdomadaires des séances de l'Académie des Sciences* **37**, 324 f.

Cauchy, Augustin Louis 1853g. Sur la plus grande erreur à craindre dans un résultat moyen, et sur le système de facteurs qui rend cette plus grande erreur un minimum. *Comptes rendus hebdomadaires des séances de l'Académie des Sciences* **37**, 326–334. All references are to the reprint in [Cauchy 1900, pp. 114–124].

Cauchy, Augustin Louis 1853h. Mémoire sur les résultats moyens d'un très-grand nombre des observations. *Comptes rendus hebdomadaires des séances de l'Académie des Sciences* **37**, 381–385. All references are to the reprint in [Cauchy 1900, pp. 125–130].

Cauchy, Augustin Louis 1897. *Œuvres complètes (1)* **10**. Paris: Gauthier-Villars.

Cauchy, Augustin Louis 1900. *Œuvres complètes (1)* **12**. Paris: Gauthier-Villars.

Chandler, Seth Carlo 1872. On the Construction of a Graduated Table of Mortality from a Limited Experience. *Journal of the Institute of Actuaries* **17**, 161–171.

Charlier, Carl Vilhelm Ludvig 1888. Letter to Chebyshev, 23 May 1888, edited in [Chebyshev 1951, 450 f.].

Charlier, Carl Vilhelm Ludvig 1905a. Über das Fehlergesetz. *Arkiv för Matematik, Astronomi och Fysik* **2**, Nr. 8.

Charlier, Carl Vilhelm Ludvig 1905b. Die zweite Form des Fehlergesetzes. *Arkiv för Matematik, Astronomi och Fysik* **2**, Nr. 15.

Charlier, Carl Vilhelm Ludvig 1905c. Über die Darstellung willkürlicher Funktionen. *Arkiv för Matematik, Astronomi och Fysik* **2**, Nr. 20.

Chatterji, Srishti D. 1993. Life and Works of Norbert Wiener (1894–1964). In S. D. Chatterji (ed.), *Jahrbuch Überblicke Mathematik 1993*, pp. 153–184, Braunschweig, Wiesbaden: Vieweg.

Chatterji, Srishti D. 2006. Commentary on [Hausdorff 1923/2006]. In [Hausdorff 2006, pp. 723–756].

Chatterji, Srishti D. 2007. Lindeberg's Central Limit Theorem à la Hausdorff. *Expositiones Mathematicae* **25**, 215–233.

Chebyshev, Pafnutii Lvovich 1845. Opyt elementarnogo analiza teorii veroyatnostei. Moskau. All references are to the reprint in [Chebyshev 1951, pp. 26–87].

Chebyshev, Pafnutii Lvovich 1846. Démonstration élémentaire d'une proposition générale de la théorie des probabilités. *Journal für die reine und angewandte Mathematik* **33**, 259–267. Reprinted in [Chebyshev 1899, pp. 17–26].

Chebyshev, Pafnutii Lvovich 1854. Sur une formule d'analyse. *Bulletin Physico-mathématique de l'Académie des Sciences de St. Pétersbourg* **13**, 210–211. All references are to the reprint in [Chebyshev 1899, pp. 701–702].

Chebyshev, Pafnutii Lvovich 1855/58. Sur les fractions continues. Originally published in Russian in *Uchennye zapiski Akademii Nauk po pervomu i tretemu otdlenyam* **3**, 636–664 (1855). French translation in *Journal de mathématiques pures et appliquées (2)* **3**, 1858, 289–323. All references are to the reprint in [Chebyshev 1899, pp. 203–230].

Chebyshev, Pafnutii Lvovich 1859. Sur le développement des fonctions à une seule variable. *Bulletin physico-mathématique de l'Académie Impériale des sciences de St. Pétersbourg* **1**, 193–200. All references are to the reprint in [Chebyshev 1899, pp. 501–508].

Chebyshev, Pafnutii Lvovich 1867. Des valeurs moyennes. *Journal de mathématiques pures et appliquées (2)* **12**, 177–184. Reprinted in [Chebyshev 1899, pp. 687–694]. Originally published in Russian in *Matematicheskii Sbornik* **2**, 1–9 (1867). German translation in [Schneider 1988, pp. 154–161].

Chebyshev, Pafnutii Lvovich 1874a. Sur les valeurs limites des intégrales. *Journal de mathématiques pures et appliquées (2)*, **19**, 157–160. Reprinted in [Chebyshev 1907, pp. 183–185].

Chebyshev, Pafnutii Lvovich 1874b. Sur les quadratures. *Journal de mathématiques pures et appliquées (2)* **19**, 19–34. Reprinted in [Chebyshev 1907, pp. 165–180].

Chebyshev, Pafnutii Lvovich 1885/87. Sur la représentation des valeurs limites des intégrales par des résidus intégraux. Originally published in Russian in *Annales de l'Académie des Sciences de St. Pétersbourg* **51** (1885). All references are to the French translation in *Acta mathematica* **9**, 35–56 (1887). Reprinted in [Chebyshev 1907, pp. 421–440].

Chebyshev, Pafnutii Lvovich 1887/89. Sur les résidus intégraux qui donnent des valeurs approchées des intégrales. Originally published in Russian in *Zapiski Akademii Nauk* **55** (1887). All references are to the French translation in *Acta mathematica* **12**, 287–322. Reprinted in [Chebyshev 1907, pp. 443–477].

Chebyshev, Pafnutii Lvovich 1887/90. Sur deux théorèmes relatifs aux probabilités. Originally published in Russian in *Zapiski Akademii Nauk* **55** (1887). All references are to the French translation in *Acta mathematica* **14**, 1890/91, 305–315. Reprinted in [Chebyshev 1907, pp. 481–491].

Chebyshev, Pafnutii Lvovich 1891/1907. Sur les sommes composées des valeurs de monômes simples multipliés par une fonction qui reste toujours positive. Originally published in Russian in *Zapiski Akademii Nauk* **64** (1891). All references are to the French translation in [Chebyshev 1907, pp. 559–610].

Chebyshev, Pafnutii Lvovich 1899. *Œuvres de P. L. Tchebychef*, Bd. 1. A. Markov & M. Sonin (eds.). St. Pétersbourg: Imprimerie de l'Académie Impériale de Sciences. Unchanged reprint New York: Chelsea, no year.

Chebyshev, Pafnutii Lvovich 1907. *Œuvres de P. L. Tchebychef*, Bd. 2. A. Markov & M. Sonin (eds.). St. Pétersbourg: Imprimerie de l'Académie Impériale des Sciences. Unchanged reprint New York: Chelsea, without year.

Chebyshev, Pafnutii Lvovich 1936/2004. *Teoriya veroyatnostei*. Moskva–Leningrad: Akademiya Nauk SSSR, 1936. All references are to the English translation *Definite Integrals, the Theory of Finite Differences, the Theory of Probability* (lectures delivered in 1879–1880 as taken down by A. M. Liapunov), O. B. Sheynin (transl.). Berlin: NG-Verlag, 2004.

Chebyshev, Pafnutii Lvovich 1946. *Izbrannye matematicheskie trudy*. Moskva–Leningrad: Akademiya Nauk SSSR.

Chebyshev, Pafnutii Lvovich 1948a. *Polnoe sobranie sochinenii*, T. 3: *Matematicheskii analiz*. Moskva–Leningrad: Akademiya Nauk SSSR.

Chebyshev, Pafnutii Lvovich 1948b. *Polnoe sobranie sochinenii*, T. 4: *Teoriya mekhanizmov*. Moskva–Leningrad: Akademiya Nauk SSSR.

Chebyshev, Pafnutii Lvovich 1951. *Polnoe sobranie sochinenii*, T. 5: *Prochie sochineniya, biograficheskie materialy*. Moskva–Leningrad: Akademiya Nauk SSSR.

Christoffel, Elwin Bruno 1877. Sur une classe particulière de fonctions entières et de fractions continues. *Annali di matematica pura et applicata (2)* **8**, 1–10. All references are to the reprint in *Gesammelte mathematische Abhandlungen*, Bd. 1, L. Maurer (ed.), pp. 65–87. Leipzig–Berlin: Teubner, 1910.

Chung, Kai Lai 1949. An Estimate concerning the Kolmogorov Limit Distribution. *Transactions of the American Mathematical Society* **67**, 36–50.

Cooke, Roger S. 2005. Bochner, lectures on Fourier integrals (1932). In [Grattan-Guinness 2005, pp. 945–959].

Copson, Edward T. 1965. *Asymptotic Expansions*. Cambridge: University Press.

Costabel, Pierre, Dugac, Pierre, & Métivier, Michel (eds.) 1981. *Siméon-Denis Poisson et la science de son temps*. Palaiseau: École Polytechnique.

Cotes, Roger 1722. *Harmonia Mensurarum*. Cambridge. A German translation of the portion "De Methode Differentiali Newtoniana" is in [Kowalewski 1917, pp. 12–25].

Cournot, Antoine Augustin 1843. *Exposition de la théorie des chances et des probabilités*. Paris: Hachette.

Courtault, Jean-Michel (ed.) 2002. *Louis Bachelier: aux origines de la finance mathématique*. Besançon : Presses Univ. Franc-Comtoises.

Cramér, Harald 1923. Das Gesetz von Gauss und die Theorie des Risikos. *Skandinavisk Aktuarietidskrift* **6**, 209–237. Reprinted with original page numbers in [Cramér 1994, Vol. 1, pp. 260–288].

Cramér, Harald 1925. On Some Classes of Series Used in Mathematical Statistics. *Proceedings of the 6th Scandinavian Mathematical Congress, Copenhagen*, 399–425. Reprinted with original page numbers in [Cramér 1994, Vol. 1, pp. 438–464].

Cramér, Harald 1927. On an Asymptotic Expansion Occuring in the Theory of Probability. *Journal of the London Mathematical Society* **2**, 262–265. Reprinted with original page numbers in [Cramér 1994, Vol. 1, pp. 495–498].

Cramér, Harald 1928. On the Composition of Elementary Errors. *Skandinavisk Aktuarietidskrift* **11**, 13–74, 141–180. Reprinted with original page numbers in [Cramér 1994, Vol. 1, pp. 499–600].

Cramér, Harald 1930. On the Mathematical Theory of Risk. In *Försäkringsaktiebolaget Skandia 1855–1930*, pp. 7–84, Stockholm. Reprinted with original page numbers in [Cramér 1994, Vol. 1, pp. 601–678].

Cramér, Harald 1935. Sugli sviluppi asintotici di funzioni di ripartizione in serie di polinomi di Hermite. *Giornalio di Istituto Italiano di Attuari* **6**, 141–157. Reprinted with original page numbers in [Cramér 1994, Vol. 2, pp. 729–745].

Cramér, Harald 1936. Über eine Eigenschaft der normalen Verteilungsfunktion. *Mathematische Zeitschrift* **41**, 405–414. Reprinted with original page numbers in [Cramér 1994, Vol. 2, pp. 856–865].

Cramér, Harald 1937. *Random Variables and Probability Distributions*. Cambridge: University Press.

Cramér, Harald 1937/70. 3rd edition of [Cramér 1937]. Cambridge: University Press.

Cramér, Harald 1938. Sur une nouveau théorème-limite de la théorie des probabilités. *Actual. Sci. Indust.* **736**, 5–23. Reprinted with original page numbers in [Cramér 1994, Vol. 2, pp. 895–913].

Cramér, Harald 1946. *Mathematical Methods of Statistics*. Princeton: University Press.

Cramér, Harald 1962. A. I. Khinchin's Work in Mathematical Probability. *Annals of Mathematical Statistics* **33**, 1227–1237.

Cramér, Harald 1972. On the History of Certain Expansions Used in Mathematical Statistics. *Biometrika* **59**, 205–207. Reprinted with original page numbers in [Kendall & Plackett 1977, pp. 437–439] and [Cramér 1994, Vol. 2, pp. 1341–1343].

Cramér, Harald 1976. Half a Century with Probability Theory. Some Personal Recollections. *Annals of Probability* **4**, 509–546. Reprinted with original page numbers in [Cramér 1994, Vol. 2, pp. 1352–1389].

Cramér, Harald 1994. *Collected Works*, 2 Vols., A. Martin-Löf (ed.). New York: Springer.

Cramér, Harald & Wegman, Edward J. 1986. Some Personal Recollections of Harald Cramér on the Development of Statistics and Probability. *Statistical Science* **1**, 528–535.

Crépel, Pierre 1984. Quelques materiaux pour l'histoire de la théorie des martingales (1920–1940). Université de Rennes I, Seminaire de probabilités. `http://www.probabilityandfinance.com/misc/crepel.pdf`

Crofton, Morgan William 1870. On the Proof of the Law of Errors of Observation. *Philosophical Transactions of the Royal Society of London* **160**, 175–187.

Crofton, Morgan William 1885. Probability. In *Encyclopædia Brittanica*, 9th edn., Vol. 19. Edinburgh: Black.

Czuber, Emanuel 1891. *Theorie der Beobachtungsfehler.* Leipzig: Teubner.

Czuber, Emanuel 1899. Die Entwicklung der Wahrscheinlichkeitsrechnung und ihrer Anwendungen. *Jahresbericht der Deutschen Mathematiker-Vereinigung* **7**, Teil 2.

Czuber, Emanuel 1902. Wahrscheinlichkeitsrechnung, erster Band. Leipzig: Teubner. All references are to the 2nd edition 1908.

Dale, Andrew I. 1991. *A History of Inverse Probability*. New York: Springer.

Daniell, Percy John 1919. Integrals in an Infinite Number of Dimensions. *Annals of Mathematics* **20**, 281–288.

Daston, Lorraine 1988. *Classical Probability in the Enlightenment.* Princeton: Princeton University Press.

David, Jean Marie 1909. *Zur Dirichlet'schen Methode des Diskontinuitätsfaktors (Inaugural-Dissertation).* Zürich: Leemann.

Debeye, Peter 1909. Näherungsformeln für die Zylinderfunktionen für große Werte des Arguments und unbeschränkt veränderliche Werte des Index. *Mathematische Annalen* **67**, 535–558.

de Finetti, Bruno 1929a. Sulle funzioni a incremento aleatorio. *Rendiconti della Reale Academia Nazionale dei Lincei* **10**, 163–168.

de Finetti, Bruno 1929b. Sulla possibilità di valori eccezionali per una legge di incrementi aleatori. *Rendiconti della Reale Academia Nazionale dei Lincei* **10**, 325–329.

de Finetti, Bruno 1929c. Integrazione delle funzioni a incremento aleatorio. *Rendiconti della Reale Academia Nazionale dei Lincei* **10**, 548–553.

de la Vallée Poussin, Christian 1906. Démonstration nouvelle du théorème de Bernoulli. *Annales de la Societé scientifique des Bruxelles* **31**, 220–236.

de Moivre, Abraham 1730. *Miscellanea analytica de seriebus et quadraturis.* London: Touson & Watts.

de Moivre, Abraham 1733. *Approximatio ad summam terminorum binomii* $(a + b)^n$ *in seriem expansi.* 7 pages offprint.

de Moivre, Abraham 1756. *The Doctrine of Chances*, 3rd edn. London: Millar. 1st edition 1718, still without "de Moivre's theorem," 2nd edition 1738.

Dienger, Joseph 1852. Ueber die Ausgleichung der Beobachtungsfehler. *Archiv der Mathematik und Physik* **18**, 149–193.

Dieudonné, Jean (ed.) 1978. *Abrégé d'histoire des mathématiques 1700–1900*, 2 vols. Paris: Hermann.

Dirichlet, Peter Gustav Lejeune 1836. Ueber die Methode der kleinsten Quadrate. *Bericht über die Verhandlungen der Königlich Preußischen Akademie der Wissenschaften*, 67 f. All references are to the reprint in [Dirichlet 1889, pp. 279–282].

Dirichlet, Peter Gustav Lejeune 1838. *Wahrscheinlichkeitsrechnung nach Dirichlet.* Unpublished lecture notes in two parts, the first mainly on general probability calculus, the second on the method of least squares. 132 pp. + 35 pp., without year, certainly SS 1838, written by C. W. Borchardt. Staatsbibliothek Preußischer Kulturbesitz zu Berlin, Nachlass Borchardt 3.

Dirichlet, Peter Gustav Lejeune 1839a. Sur une nouvelle méthode pour la détermination des intégrales multiples. *Comptes rendus hebdomadaires des séances de l'Académie des Sciences* **8**, 156–160. All references are to the reprint in [Dirichlet 1889, pp. 375–380].

Dirichlet, Peter Gustav Lejeune 1839b. Ueber eine neue Methode zur Bestimmung vielfacher Integrale. *Bericht über die Verhandlungen der Königlich Preußischen Akademie der Wissenschaften,* 18–25. All references are to the reprint in [Dirichlet 1889, pp. 381–390].

Dirichlet, Peter Gustav Lejeune 1839c. Ueber eine neue Methode zur Bestimmung vielfacher Integrale. *Abhandlungen der Königlich Preußischen Akademie der Wissenschaften,* 61–79. All references are to the reprint in [Dirichlet 1889, pp. 391–410].

Dirichlet, Peter Gustav Lejeune 1841/42. *Anwendung der bestimmten Integrale auf die Wahrscheinlichkeitsrechnung.* Unpublished lecture notes, 69 pp., written by Ph. L. Seidel. Institut für Geschichte der Naturwissenschaften der Universität München, Nachlass Seidel.

Dirichlet, Peter Gustav Lejeune 1846. *Anwendungen der bestimmten Integrale auf Wahrscheinlichkeitsbestimmungen, besonders auf die Methode der kleinsten Quadrate. Lejeune Dirichlet.* Unpublished lecture notes written by an unknown author, 37 pp., undated, most probably SS 1846. Institut für Geschichte der Naturwissenschaften der Universität München, Nachlass Seidel.

Dirichlet, Peter Gustav Lejeune 1889. *Werke*, Vol. I, L. Kronecker & L. Fuchs (eds.). Berlin: Georg Reimer.

Dirichlet, Peter Gustav Lejeune 1897a. *Werke*, Vol. II, L. Kronecker & L. Fuchs (eds.). Berlin: Georg Reimer. Vols. I and II together reprinted with original page numbers. New York: Chelsea, 1969.

Dirichlet, Peter Gustav Lejeune 1897b. Bemerkungen über die zweckmäßigste Art, Beobachtungen zur Bestimmung unbekannter Elemente zu verbinden. Posthumously edited in [Dirichlet 1897a, pp. 347–351].

Doeblin, Wolfgang 1939. Sur les sommes d'un grand nombre de variables aléatoires indépendantes. *Bulletin des sciences mathématiques* **63**, 23–32, 35–64.

Doeblin, Wolfgang 1940. Sur l'ensemble de puissance d'une loi de probabilité. *Studia mathematica* **9**, 71–96.

Doeblin, Wolfgang & Lévy, Paul 1936. Sur les sommes de variables aléatoires indépendantes à dispersions bornées inférieurement. *Comptes rendus hebdomadaires des séances de l'Académie des Sciences de Paris* **202**, 2027–2029.

Dötsch, Gustav 1935. Thetarelationen als Konsequenzen des Huygensschen und Eulerschen Prinzips in der Theorie der Wärmeleitung. *Mathematische Zeitschrift* **40**, 613–628.

Donsker, Monroe D. 1949. *The Invariance Principle for Wiener Functionals.* Ph.D. thesis, University of Minnesota.

Donsker, Monroe D. 1951. An Invariance Principle for Certain Probability Limit Theorems. *Memoirs of the American Mathematical Society* **6**, 12 pp.

Donsker, Monroe D. 1952. Justification and Extension of Doob's Heuristic Approach to the Kolmogorov–Smirnov Theorems. *Annals of Mathematical Statistics* **23**, 277–281.

Doob, Joseph Leo 1940. Regularity Properties of Certain Families of Chance Variables. *Transactions of the American Mathematical Society* **47**, 455–486.

Doob, Joseph Leo 1949. Heuristic Approach to the Kolmogorov–Smirnov Theorems. *Annals of Mathematical Statistics* **20**, 393–403.

Doob, Joseph Leo 1953. *Stochastic Processes.* New York: Wiley.

Doob, Joseph Leo 1986. Comment on [Le Cam 1986]. *Statistical Science* **1**, 93–94.

Doob, Joseph Leo 1990. Feller, William. In *Dictionary of Scientific Biography*, Supplement II, Vol. 17, F. L. Holmes (ed.), pp. 287–289. New York: Scribner's.

Dürer, Albrecht 1525. *Vnderweysung der messung mit dem zirckel vnd richtscheyt.* Nürnberg: Hieronymus Andreae. Several facsimiles and digitalized versions.

Dugac, Pierre 1978. Fondements de l'analyse. In [Dieudonné 1978, Vol. 1, pp. 359–421].

Edgeworth, Francis Ysidro 1883. The Law of Error. *The London, Edinburgh and Dublin Philosophical Magazine (5)* **16**, 300–309.

Edgeworth, Francis Ysidro 1894. The Asymmetrical Probability Curve. *Proceedings of the Royal Society of London* **57**, 563–568.

Edgeworth, Francis Ysidro 1898. On the Representation of Statistics by Mathematical Formulae. *Journal of the Royal Statistical Society* **61**, 670–700.

Edgeworth, Francis Ysidro 1905. The Law of Error. *Transactions of the Cambridge Philosophical Society* **20**, 37–65, 113–141.

Edgeworth, Francis Ysidro 1906. The Generalised Law of Error, or Law of Great Numbers. *Journal of the Royal Statistical Society* **69**, 497–539.

Edgeworth, Francis Ysidro 1917. On the Mathematical Representation of Statistical Data. *Journal of the Royal Statistical Society*, without volume, 65–83, 266–288, 411–437.

Einstein, Albert 1905. Über die von der molekularkinetischen Theorie der Wärme geforderte Bewegung von in ruhenden Flüssigkeiten suspendierten Teilchen. *Annalen der Physik (4)* **17**, 549–560.

Eisenhart, Churchill 1983. Laws of Error. In [Johnson & Kotz 1982–1989, Vol. 4, pp. 530–566].

Elfving, G. 1981. *The History of Mathematics in Finnland 1828–1918.* Helsinki: Societas Scientiarum Fennica.

Ellis, Robert Leslie 1844. On the Method of Least Squares. *Transactions of the Cambridge Philosophical Society* **8**, 204–219. Reprinted in [Ellis 1863, pp. 12–37].

Ellis, Robert Leslie 1863. *The Mathematical and other Writings*, W. Walton (ed.). Cambridge: Deighton.

Encke, Johann Franz 1850. Über die Anwendung der Wahrscheinlichkeitsrechnung auf Beobachtungen. *Berliner Astronomisches Jahrbuch für 1853*, 310–352.

Erdös, Paul & Kac, Marc 1946. On Certain Limit Theorems of the Theory of Probability. *Bulletin of the American Mathematical Society* **52**, 292–302.

Erdös, Paul & Kac, Marc 1947. On the Number of Positive Sums of Independent Random Variables. *Bulletin of the American Mathematical Society* **53**, 1011–1020.

Ermaloeva, Nataliya S. 1987. Obodnom neopublikovannom kurse terorii veroyatnostei P. L. Chebysheva. *Voprosy istorii estestvoznaniya i tekhniki* No. 4, 106–112.

Esseen, Carl Gustav 1942. On the Liapounoff Limit of Error in the Theory of Probability. *Arkiv för Matematik, Astronomi och Fysik* A **28**, Nr. 9.

Esseen, Carl Gustav 1943. Determination of the Maximum Deviation from the Gaussian Law. *Arkiv för Matematik, Astronomi och Fysik* A **29**, Nr. 20.

Esseen, Carl Gustav 1945. Fourier Analysis of Distribution Functions. A Mathematical Study of the Laplace–Gaussian Law. *Acta Mathematica* **77**, 1–125.

Euler, Leonhard 1748. *Introductio in analysin infinitorum, tomus I.* Lausanne: Bousquet. All references are to the reprint in *Opera omnia (1)* **8**, A. Kreuzer & F. Rudio (eds.). Berlin: Teubner, 1922.

Euler, Leonhard 1750. De fractionibus continuis observationes. *Commentarii academiae scientiarum Petropolitanae* **11**, 32–81. Submitted in 1739. Reprinted in *Opera omnia (1)* **14**, C. Böhm & G. Faber (eds.), pp. 291–349. Berlin: Teubner, 1925.

Faà-di-Bruno, Francesco 1869. *Traité élémentaire du calcul des erreurs, avec des tables stéréotypes, ouvrage utile à ceux qui cultivent les sciences d'observation.* Paris: Gauthier-Villars.

Farebrother, Richard W. 1999. *Fitting Linear Relationships: A History of the Calculus of Observations (1750–1900).* New York: Springer.

Fechner, Gustav Theodor 1860. *Elemente der Psychophysik*, 2 vols. Leipzig: Breitkopf und Härtel.

Fechner, Gustav Theodor 1897. *Kollektivmaßlehre.* Edited by F. Lipps. Leipzig: Teubner.

Feller, Willy 1935. Über den zentralen Grenzwertsatz der Wahrscheinlichkeitsrechnung. *Mathematische Zeitschrift* **40**, 521–559.

Feller, Willy 1937a. Über den zentralen Grenzwertsatz der Wahrscheinlichkeitsrechnung II. *Mathematische Zeitschrift* **42**, 301–312.

Feller, Willy 1937b. Über das Gesetz der großen Zahlen. *Acta litterarum ac scientiarum regiae universitatis hungaricae Francisco-Iosephinae, sectio scientiarum mathematicarum* **8**, 191–201.

Feller, Willy 1945. The Fundamental Limit Theorems in Probability. *Bulletin of the American Mathematical Society* **51**, 800–832. Reprinted in [Adams 2009, pp. 80–113].

Feller, Willy 1971. *An Introduction to Probability Theory and its Applications*, Vol. 2, 2nd edn. (1st edn. 1966). New York: Wiley.

Fischer, Hans 1994. Dirichlet's Contributions to Mathematical Probability Theory. *Historia Mathematica* **21**, 39–63.

Fischer, Hans 2000. *Die verschiedenen Formen und Funktionen des zentralen Grenzwertsatzes in der Entwicklung von der klassischen zur modernen Wahrscheinlichkeitsrechnung.* Aachen: Shaker.
`http://www.shaker.de/Online-Gesamtkatalog/Details.asp?ISBN=978-3-8265-7767-3`

Fischer, Hans 2004. Jakob Friedrich Fries und die Grenzen der Wahrscheinlichkeitsrechnung. In: *Form, Zahl, Ordnung*, R. Seising, M. Folkerts, & U. Hashagen (eds.), pp. 277–299. Stuttgart: Franz Steiner Verlag.

Fischer, Hans 2006. Laplace's Approximation of the Gamma Function, a Direct Approach. Katholische Universität Eichstätt–Ingolstadt, Mathematik, Preprint-Reihe, 2006-01. `http://www.ku-eichstaett.de/Fakultaeten/MGF/Mathematik/Didmath/Didmath.Fischer`

Fischer, Hans 2007. Die Geschichte des Integrals $\int_0^\infty \frac{\sin x}{x} dx$, eine Geschichte der Analysis in der Nußschale. *Mathematische Semesterberichte* **54**, 13–30.

Förster, Gustav 1915. Das Fehlergesetz. *Zeitschrift für Vermessungswesen* **44**, 65–72.

Fortet, Robert 1949. Quelques travaux recent sur le mouvement Brownian. *Annales d'l'Institut Henri Poincaré* **11**, 175–226.

Fortet, Robert & Mourier, Edith 1952. Loi des grands nombres et théorie ergodique. *Comptes rendus hebdomadaires de l' Académie des Sciences, Paris* **234**, 699–700.

Fortet, Robert & Mourier, Edith 1953. Lois des grands nombres pour des éléments aléatoires prenant leurs valeurs dans un espace de Banach. *Comptes rendus hebdomadaires de l' Académie des Sciences, Paris* **237**, 18–20.

Fortet, Robert & Mourier, Edith 1954. Résultats complémentaires sur les éléments aléatoires prenant leurs valeurs dans un espace de Banach. *Bulletin des Sciences Mathématiques (2)* **78**, 14–30.

Fortet, Robert & Mourier, Edith 1955. Les fonctions aléatoires comme éléments aléatoires dans les espaces de Banach. *Studia mathematica* **15**, 62–79.

Fréchet, Maurice 1915. Sur l'intégrale d'une fonctionelle étendue a un ensemble abstrait. *Bulletin de la Societé Mathématique des France* **43**, 248–265.

Fréchet, Maurice 1928. Sur l'hypothèse de l'additivité des erreurs partielles. *Bulletin des sciences mathématiques (2)* **52**, 203–216.

Fréchet, Maurice 1930. Sur la convergence en probabilité. *Metron* **8**, 1–48.

Fréchet, Maurice 1936/38. *Recherches théoriques modernes sur le calcul des probabilités (= Traité du calcul des probabilités et de ses applications I, 3).*
T. 1: *Géneralités sur les probabilités. Éléments aléatoires.* Paris: Gauthier-Villars, 1936.
T. 2: *La méthode des fonctions arbitraires. Les événements en chaîne dans le cas d'un nombre fini d'états possibles.* Paris: Gauthier-Villars, 1938.

Fréchet, Maurice 1948. Les éléments aléatoires de nature quelconque dans un espace distancié. *Annales de l' Institut Henri Poincaré* **10**, 215–310.

Fréchet, Maurice 1951. Généralisations de la loi de probabilité de Laplace. *Annales de l'Institut Henri Poincaré* **12**, 1–29.

Freudenthal, Hans 1970–76. Cauchy, Augustin-Louis. In [Gillispie 1970–1976, Vol. 3, pp. 131–148].

Fries, Jakob Friedrich 1842. *Versuch einer Kritik der Prinzipien der Wahrscheinlichkeitsrechnung.* Braunschweig: Vieweg.

Fuchs, Werner 1973. Der Beitrag von C. F. Gauß zur numerischen Integration. *Mitteilungen der Gauss-Gesellschaft*, No. 10, 8–31.

Fürth, Reinhold 1917. Einige Untersuchungen über Brownsche Bewegung an einem Einzelteilchen. *Annalen der Physik* **53**, 177–214.

Gänssler, Peter & Stute, Winfried 1977. *Wahrscheinlichkeitstheorie.* New York: Springer.

Galton, Francis 1875. Statistics by Intercomparison, with Remarks on the Law of Frequency and Error. *Philosophical Magazine (4)* **49**, 33–46.

Galton, Francis 1879. The Geometric Mean, in Vital and Social Statistics. *Proceedings of the Royal Society of London* **29**, 365–367.

Gauss, Carl Friedrich 1809. Theoria motus corporum coelestium. Hamburg: Perthes & Besser. All references are to the reprint in *Werke*, Bd. 7, pp. 3–280, Teubner: Leipzig, 1906.

Gauss, Carl Friedrich 1811. Disquisitio de elementis ellipticis Palladis es oppositionibus annorum 1803, 1804, 1805, 1807, 1808, 1809. *Commentationes societatis regiae scientiarum Gottingensis recentiores* **1**. All references are to the reprint in *Werke*, Bd. 6, pp. 2–64, Göttingen, 1874.

Gauss, Carl Friedrich 1813. Disquisitiones generales circa seriam infinitam

$$1 + \frac{\alpha\beta}{1\cdot\gamma}x + \frac{\alpha(\alpha+1)\beta(\beta+1)}{1\cdot2\cdot\gamma(\gamma+1)}xx + \frac{\alpha(\alpha+1)(\alpha+2)\beta(\beta+1)(\beta+2)}{1\cdot2\cdot3\cdot\gamma(\gamma+1)(\gamma+2)}x^3 + \text{etc.},$$

pars prior. *Commentationes societatis regiae scientiarum Goettingensis recentiores* **2**, 125–162. All references are to the reprint in *Werke*, Bd. 3, pp. 124–162, Göttingen, 1876.

Gauss, Carl Friedrich 1814. Methodus nova integralium valores per approximationem inveniendi. *Commentationes societatis regiae scientiarum Goettingensis recentiores* **3**, 165–196. All references are to the reprint in *Werke*, Bd. 3, pp. 163–196, Göttingen, 1876.

Gauss, Carl Friedrich 1816. Bestimmung der Genauigkeit von Beobachtungen. *Zeitschrift für Astronomie und verwandte Wissenschaften* **1**, 185–197. All references are to the reprint in *Werke*, Bd. 4, pp. 109–117, Göttingen, 1880.

Gauss, Carl Friedrich 1821. Summary of [Gauss 1823]. *Göttingische gelehrte Anzeigen,* 1821 Februar 26. All references are to the reprint in *Werke*, Bd. 4, pp. 95–100, Göttingen, 1880.

Gauss, Carl Friedrich 1823. Theoria combinationis observationum erroribus minimis obnoxiae, pars prior. *Commentationes societatis Regiae scientiarum Goettingensis recentiores* **5**. All references are to the reprint in *Werke*, Bd. 4, pp. 3–26, Göttingen, 1880.

Gauss, Carl Friedrich 1880. *Werke*, Ergänzungsreihe Bd. 1, Briefwechsel C. F. Gauss – F. W. Bessel. Leipzig: Wilhelm Engelmann. Reprint Hildesheim–New York: Olms, 1975.

Gauss, Carl Friedrich 1900. *Werke*, Bd. 8. Leipzig: Teubner. Nachdruck Hildesheim–New York: Olms, 1981.

Gautschi, Walter 1981. A Survey of Gauss–Christoffel Quadrature Formulae. In *E. B. Christoffel, The Influence of his Work on Mathematics and the Physical Sciences*, P. L. Butzer & F. Fehér (eds.), pp. 72–147. Basel: Birkhäuser.

Gerling, Christian 1843. *Die Ausgleichungs-Rechnung der practischen Geometrie oder die Methode der kleinsten Quadrate mit ihren Anwendungen auf geodätische Aufgaben.* Hamburg–Gotha: Perthes.

Gillispie, Charles C. (ed.) 1970–1976. *Dictionary of Scientific Biography*, 16 Vols. New York: Scribner.

Gillispie, Charles C. (ed.) 1997. *Pierre-Simon Laplace 1749–1827—A Life in Exact Sciences.* Princeton: University Press.

Girlich, Hans-Joachim 1996. Hausdorffs Beiträge zur Wahrscheinlichkeitsrechnung. In E. Brieskorn (ed.), *Felix Hausdorff zum Gedächtnis. Aspekte seines Werkes*, pp. 31–70. Braunschweig/Wiesbaden: Vieweg.

Glaisher, James Whitbread Lee 1872a. Remarks on Certain Portions of Laplace's Proof of the Method of Least Squares. *The London, Edinburgh and Dublin Philosophical Magazine (4)* **43**, 194–201.

Glaisher, James Whitbread Lee 1872b. On the Law of Facility of Errors, and on the Method of Least Squares. *Memoirs of the Royal Astronomical Society (London)* **39**, 75–124.

Gnedenko, Boris Vladimirovich 1939a. K teorii predelnykh teorem dlya summ nezavisimykh sluchainykh velichin. *Izvestiya Akademii Nauk SSSR, ser. mat.*, without volume, 181–232, 643–647.

Gnedenko, Boris Vladimirovich 1939b. K teorii oblastei prityazheniya ustoichivykh zakonov. *Uchenye zapiski Moskovskogo universiteta* **30**, 61–82.

Gnedenko, Boris Vladimirovich 1959/2004. O rabotakh A. M. Lyapunova po teorii veroyatnostei. *Istoriko-matematicheskie Issledovaniya*, **12**, 135–160. All references are to the English translation "On the Work of Liapunov in the Theory of Probability" by O. B. Sheynin in [Nekrasov 2004, 156–175].

Gnedenko, Boris Vladimirovich 1997. *Theory of Probability,* 6th edn. Amsterdam: Gordon & Breach Science Publishers.

Gnedenko, Boris Vladimirovich & Kolmogorov, Andrei Nikolaevich 1949. *Predelnye raspredelniya dlya summ nezavisimykh sluchainykh velichin.* Moskva–Leningrad.

Gnedenko, Boris Vladimirovich & Kolmogorov, Andrei Nikolaevich 1949/68. *Limit Distributions of Sums of Independent Random Variables.* 2nd English edn. of [Gnedenko & Kolmogorov 1949]. New York: Wiley.

Gnedenko, Boris Vladimirovich & Sheynin, Oscar B. 1992. The Theory of Probability. In *Mathematics of the 19th Century*, Vol. 1, A. N. Kolmogorov & A. P. Yushkevich (eds.), pp. 211–288. Basel: Birkhäuser.

Goldstine, Herman H. 1977. *A History of Numerical Analysis from the 16th through the 19th Century*. New York: Springer.

Gourier, G. 1883. Sur une méthode capable de fournir une valeur approchée de l'intégrale $\int_{-\infty}^{\infty} F(x)dx$. *Comptes rendus hebdomadaires des séances de l'Académie des Sciences de Paris* **97**, 79–82.

Gram, Jørgen Pedersen 1883. Über die Entwickelung reeller Functionen in Reihen mittelst der Methode der kleinsten Quadrate. *Journal für die reine und angewandte Mathematik* **94**, 41–73.

Grattan-Guinness, Ivor (ed.) 2005. *Landmark Writings in Western Mathematics, 1640–1940*. Amsterdam: Elsevier.

Grattan-Guinness, Ivor & Ravetz, Jerome R. 1972. *Joseph Fourier 1768–1830*. Cambridge (Mass.): MIT Press.

Grigorian, A. T. 1970–76. Lyapunov, Aleksandr Mikhailovich. In [Gillispie 1970–1976, Vol. 8, pp. 559–563].

Grodzenskii, Sergei Ya. 1987. *Andrei Andreevich Markov*. Moskva: Nauka.

Hagen, Gotthilf 1831. Brief an Bessel, 3 August 1831. Archiv der Berlin–Brandenburgischen Akademie der Wissenschaften, Nachlass Bessel 242.

Hagen, Gotthilf 1836a. Brief an Bessel, 2 February 1836. Archiv der Berlin–Brandenburgischen Akademie der Wissenschaften, Nachlass Bessel 242.

Hagen, Gotthilf 1836b. Brief an Bessel, 28 July 1836. Archiv der Berlin–Brandenburgischen Akademie der Wissenschaften, Nachlass Bessel 242.

Hagen, Gotthilf 1837. *Grundzüge der Wahrscheinlichkeitsrechnung*. Berlin: Dümmler.

Hald, Anders 1981. T. N. Thiele's Contributions to Statistics. *International Statistical Review* **49**, 1–20.

Hald, Anders 1990. *A History of Probability and Statistics and their Applications before 1750*. New York: Wiley.

Hald, Anders 1998. *A History of Mathematical Statistics from 1750 to 1930*. New York: Wiley.

Hald, Anders 2000. The Early History of the Cumulants and the Gram–Charlier Series. *International Statistical Review* **68**, 137–153. All references are to the reprint in [Lauritzen 2002, 232–248].

Hald, Anders 2002. *On the History of Series Expansions of Frequency Functions and Sampling Distributions, 1873–1944* (Matematisk-fysiske meddelelser, 49). Copenhagen: Reitzel.

Hald, Anders 2007. *A History of Parametric Statistical Inference from Bernoulli to Fisher, 1713–1935*. New York: Springer.

Hamburger, Hans Ludwig 1919. Beiträge zur Konvergenztheorie der Stieltjesschen Kettenbrüche. *Mathematische Zeitschrift* **4**, 186–222.

Hamburger, Hans Ludwig 1920–21. Über eine Erweiterung des Stieltjesschen Momentenproblems. *Mathematische Annalen* **81**, 235–319; **82**, 120–164, 168–187.

Hardy, Godfrey Harold 1917. On Stieltjes' «problème des moments». *Messenger of Mathematics* **46**, 175–182.

Hardy, Godfrey Harold, Littlewood, John Edensor, & Pólya, Georg 1934. *Inequalities*. All references are to the 2nd edn. 1952. Cambridge: University Press.

Harter, H. Leon 1988. History and Role of Order Statistics. *Communications in Statistics/Theory and Methods* **17**, 2091–2107.

Hauber, Carl Friedrich 1830. Theorie der mittleren Werthe (Theil 1). *Zeitschrift für Mathematik und Physik* **8**, 25–56.

Hausdorff, Felix 1897. Das Risico bei Zufallspielen. *Berichte über die Verhandlungen der Königlich-Sächsischen Gesellschaft der Wissenschaften zu Leipzig, Mathematisch-Physikalische Classe* **49**, 497–548. Reprinted with original page numbers in [Hausdorff 2006, pp. 443–496].

Hausdorff, Felix 1901. Beiträge zur Wahrscheinlichkeitsrechnung. *Berichte über die Verhandlungen der Königlich-Sächsischen Gesellschaft der Wissenschaften zu Leipzig, Mathematisch-*

Physikalische Classe **53**, 152–178. Reprinted with original page numbers in [Hausdorff 2006, pp. 529–555].

Hausdorff, Felix 1914. *Grundzüge der Mengenlehre.* Leipzig: Veit. Reprinted in [Hausdorff 2002].

Hausdorff, Felix 1923/2006. Wahrscheinlichkeitsrechnung (Vorlesung Univ. Bonn SS 1923, SS 1931). Manuscript, edited in [Hausdorff 2006, 595–723].

Hausdorff, Felix 2002. *Gesammelte Werke*, Band II, S. D. Chatterji et al. (eds.). New York: Springer.

Hausdorff, Felix 2006. *Gesammelte Werke*, Band V, J. Bemelmans et al. (eds.). New York: Springer.

Hawkins, Thomas 1975. *Lebesgue's Theory of Integration: Its Origins and Development.* 2nd edn. The Bronx, NY: Chelsea.

Heine, Eduard Heinrich 1866. Ueber Kettenbrüche. *Monatsbericht der Königlich Preußischen Akademie der Wissenschaften zu Berlin*, without volume, 436–451.

Heine, Eduard Heinrich 1878. *Handbuch der Kugelfunctionen*, Band 1, 2. Aufl. Berlin: Georg Reimer.

Heine, Eduard Heinrich 1881. *Handbuch der Kugelfunctionen*, Band 2, 2. Aufl. Berlin: Georg Reimer. The 1st edition of "Kugelfunctionen" appeared in 1861 in one volume at Reimer.

Helly, Eduard 1930. Über Systeme von abgeschlossenen Mengen mit gemeinschaftlichen Punkten. *Monatshefte für Mathematik* **37**, 281–302.

Hengartner, Walter & Theodorescu, Radu 1973. *Concentration Functions.* London: Academic Press.

Hensel, Susan 1989. Die Auseinandersetzung um die mathematische Ausbildung der Ingenieure an den Technischen Hochschulen in Deutschland Ende des 19. Jahrhunderts. In *Mathematik und Technik im 19. Jahrhundert in Deutschland*, Hensel, S., Ihmig K.-N., & Otte, M. (eds.), pp. 1–111. Göttingen: Vandenhoek & Rupprecht.

Hermite, Charles 1864. Sur un nouveau développement en série des fonctions. *Comptes rendus hebdomadaires des séances de l' Académie des Sciences de Paris* **58**, 93–100, 266–273. Reprinted in *Œuvres*, T. II, E. Picard (ed.), pp. 293–308. Paris: Gauthier-Villars, 1908.

Herschel, John 1850. Quetelet on Probabilities. *Edinburgh Review* **92**, 1–57.

Heyde, Chris C. & Seneta, Eugene 1977. *I. J. Bienaymé: Statistical Theory Anticipated.* New York: Springer.

Hilbert, David 1899. *Grundlagen der Geometrie*. Leipzig: Teubner.

Hilbert, David 1900. Mathematische Probleme, Vortrag, gehalten auf dem internationalen Mathematiker Kongress zu Paris 1900. *Nachrichten der Königlichen Gesellschaft der Wissenschaften zu Göttingen, math.-phys. Klasse* **3**, 253–297. All references are to the reprint in *Die Hilbertschen Probleme,* P. S. Alexandrov (ed.), 3rd edn., pp. 22–80. Leipzig: Akademische Verlagsgesellschaft.

Hochkirchen, Thomas 1999. *Die Axiomatisierung der Wahrscheinlichkeitsrechnung und ihre Kontexte.* Göttingen: Vandenhoeck & Ruprecht.

Hölder, Otto 1889. Über einen Mittelwerthsatz. *Nachrichten der Königlichen Gesellschaft der Wissenschaften zu Göttingen*, without volume, 38–47.

Hoem, Jan M. 1983. The Reticent Trio: Some Little-Known Early Discoveries in Life Insurance Mathematics by L. H. F. Oppermann, T. N. Thiele and J. P. Gram. *International Statistical Review* **51**, 213–221.

Hoppe, Edmund 1920. Die Bedeutung der $\nu\varepsilon\acute{\upsilon}\sigma\varepsilon\iota\varsigma$ in der griechischen Mathematik. *Mitteilungen der Mathematischen Gesellschaft in Hamburg* **5** (1911–20), 289–304.

Hsu, Pao Lu 1945. The Approximate Distribution of the Mean and of the Variance of Independent Variates. *Annals of Mathematical Statistics* **16**, 1–29.

Jacobi, Carl Gustav 1826. Über Gauss' neue Methode, die Werte der Integrale näherungsweise zu finden. *Journal für die reine und angewandte Mathematik* **1**, 301–308. All references are to the reprint in [Jacobi 1891, pp. 3–11].

Jacobi, Carl Gustav 1827. Über eine besondere Gattung algebraischer Funktionen, die aus der Entwicklung der Function $(1 - 2xz + z^2)^{\frac{1}{2}}$ entstehen. *Journal für die reine und angewandte Mathematik* **2**, 223–226. All references are to the reprint in [Jacobi 1891, pp. 21–25].

Jacobi, Carl Gustav 1830. Brief an F. W. Bessel, 8 January 1830. Archiv der Berlin–Brandenburgischen Akademie der Wissenschaften, F. W. Bessel, Briefband 15.

Jacobi, Carl Gustav 1859. Untersuchungen über die Differentialgleichung der hypergeometrischen Reihe. Aus den hinterlassenen Papieren C. G. J. Jacobi's mitgetheilt durch E. Heine. *Journal für die reine und angewandte Mathematik* **56**, 149–165. All references are to the reprint in [Jacobi 1891, pp. 184–202.]

Jacobi, Carl Gustav 1891. *Gesammelte Werke*, Bd. 6, K. Weierstraß (ed.). Berlin: Georg Reimer.

Jahnke, Hans Niels (ed.) 2003a. *A History of Analysis*. Rhode Island: American Mathematical Society.

Jahnke, Hans Niels 2003b. Algebraic Analysis in the 18th Century. In [Jahnke 2003a, pp. 105–136].

Jessen, Børge 1929. Über eine Lebesguesche Integrationstheorie für Funktionen unendlich vieler Veränderlichen. *VII. Skandinavischer Mathematikerkongress*, pp. 127–138.

Johnson, Norman L. & Kotz, Samuel 1982–1989. *Encyclopedia of Statistical Sciences*, Vol. 1–9, 1 supplement volume. New York: Wiley.

Kac, Marc 1946. On the Average of a Certain Wiener Functional and a Related Limit Theorem in Calculus of Probability. *Transactions of the American Mathematical Society* **59**, 401–414.

Kac, Marc 1949a. On Deviations between Theoretical and Empirical Distribution Functions. *Proceedings of the National Academy of Sciences* **35**, 252–257.

Kac, Marc 1949b. On Distributions of Certain Wiener Functionals. *Transactions of the American Mathematical Society* **65**, 1–13.

Kac, Marc 1985. *Enigmas of Chance. An Autobiography.* Berkeley: University of California Press.

Kahane, Jean-Pierre 1998. Le mouvement brownien. Un essai sur les origines de la théorie mathématique. In *Matériaux pour l'histoire des mathématiques au XXieme siècle*, J. A. Dieudonné & M. Audin (eds.), pp. 123–155. Marseille: Société Mathématique de France.

Kameda, Toyojirô 1915. Theorie der erzeugenden Funktionen und ihre Anwendung auf die Wahrscheinlichkeitsrechnung. *Proceedings of the Tokyo Mathematico-Physical Society (2)* **8**, 262–295, 336–360.

Kameda, Toyojirô 1925. Theory of Generating Functions and Its Application to the Theory of Probability. *Journal of the Faculty of Science, Imperial University of Tokyo, Section I* **1**, 1–62.

Kappler, Eugen 1931. Versuche zur Messung der Avogadro–Loschmidtschen Zahl aus der Brownschen Bewegung einer Drehwaage. *Annalen der Physik (5)* **11**, 233–256.

Kapteyn, Jacobus Cornelius 1903. *Skew Frequency Curves in Biology and Statistics* (= *Publications of the Astronomical Laboratory at Groningen* (1903)).

Kendall, Maurice G. & Plackett, Robert L. 1977. *Studies in the History of Statistics and Probability*, Vol. 2. London: Griffin.

Khinchin, Aleksandr Yakovlevich 1929. Ueber einen neuen Grenzwertsatz der Wahrscheinlichkeitsrechnung. *Mathematische Annalen* **101**, 745–752.

Khinchin, Aleksandr Yakovlevich 1933. *Asymptotische Gesetze der Wahrscheinlichkeitsrechnung* (=*Ergebnisse der Mathematik und ihrer Grenzgebiete* **2**, Heft 4). Berlin: Springer.

Khinchin, Aleksandr Yakovlevich 1935. Sul dominio di attrazione della legge di Gauss. *Giornale dell'Istituto Italiano degli Attuari* **6**, 378–393.

Khinchin, Aleksandr Yakovlevich 1936. Su una legge dei grandi numeri generalizzata. *Giornale d'istituto italiano d'attuari* **7**, 365–377.

Khinchin, Aleksandr Yakovlevich 1937. Zur Theorie der unbegrenzt teilbaren Verteilungen. *Matematicheskii Sbornik (Moskva) (2)* **44**, 79–120.

Khinchin, Aleksandr Yakovlevich 1937/2005. The Theory of Probability in Pre-Revolutionary Russia and in the Soviet Union. Published originally in Russian in *Front Nauki i Tekhniki*, No. 7, 36–46 (1937). All references are to the English translation in [Sheynin 2005a, pp. 40–55].

Khinchin, Aleksandr Yakovlevich 1938. *Predelnye teoremy dlya summ nezavissimykh sluchainykh velichin.* Moskva–Leningrad: GONTI.

Khinchin, Aleksandr Yakovlevich & Kolmogorov Andrei Nikolaevich 1925. Über die Konvergenz von Reihen, deren Glieder durch den Zufall bestimmt werden. *Matematicheskii Sbornik* **32**, 668–677. English translation: [Kolmogorov 1992, pp. 1–10].

Khinchin, Aleksandr Yakovlevich & Lévy, Paul 1936. Sur les lois stables. *Comptes rendus hebdomadaires des séances de l'Académie des Sciences de Paris* **202**, 374–376. All references are to the reprint in [Lévy 1976, pp. 345 f.].

Kjeldsen, Tine Hoff 1993. The Early History of the Moment Problem. *Historia Mathematica* **20**, 19–44.

Kline, Morris 1972. *Mathematical Thought from Ancient to Modern Times.* New York: Oxford University Press.

Knobloch, Eberhard 1987. Émile Borel as a Probabilist. In [Krüger, Daston, & Heidelberger 1987, pp. 215–233].

Knobloch, Eberhard 1992. Historical Aspects of the Foundations of Error Theory. In *The Space of Mathematics*, J. Echeverria, A. Ibarra, & Th. Mormann (eds.), pp. 253–279. Berlin–New York: Walter de Gruyter.

Kolmogorov Andrei Nikolaevich 1928. Über die Summen durch den Zufall bestimmter unabhängiger Größen. *Mathematische Annalen* **99**, 309–319. English translation: [Kolmogorov 1992, pp. 15–31].

Kolmogorov, Andrei Nikolaevich 1931a. Über die analytischen Methoden in der Wahrscheinlichkeitsrechnung. *Mathematische Annalen* **104**, 415–458. English translation: [Kolmogorov 1992, 62–108].

Kolmogorov, Andrei Nikolaevich 1931b. Eine Verallgemeinerung des Laplace–Liapounoffschen Satzes. *Izvestiya Akademii Nauk SSSR, ser. mat.*, without volume, 959–962. English translation: [Kolmogorov 1992, 118–121].

Kolmogorov, Andrei Nikolaevich 1932. Sulla forma generale di un processo stocastico omogeneo. *Rendiconti della Reale Academia dei Lincei* **15**, 805–808, 866–869. English translation: [Kolmogorov 1992, 121–127].

Kolmogorov, Andrei Nikolaevich 1933a. *Grundbegriffe der Wahrscheinlichkeitsrechnung* (= *Ergebnisse der Mathematik und ihrer Grenzgebiete* **2**, Heft 3). Berlin: Springer.

Kolmogorov, Andrei Nikolaevich 1933b. Sulla determinazione empirica di una legge di distributione. *Giornale dell'Istituto Italiano degli Attuari* **4**, 83–91. English translation: [Kolmogorov 1992, 139–146].

Kolmogorov, Andrei Nikolaevich 1933c. Über die Grenzwertsätze der Wahrscheinlichkeitsrechnung. *Izvestiya Akademii Nauk SSSR, ser. math.*, without volume, 366–372. English translation: [Kolmogorov 1992, pp. 147–155].

Kolmogorov, Andrei Nikolaevich 1933/50. *Foundations of the Theory of Probability*. New York: Chelsea. English translation of [Kolmogorov 1933a].

Kolmogorov, Andrei Nikolaevich 1935. La transformation de Laplace dans les espaces linéaires. *Comptes rendus hebdomadaires des séances de l' Académie des Sciences de Paris* **200**, 1717–1718. English translation: [Kolmogorov 1991, 194–195].

Kolmogorov, Andrei Nikolaevich 1947/2005. The Role of Russian Science in the Development of the Theory of Probability. Published originally in Russian in *Uchenye Zapiski Moskovskogo Universiteta* **91**, 53–64 (1947). All references are to the English translation in [Sheynin 2005a, pp. 68–84].

Kolmogorov, Andrei Nikolaevich 1948. Commentary on [Chebyshev 1887/90]. In [Chebyshev 1948a, pp. 404–409].

Kolmogorov, Andrei Nikolaevich 1956. Two Uniform Limit Theorems for Sums of Independent Terms. *Theory of Probability and its Applications* **1**, 384–394. Reprinted in [Kolmogorov 1992, 429–441].

Kolmogorov, Andrei Nikolaevich 1958. Sur les propriétés des fonctions de concentration de M. P. Lévy. *Annales de l'Institut Henri Poincaré* **16**, 27–34. English translation: [Kolmogorov 1992, pp. 459–464].

Kolmogorov, Andrei Nikolaevich 1963. Approximation to Distributions of Sums of Independent Terms by Means of Infinitely Divisible Distributions. *Transactions of the Moscow Mathematical Society* **12**, 492–509. Reprinted in [Kolmogorov 1992, pp. 484–504].

Kolmogorov, Andrei Nikolaevich 1991. *Selected Works, Vol. I: Mathematics and Mechanics*, V. M. Tikhomirov (ed.). Dordrecht: Kluwer.

Kolmogorov, Andrei Nikolaevich 1992. *Selected Works, Vol. II, Probability Theory and Mathematical Statistics*, A. N. Shiryayev (ed.). Dordrecht: Kluwer.

Kolmogorov, Andrei Nikolaevich & Prokhorov, Yuri 1956. Zufällige Funktionen und Grenzverteilungssätze. In B. V. Gnedenko (ed.), *Bericht über die Tagung Wahrscheinlichkeitsrechnung und mathematische Statistik in Berlin vom 19. bis 22. Oktober 1954*, pp. 113–126. Berlin: VEB Deutscher Verlag der Wissenschaften. English translation: [Kolmogorov 1992, pp. 442–458].

Kowalewski, Arnold 1917. *Newton, Cotes, Gauss, Jacobi. Vier grundlegende Abhandlungen über Interpolation und genäherte Quadratur.* Leipzig: Veit.

Krein, Mark Grigorevich 1951/59. The Ideas of P. L. Chebyshev and A. A. Markov in the Theory of Limiting Values of Integrals and their Further Development. Originally published in Russian in Uspekhi mat. nauk (n. s.) **6**, 3–120 (1951). All references are to the English translation in *American Mathematical Society Translations (2)* **12**, 1959, 1–121.

Krüger, Lorenz, Daston, Lorraine, & Heidelberger, Michael (eds.) 1987. *The Probabilistic Revolution*, Bd. 1. Cambridge (Mass.): MIT Press.

Kummell, Charles H. 1876. New Investigation of the Law of Errors of Observation. *The Analyst: A Monthly Journal of Pure and Applied Mathematics* **3**, 133–140, 165–171.

Kummell, Charles H. 1877. Remarks on Mr. Merrimans Article, Entitled "An Elementary Discussion of the Principle of Least Squares." *Journal of the Franklin Institute (3)* **74**, 270–274.

Kummell, Charles H. 1879. Revision of Proof of the Formula for the Error of Observation. *The Analyst: A Monthly Journal of Pure and Applied Mathematics* **6**, 80 f.

Kummell, Charles H. 1882. On the Composition of Errors from Single Causes of Error. *Astronomische Nachrichten* **103**, Col. 177–206.

Lagrange, Joseph Louis 177?. Mémoire sur l'utilité de la méthode de prendre le milieu entre les résultats de plusieurs observations. *Miscellanea Taurinensia* **5**, 167–232. All references are to the reprint in *Œuvres*, T. 2, J. A. Serret (ed.), pp. 173–234. Paris: Gauthier-Villars, 1868. The precise year of publication is unknown, and is suspected to be between 1773 [Seal 1949, 214] and 1776 [Stigler 1986, 387].

Lagrange, Joseph Louis 1795. Leçons élémentaires sur les mathématiques données à l'École Normale en 1795. *Séances des Écoles normales* an III (1794/95). Reprinted in *Journal de l'École Polytechnique* **2**, 173–278. All references are to the reprint in *Œuvres*, T. 7, J. A. Serret (ed.), pp. 183–288. Paris: Gauthier-Villars, 1877.

Laguerre, Edmond Nikolas 1879. Sur l'intégrale $\int_x^\infty e^{-x}\frac{dx}{x}$. *Bulletin de la Societé mathématique de France* **7**, 72–81. Reprinted in *Œuvres*, T. I, Ch. Hermite (ed.), pp. 428–437. Paris: Gauthier-Villars, 1898.

Lakatos, Imre 1966. Cauchy and the Continuum: The Significance of Non-Standard Analysis for the History of Mathematics. In *Philosophical Papers*, Vol. 2, pp. 43–60. Cambridge: University Press, 1978.

Lambert, Johann Heinrich 1760. *Photometria.* Augsburg: Klett. A German translation of the stochastic parts is in [Schneider 1988, pp. 228–233].

Lancaster, Henry O. 1971. Development of the Notion of Statistical Dependence. *Mathematical Chronicle* **2**, 1–16. All references are to the reprint in [Kendall & Plackett 1977, pp. 293–308].

Laplace, Pierre-Simon 1774. Mémoire sur la probabilité des causes par les événements. *Mémoires de l'Académie Royale des Sciences de Paris* **6**, 621–656. All references are to the reprint in [Laplace 1891, pp. 27–65].

Laplace, Pierre-Simon 1776. Mémoire sur l'inclinaison moyenne des orbites des comètes, sur la figure de la terre et sur les fonctions. *Mémoires de l'Académie Royale des Sciences de Paris*, année 1773. All references are to the reprint in [Laplace 1891, pp. 279–324].

Laplace, Pierre-Simon 1781. Mémoire sur la probabilités. *Mémoires de l'Académie Royale des Sciences de Paris*, année 1778, 227–332. All references are to the reprint in [Laplace 1893, pp. 383–485].

Laplace, Pierre-Simon 1785. Mémoire sur les approximations des formules qui sont fonctions de très grands nombres. *Mémoires de l'Académie Royale des Sciences de Paris*, année 1782, 1–88. All references are to the reprint in [Laplace 1894, pp. 209–291].

Laplace, Pierre-Simon 1810a. Mémoire sur les approximations des formules qui sont fonctions de très grands nombres et sur leur application aux probabilités. *Mémoires de l'Académie Royale des Sciences de Paris*, année 1809, 353–415. All references are to the reprint in [Laplace 1898, pp. 301–345].
Laplace, Pierre-Simon 1810b. Supplément au Mémoire sur les approximations des formules qui sont fonctions de très grands nombres. *Mémoires de l'Académie Royale des Sciences de Paris*, année 1809, 559–565. All references are to the reprint in [Laplace 1898, pp. 349–353].
Laplace, Pierre-Simon 1811. Mémoire sur les intégrales définies et leur applications aux probabilités, et specialement à la recherche du milieu qu'il faut choisir entre les résultats des observations. *Mémoires de l'Académie Royale des Sciences de Paris*, année 1810, 279–347. All references are to the reprint in [Laplace 1898, pp. 357–412].
Laplace, Pierre-Simon 1812/20/86. *Théorie analytique des probabilités*. 1st edn. 1812, 2nd edn. 1814, 3rd enlarged edn. 1820. Paris: Courcier. All references are to the reprint of the 3rd edn. in [Laplace 1886].
Laplace, Pierre-Simon 1814/20/86. *Essai philosophique sur les probabilités*. Paris: Courcier. This text served as an introduction of the *TAP* from its second edition (1814), and was also published separately in the same year. All references are to introduction of the third edition of the *TAP*, as reprinted in [Laplace 1886, pp. III–CLIII].
Laplace, Pierre-Simon 1886. *Œuvres complètes de Laplace* VII. Paris: Gauthier-Villars.
Laplace, Pierre-Simon 1891. *Œuvres complètes de Laplace* VIII. Paris: Gauthier-Villars.
Laplace, Pierre-Simon 1893. *Œuvres complètes de Laplace* IX. Paris: Gauthier-Villars.
Laplace, Pierre-Simon 1894. *Œuvres complètes de Laplace* X. Paris: Gauthier-Villars.
Laplace, Pierre-Simon 1895. *Œuvres complètes de Laplace* XI. Paris: Gauthier-Villars.
Laplace, Pierre-Simon 1898. *Œuvres complètes de Laplace* XII. Paris: Gauthier-Villars.
Laplace, Pierre-Simon 1912. *Œuvres complètes de Laplace* XIV. Paris: Gauthier-Villars.
Laugwitz, Detlev 1986. *Zahlen und Kontinuum*. Mannheim: BI-Wissenschaftsverlag.
Laugwitz, Detlev 1990. Frühe Delta-Funktionen — Eine Fallstudie zu den Beziehungen zwischen Nichtstandard-Analysis und mathematischer Geschichtsschreibung. In [Spalt 1990, pp. 23–41].
Laugwitz, Detlev 1996. *Bernhard Riemann, 1826–1866; Wendepunkte in der Auffassung der Mathematik* (= *Vita mathematica* **10**). Basel: Birkhäuser.
Laugwitz, Detlev 1999. *Bernhard Riemann, 1826–1866; Turning Points in the Conception of Mathematics*. Basel: Birkhäuser. English translation (Abe Shenitzer) of [Laugwitz 1996].
Lauritzen, Steffen L. 2002. *Thiele, Pioneer in Statistics*. Oxford: University Press.
Lavrynovich (Lawrynovicz), Kasimir 1995. *Friedrich Wilhelm Bessel 1784–1846* (= *Vita mathematica* **9**). Basel: Birkhäuser.
Lebesgue, Henri 1928. *Leçons sur l'intégration et la recherche des fonctions primitives: professées au Collège de France*. Paris: Gauthier-Villars.
Le Cam, Lucien M. 1947. Un instrument d'étude des fonctions aléatoires: la fonctionnelle caractéristique. *Comptes rendus hebdomadaires des séances de l' Académie des Sciences de Paris* **224**, 710–711.
Le Cam, Lucien M. 1965. On the distribution of sums of independent random variables. In *Bernoulli (1713) Bayes (1763) Laplace (1813)*, L. M. Le Cam & J. Neyman (eds.), pp. 179–202. New York: Springer.
Le Cam, Lucien M. 1970. Remarques sur le théorème limite central dans les espaces localement convexes. In *Les probabilités sur les structures algébriques*, A. Badrikian & P.-L. Hennequin (eds.), pp. 233–249. Paris: CNRS.
Le Cam, Lucien M. 1986. The Central Limit Theorem around 1935. *Statistical Science* **1**, 78–96. Reprinted in [Adams 2009, pp. 115–137].
Legendre, Adrien Marie 1805. *Nouvelles méthodes pour la détermination des orbites des comètes*. Paris: Courcier.
Legendre, Adrien Marie 1811. *Exercises du calcul intégral*, Vol. 1. Paris: Courcier.
Legendre, Adrien Marie 1817. *Exercises du calcul intégral*, Vol. 2. Paris: Courcier.

Lévy, Paul 1919/2008. *Compléments au cours d'Analyse par M. L. Lévy.* Lecture notes by an unknown author. Facsimile in *Electronic Journal for History of Probability and Statistics* **4**, June 2008.

Lévy, Paul 1922a. Sur la rôle de la loi de Gauss dans la théorie des erreurs. *Comptes rendus hebdomadaires de l'Académie des Sciences de Paris* **174**, 855–857. Reprinted in [Lévy 1976, pp. 9–11].

Lévy, Paul 1922b. Sur la loi de Gauss. *Comptes rendus hebdomadaires de l'Académie des Sciences de Paris* **174**, 1682–1684. Reprinted in [Lévy 1976, pp. 12–13].

Lévy, Paul 1922c. Sur la détermination des lois de probabilité par leurs fonctions charactéristiques. *Comptes rendus hebdomadaires de l'Académie des Sciences de Paris* **175**, 854–856. Reprinted in [Lévy 1976, pp. 333–335].

Lévy, Paul 1923a. Sur une application de la dérivée d'ordre non entier au calcul des probabilités. *Comptes rendus hebdomadaires de l'Académie des Sciences de Paris* **176**, 1118–1120. Reprinted in [Lévy 1976, pp. 339–341].

Lévy, Paul 1923b. Sur les lois stables en calcul des probabilités. *Comptes rendus hebdomadaires de l'Académie des Sciences de Paris* **176**, 1284–1286. Reprinted in [Lévy 1976, pp. 342–344].

Lévy, Paul 1923c. Sur une opération fonctionelle généralisant la dérivation d'ordre non entier. *Comptes rendus hebdomadaires de l'Académie des Sciences de Paris* **176**, 1441–1443. Reprinted in [Lévy 1976, pp. 336–338].

Lévy, Paul 1924. Théorie des erreurs. La loi de Gauss et les lois exceptionelles. *Bulletin de la Societé mathématique de France* **52**, 49–85. All references are to the reprint in [Lévy 1976, pp. 14–49].

Lévy, Paul 1925a. Les lois de probabilité dans les ensembles abstraits. *Revue de Métaphysique et de Morale* **32**, 149–174. All references are to the reprint in [Lévy 1925b, pp. 325–345].

Lévy, Paul 1925b. *Calcul des probabilités*, Paris: Gauthier–Villars.

Lévy, Paul 1929. Sur quelques travaux relatifs à la théorie des erreurs. *Bulletin des sciences mathématiques (2)* **53**, 1–21. All references are to the reprint in [Lévy 1976, pp. 50–70].

Lévy, Paul 1931. Sur les séries dont les termes sont des variables éventuelles indépendantes. *Studia mathematica* **3**, 119–155. All references are to the reprint in [Lévy 1976, pp. 123–159].

Lévy, Paul 1934a. Sur les intégrales dont les éléments sont des variables aléatoires indépendantes. *Annali della Reale Scuola Normale Superiore di Pisa (2)* **3**, 337–366. Reprinted in [Lévy 1980, pp. 9–38].

Lévy, Paul 1934b. L'addition de variables aléatoires enchaînées. *Bulletin de la Societé Mathématique de France* **62**, 42 f. All references are to the reprint in [Lévy 1976, 160 f.].

Lévy, Paul 1934c. Propriétés asymptotiques des sommes de variables aléatoires enchaînées. *Comptes rendus hebdomadaires de l'Académie des Sciences de Paris* **199**, 627–629. All references are to the reprint in [Lévy 1976, 162 f.].

Lévy, Paul 1935a. Propriétés asymptotiques des sommes des variables aléatoires enchaînées. *Bulletin de la Societé mathématique de France (2)* **59**, 1–32. All references are to the reprint in [Lévy 1976, pp. 201–232].

Lévy, Paul 1935b. Propriétés asymptotiques des sommes des variables aléatoires indépendantes ou enchaînées. *Journal de mathématiques pures et appliquées* **14**, 347–402.

Lévy, Paul 1936a. Détermination générale des lois limites. *Comptes rendus hebdomadaires des séances de l' Académie des Sciences* **202**, 2027–2029. All references are to the reprint in [Lévy 1976, pp. 265 f.].

Lévy, Paul 1936b. Sur la notion de probabilité conditionelle. *Bulletin des Sciences Mathématiques* **60**, 66–71.

Lévy, Paul 1937a. *Théorie de l'addition des variables aléatoires* (= *Collection de monographies sur la théorie des probabilités*, Fasc. I). Paris: Gauthier–Villars.

Lévy, Paul 1937b. Complément à un théorème sur la loi de Gauss. *Bulletin des sciences mathématiques* **61**, 115–128. All references are to the reprint in [Lévy 1976, pp. 267–280].

Lévy, Paul 1939. Sur certains processus stochastiques homogènes. *Compositio mathematica* **7**, fasc. 2, 283–339. All references are to the reprint in [Lévy 1980, pp. 46–102].

Lévy, Paul 1970. *Quelques aspects de la pensée d'un mathématicien.* Paris: Blanchard.

Lévy, Paul 1976. *Œuvres* III, D. Dugué (ed.). Paris: Gauthier–Villars.

Lévy, Paul 1980. *Œuvres* IV, D. Dugué (ed.). Paris: Gauthier–Villars.

Lindeberg, Jarl Waldemar 1920. Über das Exponentialgesetz in der Wahrscheinlichkeitsrechnung. *Annales academiae scientiarum Fennicae* (mathematisch-physikalische Klasse) **16**, 1–23.

Lindeberg, Jarl Waldemar 1922a. Sur la loi de Gauss. *Comptes rendus hebdomadaires de l' Académie des Sciences de Paris* **174**, 1400–1402.

Lindeberg, Jarl Waldemar 1922b. Über das Gauss'sche Fehlergesetz. *Skandinavisk Aktuarietidskrift* **5**, 217–234.

Lindeberg, Jarl Waldemar 1922c. Eine neue Herleitung des Exponentialgesetzes in der Wahrscheinlichkeitsrechnung. *Mathematische Zeitschrift* **15**, 211–225.

Lindelöf, E. 1934. Jarl Waldemar Lindeberg. *Academia scientiarum fennica: Årsbok-Vuosikirja* **12**, Nr. 3.

Lipps, Gottlob Friedrich 1897. Ueber Fechner's Collectivmaßlehre und die Vertheilungsgesetze der Collectivgegenstände. *Philosophische Studien* **13**, 579–612.

Lipps, Gottlob Friedrich 1901. Die Theorie der Collectivgegenstände. *Philosophische Studien* **17**, 79–184, 467–575.

Loève, Michel 1950. Fundamental Limit Theorems of Probability Theory. *Annals of Mathematical Statistics* **21**, 321–338.

Loève, Michel 1955. *Probability Theory*. Princeton: Van Nostrand.

Loève, Michel 1978. Théorie des probabilités. In [Dieudonné 1978, Vol. 2, pp. 708–747].

Loud, W. S. 2005. Mathematics Fifty Years ago at Minnesota. *University of Minnesota, School of Mathematics Newsletter*, Nr. 11.
`http://www.math.umn.edu/newsletter/2005/loudrecollections.htm`

Lorentz, George G. 2002. Mathematics and Politics in the Soviet Union from 1928 to 1953. *Journal of Approximation Theory* **116**, 169–223.

Lützen, Jesper 2003. The Foundation of Analysis in the 19th Century. In [Jahnke 2003a, pp. 155–196].

Luhmann, Niklas 1992. *Beobachtungen der Moderne*. Opladen: Westdeutscher Verlag.

Lyapunov, Aleksandr Mikhailovich 1895. *Pafnutii Lvovich Chebyshev*. Kharkov. Reprinted in parts in [Chebyshev 1946, pp. 9–21].

Lyapunov, Aleksandr Mikhailovich 1900. Sur une proposition de la théorie des probabilités. *Bulletin de l'Académie Impériale des Sciences de St.-Pétersbourg (5)* **13**, 359–386. English translation in [Adams 2009, pp. 151–171].

Lyapunov, Aleksandr Mikhailovich 1901a. Sur un théorème du calcul des probabilités. *Comptes rendus hebdomadaires des séances de l' Académie des Sciences de Paris* **132**, 126–128. English translation in [Adams 2009, pp. 149–150].

Lyapunov, Aleksandr Mikhailovich 1901b. Nouvelle forme du théorème sur la limite de probabilité. *Mémoires de l'Académie Impériale des Sciences de St.-Pétersbourg VIIIe Série, Classe Physico-Mathématique* **12**, 1–24. English translation in [Adams 2009, pp. 175–191].

Lyapunov, Aleksandr Mikhailovich 1901c. Une proposition générale du calcul des probabilités. *Comptes rendus hebdomadaires de l' Académie des Sciences de Paris* **132**, 814 f. English translation in [Adams 2009, pp. 173–174].

Lyotard, Jean-François 1979/84. *The Postmodern Condition*. Manchester: Manchester University Press, 1984. Originally published in French with title *La condition postmoderne*, Paris: Éditions de Minuit, 1979.

Maistrov, Leonid E. 1974. *Probability Theory—A Historical Sketch*. New York: Academic Press.

Makropoulos, Michael 1997. *Modernität und Kontingenz*. München: Fink.

Malmquist, G. 1960. C. V. L. Charlier. *Kungliga Svenska Vetenskapsakademiens, Årsbok*, without volume, 385–405.

Mark, Abraham M. 1949. Some Probability Limit Theorems. *Bulletin of the American Mathematical Society* **55**, 885–900.

Markov, Andrei Andreevich 1884a. Démonstration de certaines inégalités de M. Chebyshev. *Mathematische Annalen* **24**, 172–180.

Markov, Andrei Andreevich 1884b. *O nekotorykh prilozhenyakh algebraicheskikh nepreryvnykh drobei. Rassuzhdenie A. Markova.* St. Petersburg.

Markov, Andrei Andreevich 1886. Lettre adressée à M. Hermite. *Annales scientifiques de l'École Normale Superieure (3)* **3**, 81–88.

Markov, Andrei Andreevich 1896 *Differenzenrechnung*. Leipzig: Teubner.

Markov, Andrei Andreevich 1898. Sur les racines de l'équation $e^{x^2} \frac{d^m e^{-x^2}}{dx^m} = 0$. *Bulletin de l'Académie Impériale des Sciences de St.-Pétersbourg (5)* **9**, 435–446. Russian translation in [Markov 1951, pp. 253–269].

Markov, Andrei Andreevich 1898/1912. Über die Wurzeln der Gleichung $e^{x^2} \frac{d^m e^{-x^2}}{dx^m} = 0$. German translation of [Markov 1898], slightly modified compared with the original, in [Markov 1912, pp. 259–271].

Markov, Andrei Andreevich 1899. Zakon bolshikh chisel i sposob naimenshikh kvadratov. (Izvlechenie iz pisem A. A. Markova k A. V. Vasilevu). *Izvestiya fiz.-mat. obschestva Kazan univ. (2)* **8**, 110–128. Reprinted in [Markov 1951, pp. 231–251].

Markov, Andrei Andreevich 1899/2004. The Law of Large Numbers and the Method of Least Squares. English translation (by O. B. Sheynin) of [Markov 1899], in [Sheynin 2004a, pp. 130–142].

Markov, Andrei Andreevich 1900. *Ischislenie veroyatnostei* (2nd edn. 1908, 3rd edn. 1913, 4th edn. 1924). St. Petersburg.

Markov, Andrei Andreevich 1906/2004. Rasprostranenie zakona bolshikh chisel na velichin, zavisyashchikh drug ot druga. *Izvestiya fiz.-mat. obshchestva Kazan univ. (2)* **15**, 135–156 (1906). Reprinted in [Markov 1951, pp. 339–362]. All references are to the English translation in [Sheynin 2004a, 143–158].

Markov, Andrei Andreevich 1907/10. Recherches sur un cas remarquable d'épreuves dépendantes. *Acta mathematica* **33**, 87–104 (1910). Originally published under the title "Issledovanie zamechatelnogo sluchaya zavisimykh ispytanii" in *Bulletin de l'Académie Impériale des Sciences de St.-Pétersbourg (6)* **1**, 61–80 (1907).

Markov, Andrei Andreevich 1908. Rasprostranenie predelnykh teorem ischisleniya veroyatnostei na summu velichin v tsep. *Mémoires de l'Académie Impériale des Sciences de St.-Pétersbourg, Classe physico-mathématique (8)* **22**, 1–29 (1908). Reprinted in [Markov 1951, pp. 363–398]. English translation in [Sheynin 2004a, pp. 159–180].

Markov, Andrei Andreevich 1908/12. Ausdehnung der Sätze über die Grenzwerte in der Wahrscheinlichkeitsrechnung auf eine Summe verketteter Größen. German translation of [Markov 1908] in [Markov 1912, pp. 272–298].

Markov, Andrei Andreevich 1908/13/2004. O nekotorykh sluchayakh teoremy o predele veroyatnosti. *Bulletin de l'Académie Impériale des Sciences de St.-Pétersbourg (6)* **2**, 483–496 (1908). An amended version "Teorema o predele veroyatnosti dlya sluchaev akademika A. M. Lyapunova" appeared as a supplement to the 3rd edn. of *Ischislenie veroyatnostei* (1913) and was reprinted in [Markov 1951, pp. 319–338]; also separately published in French translation in 1913 with title *Démonstration du second théorème-limite du calcul des probabilités par la méthode des moments*, St.-Pétersbourg. All references are to the English translation "The Theorem on the Limit of Probability for the Liapunov case" by O. B. Sheynin in [Nekrasov 2004, pp. 141–155].

Markov, Andrei Andreevich 1910. Issledovanie obshchego sluchaya ispytanii, svyazannykh v tsep. *Zapiski Akademii Nauk po Fiziko-matematicheskomu otdeleniyu (6)*, 1–33. Reprinted in [Markov 1951, pp. 465–508]. English translation ("An Investigation of the General Case of Trials Connected into a Chain") in [Sheynin 2004a, pp. 181–205].

Markov, Andrei Andreevich 1911. O svyazannykh velichinakh, ne obrazuyushchikh nastoyashchei tsepi. *Bulletin de l'Académie Impériale des Sciences de St.-Pétersbourg (6)* **5**, 113–126. Reprinted in [Markov 1951, pp. 399–416].

Markov, Andrei Andreevich 1911/12. Über verbundene Größen, die keine eigentlichen Ketten bilden. German translation of [Markov 1911] in [Markov 1912, pp. 299–311].

Markov, Andrei Andreevich 1912. Wahrscheinlichkeitsrechnung. Leipzig: Teubner. German translation of the 2nd Russian edn.

Markov, Andrei Andreevich 1912/2004. A Rebuke to P. A. Nekrasov. In [Nekrasov 2004, pp. 73–79]. English translation of "Otpoved P. A. Nekrasovu," *Matematicheskii Sbornik* **28**, 215–227 (1912).

Markov, Andrei Andreevich 1951. *Izbrannye trudy*. Yu. V. Linnik (ed.). Leningrad: Akademiya Nauk SSSR.

Masani, Pesi R. 1990. *Norbert Wiener* (= *Vita mathematica* **5**). Basel: Birkhäuser.

Maurer, Ludwig 1896. Ueber die Mittelwerthe der Functionen einer reellen Variabeln. *Mathematische Annalen* **47**, 263–280.

Mazliak, Laurent 2009. How Paul Lévy Saw Jean Ville and Martingales. *Electronic Journal for History of Probability and Statistics* **5**, June 2009.

McAlister, Donald 1879. The Law of the Geometric Mean. *Proceedings of the Royal Society of London* **29**, 367–376. Supplements [Galton 1879].

Mehler, Ferdinand Gustav 1864. Bemerkungen zur Theorie der mechanischen Quadraturen. *Journal für die reine und angewandte Mathematik* **63**, 152–157.

Mehrtens, Herbert 1990. *Moderne—Sprache—Mathematik*. Frankfurt: Suhrkamp.

Mehrtens, Herbert, Bos, Henk, & Schneider, Ivo (eds.) 1981. *Social History of Nineteenth Century Mathematics*. Basel: Birkhäuser.

Merriman, Mansfield 1877a. An Elementary Discussion of the Principle of Least Squares. *Journal of the Franklin Institute (3)* **74**, 173–187.

Merriman, Mansfield 1877b. Remarks on Hagen's Proof of the Method of Least Squares. *Journal of the Franklin Institute (3)* **74**, 330–334.

Merriman, Mansfield 1877c. A List of Writings Relating to the Method of Least Squares, with Historical and Critical Notes. *Transactions of the Connecticut Academy of Arts and Sciences* **4**, 151–227.

Meyer, Anton 1874. Cours de Calcul des Probabilités fait à l'université de Liège depuis 1849 à 1857. *Mémoires de la Société des Sciences de Liège (2)* **4**, 1–458.

Meyer, Anton 1874/79. *Vorlesungen über Wahrscheinlichkeitsrechnung*. German translation by E. Czuber of [Meyer 1874]. Leipzig: Teubner.

MIT 2004. Longtime math department head Ted Martin dies at 92. *Massachusetts Institute of Technology News*, June 4, 2004.
`http://web.mit.edu/newsoffice/2004/martin.html`

Molina, Edward Charles 1930. The Theory of Probability: Some Comments on Laplace's Théorie Analytique. *Bulletin of the American Mathematical Society* **36**, 369–392.

Mourier, Edith 1953a. Éléments aléatoires dans un espace de Banach. *Annales de l'Institut Henri Poincaré* **13**, 161–244.

Mourier, Edith 1953b. Éléments aléatoires laplaciens dans un espace de Banach. *Comptes rendus hebdomadaires de l' Académie des Sciences, Paris* **236**, 575–576.

Natani, Leopold 1866. Quadrate (Methode der kleinsten). In *Mathematisches Wörterbuch*, Bd. 5, L. Hoffmann & L. Natani (eds.). Berlin: Bosselmann.

Nekrasov, Pavel Alekseevich 1898. Obshchie svoistva massovykh nezavicimykh yavlenii v svyazi s priblizhennym vychicleniem funktsii vesma bolshikh chisel. *Matematicheskii Sbornik (Moskva)* **20**, 431–442. English translation ("The General Properties of Mass Independent Phenomena in Connection with Approximate Calculation of Functions of Very Large Numbers") by O. B. Sheynin in [Nekrasov 2004, pp. 12–21].

Nekrasov, Pavel Alekseevich 1902. Filosofiya i logika nauki massovykh proyavleniyakh chelovechkoi deyatelnosti. *Matematicheskii Sbornik (Moskva)* **23**, 463–600.

Nekrasov, Pavel A. 2004. *The Theory of Probability*, O. B. Sheynin (ed.). Berlin: NG Verlag.

Newton, Isaac 1687. *Philosophiae Naturalis Principia Mathematica*. Cambridge.

Newton, Isaac 1711. *Methodus Differentialis*. London.

Nikodym, Otton 1930. Sur une généralisation des intégrales de M. J. Radon. *Fundamenta Mathematicae* **15**, 131–179, 358.

NYT 1991. Monroe Donsker, 66, N.Y.U. Math Professor. *The New York Times*, June 12.

Olesko, Kathryn M. 1995. The Meaning of Precision: The Exact Sensibility in Early Nineteenth-Century Germany. In *The Values of Precision*, N. M. Wise (ed.), pp. 103–134. Princeton: University Press.

Ondar, Kh. O. 1981. *The Correspondence between A. A. Markov and A. A. Chuprov on the Theory of Probability and Mathematical Statistics.* New York: Springer. Originally published in 1977 in Russian.

Ottmann, Ernst 1934. *Gotthilf Hagen.* Berlin: Ernst & Sohn.

Paley, Raymond E. & Wiener, Norbert 1934. *Fourier Transforms in the Complex Domain.* Providence (R.I.): American Mathematical Society.

Parthasarathy, Kalyanapuram Rangachari 1967. *Probability Measures on Metric Spaces.* New York: Academic Press.

Pearson, Egon S. & Kendall, Maurice G. (eds.) 1970. *Studies in the History of Statistics and Probability.* London: Griffin.

Pearson, Karl 1978. The History of Statistics in the 17th and 18th Centuries. London: Griffin.

Perron, Oskar 1913. *Die Lehre von den Kettenbrüchen.* Leipzig: Teubner. (2nd edn. 1929, 3rd edn. in two volumes 1953/57). All references are to the 1st edn.

Petrov, Valentin V. 1995. *Limit Theorems of Probability Theory: Sequences of Independent Random Variables.* Oxford: Clarendon Press.

Petrovskii, Ivan Georgievich 1934. Über das Irrfahrtsproblem. *Mathematische Annalen* **109**, 425–444.

Pettis, Billy James 1938. On Integration in Vector Spaces. *Transactions of the American Mathematical Society* **44**, 277–304.

Pizzetti, Paolo 1892. I fondamenti matematici per la critica dei risultati spermentali. *Atti della Regia Università di Genova*, 113–333.

Plarr, Gustave 1857. Note sur une proprieté commune aux séries dont le terme général dépend des fonctions X_n de Legendre, ou des cosinus et sinus des multiples de la variable. *Comptes rendus hebdomadaires des séances de l'Académie des Sciences* **44**, 984–986.

Poincaré, Henri 1896. *Calcul des probabilités.* Paris: Georges Carré.

Poincaré, Henri 1908. *Science et Méthode.* Paris: Flammarion.

Poincaré, Henri 1912. 2nd edn. of [Poincaré 1896]. Paris: Gauthier–Villars.

Poisson, Siméon Denis 1818. Mémoire sur la théorie des ondes. *Mémoires presentées à l'Académie Royale des Sciences par divers savants* **1**, 71–186.

Poisson, Siméon Denis 1824. Sur la probabilité des résultats moyens des observations. *Connaissance des tems*[1] pour l'an 1827, 273–302.

Poisson, Siméon Denis 1829. Suite du mémoire sur la probabilité des résultats moyens des observations, inséré dans la connaissance des tems de l'année 1827. *Connaissance des tems* pour l'an 1832, 3–22.

Poisson, Siméon Denis 1837. *Recherches sur la probabilité des jugements en matière criminelle et en matière civile, précédés des règles générales du calcul des probabilités.* Paris: Bachelier.

Pólya, Georg 1914/15. Review of [Markov 1908/13/2004]. *Jahrbuch über die Fortschritte der Mathematik* **45**, 1262 f. (JFM 45.1262.01).

Pólya, Georg 1918. Über die Nullstellen gewisser ganzer Funktionen. *Mathematische Zeitschrift* **2**, 352–383. Reprinted with original page numbers in [Pólya 1974].

Pólya, Georg 1919a. Über das Gaußsche Fehlergesetz. *Astronomische Nachrichten* **208**, 185–192.

Pólya, Georg 1919b. Über das Gaußsche Fehlergesetz. *Astronomische Nachrichten* **209**, 111 f.

Pólya, Georg 1920. Über den zentralen Grenzwertsatz der Wahrscheinlichkeitsrechnung und das Momentenproblem. *Mathematische Zeitschrift* **8**, 171–181. Reprinted with original page numbers in [Pólya 1984].

Pólya, Georg 1923. Herleitung des Gaußschen Fehlergesetzes aus einer Funktionalgleichung. *Mathematische Zeitschrift* **18**, 96–108. Reprinted with original page numbers in [Pólya 1984].

Pólya, Georg 1974. *Collected Papers*, Vol. II, R. P. Boas (ed.). Cambridge (Mass.): MIT Press.

[1] Up to the middle of the 19th century, the spelling"tems" was quite common beside the now usual spelling "temps."

Pólya, Georg 1984. *Collected Papers*, Vol. IV, G.-C. Rota (ed.). Cambridge (Mass.): MIT Press.

Porter, Theodore M. 1986. *The Rise of Statistical Thinking, 1820–1900*. Princeton: Princeton University Press.

Possé, Konstantin Aleksandrovich 1875. Sur les quadratures. *Nouvelles annales des mathématiques (2)* **14**, 49–62.

Possé, Konstantin Aleksandrovich 1886. *Sur quelques applications des fractions continues algébriques*. St. Pétersbourg: Académie Impériale des Sciences.

Price, Bartholomew 1865. *A Treatise on Infinitesimal Calculus*, Vol. 2., 2nd edn. Oxford: University Press.

Pringsheim, Alfred 1907. Über das Fouriersche Integraltheorem. *Jahresbericht der Deutschen Mathematiker-Vereinigung* **16**, 2–16.

Prokhorov, Yuri V. 1953. Raspredelenie veroyatnostei v funktsionalnykh prostranstvakh. *Uspekhi Matematicheskikh Nauk* **8**, 165–167.

Prokhorov, Yuri V. 1956. Skhodimost sluchainykh protsessov i predelnye teoremy teorii veroyatnostei. *Teoriya Veroyatnostei i Ee Primeneniya* **1**, 177–238. English translation: "Convergence of Random Processes and Limit Theorems in Probability Theory," *Theory of Probability and its Applications* **1** (1956), 157–214.

Prudnikov, Vasilii Efimovich 1964. *P. L. Chebyshev, Uchenyi i pedagog*, 2nd edn. Moskau.

Purkert, Walter 2002. Grundzüge der Mengenlehre — Historische Einführung. In [Hausdorff 2002, pp. 2–89].

Purkert, Walter 2006a. Kommentar zu [Hausdorff 1897]. In [Hausdorff 2006, pp. 497–526].

Purkert, Walter 2006b. Kommentar zu [Hausdorff 1901]. In [Hausdorff 2006, pp. 556–590].

Quetelet, Adolphe 1846. *Lettres à S.A.R. le Duc Régnant de Saxe-Cobourg et Gotha, sur la théorie des probabilités, appliquée aux sciences morales et politiques*. Brüssel: Havez.

Radau, Rodolphe 1883. Remarque sur le calcul d'une intégrale définie. *Comptes rendus hebdomadaires des séances de l'Académie des Sciences de Paris* **97**, 157 f.

Radon, Johann 1913. Theorie und Anwendungen der absolut additiven Mengenfunktionen. *Sitzungsberichte der Kaiserlichen Akademie der Wissenschaften in Wien, Mathematisch-naturwissenschaftliche Klasse, Abt. IIa* **122**. All references are to the reprint in *Gesammelte Abhandlungen*, Band 1, P. Gruber et al. (eds.), pp. 45–189. Basel: Birkhäuser.

Ramaer, J. C. 1924. Schols (Dr. Charles Mathieu). In *Nieuw Nederlandsch Biographisch Woordenboek* **6**, P. C. Molhuysen, P. J. Blok, & Fr. K. Kossmann (eds.), pp. 1228 f. Leiden: Sijthoff's.

Riesz, Friedrich 1910. Untersuchungen über Systeme integrierbarer Funktionen. *Mathematische Annalen* **69**, 449–497.

Rodrigues, Olinde 1816. Mémoire sur l'attraction des sphéroides. *Correspondance sur l'École Polytechnique* **3**, 361–385.

Rogers, Leonhard James 1888. An Extension of a Certain Theorem in Inequalities. *Messenger of Mathematics* **17**, 145–150.

Roll-Hansen, Nils 2008. Wishful Science: The Persistence of T. D. Lysenko's Agrobiology in the Politics of Science. *Osiris (2)* **23**, 166–188.

Rossberg, Hans Joachim, Jesiak, Bernd, & Siegel, Gerhard 1985. *Analytic Methods of Probability Theory*. Berlin: Akademie Verlag.

Särndal, Carl-Erik 1971. The Hypothesis of Elementary Errors and the Scandinavian School in Statistical Theory. *Biometrika* **58**, 375–391. All references are to the reprint in [Kendall & Plackett 1977, pp. 419–435].

Schneider, Ivo 1968. Der Mathematiker Abraham De Moivre. *Archive for History of Exact Sciences* **5**, 177–317.

Schneider, Ivo 1981a. Die Situation der mathematischen Wissenschaften vor und zu Beginn der wissenschaftlichen Laufbahn von Gauss. In *Carl Friedrich Gauss (1775–1855). Sammelband von Beiträgen zum 200. Geburtstag von C. F. Gauss*, I. Schneider (ed.), pp. 9–36. München: Minerva.

Schneider, Ivo 1987a. Laplace and Thereafter: The Status of Probability Calculus in the Nineteenth Century. In [Krüger, Daston, & Heidelberger 1987, pp. 191–214].

Schneider, Ivo 1987b. The Intellectual and Mathematical Background of the Law of Large Numbers and the Central Limit Theorem in the 18th and 19th Centuries. *Cahiers d'histoire et de philosophie des sciences* **20**, 214–231.

Schneider, Ivo 1988. *Die Entwicklung der Wahrscheinlichkeitstheorie von den Anfängen bis 1933. Einführungen und Texte*. Darmstadt: Wissenschaftliche Buchgesellschaft.

Schneider, Ivo 1995. Die Rückführung des allgemeinen auf den Sonderfall — eine Neubetrachtung des Grenzwertsatzes für binomiale Verteilungen von Abraham de Moivre. In *History of Mathematics: The State of the Art*, J. W. Dauben et al. (eds.), pp. 263–273. Orlando: Academic Press.

Schneider, Ivo 1998. Cases of Abstraction in 19th Century Theory of Probability in the Interplay between Users and Mathematicians. Unpublished contribution to the conference *L'abstraction mathématique entre dynamiques théoriques et inscriptions sociales et professionelles*, Centre Koyré, Paris, March 1998.

Schols, Charles Matthieu 1875/86. Théorie des erreurs dans le plan et dans l'espace. *Annales de l'École Polytechnique de Delft* **2**, 1886, 123–178. Originally published in Dutch in *Verhandelingen van de Koniglijke Akademie van Wetenschappen* **15** (1875).

Schols, Charles Matthieu 1887a. La loi de l'erreur résultante. *Annales de l'École Polytechnique de Delft* **3**, 140–150.

Schols, Charles Matthieu 1887b. Démonstration directe de la loi limite pour les erreurs dans le plan et dans l'espace. *Annales de l'École Polytechnique de Delft* **3**, 195–200.

Schubring, Gert 2005. *Conflicts Between Generalization, Rigor, and Intuition.* New York: Springer.

Seal, Hilary L. 1949. The Historical Development of the Use of Generating Functions in Probability Theory. *Bulletin de l'Association des Actuaires Suisses* **49**, 209–228. Reprinted with original page numbers in [Kendall & Plackett 1977, pp. 67–86].

Seal, Hilary L. 1979. The Fitting of a Mathematical Graduation Formula: A Historical Review with Illustrations. *Blätter (Deutsche Gesellschaft für Versicherungsmathematik)* **14**, 237–253.

Seidel, Philipp Ludwig 1855. Bemerkungen über den Zusammenhang zwischen dem Bildungsgesetze eines Kettenbruches und der Art des Fortgangs seiner Näherungsbrüche. *Abhandlungen der Königlich Bayerischen Akademie der Wissenschaften, Mathematisch-Physikalische Klasse* **7**, 559–602.

Seidel, Philipp Ludwig 1863. Resultate photometrischer Messungen an 208 der vorzüglichen Fixsterne. *Abhandlungen der Königlich Bayerischen Akademie der Wissenschaften, Mathematisch-Physikalische Klasse* **9**, 3. Abt., 419–609.

Séjour, Achille Pierre, Dionis du 1775. *Essai sur les comètes en général et particulièrement sur celles qui peuvent approcher de l'orbite de la terre*. Paris: Valade.

Seneta, Eugene 1982. Bernstein (!), Sergei Natanovich. In [Johnson & Kotz 1982–1989, Vol. 1, pp. 221–223].

Seneta, Eugene 1984. The Central Limit Problem and Linear Least Squares in Pre-Revolutionary Russia: The Background. *Mathematical Scientist* **9**, 37–77.

Seneta, Eugene 2006. Markov and the Creation of Markov Chains. In *Markov Anniversary Meeting 2006*, A. M. Langwille & W. J. Steward (eds.), pp. 1–20. Raleigh: Boson Books.

Shafer, Glenn & Vovk, Vladimir 2005. The Origins and Legacy of Kolmogorov's "Grundbegriffe." `http://www.probabilityandfinance.com/articles/04.pdf`

Shafer, Glenn & Vovk, Vladimir 2006. The Sources of Kolmogorov's "Grundbegriffe." *Statistical Science* **21**, 70–98.

Sheynin, Oscar B. 1970. Daniel Bernoulli on the Normal Law. *Biometrika* **57**, 199–202. Reprinted with original page numbers in [Kendall & Plackett 1977, pp. 101–104].

Sheynin, Oscar B. 1972. D. Bernoulli's Work on Probability. *RETE Strukturgeschichte der Naturwissenschaften* **1**, 273–300. Reprinted with original page numbers in [Kendall & Plackett 1977, pp. 105–131].

Sheynin, Oscar B. 1973. Finite Random Sums (a Historical Essay). *Archive for History of Exact Sciences* **9**, 275–305.

Sheynin, Oscar B. 1976. S. D. Laplace's Work on Probability. *Archive for History of Exact Sciences* **16**, 137–187.

Sheynin, Oscar B. 1977. Laplace's Theory of Errors. *Archive for History of Exact Sciences* **17**, 1–61.

Sheynin, Oscar B. 1978. S. D. Poisson's Work in Probability. *Archive for History of Exact Sciences* **18**, 245–300.

Sheynin, Oscar B. 1979. C. F. Gauss and the Theory of Errors. *Archive for History of Exact Sciences* **20**, 21–72.

Sheynin, Oscar B. 1989. A. A. Markov's Work on Probability. *Archive for History of Exact Sciences* **39**, 337–377.

Sheynin, Oscar B. 1990. H. Poincaré's Work on Probability. *Archive for History of Exact Sciences* **41**, 137–171.

Sheynin, Oscar B. 1994. Chebyshev's Lectures on the Theory of Probability. *Archive for History of Exact Sciences* **46**, 321–340.

Sheynin, Oscar B. 1996a. *The History of the Theory of Errors*. Egelsbach–Frankfurt–St. Peter Port: Hänsel-Hohenhausen (= Deutsche Hochschulschriften 1118).

Sheynin, Oscar B. 1996b. *Aleksandr A. Chuprov: Life, Work, Correspondence: The Making of Mathematical Statistics*. Göttingen: Vandenhoek & Ruprecht.

Sheynin, Oscar B. 2003. Nekrasov's Work on Probability: The Background. *Archive for History of Exact Sciences* **57**, 337–353.

Sheynin, Oscar B. (ed.) 2004a. *Probability and Statistics, Russian Papers*. Berlin: NG Verlag.

Sheynin, Oscar B. (ed.) 2004b. *Russian Papers on the History of Probability and Statistics*. Berlin: NG Verlag.

Sheynin, Oscar B. (ed.) 2005a. *Probability and Statistics, Soviet Essays*. Berlin: NG Verlag.

Sheynin, Oscar B. 2005b. *Theory of Probability. A Historical Essay*, 1st edn. (2nd edn. 2009). Berlin: NG Verlag.

Sheynin, Oscar B. 2009. *Theory of Probability and Statistics, as Exemplified in Short Dictums*, 2nd edn. Berlin: NG Verlag.

Shiganov, I. S. 1982. Making more precise the upper estimate of the constant in the remainder term of the central limit theorem (Russ.). In *Problema ustoichivosti stokhasticheskikh modelei, trudy seminara*, V. M. Zolotarev & V. V. Kalashnikov (eds.), pp. 109–115. Moskva: Vsesoyuznyi Nauchno-Issledovatelskii Institut Sistemnykh Issledovanii. English translation: "Refinement of the Upper Bound of the Constant in the Central Limit Theorem," *Journal of Soviet Mathematics* **35** (1986), 2545–2550.

Siegmund-Schultze, Reinhard 2001. *Rockefeller and the Internationalization of Mathematics Between the Two World Wars*. Basel: Birkhäuser.

Siegmund-Schultze, Reinhard 2004. A Non-conformist Longing for Unity in the Fractures of Modernity: Towards a Scientific Biography of Richard von Mises (1883–1953). *Science in Context* **17**, 333–370.

Siegmund-Schultze, Reinhard 2006. Probability in 1919/20: the von Mises–Pólya-Controversy. *Archive for History of Exact Sciences* **60**, 431–515.

Siegmund-Schultze, Reinhard 2010. Sets Versus Trial Sequences, Hausdorff Versus von Mises: "Pure" Mathematics Prevails in the Foundations of Probability Around 1920. *Historia Mathematica*. In press.

Sleshinskii, Ivan Vladislavovich 1892. K teorii sposoba naimenshikh kvadratov. *Zapiski mat. otdeleniya novorossiskogo obshchestva estestvoispitatelei* **14**, 201–264.

Smithies, Frank 1997. *Cauchy and the Creation of Complex Function Theory*. Cambridge: University Press.

Solovev, A. A. 1997. P. A. Nekrasov i tsentralnaya predelnaya teorema teorii veroyatnostei. *Istoriko-Matematicheskie Issledovaniya* **37**, 9–22.

Sommerfeld, Arnold 1904. Eine besondere anschauliche Ableitung des Gaussischen Fehlergesetzes. In *Festschrift, gewidmet L. Bolzmann zum sechzigsten Geburtstage*, pp. 848–859. Leipzig: Teubner.

Spalt, Detlev 1981. *Vom Mythos der mathematischen Vernunft*. Darmstadt: Wissenschaftliche Buchgesellschaft.

Spalt, Detlev (ed.) 1990. *Rechnen mit dem Unendlichen*. Basel: Birkhäuser.

Steele, Arthur Donald 1936. Über die Rolle von Zirkel und Lineal in der griechischen Mathematik. *Quellen und Studien zur Geschichte der Mathematik, Astronomie und Physik, Abteilung B* **3**, 287–369.

Steffens, Karl-Georg 2006. *The History of Approximation Theory: from Euler to Bernstein.* Boston: Birkhäuser.

Steinhaus, Hugo 1930. Sur la probabilité de la convergence de séries. *Studia mathematica* **2**, 21–39.

Stieltjes, Thomas Jan 1883. Sur l'évaluation approchée des intégrales. *Comptes rendus hebdomadaires des séances de l'Académie des Sciences de Paris* **97**, 740–742; 798 f. All references are to the reprint in [Stieltjes 1914, pp. 314–318].

Stieltjes, Thomas Jan 1884a. Quelques recherches sur la théorie des quadratures dites mécaniques. *Annales scientifique de l'École Normale de Paris (3)* **1**, 409–426. All references are to the reprint in [Stieltjes 1914, pp. 377–394].

Stieltjes, Thomas Jan 1884b. Sur un développement en fraction continue. *Comptes rendus hebdomadaires des séances de l'Académie des Sciences de Paris* **99**, 508–509. All references are to the reprint in [Stieltjes 1914, pp. 395–396].

Stieltjes, Thomas Jan 1884c. Note sur la densité de la terre. *Bulletin astronomique* **1**, 465–467. All references are to the reprint in [Stieltjes 1914, pp. 397–399].

Stieltjes, Thomas Jan 1884d. Sur une généralisation de la théorie des quadratures mécaniques. *Comptes rendus hebdomadaires des séances de l'Académie des Sciences de Paris* **99**, 850–851. All references are to the reprint in [Stieltjes 1914, pp. 428–429].

Stieltjes, Thomas Jan 1885a. Quelques remarques sur la variation de la densité dans l'interieur de la terre. *Verslagen en mededeelingen der Koninklijke Akademie van Wetenschappen (3)* **1**, 272–297. All references are to the reprint in [Stieltjes 1914, pp. 400–425].

Stieltjes, Thomas Jan 1885b. Note à l'occasion de la réclamation de M. Markov. *Annales scientifique de l'École Normale de Paris (3)* **2**, 183–184. All references are to the reprint in [Stieltjes 1914, pp. 430–431].

Stieltjes, Thomas Jan 1894/95. Recherches sur les fractions continues. *Annales de la Faculté des sciences de l'Université de Toulouse pour les sciences mathématiques et les sciences physiques* **8**, 1–122; **9**, 1–47. Reprinted in [Stieltjes 1918b, pp. 402–566].

Stieltjes, Thomas Jan 1914. *Œuvres complètes*, T. 1, W. Kapteyn & J. C. Kluyver (eds.). Groningen: Noordhoff.

Stieltjes, Thomas Jan 1918a. Sur certaines inégalités dues à M. P. Tchébychef. Posthumously edited in [Stieltjes 1918b, pp. 586–593].

Stieltjes, Thomas Jan 1918b. *Œuvres complètes*, T. 2, W. Kapteyn, J. C. Kluyver, & E. F. van de Sande Bakhuysen (eds.). Groningen: Noordhoff.

Stigler, Stephen M. 1973. Simon Newcomb, Percy Daniell, and the History of Robust Estimation, 1885–1920. *Journal of the American Statistical Association* **68**, 872–879.

Stigler, Stephen 1974/1999. Cauchy and the Witch of Agnesi: an Historical Note on the Cauchy Distribution. *Biometrika* **61**, 375–380. All references are to the revised version in [Stigler 1999].

Stigler, Stephen M. 1978. Francis Ysidro Edgeworth, Statistician (with Discussion). *Journal of the Royal Statistical Association (A)* **141**, 287–322. See also [Stigler 1999].

Stigler, Stephen M. 1986. *History of Statistics.* London: Belknap.

Stigler, Stephen M. 1999. *Statistics on the Table. The History of Statistical Concepts and Methods.* Cambridge (Mass.): Harvard University Press.

Stigler, Stephen M. 2005. P. S. Laplace, *Théorie analytique des probabilités*, 1st edn. (1812); *Essai philosophique sur les probabilités*, 1st edn. (1814). In [Grattan-Guinness 2005, pp. 329–340].

Stirling, James 1730. *Methodus differentialis.* London: Strahan.

Tait, Peter Guthrie 1865. On the Law of Frequency of Error. *Transactions of the Royal Society of Edinburgh* **24**, 139–145.

Tait, Peter Guthrie & Thompson, William 1867. *Treatise on Natural Philosophy*, Vol. 1. Oxford: Clarendon Press.

Thiele, Thorvald Nikolai 1873. Om en tilnærmelsesformel. *Tidsskrift for Mathematik* **3**, 22–31.

Thiele, Thorvald Nikolai 1889/2002. *Forelæsninger over Almindeling Iagttagelseslære: Sandsynlighedsregning og mindste Kvadraters Methode.* Kjøbenhavn: Reitzel. All references are to the

English translation in [Lauritzen 2002, pp. 57–197] with title "The General Theory of Observations: Calculus of Probability and the Method of Least Squares."

Thiele, Thorvald Nikolai 1899/2002. Om Iagttagelseslæres Halvinvarianter. *Oversigt over det koneglige danske Videnskabernes Selskabs Forhandlinger* **3**, 135–141. All references are to the English translation in [Lauritzen 2002, pp. 226–231] with title "On the Halfinvariants in the Theory of Observations".

Tsykalo, Alfred L. 1988. *Aleksandr Mikhailovich Lyapunov*. Moskva: Nauka.

Uspensky, James Victor 1937. *Introduction to Mathematical Probability*. New York: McGraw-Hill.

Vasilev (Wassilief), Aleksandr Vasiletivich 1898/1900. P. Tchébychef et son œuvre scientifique. *Bollettino di bibliografia e storia delle scienze matematiche* **1**, 1898. All references are to the German translation *P. L. Tschebyschef und seine wissenschaftlichen Leistungen*, Leipzig: Teubner, 1900.

Ville, Jean 1939. *Etude critique de la notion de collectif.* Paris: Gauthier–Villars.

von Mises, Richard 1912. Über die Grundbegriffe der Kollektivmaßlehre. *Jahresbericht der Deutschen Mathematiker-Vereinigung* **21**, 9–20.

von Mises, Richard 1919a. Fundamentalsätze der Wahrscheinlichkeitsrechnung. *Mathematische Zeitschrift* **4**, 1–97.

von Mises, Richard 1919b. Grundlagen der Wahrscheinlichkeitsrechnung. *Mathematische Zeitschrift* **5**, 52–99.

von Mises, Richard 1931. *Wahrscheinlichkeitsrechnung und ihre Anwendung in der Statistik und theoretischen Physik.* Leipzig: Franz Deuticke.

von Mises, Richard 1935. Deux nouveaux théorèmes de limite dans le calcul des probabilités. *Revue de la faculté des sciences de l'université d'Istanbul* **1**, 61–80.

von Mises, Richard 1936. Les lois de probabilité pour les fonctions statistiques. *Annales de l'Institut Henri Poincaré* **6**, 185–212.

von Plato, Jan 1983. The Method of Arbitrary Functions. *The British Journal for the Philosophy of Science* **34**, 37–47.

von Plato, Jan 1994 *Creating Modern Probability*. Cambridge: University Press.

von Plato, Jan 2005. A. N. Kolmogorov, *Grundbegriffe der Wahrscheinlichkeitsrechnung* (1933). In [Grattan-Guinness 2005, pp. 960–969].

Wattenberg, Diedrich 1976. Nach Bessels Tod: Eine Sammlung von Dokumenten. *Veröffentlichungen der Archenhold-Sternwarte Berlin-Treptow* **7**.

Wicksell, Sven Dag 1917. On the Genetic Theory of Frequency. *Arkiv för Matematik, Astronomi och Fysik* **12**, Nr. 20.

Wiener, Norbert 1921. The Average of an Analytic Functional and the Brownian Movement. *Proceedings of the National Academy of Sciences* **7**, 294–298. Reprinted with original page numbers in [Wiener 1976].

Wiener, Norbert 1923. Differential Space. *Journal of Mathematics and Physics* **58**, 131–174. Reprinted with original page numbers in [Wiener 1976].

Wiener, Norbert 1924. The Average Value of a Functional. *Proceedings of the London Mathematical Society* **22**, 454–467. Reprinted with original page numbers in [Wiener 1976].

Wiener, Norbert 1930. Generalized Harmonic Analysis. *Acta Mathematica* **55**, 117–258. Reprinted with original page numbers in [Wiener 1976].

Wiener, Norbert 1976. *Collected Works*, Vol. 1, P. Masani (ed.). Cambridge (Mass.): MIT Press.

Wiener, Norbert, Siegel Armand, Rankin, Bayard, & Martin, William 1966. *Differential Space, Quantum Systems, and Prediction*. Cambridge (Mass.): MIT Press.

Winckler, Anton 1866. Allgemeine Sätze zur Theorie der unregelmäßigen Beobachtungsfehler. *Sitzungsberichte der Kaiserlichen Akademie der Wissenschaften in Wien, mathematisch-naturwissenschaftliche Klasse* **53**, 6–41.

Wittstein, Theodor 1849. Die Methode der kleinsten Quadrate. In Navier, L. *Lehrbuch der Differential- und Integralrechnung*, 2nd Vol., pp. 343–442. Hannover: Halm'sche Hofbuchhandlung.

Young, Thomas 1819. Remarks on the Probabilities of Error in Physical Observations, and on the Density of the Earth, Considered, Especially with Regard to the Reduction of Experiments on

the Pendulum, in a Letter to Capt. Henry Kater, F. R. S. *Philosophical Transactions of the Royal Society of London*, 70–95.
Yushkevich (Youschkevitch), Adolf-Andrei Pavlovich 1970–76a. Chebyshev, Pafnuty (!) Lvovich. In [Gillispie 1970–1976, Vol. 3, pp. 222–231].
Yushkevich (Youschkevitch), Adolf-Andrei Pavlovich 1970–76b. Markov, Andrei Andreevich. In [Gillispie 1970–1976, Vol. 9, pp. 124–130].
Yushkevich (Youschkevitch), Adolf-Andrei Pavlovich 1970–76c. Bernstein (!), Sergey (!) Natanovich. In [Gillispie 1970–1976, Vol. 15, pp. 22–24].
Zabell, Sandy 1995. Alan Turing and the Central Limit Theorem. *American Mathematical Monthly* **102**, 483–494.
Zachariae, Karl Christian 1871. *De mindste Kvadraters Methode.* Nyborg: Schönemann.

Name Index

Subject Index